THE DEFORMATION

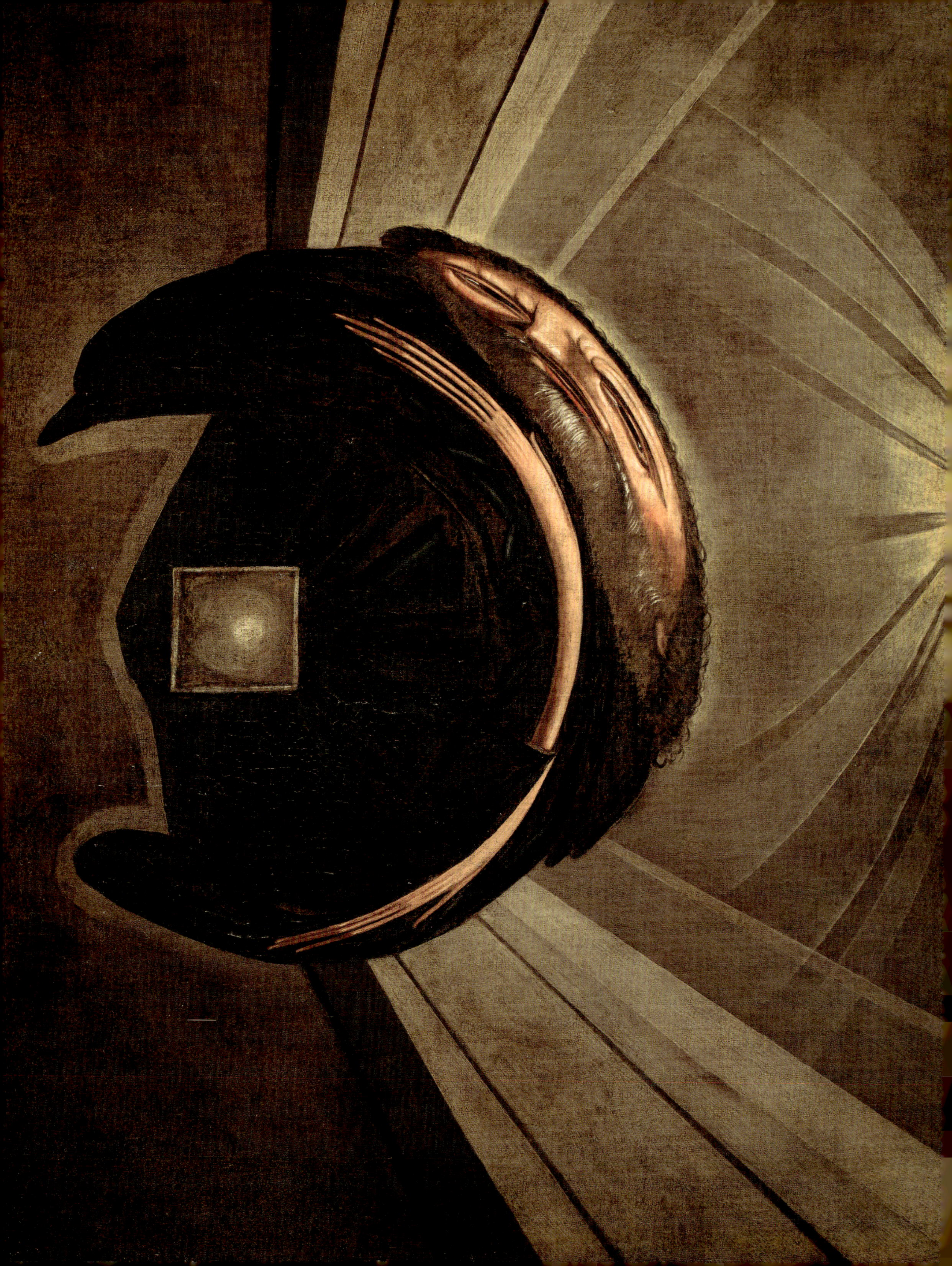

THE DEFORMATION

ATTENTION AND DISCERNMENT IN CATHOLIC REFORMATION ART AND ARCHITECTURE

SUSANNA BERGER

PRINCETON UNIVERSITY PRESS · Princeton and Oxford

For Torbjørn, Theodore, and Thomas, with love

CONTENTS

ABBREVIATIONS

ASR	Archivio di Stato di Roma
ASV	Archivio Segreto Vaticano
BAV	Biblioteca Apostolica Vaticana
BnF	Bibliothèque nationale de France
FSV	Archivio di Stato di Roma, Fondo Spada Veralli

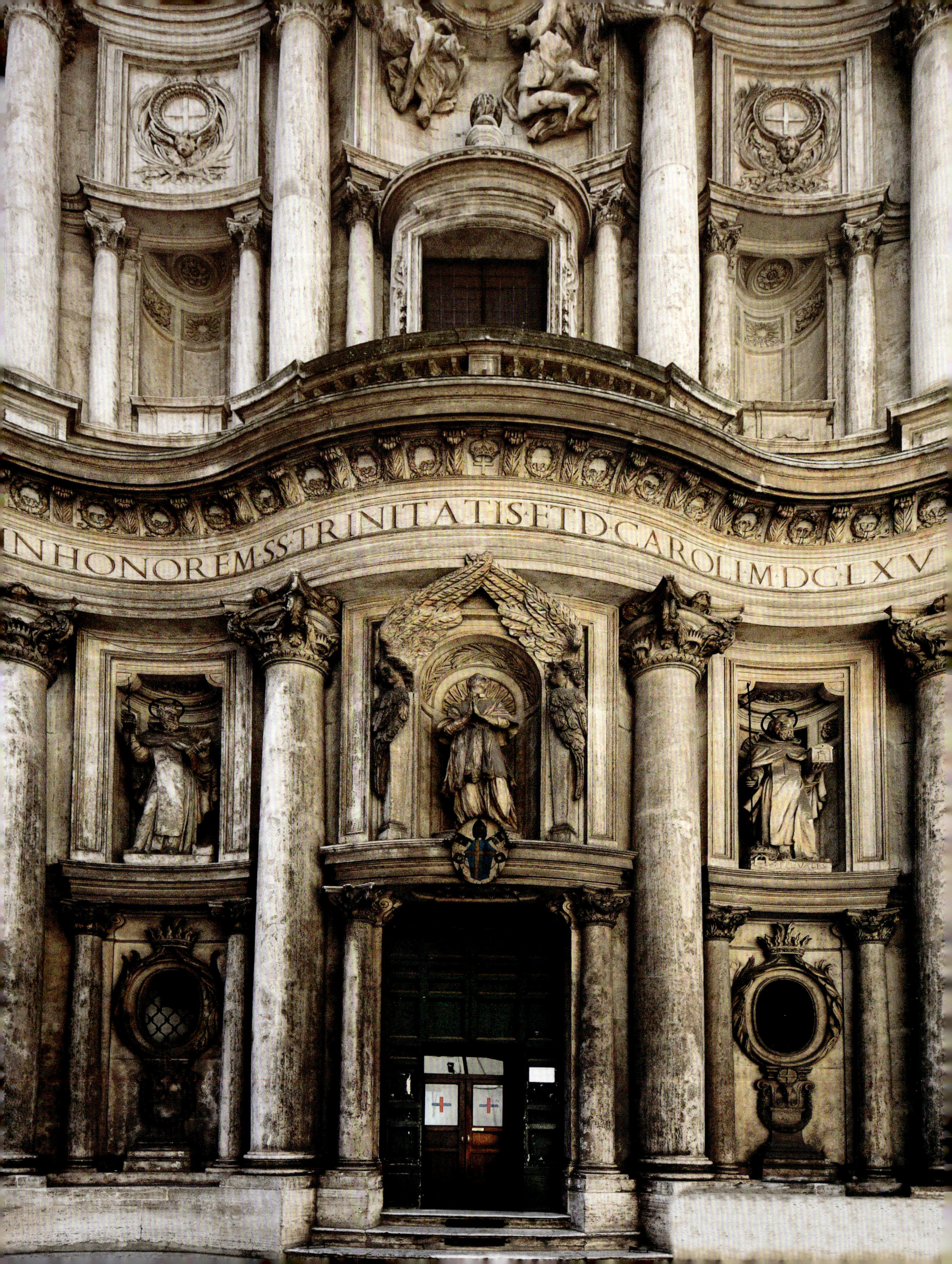
IN HONOREM·SS·TRINITATIS·ET·D·CAROLI·M·DC·LXV

INTRODUCTION

AROUND 1665–1667, biographer Giovanni Pietro Bellori (1613–1696) described San Carlo alle Quattro Fontane in Rome as both "ugly and deformed" (*brutta et deforme*). Scribbling in the margins of a book, he dismissed the church's architect as "a most ignorant goth and a corrupter of architecture, the infamy of our century" (*gotico ignorantissimo et corrutore dell'architetura, infamia dal nostro secolo*) (Figs. 1 and 2).[1] Bellori's objections to Francesco Borromini's (1599–1667) architecture would be sustained by other critics in subsequent centuries: the Italian art and architectural historian, theorist, and biographer Francesco Milizia (1725–1798), speaking on behalf of neoclassicism, would later characterize San Carlo as "Borromini's greatest delirium" (*Il delirio maggiore del Borromini*).[2] San Carlo's double S-curved facade—the first element of the church that visitors confront—especially bothered Milizia: "So many straight, concave, and convex lines, with so many columns upon columns of different shapes, and windows and niches and sculptures in such a small facade, are pitiful things."[3]

Bellori and others objected to San Carlo because they believed it violated norms of classical architecture and, in the process, cast aside the classical ideal of clarity. An unnamed grand cleric criticized Borromini for showing "too much desire to go beyond the rule" (*troppa voglia di uscir di regola*).[4] With the undulating facade, Borromini and his nephew—Bernardo Castelli-Borromini (1643–1709), who completed the facade's upper area—render deformed, to borrow Bellori's term, the morphological clarity and sense of apparent structural stability that classical buildings convey with rectangular grid schema and parallel planes.[5] San Carlo's lower convex center bay is wider than the concave framing bays in the same story, which produces a dynamic sense of rhythm in the facade's lower level. Whereas the lower story alternates between concave and convex curves, the upper level features only concave bays—and this divergence from the pattern below creates a strain that magnifies the facade's undulating movement. The angular cornice that unites the concave bays above the upper story is broken

Detail of fig. 2

Fig. 1. Giovanni Pietro Bellori's annotation in the margins of Giovanni Baglione's *Vite*, 1642. 9.21 × 6.38 in (23.4 × 16.2 cm). Accademia Nazionale dei Lincei, Rome [31.E.15].

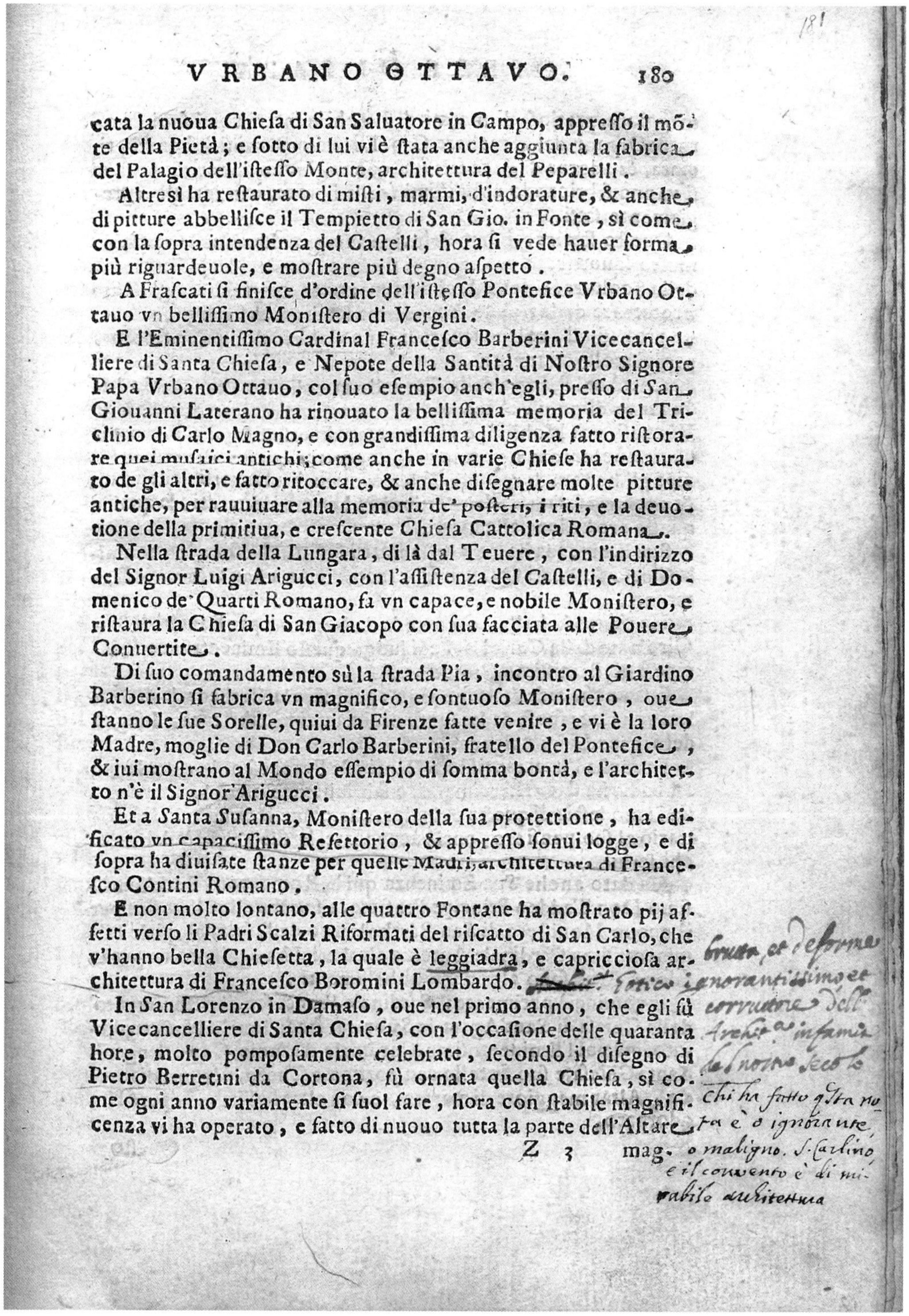

VRBANO OTTAVO. 180

cata la nuoua Chieſa di San Saluatore in Campo, appreſſo il mõte della Pietà; e ſotto di lui vi è ſtata anche aggiunta la fabrica del Palagio dell'iſteſſo Monte, architettura del Peparelli.

Altresì ha reſtaurato di miſti, marmi, d'indorature, & anche di pitture abbelliſce il Tempietto di San Gio. in Fonte, sì come con la ſopra intendenza del Caſtelli, hora ſi vede hauer forma più riguardeuole, e moſtrare più degno aſpetto.

A Fraſcati ſi finiſce d'ordine dell'iſteſſo Pontefice Vrbano Ottauo vn belliſſimo Moniſtero di Vergini.

E l'Eminentiſſimo Cardinal Franceſco Barberini Vicecancelliere di Santa Chieſa, e Nepote della Santità di Noſtro Signore Papa Vrbano Ottauo, col ſuo eſempio anch'egli, preſſo di San Giouanni Laterano ha rinouato la belliſſima memoria del Triclinio di Carlo Magno, e con grandiſſima diligenza fatto riſtorare quei muſaici antichi; come anche in varie Chieſe ha reſtaurato de gli altri, e fatto ritoccare, & anche diſegnare molte pitture antiche, per rauuiuare alla memoria de' poſteri, i riti, e la deuotione della primitiua, e creſcente Chieſa Cattolica Romana.

Nella ſtrada della Lungara, di là dal Teuere, con l'indirizzo del Signor Luigi Arigucci, con l'aſſiſtenza del Caſtelli, e di Domenico de' Quarti Romano, fa vn capace, e nobile Moniſtero, e riſtaura la Chieſa di San Giacopo con ſua facciata alle Pouere Conuertite.

Di ſuo comandamento sù la ſtrada Pia, incontro al Giardino Barberino ſi fabrica vn magnifico, e ſontuoſo Moniſtero, oue ſtanno le ſue Sorelle, quiui da Firenze fatte venire, e vi è la loro Madre, moglie di Don Carlo Barberini, fratello del Pontefice, & iui moſtrano al Mondo eſſempio di ſomma bontà, e l'architetto n'è il Signor Arigucci.

Et a Santa Suſanna, Moniſtero della ſua protettione, ha edificato vn capaciſſimo Refettorio, & appreſſo ſonui logge, e di ſopra ha diuiſate ſtanze per quelle Madri, architettura di Franceſco Contini Romano.

E non molto lontano, alle quattro Fontane ha moſtrato pij affetti verſo li Padri Scalzi Riformati del riſcatto di San Carlo, che v'hanno bella Chieſetta, la quale è leggiadra, e capricciosa architettura di Franceſco Boromini Lombardo.

In San Lorenzo in Damaſo, oue nel primo anno, che egli fù Vicecancelliere di Santa Chieſa, con l'occaſione delle quaranta hore, molto pompoſamente celebrate, ſecondo il diſegno di Pietro Berretini da Cortona, fù ornata quella Chieſa, sì come ogni anno variamente ſi ſuol fare, hora con ſtabile magnificenza vi ha operato, e fatto di nuouo tutta la parte dell'Altare

Z 3 mag.

brutta, et deforme Gotico ignorantissimo et corruttore dell' Archit.a infamia del nostro secolo

Chi ha fatto q.sta nota è o ignorante, o maligno. S. Carlino e il convento è di mirabile architettura

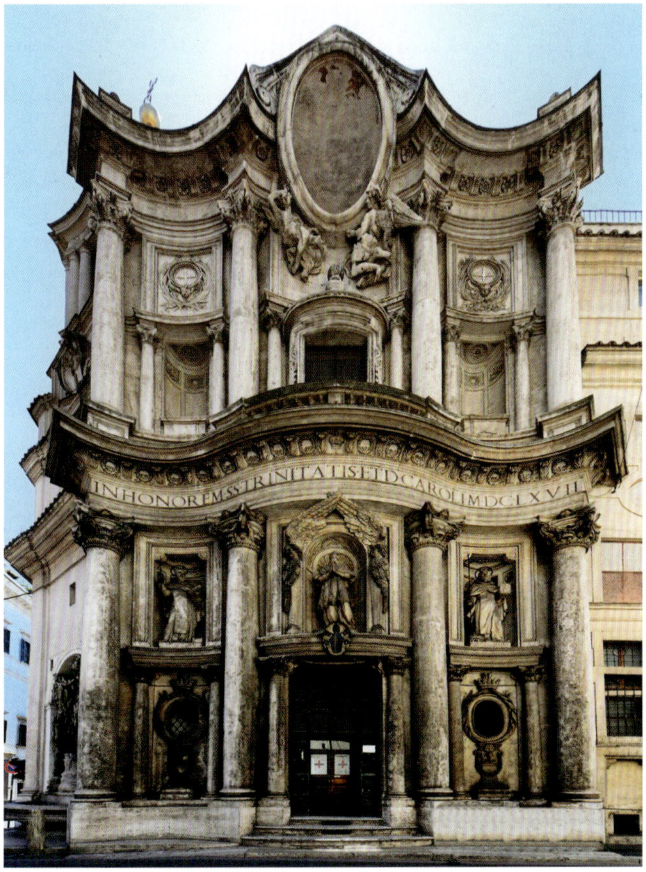

Fig. 2. San Carlo alle Quattro Fontane, lower area built in 1665–1667 by Francesco Borromini and upper area built in 1675–1677 by Bernardo Castelli-Borromini, exterior view of facade.

by a concave medallion crowned with two scrolls; these join at an acute angle, imparting to the cornice a jagged profile that adds drama to the skyline.

Although facades on single concave curves or juxtaposing concave and convex curves could be found on structures dating to antiquity (including the so-called Temple of Romulus on the Via Sacra, which featured a single concave curve), and were known in seventeenth-century Italy, early modern critics noted the incomprehensibility and the extreme apparent instability of Borromini's dynamic forms as conflicting with classical architecture's perceived regularity and steadiness.[6] A guidebook to Rome written between 1677 and 1681 describes another enigmatic facade by the architect as "distant" from antique simplicity: "Borromini, who has always sought to distance himself from the regular simplicity of the antique, has done so much [in the facade of Sant'Agnese in Agone] to turn and to turn again convex and concave lines that eyesight, which tires itself in searching therein a path, is not entirely pleased, even if at first this style has a certain something [*un je ne scais quoi*] that surprises and that prevents the sight from disapproving of it entirely."[7] After accentuating Borromini's dissociation from more legible antique models, the anonymous author notes the facade's dizzying effect on beholders—the quest for clarity and guidance amid lines that bend in opposite directions exhausts and ultimately frustrates the eye, although it is not completely displeasing.

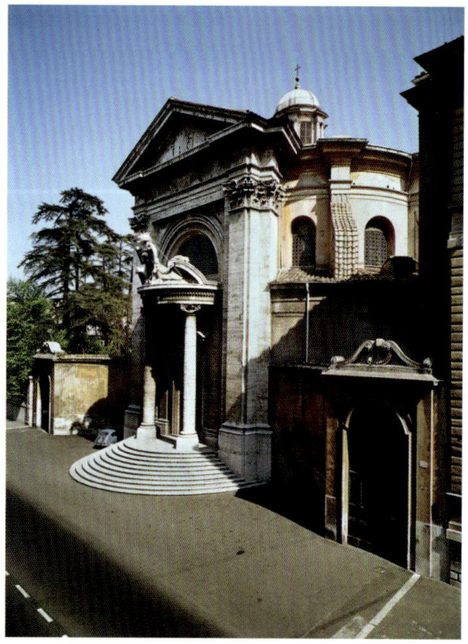

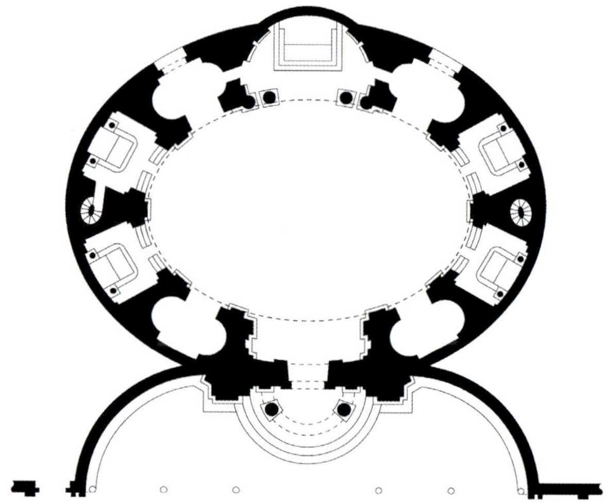

Fig. 3. Gian Lorenzo Bernini, Sant'Andrea al Quirinale, 1670–1672, exterior view of facade.

Fig. 4. Gian Lorenzo Bernini, plan of Sant'Andrea al Quirinale, 1658–1670.

San Carlo is just one example of the wide-ranging visual phenomenon, known in the seventeenth century as "deformation," that is at the center of this book. Deformation—defined in the period, as we have seen, against an ideal that an entity fails to meet and balanced against the related term "reformation" that restores the ideal—characterized architecture, images, sculptures, and other works across media. Deformations depart from ideals—not necessarily to challenge those ideals but to at least invite consideration of how we perceive and value them. During the periods that German Lutheran historians of the eighteenth century would label the "Reformation" and the "Counter Reformation," deformations became prevalent across Europe, appearing in paintings, prints, sculptures, as well as in churches and secular structures.[8] Artists and theoreticians like Borromini, Mario Bettini (1582–1657), Jean-François Niceron (1613–1646), Emanuel Maignan (1601–1676), and Andrea Pozzo (1642–1709) experimented with deformation in church architecture, in the gardens, libraries, and corridors of palaces and convents, and in printed books and pedagogical materials. This book will consider in particular Catholic Reformation paintings, prints, sculptures, and architectures produced by Italian and French artists and architects in seventeenth-century Italy, the center of the Catholic Church, although I also discuss works created in broader Catholic geographies within and beyond Europe.

In undermining the ideals of classical and Renaissance art and architecture—and in particular the classical rhetorical ideals of clarity and comprehensibility—these deformations evoked states of confusion or mental perturbation that precluded the viewer from fully exercising his or her faculties. Although (at least momentary) incomprehensibility was a common feature of deformations—in the sense that they could not be grasped immediately by the understanding—there was no single experience of deformation. A deformation could challenge verbal capacity, as observers struggled to articulate their visual experience in language,

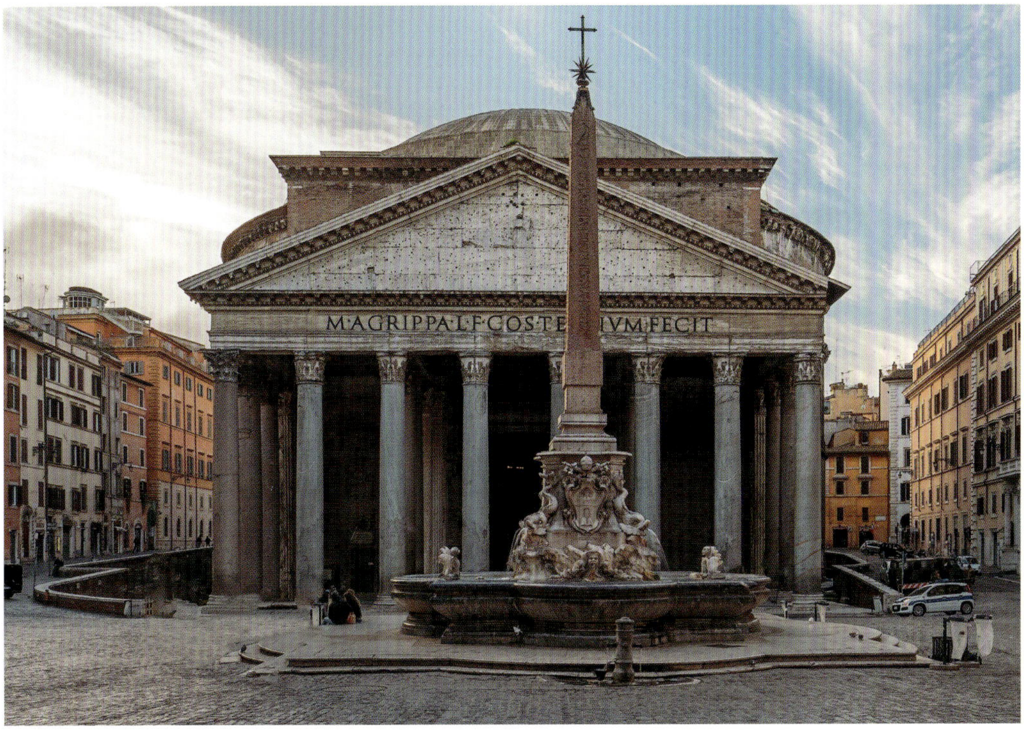

Fig. 5. Pantheon, c. 125–128 CE, exterior view of facade, piazza, and fountain.

or it could defy understandings of facture, as observers—even sophisticated observers—struggled to imagine how an object or visual effect could have been generated.

Some seventeenth-century visitors to San Carlo recorded their experiences of this deliberate incomprehensibility. Jean-Baptiste Antoine Colbert (1651–1690), Marquis de Seignelay and son of the French statesman and chief minister to Louis XIV (1638–1715), was among the visitors that San Carlo perplexed. He arrived at the structure on April 10, 1671, after having visited Sant'Andrea al Quirinale, located further down the Via Pia. Sant'Andrea did not strike the marquis as strange (Figs. 3 and 4). He describes in straightforward terms this church by Gian Lorenzo Bernini (1598–1680), whose plan is easily recognized as an oval with entrance and altar opposed across the oval's short axis: "a very small church, but very pleasant and of a figure that pleases" (*une très-petite église, mais fort agréable et d'une figure qui plait*).[9] Bernini's design respected the dignity of classical shapes; even if he elongated the classical circle into an oval in Sant'Andrea, he still employed centric geometry.[10] The comparative simplicity and intelligibility of the plans for Sant'Andrea, begun seventeen years after San Carlo had been built, has been read as "a kind of mute architectural criticism" of Borromini's more convoluted designs, which are not as clear and easy to understand when one visits the church.[11]

While the marquis might have arrived at San Carlo most immediately from Sant'Andrea, his expectations for clear and legible plans—like the expectations of other seventeenth-century visitors—were ultimately informed by his knowledge of classical structures like the Pantheon, located twenty minutes by foot from San Carlo (Fig. 5). Three days after visiting San Carlo and Sant'Andrea, on Monday, April 13, the marquis went to the Pantheon, which alongside the

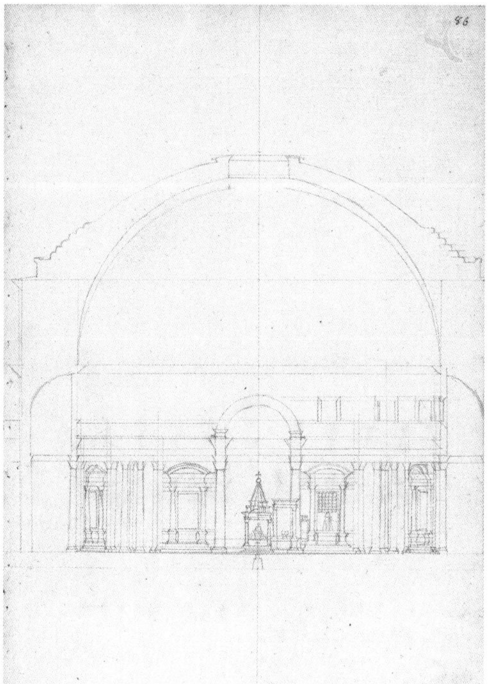

Fig. 6. Anonymous seventeenth-century drawing of an orthogonal section through the Pantheon. Graphite drawing on paper, 16.1 × 10.8 in (40.9 × 27.4 cm). Gabinetto dei Disegni, Castello Sforzesco, Milan [SM 5,99].

so-called Maison Carrée of Nîmes is regarded as the best-preserved temple of Roman antiquity. The three-dimensional profile of this building, which features an implied sphere nestled within a cylinder, was directly intelligible to the marquis, who observed in his diary that "the height of this vault is equal to its diameter" (*la hauteur de cette voûte est égale à son diamètre*).[12] The geometric forms of the Pantheon—its sphere and cylinder—can be appreciated in an orthogonal section through the building by an anonymous seventeenth-century draughtsman (Fig. 6).

Whereas the marquis finds both Sant'Andrea and San Carlo "pleasing," there are no oddities in his description of Sant'Andrea, while his characterization of San Carlo stresses the bizarre: "although this church is very bizarrely built, it is surprising and pleasing at first sight" (*quoique cette église soit très-bizarrement bâtie, elle surprend et plaît d'abord*).[13] That Borromini was sensitive to San Carlo's distance from the comprehensibility of Italian classicism is suggested by the plans he drew for San Carlo in the early 1660s, as he prepared to publish his works. Albertina 173 reduces and idealizes the church (Fig. 7).[14] The plan indicates that the church's design is based on two triangles, positioned in such a way that they share one side and make a diamond, and it features apses created from half circles and half ovals that extend from the diamond's points.[15]

The marquis was not alone in expressing his perplexity at San Carlo, whose enigmatic layout exercised visitors from all over the world.[16] The church's chronicler, a friar named Juan di San Bonaventura, reports that once the church was built, German, Flemish, French, Italian, Spanish, and Indian visitors requested plans, hoping to make sense of its confusing forms.[17] And yet for a long time Borromini kept his plans for San Carlo a secret, even from Fra Juan, and unfortunate circumstances meant they would remain largely unknown until they were finally

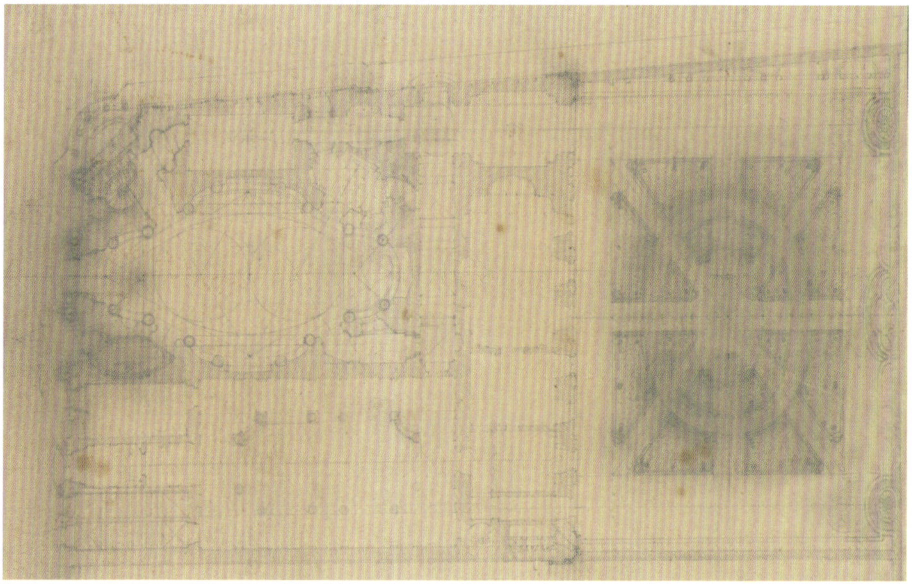

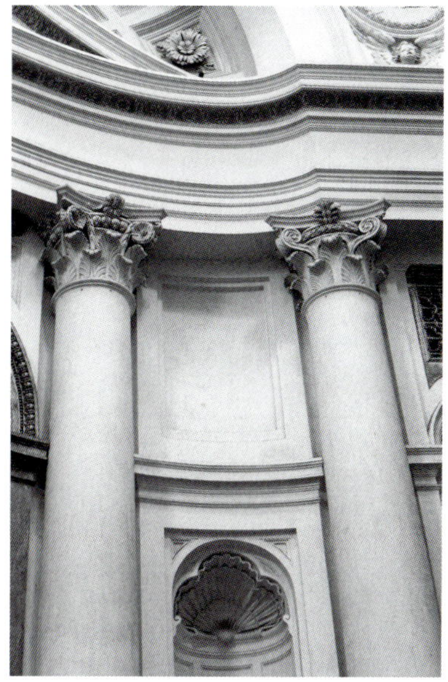

Fig. 7. Francesco Borromini, plan of San Carlo alle Quattro Fontane. Pencil on paper, 12.2 × 18.5 in (31 × 47 cm). The Albertina Museum, Vienna [Alb. 173].

Fig. 8. Francesco Borromini, San Carlo alle Quattro Fontane, 1638–1641, interior view of capitals of columns.

published in the 1950s and '60s.[18] As a result, San Carlo's form continued to bewilder visitors well into modern times. Leo Steinberg recounts that twelve respected scholars in the nineteenth and twentieth centuries would examine San Carlo and provide twelve distinct rationales supposedly underlying the structure's forms.[19]

If observers' most immediate response to San Carlo's facade and plan is perplexity, their confusion increases as they more closely examine the church's interior details. A visitor whose understanding of architecture has been informed by classical structures might be especially confused on noticing Borromini's treatment of San Carlo's columns: whereas the capitals of half the columns inside the church feature volutes positioned in the conventional fashion, with scrolls that spiral outward, the capitals of the other columns have inverted volutes, their scrolls spiraling inward (Fig. 8). The Marquis de Seignelay was among those perturbed by this aspect of Borromini's design, to the point of forgetting to mention that half the volutes follow the conventional form: "In this church there are sixteen columns whose capitals are composed so strangely that the volutes turn inward instead of outward."[20] Capitals with inverted volutes were in fact produced in antiquity—for instance, in the villa of Hadrian (76–138 CE) at Tivoli (118–138 CE)—and then well into the fifteenth century, with a renewed interest in the motif after 1590—for instance, in the facade by Giacomo della Porta (1532–1602) for San Giuseppe dei Falegnami.[21] This long history of inverted volutes underlines the complexity of Borromini's violation of classical norms, insofar as some of his alleged transgressions involved the transposition of classical motifs, and often in a systematic fashion.[22] And yet, even a viewer familiar with this history would have confronted Borromini's refusal of immediate comprehensibility. Despite a long tradition justifying the use of inverted volutes, the contrast Borromini creates by using two, apparently competing, types of classical volutes in the same structure invites puzzlement and signals that something is amiss, that there is a gap between the norms of classicism and Borromini's confusing ornamental juxtapositions.[23]

Even so, within this historical moment, the very things that made some contemporaries label San Carlo as deformed, ugly, and confusing could also make it compelling. Here we might recall that the marquis described San Carlo as "pleasing," as well as bizarre. If Bellori, Bernini, and Milizia disliked Borromini's deformations in San Carlo, Fra Juan reported that foreign visitors flocked to the church:

> Every day we see these foreign people entering this church, moved to see it by the fame and reputation that it gained in their countries. And when they are in the church, they do nothing but look upward and turn themselves around throughout the whole church: all its objects are, in fact, placed in such a way that each one calls to the other, and the one stimulates the viewer to look at the other. And so, from the tribunes and balustrades of the church, we often see those foreigners act this way without having the force to leave nor to say anything for a while. And what is most amazing is that the more one looks at this church, the more pleasure it gives and seems like it is seen for the first time, and leaves a desire to come back to see it again. Why? Because they notice that it gives so much pleasure without ever boring them, but leaving them with a desire to see it again. It seems to me that this has a kind of similarity (insofar as one can say) to Divinity. … The construction of this church when it is seen by these foreigners seems to suspend their intellect, because for a long time, you can see them incapable of doing anything but look at it. And after the visit, they return to see it, and this not only many times on one occasion, but a great number of times on many different days. If the church had either satisfied or bored them the first time (as earthly things do after having been seen), they would not return many times to see it again.[24]

San Carlo demanded—and did so more intensely than would a classical structure—the temporal expansion of the beholding and comprehending process. Fra Juan compares the observation of San Carlo to the beatific vision: the building put observers in touch with something transcendent (akin to the experience of angels contemplating the Holy Spirit in heaven). The intellectual bafflement evoked by this building generated a near-mystical experience that observers—unable "to say anything for a while"—could not articulate with language. San Carlo induced a state of confusion and—insofar as it "seems to suspend their intellect"—of stupor, an intoxicating mental paralysis in which visitors could not explain what they experienced.

To experience the alleged deformations in Borromini's architecture, then, was to move in some sense beyond the grasp of language. That deformations were incomprehensible, and that articulating one's experience of a deformation with language was challenging, are manifest in the phrase *je ne sais quoi* used to characterize Borromini's style in the late seventeenth-century guidebook to Rome I mentioned above. Translated literally as "I don't know what," the term had by then acquired a substantive role—denoting "a certain something."[25] The words labeled an experience of the ineffable, as conceptualized in France from the late sixteenth through the late seventeenth centuries, an intense experience or perception of specific phenomena whose cause or explanation a subject finds difficult to express linguistically.

In a popular text that appeared in 1671, the critic and grammarian Dominique Bouhours (1628–1702), who taught at the Jesuit collège de Clermont in Paris, makes the *je ne sais quoi* the subject of a dialogue. According to his character Eugène, the *je ne sais quoi* "is much easier to perceive than to know … it is in its nature to be incomprehensible and inexplicable."[26] This remark highlights the limits on our capacity to express discernment: the *je ne sais quoi* can be recognized without being understood. The term could refer to mental refinement, tastes, and modes of behaving. And it could be employed in relation to a broad range of realms, including aesthetic and literary discourse, theology, the natural world, the passions, and culture.[27] Eugène's interlocutor Ariste adds the difficulty in recognizing the *je ne sais quoi*: "Indeed, it is something so delicate, and so imperceptible, that it escapes the most penetrating and subtle intelligence."[28] Ariste's implication is clear: those who recognize the *je ne sais quoi* distinguish themselves as belonging to an elite group of discriminating observers. The suitability of applying the term to the divine was debated. The French lawyer and Jansenist Jean Barbier d'Aucour (1641–1694) and the French diplomat and civil servant Antoine de Courtin (1622–1685) criticized Bouhours for using the notion to refer to God.[29] The idea that God and grace could not be comprehended did not concern them; rather, they objected to marking God with a term they considered base. Thus, although the *je ne sais quoi* denotes the inexplicable and the visitors to San Carlo described by Fra Juan could not articulate with language their experiences in the church, the divine setting of the visitors' mystical encounters of the ineffable indicates that Barbier d'Aucour and de Courtin would likely not have approved of using this term. They would presumably also have scoffed at the reaction recorded by the observer in the guidebook to Rome written between 1677 and 1681 on the style of the church of Sant'Agnese in Agone as having a certain something (*un je ne scais quoi*). Notwithstanding these disagreements over its use, the phrase's popularity in seventeenth-century Europe offers evidence that this period witnessed a sensitivity to deliberate incomprehensibilities that functioned to solidify distinctions between the apparently discerning elites and more ordinary people, who could be stupefied by the incomprehensible but lacked the discernment to eventually describe their experience.

As Bouhours's dialogue suggests, the *je ne sais quoi* gestures toward an understanding that a structure that broke the rules of classical rhetoric might not in fact fall short of those rules but could instead transcend them. If the deformation of San Carlo outraged some, it also helped draw visitors from all over the world to enter the church in stupefaction and awe. Seventeenth-century visual artists and architects, like their predecessors, relished in collaging formal motifs from disparate sources. To an extent unlike their predecessors, they also sought meaning in presenting shared visual languages—whether of classicism, single-point perspective, or other models—and engaging in a formal undoing or testing of those languages. By manipulating these visual languages, they achieved a heightened formal tension between the undoing or deformation and the reformation or conformation of formal motifs—thus arousing confusion among observers, cueing their attention, and leading them to appreciate and analyze the devices that separated the deformed from the ideal.[30] In the gap between the ideal and the distorted left by deformations, the Catholic Church elicited interest and drew the faithful into its arms and toward God, and ambitious families incited admiration to strengthen their positions within society.

Why Deformations?

This book asks why Catholic Reformation patrons, scholars, artists, architects, artisans, and observers were so drawn to visual and spatial deformations, how they used deformations, and what we might learn from their conscious manipulations of form. At first glance, the passion for deformation may seem counterintuitive. From antiquity through the start of the twentieth century, from Aristotle onward, Western thinkers elevated clarity and opposed ambiguity in rhetorical and philosophical writing and speech.[31] Following Giulio Carlo Argan's observation in 1954 that baroque art, in order to affect and persuade observers, reconfigures representation in rhetorical terms, scholars have repeatedly called attention to the interrelations between sixteenth- and seventeenth-century Italian art theory and practice, on the one hand, and rhetorical theory, on the other.[32] If we begin from the long-standing elevation of clarity as a rhetorical virtue and the concomitant criticism of obscurity as an obstacle to persuasive discourse—and the application of rhetorical principles in baroque visual art and architecture—we may find puzzling the investment within Catholic Reformation circles of deliberately incomprehensible deformations, artworks, and buildings that (like San Carlo alle Quattro Fontane) could be criticized for plunging observers into a perceptual confusion that was overcome only with difficulty.[33] This investment becomes less puzzling in the context of the parallel interest in rhetorical obscurity from antiquity onward, in such texts as, for example, *De finibus bonorum et malorum* (*On the Ends of Good and Evil*) by Cicero (106 BCE–43 BCE), who (at 2.15) draws a distinction between well-founded and objectionable obscurity.

Nevertheless, since the demand for clarity was also deeply embedded in Catholic Reformation policies, in line with the dominant norms of classical rhetoric, the popularity of deformations in seventeenth-century Catholic geographies might surprise. The twenty-fifth session of the Council of Trent, on December 3 and 4, 1564, framed sacred images in pedagogical terms: "Let the bishops diligently teach that by means of the stories of the mysteries of our redemption portrayed in paintings and other representations the people are instructed and confirmed in the articles of faith."[34] This passage recalls the earlier, often-cited defense of images' didactic powers that appears in the letters of Gregory the Great (c. 540–604) to the Bishop of Marseilles.[35] Gabriele Paleotti (1522–1597), a controversial figure within the Church who was marginalized after Trent, would echo the Tridentine principles in a famous treatise printed in 1582 in Bologna.[36] There Paleotti—who became a cardinal in 1565 and then Bishop of Bologna the following year—unequivocally framed sacred images as rhetorical tools: "another very notable and important effect deriving from Christian pictures, which is to persuade the people, as orators do, and draw them, by means of pictures, to embrace anything pertinent to religion."[37] Paleotti's later argument that the "principal end" of Christian images "to persuade persons to piety and order them toward God" helps to explain the decision by the Council of Trent to ban imagery with perplexing forms: "zeal and care should be exhibited by the bishops with regards to these things that nothing may appear that is disorderly or unbecoming and confusedly arranged."[38]

Indeed, the ban on perplexing forms—on deformation itself—was explicitly extended to churches. Saint Carlo Borromeo (1538–1584)—the cardinal and archbishop of Milan who actively promoted the Catholic Reformation and, perhaps ironically, to whom San Carlo alle Quattro Fontane is dedicated—prohibited deformations in all ecclesiastical buildings: "No work, whatever it may be, shall be established, made, inscribed, fashioned, or expressed, that is … deformed."[39] Carlo Borromeo's younger cousin, Federico Borromeo (1564–1631), who was Cardinal-Archbishop of Milan, likewise decried visual obfuscation in his 1624 treatise on sacred painting: "Although painters or sculptors can embellish and illustrate history as much as they like, they must not try to joust with historical truth and obscure or corrupt long-standing Christian traditions on any issue."[40] Paleotti's 1582 work had also stressed the importance of clarity in both textual and imagistic communications:

> One of the principal commendations that any author or professor of any science can receive is that he is able to explain his concepts clearly and render even recondite and difficult matters intelligible and plain to all with his accessible use of language. We may assert that the same is true of painters in general, especially since their works serve mainly as books for the illiterate, whom it is always imperative to address openly and clearly.[41]

And nevertheless, the statements of Catholic Reformation leaders like Paleotti should not be read as prescriptive support for current practice, but rather as proscriptive. As their counterexamples make clear, these leaders speak against deformation as a widespread artistic trend they do not like. It was his contemporaries' "obscure and ambiguous" paintings that Paleotti condemned as contrary to religious devotion:

> There are some who refuse to heed this precept [to address the illiterate openly and plainly], which is why we are constantly encountering paintings so obscure and ambiguous, especially in churches; rather than illuminating the intellect, as they should, and arousing devotion and heartfelt contrition, their obscurity so confounds the mind that it is pulled in a thousand ways as it tries to make out what the figure before it is, while devotion drains away.[42]

Here confusing images distract the mind, causing it to lose focus and thereby undermining religious worship. Although Paleotti is talking about paintings and not architecture in this passage, his remarks also run counter to those of friar Juan about seventy years later, for whom the stupefaction elicited by Borromini's San Carlo served to enhance, not undermine, devotion: if some in the Church feared obscurity, the artistic output of the Catholic Reformation period demonstrates that others embraced its potential to capture believers' attention. These differing perspectives on the rhetorical benefits and drawbacks of obscurity existed in sustained tension with each other throughout the early modern period. If some thought unclarity could only hinder, others hoped it could bridge toward understanding the great mysteries of faith.

For, of course, the prescriptions (or proscriptions) of the Council of Trent were not slavishly followed.[43] Nor were "deformed" works necessarily inconsistent with Church appeals for clarity, which were less concerned with form or rhetoric than with theology or doctrine, such as depictions of non-officially canonized saints with the traditional figurative apparatus of saints (such as halos). San Carlo is not a genuine departure from Church authority; it merely reflects a different understanding of rhetorical architecture, according to which deformation and perplexity may be enlisted in the project of persuasion. In general, we should recognize deformations as a departure from aesthetic tradition, but not assume that implies a departure from Catholic orthodoxy.

Seventeenth-century readers and observers in fact brought many tools and many pieces of information to those sixteenth-century Church statements to recognize that the Church's position on visual clarity allowed widespread embrace of the works then described as deformations—though where matters of faith were concerned, deformations were to be encountered within a defined journey of exposition that ultimately resolved in a doctrinally orthodox image or object of contemplation. In the chapters that follow, we will see that knowledge, salvation, truth, and understanding of the universe and divine purpose within it were recognized to take place in the space between—and most particularly in the transition between—obscurity and clarity. In numerous works that seem to contradict a stated Catholic proscription against "unclear" art, we can find that the strain these works created between ideal form and deformation was resolvable with a reform that supported the goals of the Church. That strain—and its resolution—was rhetorical in its power to persuade, if not immediately.

While Fra Juan observed that deformation was able to amaze and attract, the French mathematician, physicist, and religious philosopher Blaise Pascal (1623–1662) proposed a possible explanation for why deformation could be alluring. Discussing the art of persuasion around 1658 or 1659, Pascal highlighted certain logicians' valuation (or, in his view, overvaluation) of the inaccessible, the incomprehensible. In particular, he censures the mnemonic poem of scholastic logicians that has been identified as a possible source for the term "baroque": "*Barbara* and *baralipton* do not mold reasoning."[44] He suggests individuals are drawn to incomprehensible forms because they attach value to inaccessibility:

> One of the principal reasons why those who enter the realm of knowledge are carried so far from the true course they should follow is their first notion that good things are inaccessible. That is why they call them grand, high, elevated, sublime. This undoes everything. I should like to call them low, common, familiar; these suit them better; I hate swollen words ... [45]

Pascal's rejection of "swollen words" presents a philosophical analogue to the rejection of deformations among Borromini's critics and underscores how ancient calls for clarity in rhetoric could potentially be deployed against visual deformations. In this passage, Pascal pinpoints the appeal of the inaccessible in a belief, which he takes to be faulty, that the impenetrable is somehow superior to the penetrable. On the one hand, his remark suggests we might consider whether and how the artists and their observers introduced in the pages that follow subscribed to this belief that inaccessible things are by default good: did they deploy deformations and incomprehensibility deliberately, to manipulate others into believing that what they could not understand was in some sense elevated? On the other hand, Pascal's criticisms were not shared by all contemporaries: others, as these chapters will show, did not discourage incomprehensibility if it ultimately contributed to supporting religious mysteries and phenomena that could not be neatly outlined in words or images.

Deformations are rhetorical in the sense that they are persuasive, but they persuade via other means than strict clarity. They initially transcend the ability to articulate experience, offering a moment of transcendence within a process where comprehension is delayed, if not ultimately denied. In particular, visual deformations privilege a rhetorical approach that delays resolution, certainty, and clarity to emphasize authority and, often, to attach greater attention to a resolved image.

The Deformation

In the early modern era, the notion of reformation was generally conceived of in relation to a prior deformation. For instance, Protestants saw themselves as reforming a deformed church. In period conceptions, the notions *deformatio* ("deformation") and *reformatio* ("reformation") are inextricably intertwined, because both assume the existence of an ideal configuration. If a deformation moves away from that ideal, a reformation moves back toward it. Deformation was, then, understood as a derivative form that stands at a recognizable distance from an ideal and carries within it the stamp of that ideal, if changed and distorted. And to write of deformations calls attention to the overlapping implications of deformations and reformations in artistic and religious domains.

As we have seen, the higher-level goal of Catholic Reformation art and architecture was to function as a rhetorical instrument that would, in Paleotti's words, "persuade persons to piety and order them toward God." Catholic Reformation deformations, then, embodied religious ends, which I will discuss in the chapters that follow, even if not all of these goals were apparent to all observers. That ecclesiastical frame meant that Catholic visual artists and architects were not free to express themselves however they wished; a deformation could be original and bold, but not unorthodox in its key points or motivations. And it is Catholic deformations and reformations, in particular, that I will consider in this book. Even though objects similar to those I discuss were also employed or generated in Protestant contexts, Protestants deployed those works toward doctrinally different ends; I have kept discussion of Protestant works to a minimum rather than attempting to write a large-scale comparative work. Moreover, I write primarily in dialogue with scholarship in art and architectural history, although I situate my work in a cultural-historical context. In a period of crisis, the Catholic Church upheld the functional ideal Paleotti highlights, the goal of bringing people toward piety and ordering them toward God—whether in the production of visual art and architecture or in other endeavors. The pursuit of ideal practice also, of course, engaged the larger question that drove the period's reformations: whether and how the Catholic Church should itself be reconfigured, that is, how it might be returned to an original ideal form.

In Borromini's time, the endeavors of Martin Luther would not yet have been exclusively defined as the "Reformation"; that term would emerge by the mid-1700s, as Lutheran historians in Germany attached the label to a period that began when Luther posted the Ninety-five Theses in 1517 and ended in 1555 with the Peace of Augsburg, in which Catholics and Lutherans agreed that both faiths could exist in Germany provided that in each of its territories' subjects adopted their ruler's religion.[46] In the nineteenth century, Leopold von Ranke (1795–1886), with whom the Swiss cultural historian Jacob Burckhardt (1818–1897) studied early modern history in Berlin, helped establish the meaning of the Reformation among historians beyond Germany through his lengthy work *Deutsche Geschichte im Zeitalter der Reformation* (*German History in the Epoch of the Reformation*, 1839–1847).[47] Even by the late eighteenth century, the name was sufficiently ingrained for the related idea of a "Counter Reformation" (*Gegenreformation*) to emerge and gain approval from German Lutherans; the Lutheran historian and law professor Johann Stephan Pütter (1725–1807) first used the term *Gegenreformation* in 1776.[48] As Evonne Levy has stressed, when German historians employed the term "Counter Reformation" during the mid 1800s, they referred to the Jesuit recovery by conquest of areas that had been Reformed.[49]

If today the Reformation label commonly denotes early modern Protestantism, in the seventeenth century *reformatio* was also associated with Catholic endeavors, and for centuries before Luther the term was actively used in discussions of religious practice.[50] As John W. O'Malley has noted, *reformatio* features in ecclesiastical writings from the eleventh century on; it appears with increased frequency in works relating to the Council of Constance (1414–1418), and 50 percent of the directives issued by the Council of Trent (1545–1563) pertain to reform (*de reformatione*).[51] In these ecclesiastical writings, *reformatio* generally signifies corrective action, presented with acute self-awareness by ecclesiastical leaders who sought to return clerical conduct to the moral and doctrinal purity of the primitive church.

In the early modern period, whether Luther was seen to have led a "reformation" that would bring his followers closer to salvation or a "deformation" that would turn them away from God hinged on what observers made of his protests against the papacy and the Catholic Church. These were the terms the German theologian Jerome Emser (1478–1527) used in a combative response to Luther's fierce attack on the papacy: "Is this a reformation or a deformation?" (*Ist das ein reformation oder deformation?*)[52] In posing the question in these terms, Emser implied that those who identified Luther's message with a process of reforming, those who thought he was bringing Christianity closer to its ideal form, were mistaken. In Emser's view, Luther was leading his followers astray.

Some Catholics agreed that the Church had undergone a deformation that demanded its reformation. But they diverged from Protestants in insisting that the papacy should lead this reform effort. In 1522, Pope Adrian VI (1459–1523), instructing the papal nuncio Francesco Chieregati (1478/1479–1539) on what to say about Luther and his ideas to people assembled in Nuremberg, called the Church a "deformed bride" and urged patient reform:

> We ... have never sought to gain this pontificate ... we submitted, therefore, with neck held high for the office, not out of a desire to dominate, and not in order to enrich our relatives, but in order to obey divine will, in order to reform its deformed bride the Catholic Church, in order to aid the oppressed, and to raise up and praise the learned and those gifted with virtue, who have lain neglected already for a long time, and indeed to do everything else that is appropriate for a good pope and the most legitimate successor of the blessed Peter to do; although no one ought to marvel, if he does not immediately see all the errors and all the abuses fixed by us; the disease is too old, and not simple, but various and manifold, and we must proceed step by step in its cure; first we must encounter graver and greater dangers, lest we, who wish to reform everything equally, stir up everything; all sudden changes (says Aristotle) are dangerous for the republic; he who cleans his nose too much draws blood.[53]

The Pope suggests here that once the deformed Church is reformed, it will be able to accomplish its goals more effectively. One of those goals—perhaps its most important overarching goal—is leading the people toward piety.

Further disagreements over whether the Catholic Church had led a reformation or deformation emerged in the wake of the Council of Trent. In the introduction to his history of the Council of Trent, written in direct response to Paolo Sarpi's (1552–1623) history of the council, the cardinal and historian Francesco Maria Sforza Pallavicino (1607–1667) titles one section "Whether the Council Reformed or Deformed the Ecclesiastical Order" (*Se il Concilio habbia riformato, o deformato l'Ordine Ecclesiastico*).[54] Pallavicino unequivocally characterizes the council as a reformation that brought Catholicism closer to its original ideal configuration; the council, he argues, returned the Catholic Church to a state in which it could fulfill its aims and increase worship and goodness in Christianity.[55] He claims that anyone who has eyes to look to the past and the present will admire the immense improvement of ecclesiastical customs, or forms of behavior and procedure, that the council instituted. He stresses improvements in such domains as the decorum of sacred functions, the observance of the ecclesiastical rites, worship, the modesty of dress, and the way of life. He asserts that even though a hundred years have passed since the council ended, its healing and restorative virtues thrive.

Beyond these institutional discussions regarding the deformation / reformation of the Catholic Church, seventeenth-century scholars analyzed the implications of the deformation / reformation of visual forms and of analogies to visual forms. In the *Vocabolario* (*Vocabulary*) of the Accademia della Crusca, "to deform" (*difformare*) is defined as a negative rearrangement of form: "To spoil the form, take away the beauty."[56] "To reform" (*riformare*) is defined as a rearrangement of form: "To reorder, and give new, and better form."[57] Early modern theologians would not have welcomed the term "new" in this definition—and would not have used it when talking about religious reform, but the *Vocabolario* reinforces the religious connotations. Under this definition, the *Vocabolario* offers an example of spiritual reformation from the writings of Saint Gregory the Great: "But he reforms, and reinforces his soul with hope."[58]

This notion of the soul's reformation is also portrayed in Jesuit meditative handbooks that imagine the soul as a painter, who reforms its appearance through meditative prayer.[59] The *Veridicus Christianus* (*True Christian*, 1601) by the Flemish Jesuit Johannes David (1545–1613) describes meditative prayer as a mode of engraving or of painting the *imago Christi*, or "image of Christ," onto the soul of the exercitant.[60] For David, to be a "true Christian" (*veridicus Christianus*) requires picturing Christ's lessons and story inside oneself: by imitating Christ more closely, one may reform one's soul, mind, and heart with the aim of being closer to the original example of Christ. The second title page to the *Veridicus Christianus*, engraved by Cornelis Galle (1576–1650), depicts Christ bearing the cross on Golgotha (Fig. 9).[61] To his sides and below him are ten artists working at their easels. While the painter in the foreground portrays Christ carrying the cross, other easels display further important scenes from His life, such as the Adoration of the Magi and the Last Judgment. In addition, two foreground paintings to the left and right—displaying the face of the Devil and a female confronting two beasts—denote how sin can pervert the virtual imagining of Christ.[62] True Christians must become good painters of the life of Christ to form Him within themselves. Christians must not deform His image in the way shown by the two artists who represent not Christ, but Satan and a woman

Fig. 9. Cornelis Galle (engraver), second title page from Johannes David's *Veridicus Christianus*, 1601–1603. Engraving printed on paper, 8.7 × 6.1 in (22 × 15.5 cm). BnF, département Arsenal, Paris [EST-1353].

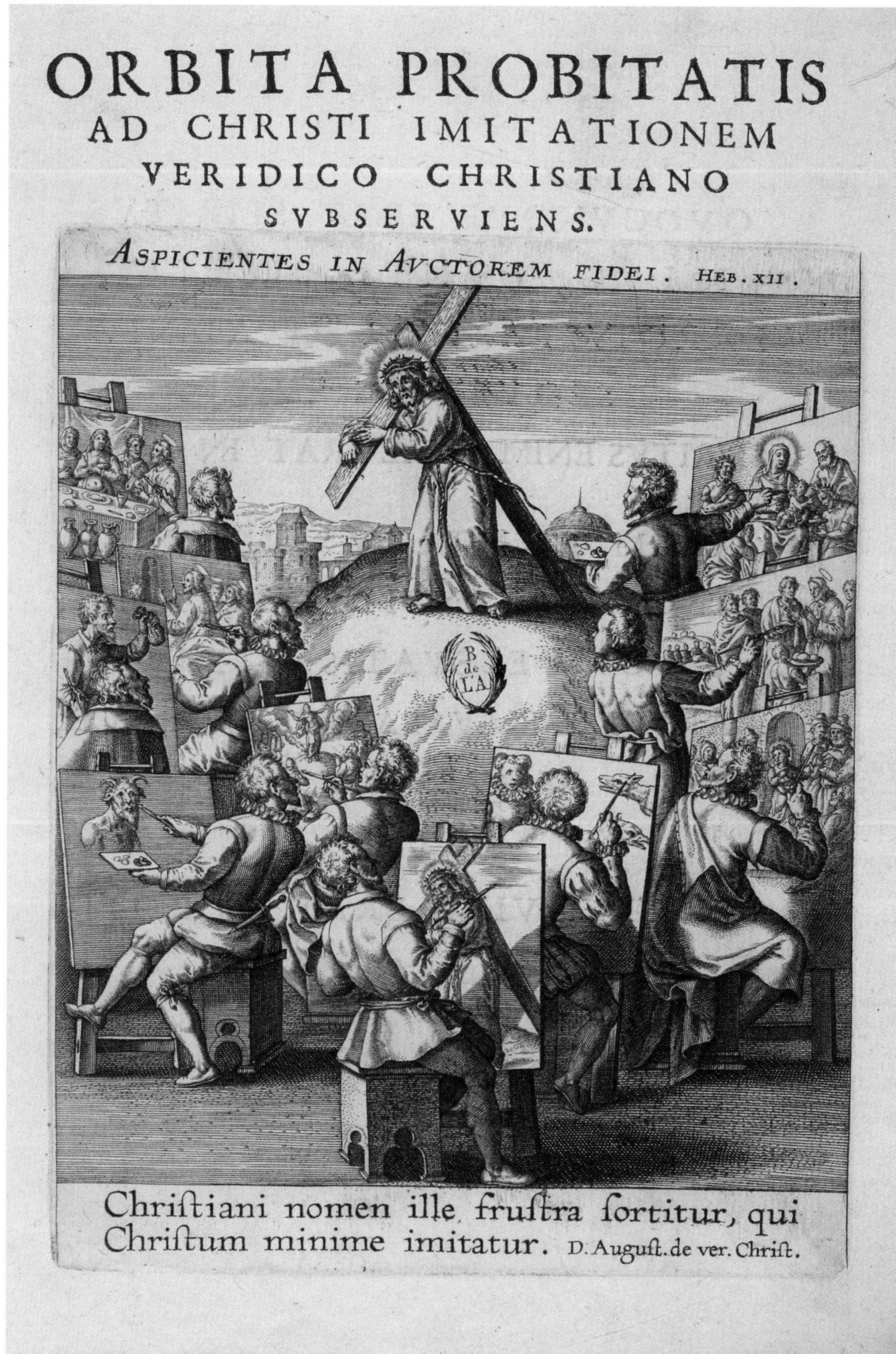

alongside two beasts. They must conform themselves to Christ. The annotation *Aspicientes in auctorem fidei* above the scene refers to Hebrews 12:2, "Looking on Jesus, the author and finisher of faith," underscoring the message to picture Christ. In the *proemium*, David clarifies the comparison between the act of painting and the Christian project of reforming one's life after the model Christ offered:

> Just as any excellent painter, with all his industriousness, tries to copy from the life that which he had presumed needed to be imitated by art, so this care falls upon Christians to imitate Christ our Savior in their life and sacred conversation, and to show Christ in themselves, as if depicted from the life. … And in this representation of Christ, in this way, at length, they make progress in every type of virtue, so that they represent Christ in themselves as if their prototype, just like in a painting.[63]

David holds that if Christians mimic Christ in their external actions, they will move toward a more Christlike interior state.

That the exercitant might reform—positively rearrange—his soul into a picture of Christ also features in the beloved treatise *Via vitae aeternae* (*Road of Eternal Life*, 1620) by the Belgian Jesuit Antonius Sucquet (1574–1627). The motif is manifest in Sucquet's definition of meditation: "To meditate is to consider in one's mind, and as it were, to paint in one's heart, the mystery or some doctrine of the Life of Christ, or even the perfection of God, by [representing] the circumstances: people, actions, words, place, and time."[64] It appears again in the *Via vitae aeternae*'s engravings: in the eleventh *imago*, for instance, the Soul of a votary paints the nativity in a series of memory places diagrammed onto a heart-shaped panel (Fig. 10). This image and others like it illustrate the soul as a painter, who pursues self-reformation through meditative prayer. In a related vein, the Jesuit theologian Jerome Nadal (1507–1580) asked Ignatius of Loyola (1491–1556) over a period of almost four years to describe how God went about forming him in spirit from the moment of his conversion, since Nadal believed that this account could be of benefit to the society.[65] In sum, Catholic thinkers of the period conceived of spiritual reformations as giving a better shape to deformed souls, by which they meant a form closer to the original model of Christ.

When early modern art theorists debated the relationships between deformations and reformations, they tended to worry about "corruption" or "sordidness," with deformation articulated as a negative change, a removal or obscuring of beauty that needed to be reformed or restored. Paleotti decries "Catholics, who, while retaining the use of images, have nevertheless corrupted and deformed their dignity in various ways," and sees the sacred Council of Trent as aiming to amend those abuses by Catholic practitioners.[66] He intends his treatise to "reform the Catholic."[67] Paleotti further summarizes his aim later, as he differentiates between things that can and cannot be depicted: "Now to begin our discourse about these abuses—so that, with the sordidness that is deforming the beauty of images laid bare, their purity may more easily be restored."[68] In describing a restoration of purity, Paleotti is describing a return to an ideal, a reformation.

Paleotti's recommendations around avoiding "sordidness" resonate with Bellori's characterization of San Carlo as "ugly and deformed" (*brutta et deforme*) or his choice to brand Borromini as a "corrupter" of architecture, debasing and perverting its forms from

Fig. 10. Boëtius à Bolswert (engraver), eleventh *imago* from Antonius Sucquet's *Via vitae aeternae*, 1620. Engraving printed on paper, 7.1 × 4.5 in (18 × 11.4 cm). Princeton University Library, Special Collections.

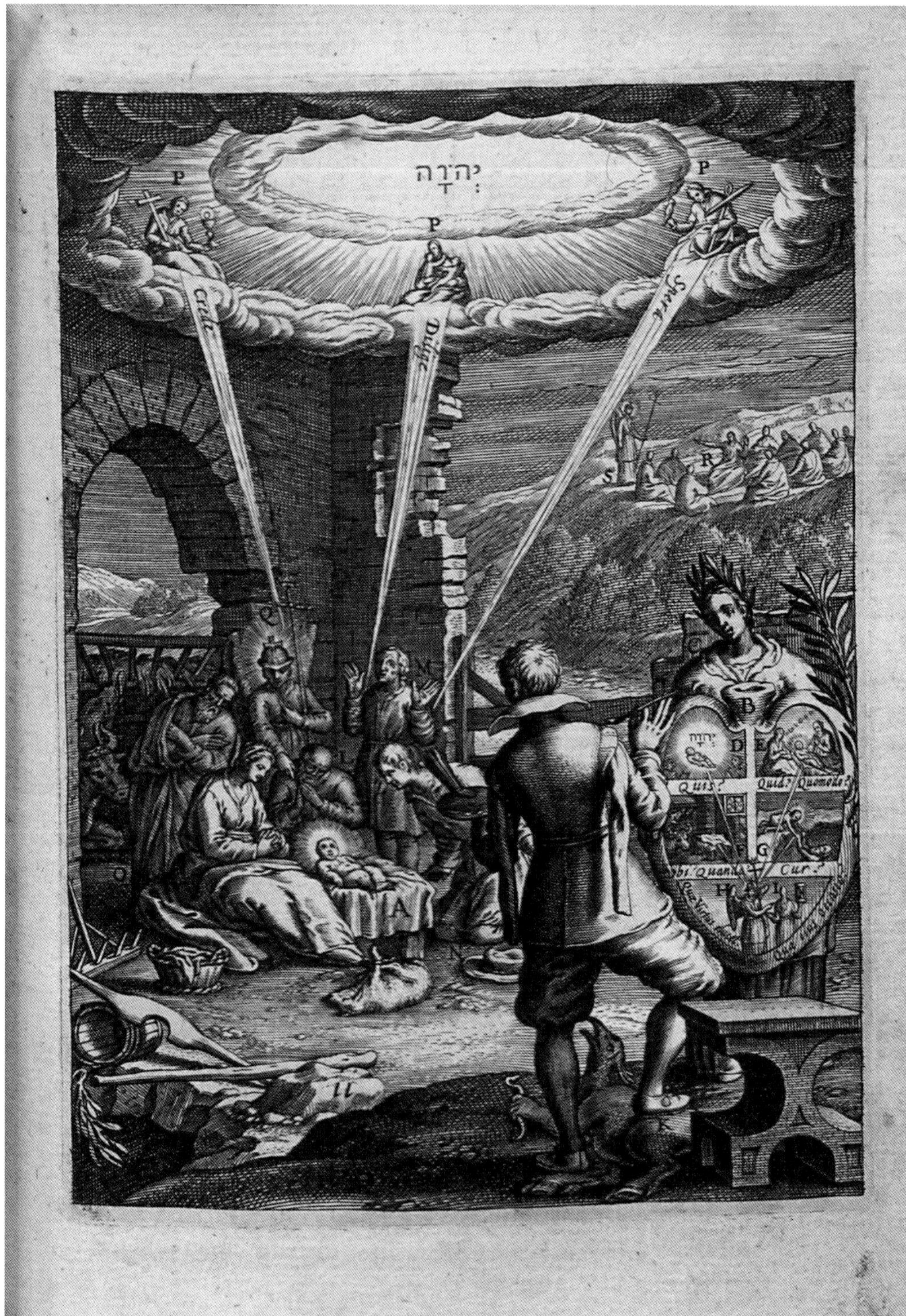

the "preferred" condition of Italian classicism. Bellori's art-theoretical writing in *L'Idea* (*The Idea*)—a discourse on painting, sculpture, and architecture first delivered in 1664 as a lecture to the Accademia di San Luca and then published in 1672 as a preface to his biographies of artists—helps us make sense of his condemnation of San Carlo and supports a reading of deformation as a declension from ideal forms. Bellori opens *L'Idea* by describing the structure of the Aristotelian-Ptolemaic cosmos, in which the sphere of the moon marks the boundary between the sublunary realm below and the superlunary realm above. If the superlunary bodies are "eternally perfect and most beautiful," Bellori describes the sublunary bodies as "subject to change and to ugliness."[69] He explains that "owing to the inequality of matter, forms are altered, and human beauty in particular is confounded, as we see in the innumerable deformities and disproportions that there are in us."[70] In the sublunary realm, the "first forms" (*prime forme*), or Platonic ideas, are deformed. Bellori argues that artists and architects should aim for the beauty of the celestial realm in their works and not the deformations of the sublunary world.[71] He holds, moreover, that those artists and architects who do not follow the examples of the ancients deform the first forms: "regrettably, those who transform [the one beauty] with innovations deform it, since ugliness stays close by beauty, just as the vices can touch the virtues."[72] And yet, Bellori recognizes here that deformations remain near to their ideal originals. A deformation, although distorted, is still connected to its ideal and still reflects that ideal in some way.

In artistic deformations, proximity to an ideal could be simultaneously part of the attraction and part of the problem. San Carlo was so stupefying because observers could intuit the ways in which it related to and deformed classical ideals. This church and other Catholic deformations of the period propose an alternative appraisal of the relationship between deformation and ideal forms, one that invests value in the work of deforming or unraveling and *re*forming a relationship to an ideal. San Carlo, in confusing visitors and requiring observers to think about the ideal rather than just process it without reflection, offered a different sort of persuasive capacity that was not less than, and potentially greater than, that offered from the conformity of classical forms.[73] Moving structures away from expected forms became not a diminishment but a chosen practice that generated a new kind of experience for the observer.

In the early modern era, deformation was multivalent: at times understood as a deformation of all creation, elsewhere as a deformation of entire categories of created beings, or in some cases as a departure from the norm at the individual level. Indeed, as Bellori implies, all sublunary entities were understood to be deformed, not only religious institutions, souls, and certain works of art and architecture. Aristotle had also already introduced the idea that a class of bodies could fall short of an ideal, as when he criticized lobsters for using their claws for locomotion rather than their natural purpose of grasping.[74] He does not single out an individual lobster but faults all lobsters for their alleged shortcomings. In a similar vein, Aristotle characterized all females as deficient when compared to males, deficient because they cannot produce pure semen that can act as the origin of an offspring's soul.[75] In treating the grasping claw or the purified semen-producing body as ideals, and the locomotive claw or the non-purified-semen-producing body as aberrant, Aristotle gestures toward what seventeenth-century thinkers would have seen as deformation.

While continuing to recognize all sublunary entities as deformed from superlunary ideals, thinkers of the early modern era also applied "deformation" or "deformity" as a label for entities and individuals they saw as markedly deviating from ideals or norms. A person or object was presumed to exist in relation to the ideal as on a spectrum, whether distance was an opportunity for comparison (as in Bellori's recognition that "ugliness stays close by beauty") or a representation of degree, as when Richard III describes himself as "Cheated of feature … Deformed, unfinished … scarce half made up" in Shakespeare's *Richard III* (1.1.19–21, c. 1593–1595). While Shakespeare and contemporaries like Francis Bacon (1561–1626), as in his essay "Of Deformity," debated the moral associations of departures from the norms of the human body, in what follows, I do not use the term deformation in relation to living bodies. I do follow the period conception of deformation as a state that characterizes all sublunary entities, from postlapsarian souls and institutions to artworks, and while recognizing the contemporary perspective that all artworks necessarily fall short of "first forms," I pay particular attention to certain works that were seen to deviate more conspicuously from ideals—and especially from classical ideals—and to the ways in which those deformations were articulated.

And yet it must be stressed: despite abundant contemporary (and even classical) theoretical opposition to deformations, patrons and artists of the Catholic Reformation invested considerable financial and creative resources in producing works that complicate classical ideals to embrace, as does San Carlo alle Quattro Fontane, what the Swiss art historian Heinrich Wölfflin (1864–1945) would define as *Unklarheit* ("unclearness"). If some practitioners avoided deformations as problematic, others openly applied the term to their work, because its confusions were understood to yield positive spiritual benefits.[76]

The Baroque

How we think about the seventeenth-century relationship to deformation is crucial to how we understand what constitutes the "baroque"—a notion that, although fundamental to the discipline of art history, is also among its most debated terms.[77] The baroque as a category remains important to various cultural productions, across artistic, literary, rhetorical, and musical styles and extending even to the description of political systems and science.[78] Levy and Jens Baumgarten have called the baroque a "miraculous cake" that is cooked repeatedly and through an unending assortment of quickly forgotten recipes: ingredients, methods, even outcomes may vary, but the cake remains "baroque."[79] These critics articulate their surprise at the large assortment of baroques in existence—and at how easily scholars prepare new ones when they have already devoured countless others. Yet despite its elusive, multifarious nature, the baroque remains a category important to understanding the cultural productions in seventeenth-century Catholic geographies.[80]

This book's account of deformation belongs to a particular tradition of defining the baroque, articulated in the past and more recently. I relate my discussion to Wölfflin's *Unklarheit,* a quality he attached to the baroque: "Diverse as the stylistic variants of the postclassical era may be, they all have this one remarkable characteristic: their appearance somehow recedes from full intelligibility."[81] Wölfflin had famously, two decades earlier, described a progression from a strict Renaissance style to a free and painterly style that he qualifies as a progression "from the formed to the formless" (*vom Geformten zum Formlosen*); in 1915, he declared that "the seventeenth century found beauty in the darkness that engulfs form."[82] Although I share Wölfflin's emphasis on unclearness and his attention to form and formal arrangements, I supplant his notion of the "formless"—which makes little sense as a descriptor of art and architecture that necessarily has a form, even if it is an irregular one—with the notion of deformation, or the process of undoing and testing forms and formal arrangements in relation to an ideal.[83]

I align my work with scholars who have more recently centered form in their conversations around the baroque, as when Victor I. Stoichita explains that seventeenth-century paintings grow "self-aware" and depict their own formation and structure.[84] More recently, Levy has noted the role of "forming and reforming" in Jesuit imagery in relation to the process of subject formation.[85] This study builds on this sense of the baroque as distinctly attentive to form, and on Aaron M. Hyman's still more recent work on "conformity"—from the Latin *conformis* ("to share forms") and from the Spanish verb *conformar,* which appears in colonial notarial contracts to signal that commissioned works had to conform to certain compositions. For Hyman, "conformity" is "a guiding principle" of baroque artistic practice in relation to copying models in print media.[86] Conformity is useful for thinking about the baroque as attentive to agreement and—implicitly—disagreement among forms. In making explicit what is implicit in Hyman's discussion, I argue that it is the terrifying tension that boils up in the gap between a deformation, or what might be termed a "nonconforming copy," and its ideal that drives baroque artistic practice and distinguishes it from earlier practices. Recognizing this gap as a constitutive component of the baroque enables us to account in a more precise fashion for artistic modes in Catholic geographies over the long seventeenth century. If creators of deformations saw the fear of deformations that drove so much of the theoretical discourse, instead of allowing that fear to drive them away from the practice of deforming, they harnessed it toward the rhetorical ends of strengthening class hierarchies and drawing stupefied individuals into the arms of the Church, where the faithful could be pointed to the divine.

Scholars disagree about the etymology of the term "baroque": It may derive from the mnemonic word *baroco,* which in scholastic logic refers to a particular form of valid argument.[87] Alternatively, it may well descend from the Spanish and Portuguese words *barocco* and *barueca,* denoting a peculiarly formed pearl and deriving from the Latin *verruca* ("wart"). The satirist and lexicographer Antoine Furetière (1619–1688) defines the French word *baroque* in his *Dictionnaire universel* (*Universal Dictionary,* 1690) as "a jeweler's term, which is said only in regard to pearls that are not perfectly round."[88] By the eighteenth century, the term was also applied to the visual arts. In 1757 the French writer Antoine-Joseph Pernety (1716–1796) described "baroque" in his dictionary as a term for "what is not according to the rules of proportion, but of caprice."[89]

If one account of the term's origins affiliates the "baroque" with a form that achieves the philosophical ideal of valid argumentation, the other associates it with the deformed. Even as style labels are being reevaluated, I use the term in this study, because this etymological tension between ideal form and the deformed defines the baroque as a conflicted approach to artistic composition that is essential to the phenomenon under discussion in this book.

Attention and Discernment

The Deformation examines a rich trove of understudied sources—from gigantic illusionistic frescoes to miniature ivory towers—to make sense of the interest some Catholic Reformation thinkers directed toward objects that evoked deliberate visual incomprehensibility and confusion. In examining that interest in the chapters that follow, I will consider deformations not only in the sense implied by the twin etymologies of the baroque—in terms of an objective relationship to an ideal, whether conforming or nonconforming—but also in subjective, experiential terms, as these varied deformations act upon the senses of observers. I show how deformations cue first the attention of observers and second their visual discernment.[90] Deformations prompt observers to notice and then to assess how their departures from the ideal challenge their expectations, as I have suggested the Marquis de Seignelay would have noticed and then evaluated the ways that San Carlo confounded the expectations he had carried with him as he traveled the 150 meters from Sant'Andrea. Individuals who possessed visual discernment could draw on accrued observational experience to perceive meaning, to distinguish, to recognize, and to describe forms that inexpert viewers might regard as meaningless. In demanding discernment, then, deformations invited an incremental process that is more expansive than perception. While the latter extends only to becoming aware of phenomena through the senses, the former combines sensory perception—seeing, hearing, touching, tasting, and smelling—with the cognitive processes of understanding, recognizing, identifying, and judging.

In tracing visual experience in the intellectual climate of the Catholic Reformation and European colonial expansion, this book thus offers a prehistory of attention in the West; it challenges an assumption that our present anxiety around attention first emerged in the late nineteenth and early twentieth centuries in relation to systems of modernization.[91] According to this line of thought, human subjectivity was radically reconstructed in the late 1800s, such that individuals could characterize and determine who they were through their ability both to pay attention to particular stimuli and to disassociate themselves from an expansive terrain of sensory data. This elevation of focused attentiveness is said to have been entwined with social and economic changes and more precisely with the shift toward centralized corporate capitalism and away from older modes and technologies of exhibition and consumption. In what follows, I identify an earlier reorientation and expansion in mental engagement closely associated with reformation in its multiple early modern manifestations, with Catholic reformers in particular concerned with holding attention in the context of deformation, as we have seen in Paleotti's discussion of paintings in churches whose "obscurity so confounds the mind that it is pulled in a thousand ways." The chapters that follow explore how Catholic Reformation leaders and artists harnessed deliberate incomprehensibility to capture observers' attention and to direct their minds and, in the end, their conduct more generally, to sustain their relation to the Catholic Church to lead the faithful to God and to build, maintain, and reinforce class identities.[92]

The contrast between deformations and reformations is a recurring theme in the visual experience of Europe in the sixteenth and seventeenth centuries. Staying focused on this tension allows us to resolve certain contradictions scholars have posited regarding viewing experiences in this period. On the one hand, some scholars, such as David Freedberg, have held that this period was characterized by the rise of naturalism, empiricism, and a commitment to using visual representations in the transmission of knowledge among networks of viewers who trusted what they saw—in other words, who trusted that they observed without distortions.[93] On the other hand, scholars including Carl Havelange and Stuart Clark have posited that this moment was characterized by skepticism and deceit, a mistrust of the visual and visual experience, as well as an enjoyment of playful images that were unstable—in other words, an enjoyment of deformations.[94] If we think about the early modern viewing experience with the scholars in the second group, we are in the territory of St. Paul's famous remark on the limits of perception: "We see now through a glass in a dark manner; but then face to face. Now I know in part; but then I shall know even as I am known."[95] And yet a related biblical text on perception—"For the invisible things of him, from the creation of the world, are clearly seen, being understood by the things that are made"—underlies the argument that viewers expected, or preferred, to see clearly.[96] The tension scholars observe in viewing experiences of the early modern period is thus present in the earliest Christian sources, and the religious viewing experience, broadly speaking, might be understood as a journey from misapprehension or partial understanding and uncertainty toward a divinely provided clarity and certainty. Juan Caramuel y Lobkowitz (1606–1682), the polymath at the center of chapter 5 below, who developed the theory of "oblique architecture," stressed human inability to discern entities' nature and argued that architects should compensate for inevitable visual misapprehensions. In other words, an expectation of visual misapprehensions—a recognition of the limits on humans' capacity for discernment—came to motivate a theory of architecture designed to reform deformations. Claude Perrault (1613–1688), a second theorist, by contrast, praised humans' perspicacity and argued that such architectural recommendations would in the end result in deformations as opposed to reformations.

The Deformation considers this tension between what we might label perspicacious viewing and misapprehensive viewing—perceptions differing in their understanding of the observer's experience. And, in turn, these differing perceptions of the beholding experience informed discussions concerning the artist's own responsibilities around truthful representation: we thus see a persistent contrast in seventeenth-century religion, philosophy, and visual culture between what Italians would call *inganno* ("deceit") and *disinganno* ("the revealing of a deceit"). This tension has also been observed in the culture of seventeenth-century Spain, where Felipe Pereda has explored the contrast between *engaño* ("deceit") and *desengaño* ("undeceit") as these terms were employed to scrutinize the paradoxical forms in which painters, writers, and theorists reflected on and framed visual experience and illusion.[97] The period that saw the rise of thought-out, carefully articulated, and communal practices of observation and representation was also the era of mystic visions and philosophical, religious, and artistic questioning of the limits of visual experience. It is precisely the multivalent nature of vision—and the multivalent explanations of visual experience that coexisted in this period—that propagated the investment in matters of visual representation and the diverging views on how far an artist should appeal to intuitive, immediate perception as opposed to a more temporally extended practice of discernment.

Jesuits, Minims, Trinitarians (who patronized San Carlo alle Quattro Fontane), and other Catholic Reformation groups were preoccupied with visual incomprehensibility, and beyond the biblical foundations for that preoccupation we can look elsewhere, to the classical tradition. That preoccupation could also be grounded in the ancient Platonic distrust of the senses or in conflicting ancient epistemologies' interpretations around the regulation and productive use of sensory evidence. But Catholic Reformation leaders' attention to the unclear is not merely a Christian recapitulation of Hellenistic philosophies.[98] New in the seventeenth century is the surge of treatises on perspective, dealing with such topics as illusionistic artworks, distortions in visual experience, and the making of visual art. These works, published in rapid succession, helped to drive intense interest in and fierce debates concerning the interrelations of theoretical instructions, the judgment of the eye, artistic practice, and the goals of visual art.[99]

In France, theorists of artistic production debated the relative importance in facture of perspective, on the one hand, or visual discernment (i.e., the eye's capacity to judge), on the other. The printmaker and writer Abraham Bosse (1602/4–1676), who published several works interpreting the views of the mathematician and engineer Girard Desargues (1591–1661), was expelled from the Académie Royale de Peinture et de Sculpture in 1661 and prohibited from teaching perspective there following a series of disagreements with the painter and art theorist Charles Le Brun (1619–1690). Bosse criticized members of the Académie—one of whom he later dubbed "the new reformer of the alleged depravities of perspective" (*le nouveau Reformateur des pretendues depravations de la Perspective*)—for supposedly teaching youths that they did not need to master geometrical rules and could, in creating artworks, rely on how the eye saw things.[100] Le Brun did in fact accept perspective as important to compositional practice, but he considered it as only one component, rather than the entirety, of what goes into painting. Bosse responded to his expulsion by founding his own drawing academy at Saint-Denis-de-la-Chartre; Le Brun in turn ensured that Bosse's school was shut down by royal edict. This seventeenth-century crisis concerning facture further motivated the fascination with form, deformation, and compositional practice that I take to be emblematic of the baroque.

Beyond the outpouring of perspective treatises and the intensity of debates over the rules and ideals that should govern the making of art, further novelty arose in the realm of material culture. New material forms and the increased production of a variety of optical contrivances—from multifocal beveled lenses to magic lanterns—integrated observers into the experience of an inverted universe. To an unprecedented degree, artworks, architecture, and the experience of two- and three-dimensional works—not unlike such scientific instruments as the microscope or telescope—made it possible for observers to become aware of how they were seeing, as they were seeing.

The perspectival colonnade in Palazzo Spada in Rome analyzed in chapter 1 raises questions around the gap between what is seen and what is in fact present. Similar questions are raised in chapters 2, 3, and 4 by anamorphoses—images deliberately constructed to appear incomprehensible, resolving into comprehensible shapes only when observed from precise vantage points or through reflection from suitable mirrors—and by the multifocal beveled lens, which generates a single image from another image consisting of multiple separate forms. Observers who were familiar with telescopes or microscopes but had not previously encountered the multifocal lens would expect, upon looking at an image with multiple separate forms through the lens, to find a magnified version of that same image. Instead, they would find a novel singular form, reformed from the several forms observed by the naked eye. By subverting expectations, this optical device's reformation undermined observers' presumptions concerning their capacities to decipher forms. Such lenses raised questions about an individual's capacity to see and to understand or to discern.

To consider the seventeenth-century theorization and cultivation of attention and visual discernment—achieved through deformations and reformations that evoked deliberate visual incomprehensibility—is to offer important examples of how the Catholic Reformation and the "scientific revolution" intersected, building on current directions in scholarship.[101] Without questioning the Catholic Church's intense use of visual culture to persuade souls to convert, and without disputing the reality that the Church persecuted Galileo (1564–1642), I join other scholars in seeking to widen the picture of the baroque to include the functions of deformations and other visual culture in the presentation of scientific knowledge—especially concerning visual experience, catoptrics (the branch of optics that deals with reflection), horology (the study of time and the art of measuring it), astronomy, and natural history—under the auspices of Catholic Reformation leaders and scholars. Conversely, I also demonstrate how scientific advancements and experiments in the visual arts and architecture with deformations and reformations functioned as crucial instruments of theological thought and teaching.

Book Structure

This book is organized around a set of questions: Why were Catholic Reformation patrons, practitioners, and observers so drawn to deformations that trigger visual and spatial confusion? How did they deploy deformations, and what do these uses show us? By considering paintings, prints, drawings, sculptures, architectures, and theoretical writings in relation to these queries, I emphasize that Catholic thinkers invested importance in visual and spatial deformations to achieve a long list of ends: they turned to deformations in search of pleasure, or to transform exhibitions of wealth into conversation pieces, to bolster their status within systems of class and power, to establish philosophical and theological truths, to represent known prophecies, to display the enigmas of visual experience, to reshape and reveal strategies of deception, to represent a revelation of God's truth, to prolong and train observation, to revive and invent artistic forms, and, ultimately, to shed light on the theological mysteries of the Church itself (such as the Incarnation and the Eucharist).

The Catholic characters who produced and patronized works that were referred to as deformations included cardinals, bishops, teachers, and students in the Society of Jesus and Minim order and other clerics as well as members of noble or ruling families or families who aspired to move up within the social pecking order. I focus on contributions of Jesuits and Minims because members of these two communities persistently emphasized visual experience, basing meditative exercises on intricate visualizations and performing elaborate experiments with optical devices and imagery.[102] I define an intellectual cohort of figures who knew each other personally—or at least are likely to have known of each other, from seeing one another's paintings, buildings, and sculptures or by reading one another's books, or simply by word of mouth. This is not, then, a systematic study of all Catholic Reformation thinkers or of all Catholic Reformation deformations. I consider a selection of early modern Catholic theorists and practitioners—some quite mainstream within the Church and others more controversial—who engaged in different ways with a set of ideas around representing reality and guiding visual perception and discernment. Moreover, many of the artworks and structures addressed in the chapters that follow are situated within walking distance of each other in Rome, as indicated by the key locations I have highlighted on a map of the city created by the Dutch publisher Pieter van der Aa (1659–1733) and the engraver Giovanni Battista Falda (c. 1643–1678) in 1705

(Fig. 11). Although I approach Catholic geography from Rome and Italy as the central home of the Catholic Church, I also gesture to broader Minim and Jesuit geographies in France, China, and Chile.

The book is loosely organized around the practice of deformation, with some chapters describing works that are explicitly so labeled (and celebrated or denounced accordingly), while others address artistic theory and practice that relate to deformation implicitly and primarily in experiential terms. Focusing on the patron Cardinal Bernardino Spada (1594–1661) and the artwork and architectural features he commissioned for his Palazzo Spada in Rome, the first chapter looks at deformation as a means to achieve social power in a secular setting. I consider Bernardino's use of architectural deformation—especially a perspectival colonnade—in a context where every move a guest made through a host's space was carefully choreographed, and one demonstrated elite status by knowing this choreography and also knowing how to make civil conversation within reception spaces. I show how Bernardino anticipated where etiquette would dictate guests should move and used that knowledge to direct visitors to predetermined viewing positions, in effect requiring them to look at something he had commissioned, admire the work, and then discuss it with him, in encounters that optimized his and his family's social power and thus reinforced their social status more broadly.

If chapter 1 focuses on a single influential patron, chapters 2, 3, and 4 examine how and why entire religious communities, specifically the Minim order and the Society of Jesus, integrated deformations, such as anamorphoses, into their practice. After documenting and explaining the Minim interest in deformations in chapter 2 and the Jesuit interest in chapter 3, the fourth chapter shows how the drive to deform and reform was most powerful when it came to the transmission of key theological mysteries, such as the Eucharist, which Catholic Reformation leaders prioritized as tenets of faith. Bernardino does not directly call his perspectival colonnade a deformation, but we can discuss it together with the anamorphic deformations of chapters 2, 3, and 4 in part because all these works are experienced via an analogous transition between the initially perceived and the ultimately comprehended. As Bernardino anticipated how guests would move in his palace to experience the confusion and then understanding of his architectural deformation, Minims and Jesuits similarly anticipated how professors, students, and visitors to their convents would move through corridors to confront initially incomprehensible images that transformed into clearer theological representations as they moved forward in space.

Whereas the first four chapters explore Catholic Reformation thinkers who intentionally sought out visual and spatial deformations, chapter 5 considers the Spanish theologian Caramuel, who ultimately served as Bishop of Vigevano, as he devised recommendations he claimed would generate architecture "with grace, and without deformity" (*con gracias, y sin deformidad*).[103] Yet, as noted above, architectural theorist Claude Perrault would have read Caramuel's recommendations as promoting deformations rather than reforming architecture; the example of Caramuel thus suggests that at least one Catholic Reformation theorist, who admittedly was outside the mainstream, would have been understood by one thinker to be drawn to deformation inadvertently. Finally, chapter 6 considers turned ivories, small-scale works crafted by cutting material into a form while rotating it in a lathe. Although not described in the early modern era as "deformations," the task of understanding or interpreting a turned ivory involved

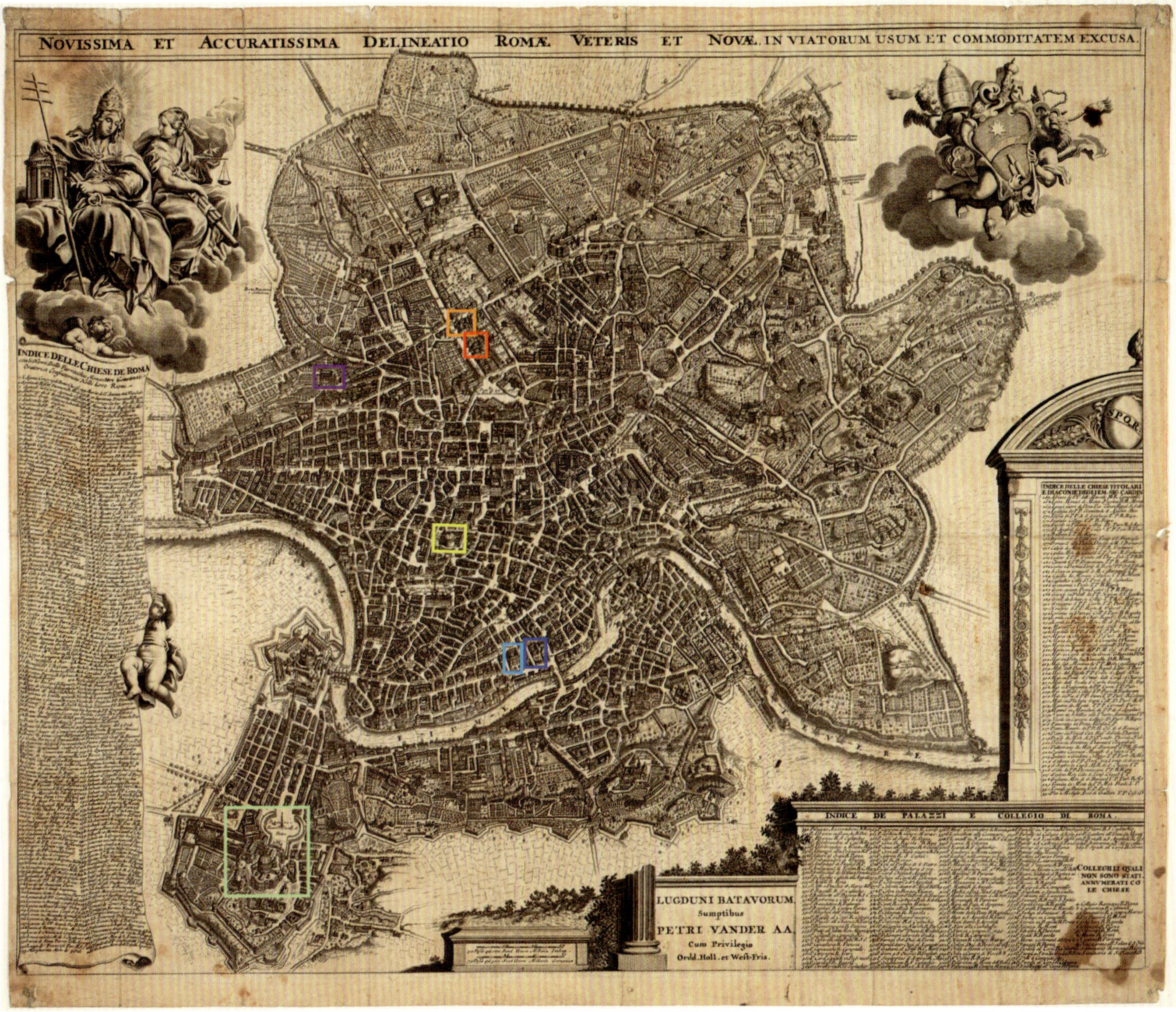

Fig. 11. Key locations discussed in this book are annotated onto Giovanni Battista Falda's *Novissima et acuratissima delineatio Romae veteris et novae, in viatorem usum et commoditatem* (Leiden: Pieter van der Aa, 1700–1721). Engraving printed on paper, 21.9 × 23.6 in (55.5 × 60 cm). University of Chicago Library, Hanna Holborn Gray Special Collections Research Center.

operations of the mind similar to those necessitated by the images and structures that were identified as such. To find similarity between turned ivories and visual and spatial deformations, the chapter draws on the French Minim Charles Plumier's (1646–1704) account of turning, ultimately demonstrating that the collectors and makers of these exquisite sculptures, like patrons and practitioners of deformations, derived social power from the incomprehensible, quasi-supernatural powers associated with this transcendent art. The transcendent incomprehensibility of turned ivories prolongs—perhaps infinitely—the time it takes to move from not-knowing to knowing, suggesting that some turned works, like San Carlo, may ultimately remain "unknowable." The chapter concludes the book's extended reflection on the intellectual puzzle of deformation by recognizing that contemporary understanding allowed for the possibility, even the inevitability, that further truths or mysteries remained beyond human comprehension. Whereas many deformations modeled the revelatory journey from obscurity to clarity, some of the baroque's most exceptional objects and structures defied comprehensibility altogether.

✚

If the forms of San Carlo alle Quattro Fontane would have been oppositional to at least one reading of contemporary church aesthetic theory, we can nevertheless seek consistency with a distinct rhetorical norm advocating for persuasion by means other than immediate clarity. Fra Juan saw San Carlo as rhetorical, even though the church does not allow for immediate comprehension, because it persuaded visitors from all over the world to enter again and again and to experience a relationship with the transcendent. Moreover, in the existence of numerous works that seemingly contradict the stated Catholic Reformation proscription against "unclear" art, we find that practitioners and observers of deformation saw the confusion emerging from the space they created between ideal form and deformation as productive, insofar as it encouraged observers to slow down, look, and pay attention. Practitioners of deformations and their patrons and audiences saw these works as offering access to a different model of spiritual persuasion, one that mimicked the limits on human perception when confronting the divine. Catholic Reformation leaders and observers embraced deformation for its capacity to hold observers in suspense—indefinitely or temporarily containing them in a state of stupefaction or bafflement—and then at times to move observers (sometimes literally) from incomprehension to comprehension, via a defined physical or mental path that represented the interpretive authority of the Catholic Church or a collector. As the Church mediated to help the chaotic human mind approach sublime, spiritual, transcendent meaning, so could deformation mediate visual experience, transport observers to a particular vantage point—and thus concentrate the beholder's attention on particular, illuminating subject matter. This capacity motivated Jesuit and Minim artists, as well as other painters and architects like Borromini, to deploy the new optical sciences and mathematics in support of Catholic Reformation claims to institutional, theological, and exegetical authority.

In other words, despite Tridentine recommendations that might seem to privilege rhetorical clarity in the visual arts, Counter Reformation artistic practice was rhetorical in a different sense. Some Catholic Reformation thinkers valued deformations because they were persuasive, ultimately privileging elements of classical rhetoric that reinforced the authority of the speaker and the persuasive power of analogy over direct rhetorical clarity. The deformation's capacity to stun and to elicit wonder heightened emotional affect, and resolving the deformation involved the observer in an exercise of interpretive power where the patron has predetermined the interpretive outcome. Through this exegetical exercise the viewer was led to recognize the Church's institutional authority as a mediator leading the believer toward God and reinforcing their faith—or, in secular contexts, the viewer was led to recognize the social power of the patron or elite creator.

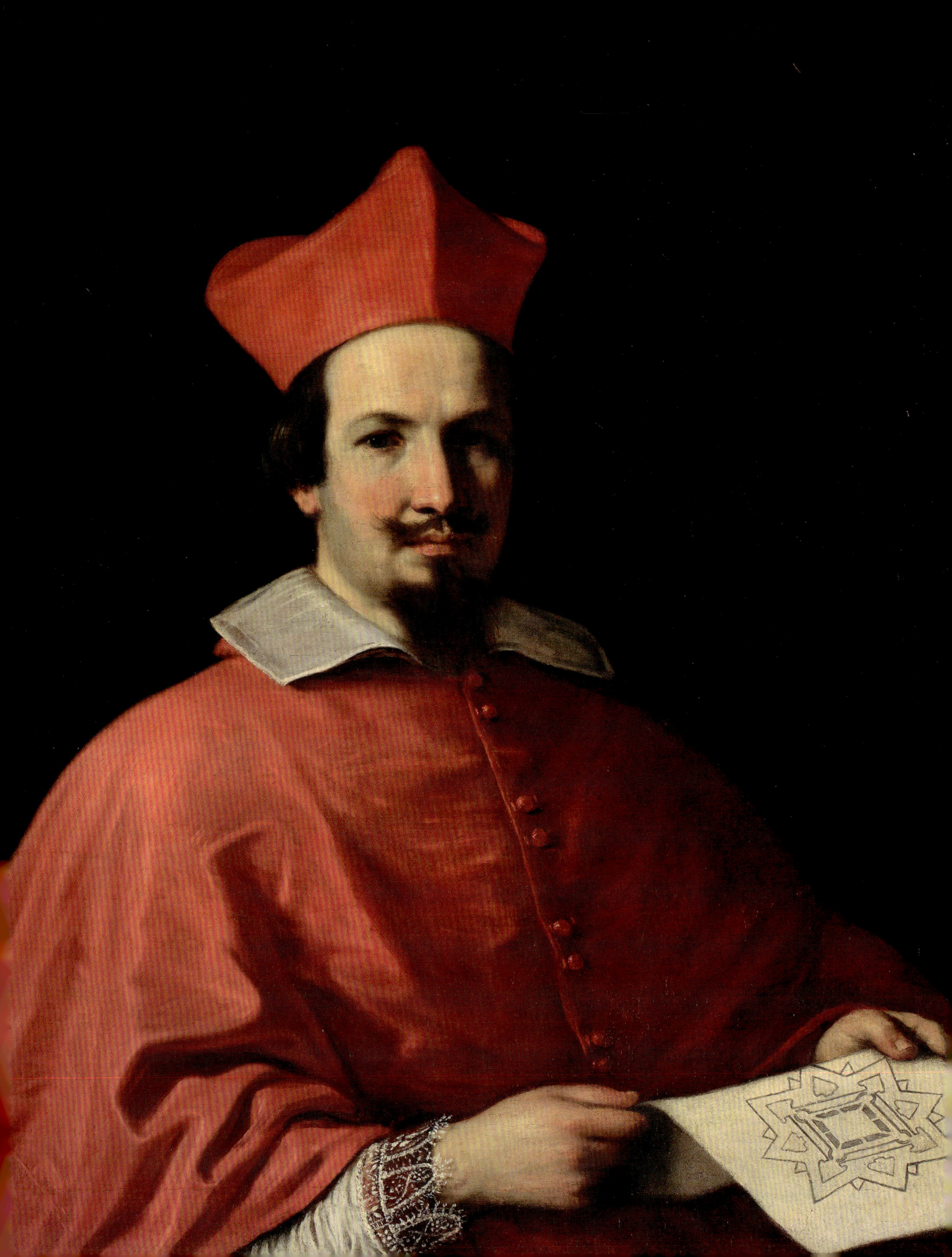

1 THE CARDINAL'S PERSPECTIVE

A VIEWER LOOKS OUT at the perspectival colonnade in the garden of Palazzo Spada (Fig. 12). She sees, at the far end of the colonnade, a cat—a cat that seems bafflingly large. The size distortion is especially convincing if the colonnade is viewed at a distance from inside the main courtyard, in front of a walnut gate with long balusters turned in 1642, or from the *Sala delle Udienze* (audience room) on the *piano nobile* (principal floor) of the palace (Figs. 13 and 14). The perspectival colonnade was added for Cardinal Bernardino Spada, one of the major patrons of baroque art in Rome, who extensively renovated the palazzo during the 1650s. To facilitate its admiration from a proper viewing distance, the Oratorian Virgilio Spada (1596–1662), Bernardino's brother, would give directions in his will that relatives should keep certain doors in the palace open, especially when entertaining eminent visitors.[1]

Entities positioned at the rear of the colonnade—whether cats, human visitors, or statues—appear disproportionately large via the technique of "accelerated perspective" (*prospettiva accelerata*), which deceives the beholder into believing that the far columns are at a greater distance than they really are.[2] The measurements of components in the colonnade's foreground—for instance, the height and width of columns and the height of the arch—are more than double the measurements of the corresponding components at the far end.

I center my opening chapter on this colonnade, also known as the *prospettiva*, because it renders deformed the most basic of classical forms, the column, insofar as it violates standard norms of columnar figurations and proportions. Moreover, it also breaches the expectation that a classical colonnade will present a series of identical columns at regular intervals. If architectural classicism introduced a set of forms and a logic around them, the *prospettiva* is deliberately incomprehensible in its unraveling of those forms and as well as their surrounding logic.[3] Colonnades emblematized the principle of regular recession—for instance, when they were

Detail of fig. 15

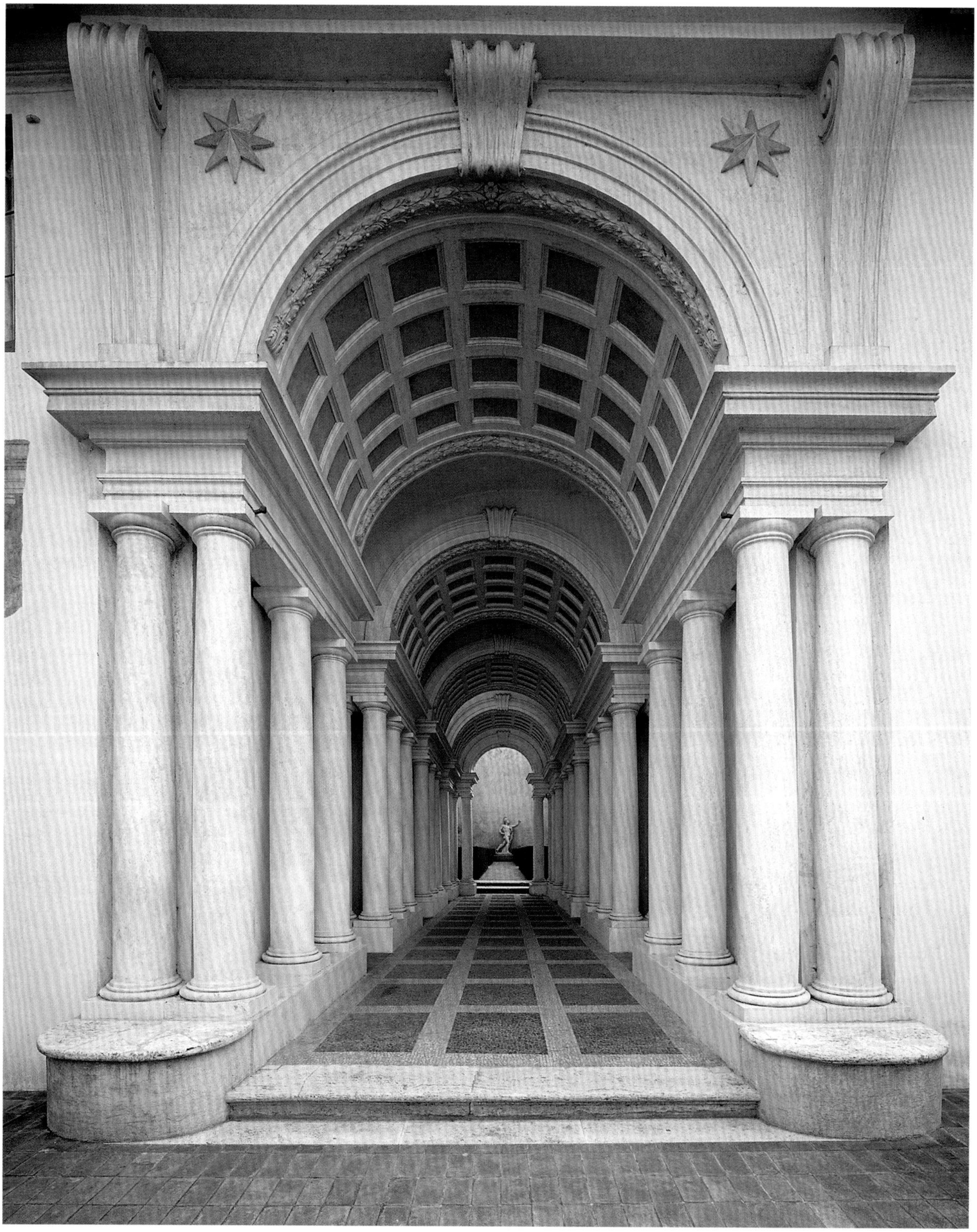

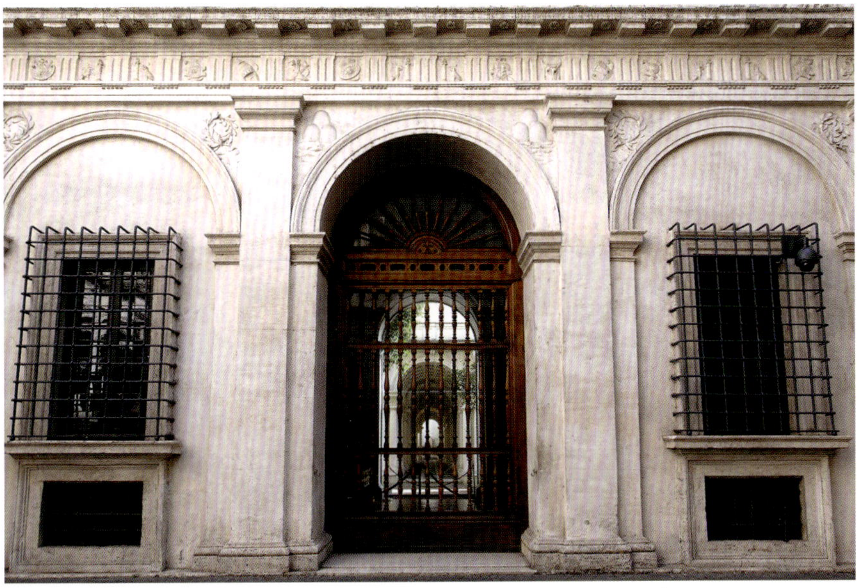

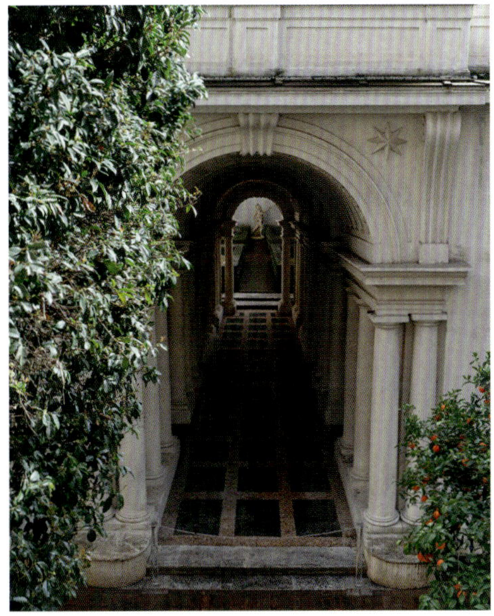

deployed alongside tiled floors in two-dimensional perspectival demonstrations of recessions into depth. Thus, being deformed in the Palazzo Spada *prospettiva* is a classical ideal that had come to represent the very idea of regularity itself. In taking apart notions of regularity, the colonnade foregrounds two deceptions. One lies in its apparent enlargement of bodies (a cat, a human, or a statue) as they recede into the depth; a second lies in its capacity to make a short span appear much longer than it is (distorting the perceived distance between the cat—or human, or statue—and the viewer).

Although in the past the Spada *prospettiva* has been attributed solely to Borromini, the structure is, in fact, based on designs by an Augustinian monk from Bitonto, Puglia, known as Fra Giovanni Maria da Bitonto (b. 1586), who, together with Borromini and his assistant Francesco Righi, oversaw its construction from 1652 to 1653.[4] This chapter is less concerned with that attribution than with the question of patronage: why did Bernardino Spada, an influential Catholic figure, patronize this deformation?[5] How did he make use of it and what do those functions reveal? In exploring Bernardino's incentives, I position him beside other Catholic Reformation patrons, practitioners, and observers who were similarly fascinated by visual and spatial deformations that evoke deliberate incomprehensibility, confusion, and uncertainty. Bernardino used visual and spatial deformation to control what guests observed when they moved through the spaces of his palace and garden; these uses capitalized on existing etiquette that governed how and where guests and hosts were supposed to perform their roles. Bernardino invited visitors, in an embodied sense, into experientially disorienting positions where they would admire his and his architects' and artists' cleverness. He was not only proffering conversation pieces that focused attention but also inviting conversations that would highlight his own discernment.

In a letter of December 7, 1653, Bernardino ascribes a primarily social function to the *prospettiva*. Writing to the poet and physician James Alban Gibbes (1611?–1677), who served

OPPOSITE

Fig. 12. Francesco Borromini and Fra Giovanni Maria da Bitonto, perspectival colonnade, c. 1652–1653. Palazzo Spada, Rome.

ABOVE

Fig. 13. Francesco Borromini and Fra Giovanni Maria da Bitonto, perspectival colonnade viewed from the main courtyard of Palazzo Spada, c. 1652–1653.

Fig. 14. Francesco Borromini and Fra Giovanni Maria da Bitonto, perspectival colonnade viewed from the *Sala delle Udienze* on the *piano nobile*, c. 1652–1653.

as Bernardino's personal doctor from 1644 and had occasionally visited Palazzo Spada, Bernardino describes the visual deception produced when the work is observed from a precise location:[6]

> I do not know if before your Lordship left from here work had started on a certain small loggia decorated with stuccos and with a number of columns, that the Augustinian Father Bitonto built here in the house facing the room where I usually receive. And truly, for the small thing that it is, it forms a very long perspective effect, hence it provided the occasion for a moral epigram that I am sending you.[7]

By situating the perspective opposite the room in which he greeted visitors in Palazzo Spada, the *Sala delle Udienze* on the *piano nobile*, Bernardino used this construction to impress guests, to capture their attention, and to arouse their curiosity. Bernardino would have also invited visitors to the ground floor room facing the *prospettiva* that was known as the *Sala dei Rilievi* (room of reliefs), because it displayed the Roman reliefs that entered the family with the dowry of Maria Veralli (1616–1686), the wife of Bernardino's nephew Orazio (1613–1687).

Bernardino's moralistic epigram survives in another letter to a Barberini Cardinal (perhaps Francesco Barberini, 1597–1679):[8]

> A huge portico is seen under a small structure.
> A long path is perceived in a short space.
> The farther away they are, the larger they seem,
> Objects that in their proper place are small.
> A work of marvelous art, the image of a deceptive world,
> Itself poor and small, it offers great things to its own.
> Great, but only in appearance, they become small when they are attained.
> Grand things under the heavens remain nothing except specters.[9]

If the wondrous *prospettiva* overturns a scalar opposition by making small entities appear large, so too does Bernardino's poem. The first two lines are structured around antitheses, such that *exigua* ("small") contrasts with *ingens* ("huge") and *longa* ("long") with *brevi* ("short"); the chiasmus in the second line points to Bernardino's concern with reversing opposites.[10] The poem calls attention to the misapprehensions the construction invites before arguing that the scalar deceptions of the colonnade emblematize grandeur misconstrued in the world at large. The epigram imbues Bitonto's and Borromini's *prospettiva*, which Bernardino characterized as a marvel, with a moral message that visitors could analyze, a message with theological undertones concerning the conflict between spiritual truth and bodily traps.

Although Bernardino does not use the term "deformation" to describe Bitonto's and Borromini's *prospettiva*, a relationship to deformation can be seen when he says that "grand things under the heavens remain nothing except specters." He evokes here an idea we have seen Bellori describe in his *L'Idea* (published a few years later), that all sublunary creation is deformed, in the sense that it departs from the ideal. Several words in the epigram, moreover, transmit a sense of the spectral or the phantasmic as a lesser imitation or deformed shadow: in addition to *spectrum* and *species, imago* is also relevant in this regard: quite literally an image but also in Latin literature it conveys the sense of a ghost—for instance, in the *Aeneid*, the ghost of Aeneas's wife Creusa is described as a *simulacrum* and an *umbra* (or shade), but also an *imago* (book 2, lines 772 and 793).

Bernardino's epigram discloses that any potential confusion evoked by the *prospettiva* serves a purpose: it reminds observers to recognize that displays of magnificence and splendor, typically understood as suggestions of wealth, are as misleading as the *prospettiva*. The moral epigram may answer in part one of the key questions that structures this book—suggesting that the didactic potential of deformation may help to explain its appeal to so many. But that answer leads to a further question: why did Bernardino invest so much money and energy into cultivating splendor in his Roman palace, if he believed that "grand things under the heavens remain nothing except specters"?[11] Most likely, this aspect of the epigram is performative rather than expressing Bernardino's "view" on wealth in a transparent way. His *prospettiva,* like other deformations, exists within a broader context of elite displays of knowledge and understanding: deformations were a particular category of luxurious display, one that demanded interlocutors interpret them and enabled Bernardino to exercise social power in controlling his visitors' experiences as they moved from incomprehension and uncertainty to comprehension and certainty.

Bernardino Spada

Bernardino was born in the town of Brisighella in Northeast Italy. His father, Paolo (1541–1631), a coal merchant, contributed to the family's rapid social elevation in the late sixteenth century. Paolo, eventually made treasurer to Pope Clement VIII (1536–1605) in the region of Romagna, not only amassed a fortune as a tax farmer but also garnered political connections that proved useful in advancing his family's interests in general and advancing his son Bernardino's career in particular.[12] Bernadino studied with the Jesuits in Bologna before becoming a priest, and thus, among other influences, would have been exposed to Jesuit drama.[13] Plays were a vital component of Jesuit education, allowing students to demonstrate their abilities publicly while learning and absorbing spiritual lessons.[14] Given the intensity and global reach of Jesuits' commitment to drama, it is tempting to trace to his educational upbringing the origins of Bernardino's interest in blocking the metaphorical performances and rituals of respect.[15]

For Bernardino and his family, architecture, collecting, and the practice of gentlemanly conversation functioned as vital means of social elevation. Conduct manuals such as Stefano Guazzo's (1530–93) *La civil conversazione* (*Civil Conversation*, 1574) offered advice on proper behavior in conversational exchanges. Bernardino, perhaps unsurprisingly, had a copy of this manual in his library and thus could have read Guazzo's interlocutor explaining that civil conversation "must be honest, praiseworthy, and virtuous."[16] Honing his powers of attention and discernment in civil conversations along Guazzo's model would, in turn, have helped Bernardino to secure his family's social standing within the Roman elite.[17] Investing in architecture, collecting art, developing educated, observant eyes and thus learning to discuss the buildings or objects before them with confidence would have signaled the family's "cultural nobility," to use Pierre Bourdieu's term. By engaging with objects of art and architecture and their interpretation, as much as by formal education, the family would leave behind its modest

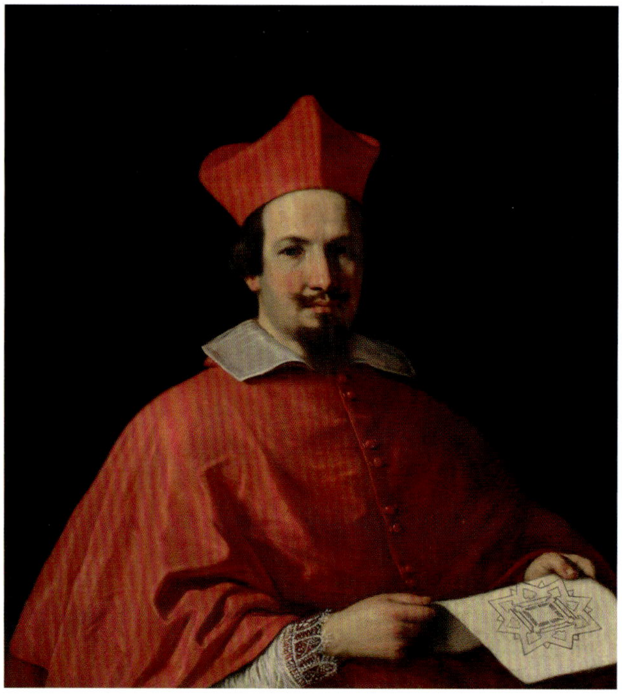

Fig. 15. Guercino, *Portrait of Bernardino Spada*, 1631. Oil on canvas, 38.11 × 34.57 in (96.8 × 87.8 cm). Galleria Spada, Rome.

origins and would come to participate in the highest echelons of Roman society. Bourdieu pinpoints the seventeenth century as the moment when the parameters of cultural nobility were first drawn in France, with those who sought this new cultural authority clashing over competing notions of culture and over the correct mode of connecting to artworks, food, music, books, and other products and practices. As Bourdieu observes, only those with cultural understanding would have been able to decipher, or discern, the code within which artworks have significance.[18] The term "edification," from the Latin *aedificare* ("to build"), points to age-old commonalities between building and mental or moral instruction, and the new attention to discernment signaled patronage as a way to build social standing as well.[19]

Bernardino was made a cardinal in 1626. In this capacity, as his father had done before him—and as had other cardinals in early modern Rome—he continued to advance his family's social standing. He devoted himself to the education of his many nephews, particularly Gregorio (1615–1686) and Orazio, ensuring they would receive training appropriate for gentlemen and would thus be prepared to participate in elevated social circles.[20] He also took care that the marriages of his nieces and nephews were politically advantageous.[21] From 1624 to 1627, Bernardino served as papal nuncio in Paris, where he brought Orazio, who was nine when his term started, and Gregorio, who was then eleven. Bernardino subsequently relocated to Bologna, where he worked as apostolic legate. During his time in Bologna, Guercino portrayed Bernardino with a furrowed brow and concentrated stare; the red of his vestments is picked up in his flushed cheeks (Fig. 15). Along with his attentive look, the sitter presents to viewers a plan of the Forte Urbano of Castelfranco Emilia, which he was involved in constructing. The portrait thus underlines his passion for military architecture, a field that involved an understanding of

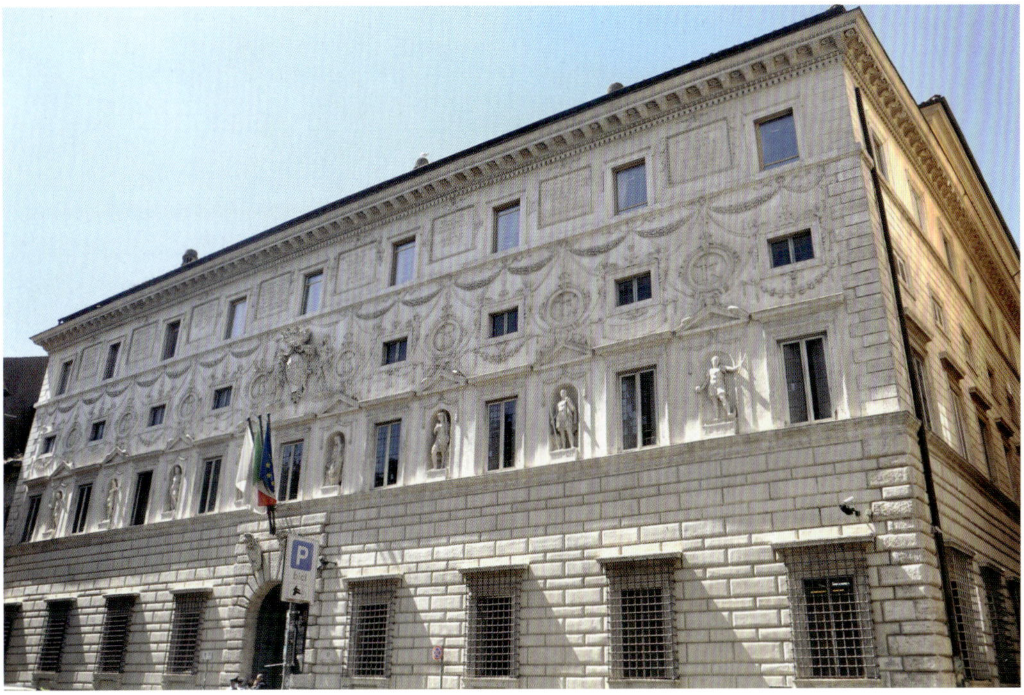

Fig. 16. Facade of Palazzo Spada.

ballistics and engineering and thus a high degree of specialized visual attention and discrimination.[22] After three years in Bologna, Bernardino settled in Rome. Orazio's diary indicates that by 1633 he and Gregorio already resided in Palazzo Spada, then known as Palazzo Capodiferro.[23] The name "Capodiferro" is derived from the palace's first proprietor, Cardinal Girolamo Capodiferro (1502–1559), who erected it in Rome, very close to the Palazzo Farnese, between 1548 and 1550; the structure passed into Bernardino's hands in 1632 and is still known today as Palazzo Spada (Fig. 16).

In his biography of Bernardino, Virgilio Spada worried that his older brother was addicted to building, a passion he had apparently inherited from their father.[24] Virgilio writes that Paolo Spada had barely refrained from destroying a building, which had perhaps eighteen rooms for each story, in order to rebuild it fifty steps away. He notes that on three separate occasions Bernardino built and entirely rebuilt the Palazzo Spada's *piano nobile*, not to mention the main staircase, which he changed at least as many times.[25] Paolo and Bernardino's enthusiasm for complete architectural destruction and reconstruction defied contemporary architectural practice, although it was not exceptional in Rome. Prior to the twentieth century, European architectural culture placed a premium on permanence, fixity, and steady change.[26] When Bernardino remodeled the *piano nobile*, main staircase, or other parts of Palazzo Spada, he could be said to have worked under a long-durational building ideology that Marvin Trachtenberg has labeled "Building-in-Time" (and has contrasted with an Albertian ideal of "Building-outside-Time," which held that in an excellent design, "nothing may be added, taken away, or altered, but for the worse").[27] Bernardino appears to have been exceptionally dedicated to a principle Trachtenberg calls "continuous redesign," whereby builders, during extended periods of construction, adjusted designs based on patrons' shifting requirements. Whereas Virgilio considered troublesome the constant need to destroy and remodel, it seems that Paolo and Bernardino found building sites to be catalysts for the imaginative and creative discernment that signaled their social aspirations.[28] And in building and rebuilding in Rome during this period, they found themselves in good company: between 1550 and 1650 an estimated 150 palaces, 300 churches, and an incalculable number of houses in the city were either erected from the ground up or were heavily renovated and extended.[29] Moreover, Rome had a vibrant culture of ephemeral architecture: stucco arches might go up and then rapidly be taken down; fictive facades were erected to cover churches temporarily. Elaborate ephemeral architectural structures were frequently raised and demolished within churches.[30] Collectively, all this activity would have added to the impression of the city as a building site.

Virgilio may have considered both Paolo and Bernardino excessive in their passion for building, but he recognized a distinction between their approaches: Paolo used inexpensive materials, while Bernardino used expensive materials, from travertine to stucco decorations.[31] The use of expensive materials may well have been part of Bernardino's plan for advancing his social standing. He earned a cutting third-person critique from his brother after he purchased, and then began to renovate, Palazzo Spada:

> Father Virgilio, his brother, took great pleasure in this purchase, because, since he knew the cardinal's sinful inclination to build and change everything, he said: "I am happy that Your Eminence has found such a fully completed palace, so that you won't be able to add one more brick, even if you wanted to." ... But as soon as the cardinal moved there, he began to fantasize about it. And he continued for more than thirty years to renovate there, to the extent that all the money he saved in having made his initial purchase with some advantage benefited him little, because by the day of his death, beyond the 31,500 scudi, he had spent 50,000 more scudi, not to mention all the debts with the workers for about another 6,000 scudi.[32]

Beyond underscoring the duration of the renovations, Virgilio focuses on the financial costs of his brother's "sinful inclination" (*humor peccante*). The lines from Bernardino's epigram concerning the illusive nature of "grand things under the heavens" suggest he was well aware that his inclination to invest in displays of wealth could be viewed as "sinful."

When Bernardino purchased the palace, Virgilio's biography reports, it "had ceilings alleged to have been gilded with the first gold taken from the Americas."[33] As this small detail suggests, the palace gratified period interest in "cabinets of wonders," or collections that juxtaposed rare entities and materials that were previously unknown or found only in distant places.[34] Like the gilded ceiling and other wonders gathered together in the palace, Bernardino's perspectival deformation can be understood as an iteration of creative endeavor that could produce wonder in the onlooker. And though wonder, famously promoted by Aristotle as a means to lead people to philosophical questions, had gone out of fashion in the Middle Ages—as European thinkers became more interested in studying general patterns rather than unusual phenomena—wonder's capacity to pique intellectual curiosity returned with force in the early modern period, as Lorraine Daston and Katharine Park observed in their classic study.[35]

The French philosopher Nicolas Malebranche (1638–1715) saw the "impulse of wonder" as essential in sustaining attention and thus overcoming the difficulties of inquiry:

> Nothing is more difficult than applying oneself for any length of time to something that fails to excite our wonder, since then the animal spirits are not so easily led into those parts of the brain necessary to represent it. Although our attention may be called for, it is not forthcoming, or at least not for any length of time, however we may be convinced in the abstract … that the thing in question deserves our attention. We must trick our imagination in order to stir our spirits, and we must represent the subject we wish to think about in a novel way so that an impulse of wonder might be excited in us.[36]

Wonder stimulated attention, and attention in turn stimulated the curiosity required for prolonged inquiry. Bernardino, four decades earlier, would have been practicing the embrace of novelty that Malebranche would later endorse—and for Bernardino, at least, thinking "in a novel way" and exciting wonder encompassed the creation we would now recognize as a deformation. The multivalent objects and structures that he would integrate into his renovations of Palazzo Spada catered to an interest in amassing entities—from an architectural deformation to paintings, books, and scientific instruments—that stimulated wonder through tricking the imagination and representing entities in new manners.[37]

That Bernardino prioritized "wonder" in his architectural projects—over other concerns, including financial prudence and utility—might be deduced from Virgilio's criticisms: he certainly thought Bernardino's projects were too expensive, and he seems also to have deemed them ill-advised on practical grounds. He claims that his brother destroyed a big room next to the antechamber of his palace by turning it into a long narrow hall with two little rooms and two cubicles, with a small open loggia, all facing south, so that they could be pleasant in winter. Then, many years later, Bernardino "ruined" the same area of the palace again: he turned the site into two big halls without the open loggia and without any comforts. Virgilio grumbles that nobody could possibly live in those rooms, meant only as a display of grandeur and splendor to those who entered. He also bemoans Bernardino's destruction of a comfortable staircase leading from the underground kitchens to the roof, a staircase Virgilio considered both ingenious and exceptional in Roman architecture.[38] Elsewhere in the biography, Virgilio complains further about Bernardino's appetite for building and lists the manifold shortcomings of his various projects, including a staircase that collapsed when it was almost finished and that, even in its final form, failed to lead where it should have.[39] Virgilio implies that Bernardino's desire to build was insatiable—"And since such a vast site [as Palazzo Spada] was not enough for his ideas and for his itch to build, he brought his mind to bear on the decoration of the walls of a neighbors' house facing the facade of his own palace"—his architectural discernment lacking, and his competence questionable at best.[40]

And yet Roman palaces were not merely structures to house family members and guests; Virgilio's emphasis on spending and basic utility (actually living in those rooms) fails to acknowledge that a building like Palazzo Spada also functioned as a public statement of family rank and as a stage set for performing critical transactions and negotiations against the backdrop of Rome's rigorous class hierarchy.[41] (Virgilio does at least nod to this role of architecture: "It is a well-established axiom that the value of men is derived from their buildings and their marriages.")[42] The status of palace guests was reflected by where in a sequence of rooms or spaces the host greeted them upon entering the structure, and where he bid farewell on their departure; Bernardino's creation of rooms that display grandeur to those who entered may have been entirely appropriate, and even shrewd.[43]

Late sixteenth- and seventeenth-century courtesy handbooks outlined proper modes of comportment within palaces: hosts and visitors needed help in navigating these complex spaces as well as the intricate social codes that palaces embodied and facilitated. The existence of these behavioral handbooks demonstrates how seriously members of the Roman elite respected—and performed—status distinctions.[44] The pseudonym "Cesare Evitascandolo"—adopted by one handbook author who served as *maestro di casa* (responsible for supervising the servants and running the house) to several Roman nobles and high-ranking clerics—literally means "Cesare avoids the blunder." Taking the correct steps—and circumventing missteps—was crucial when performing within the social space of the palace.

The Roman guidebook tradition further points to a growing interest in the cultivation of visual attention and discernment in this period—guidebooks showed readers which creative or interpretive steps to make or, presumably, avoid.[45] Guidebooks to Rome had first appeared as late medieval *Mirabilia*—recognizing that ancient buildings could themselves be wonders—and the *Mirabilia urbis Romae* (*Marvels of the City of Rome*), written in Latin by an anonymous author before 1143, was among the most popular of these works. The genre later flourished in the seventeenth century, particularly during the reign of Pope Alexander VII (r. 1655–1667).[46] Particularly suggestive (and also constitutive) of Roman elites' developing visual discernment are guidebooks that go beyond listing specific sites or facts to describe visual experiences in concrete terms, providing observers with a vocabulary for conversations about the visual arts and architecture and specifying for them what sorts of questions artworks and architecture raise; the appearance of the term *je ne sais quoi* in the Roman guidebook cited in the introduction models an appropriate response to a wonder or deformation.[47] When confronted with a deformation that overwhelms with wonder, surprise, and strong emotion (in other words, triggers stupefaction), an informed observer could turn to the *je ne sais quoi* to denote that their first-person experience could not be explicated.

Bernardino's Deformation

When a guest of high status—for instance, a cardinal or a grand duke—came to Palazzo Spada, they might have been led to its *scalone d'onore* (grand staircase), where their visit could have been broadcast with a bell and an attendant gentleman would have welcomed them.[48] Once they arrived at the *piano nobile*, the guest would have entered the *Salone* (the main hall), where Bernardino would have greeted them, or, if he particularly respected them, he might have met them in front of the entrance to this room, but not on the staircase itself. Then the guest would have been ushered into the *prima anticamera* (first anteroom) that also led to a chapel and then into the *seconda anticamera* (second anteroom) from which they would finally have been led into the *Sala delle Udienze*.

A discerning guest, who had been offered an initial impression of the *prospettiva* from this *piano nobile* audience room, might recognize from afar—or guess—that his host had generated a clever illusion in his colonnade, which he may have then discussed and analyzed with him. Bernardino recalls one discussion of the *prospettiva* with the Cistercian monk and church historian Ferdinando Ughelli (1595–1670) in a letter written to his brother Virgilio: "On Wednesday, if I recall correctly, I was talking with Father the Abbot Ughelli in my audience room about Borromini's perspective in the secret garden, and since he was praising this work, I told him what I had previously thought about, that is to say to build a similar one in a church."[49] In this discussion of his unrealized plan to build a similar deformation in a church, the mechanics of the Palazzo Spada *prospettiva* would have likely come to the fore.

A less sophisticated guest may have first perceived the *prospettiva* as an atrium with a regular recession. His host might have then invited the guest to stroll in the garden to examine the deformation more closely, approaching it from either the center (through the door in the *Sala dei Rilievi* that led to the garden) or from the left (through the door in the *Sala del Sole*).[50] Whether Bernardino would have accompanied the guest, or the guest later in the visit would have descended to the garden and then rejoined Bernardino, the *prospettiva* invited further movement into space that Bernardino had designed to communicate particular messages about his social power and sophistication. The deformation not only makes the guest think about visual deception but also, like other elements in the palace's architecture, causes the guest to behave in particular ways that ultimately redound to the power of the host and his respect for his guest.

Those who exited the *Sala dei Rilievi* and approached the structure head-on would have soon begun to notice discrepancies between it and typical atria. Any visitor who stepped into the colonnade would be confused to find that the floor, previously perceived as flat, slants upward. Walking down the colonnade, they would realize, moreover, that the floor does not consist of rectangles, but rather isosceles trapezoids. Although the guest had first taken these shapes to all be the same size, as in a standard grid, they would notice that, in fact, they are decreasing in scale and are gradually shifting from vertical to horizontal orientations as the colonnade recedes. By contrast, the individual stones in the mosaics do not decrease in size as observers move forward (Fig. 17).

Just as our imagined visitors realized that the *prospettiva*'s floor was at an incline, they would also be baffled to find that its vault, initially presumed parallel to a non-inclined floor, slants downward. As they studied up close the vault's rectangular coffering, they would realize they were also wrong to infer that it remains constant in scale. Much like the isosceles trapezoids at their feet, the elements of the coffering above them decrease in size as visitors traverse the space. When observers inspected the columns to their left and right, they would have recognized in perplexity that they too are not of uniform size but instead also decrease progressively in scale. The columns prove to be successively thinner and shorter as observers move toward the far end of the colonnade, even if they had seemed regular in size when viewed from the *piano nobile* audience room. Visitors also would have discovered that the colonnade narrows as they advance, the space of the colonnade closing in on them gradually from all sides. Even the tori—the bold projecting convex moldings at the columns' bases—are not centered beneath the shafts but protrude toward the audience room and toward the colonnade's interior (Fig. 18).[51] These deformations of the tori will grow less accentuated as the columns decrease

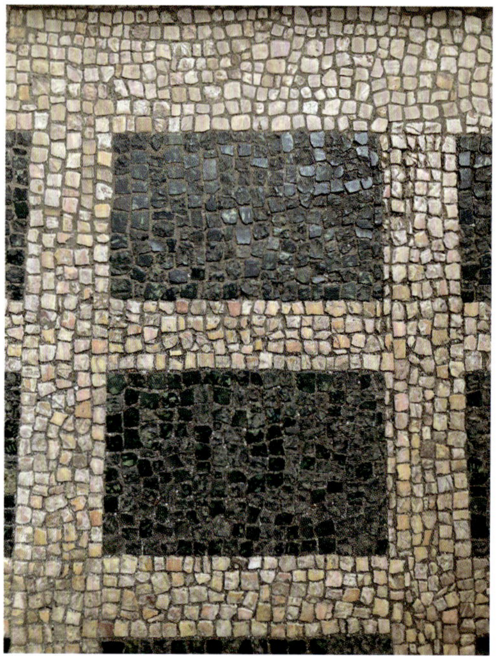

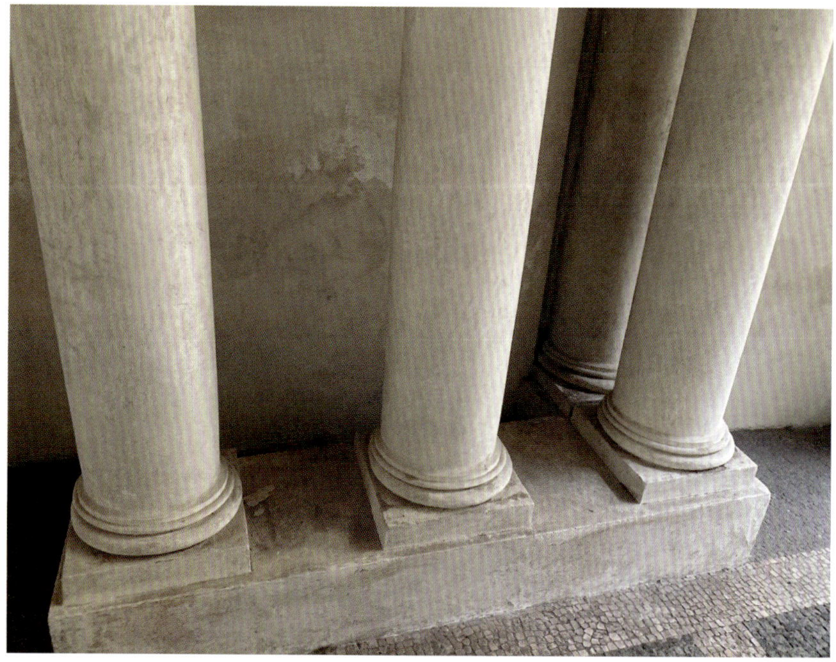

Fig. 17. Floor of perspectival colonnade. Palazzo Spada, Rome.

Fig. 18. Tori of perspectival colonnade. Palazzo Spada, Rome.

in height. Those who looked upward from the bases would notice that the capitals and architrave are oblique, which makes it seem as though they are collapsing toward the person walking through the colonnade (Fig. 19). For those who approached the colonnade from the left side (as opposed to head-on from the *Sala dei Rilievi*), those discrepancies would have been magnified, as observers immediately would find that the entablature slants downward and the columns decrease in size (Fig. 20). Contemporary observers will notice these discrepancies more readily in real life than in the photographs reproduced in Figures 19 and 20, as we are used to—and to a certain degree expect—the accentuated perspectives that wide angle lenses produce.

Of course, the viewers of Bernardino's wonders included family members and palazzo residents as well as invited guests. If multigenerational cultural nobility depended not only on cultural production by and for a head of the house—someone who patronized craftspeople and artists to create wonders such as deformations—but also on the reception of those creations by discerning younger family members, then Palazzo Spada might have been a failure. Bernardino needed his family to appreciate those wonders, their discernment recognizing or ratifying his considerable investment in producing them, and in a July 1661 letter to Maria Veralli Bernardino he lamented that his efforts were underappreciated: "If I talk about building, they start to say that it's not necessary and end up doubting whether it is enough [i.e., it is affordable]; as a result, I spend with little appreciation from others and after having spent I cannot use it [i.e., what I have built] with their satisfaction."[52] Maria's husband was among those who questioned the necessity and cost of his projects. On December 20, 1673, Orazio sent a letter to his youngest son, Bartolomeo (1657–1706), then a student at the Collegio dei Nobili in Parma, urging him and his brothers to sharpen their architectural expertise.[53] Orazio suggests that Bernardino, his uncle, spent much more than necessary on architecture because he allowed himself to be

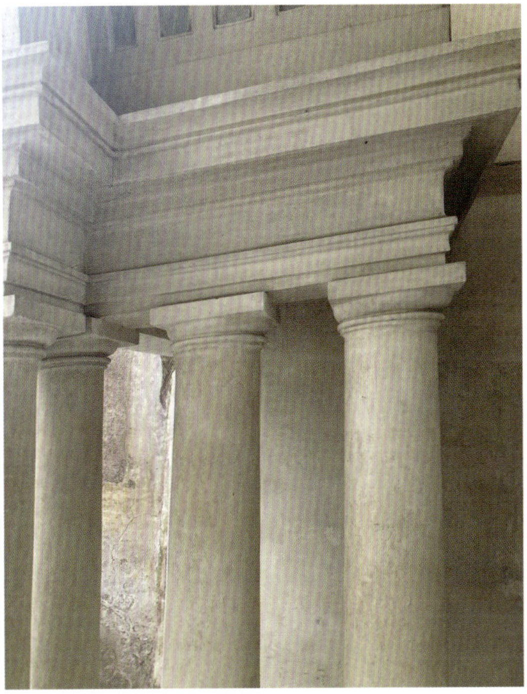

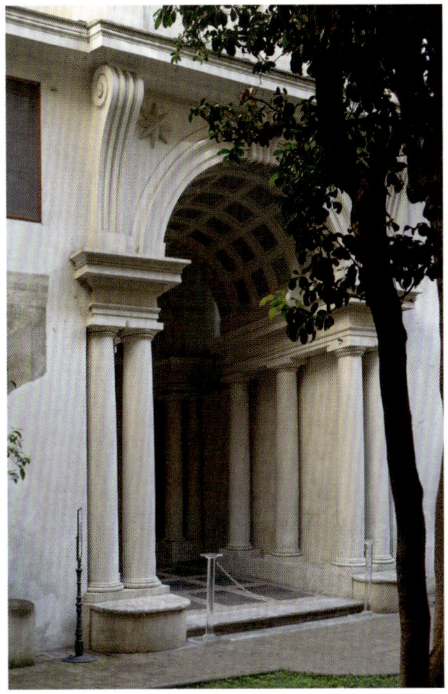

Fig. 19. Oblique capitals and architrave of perspectival colonnade. Palazzo Spada, Rome.

Fig. 20. View of perspectival colonnade from the left. Palazzo Spada, Rome.

led by the whims of the architects, who, working at his expense, proposed extravagant designs entailing exorbitant costs. Orazio recognizes that the inclination for building runs in their family and expresses his concern that he has transferred it to his children as well. He wants Bartolomeo to study architecture to avoid being cheated by architects. Orazio observes that his own architectural acumen has served him well: "And I can surely affirm that I have a very great itch to construct, but having some knowledge about this profession makes me abstain from it more than I would do otherwise."[54] Orazio's letter underscores the financial benefits tied to the acquisition of architectural expertise. He stresses that architectural discernment is valuable; the young elite must educate themselves thoroughly to enhance their perceptions. If Orazio did indeed have "some knowledge" in the field of architecture, it was likely due in part to his uncle's educational efforts while he and his brother Gregorio resided in his uncle's homes, where his family would have solidified its position in society through works of art and architecture that guided visitors through a planned set of rooms to dictate the guests' viewing experience.

The epigram Bernardino sent to Gibbes—calling the perspectival colonnade "the image of a deceptive world"—implies that this deformed architectural structure deceives only because it is in cohesion with a world that is itself understood to deceive. In that sense, the deformation mimics the deceptions that are inevitable in the sublunary realm, where, as Bellori stressed, all entities fall short of ideals. Bernardino's epigram underscores an interest in deception that is manifest throughout his palace, for which he also commissioned manifold illusionistic paintings. Like the *prospettiva*, these betray his interest in expanding the temporal experience of an artwork, from temporary stupefaction at the moment of initial perception to resolution and eventual discernment.

Fig. 21. Francesco Borromini, floor tiles in the Palazzo Spada *Salone*, 1645.

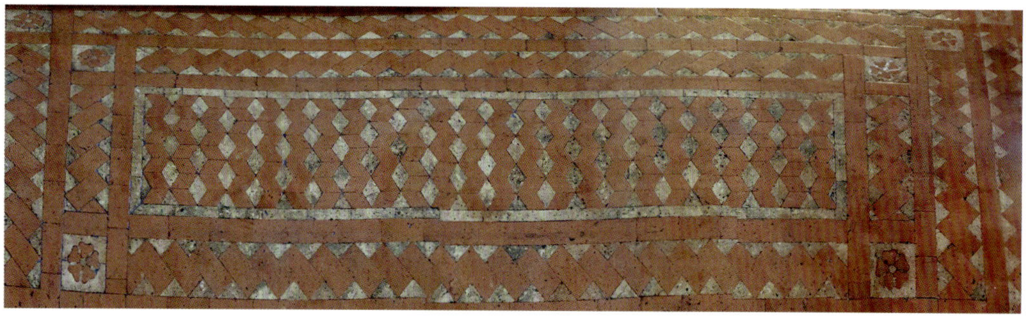

Bernardino wanted to adorn his palace in Rome with visual deceptions. And thus he summoned from Bologna the painter and stage designer Girolamo Curti (Il Dentone, 1575–1632), who specialized in *quadratura* schemes, or illusionistic architectural settings on walls and ceilings, and Angelo Michele Colonna (1600–1687), who was often employed by Dentone as an assistant.[55] Colonna also collaborated between 1634 and 1635 with Agostino Mitelli (1609–1660)—another preeminent exponent of *quadratura*—on the *Salone*, which, as we saw earlier, important guests would pass through on their way to the *Sala delle Udienze*.[56] Virgilio observes that the painters "made reliefs appear in vaults and friezes in the rooms where they were not, to the extent that many, to ascertain if they were true, even wanted to climb with ladders and touch them."[57] This account of deception and confusion before the allegedly incomprehensible perspectival frescoes is itself likely a deception, celebrating the painters' skill with a literary allusion to illusion. Much as in Pliny's (d. 79 CE) *Natural History*, in which a painted curtain fooled Zeuxis, who requested that the painted curtain be lifted, so too early modern observers of these frescoes, at least according to Virgilio's report, struggled to distinguish representation from reality and turned to the sense of touch to understand what they were viewing.[58] Virgilio underscores the fame of Curti and Colonna, who were "summoned by all those princes in northern Italy, but also in France by the king."[59] Although elsewhere in his writings Virgilio is unsupportive of Bernardino's investments in grandeur, he gestures here to the prestige that Bernardino's commission conferred on the palace. Virgilio's praise of the deceptive frescoes suggests that he recognizes the social value of the works, insofar as they present an attempt to display Bernardino's own discernment by generating a viewing experience that will amaze at least some (implicitly less sophisticated) viewers while generating a shared sense of belonging among those who recognize the visual deception without having to turn to the sense of touch.

The frescoes include monochrome paintings and fictive golden sculptures of powerful figures in Church history—Constantine (c. 272–337), Charlemagne (c. 742–814), Countess Matilda of Canossa (1046–1115), and Cardinal Gil Álvarez Carrillo de Albornoz (1310–1367)—as well as six fictive medallions that show personified Christian virtues. The expensive cut floor tiles, designed and measured by Borromini in 1645 a few years after the frescoes were completed, are laid out in concentric rectangles that formally guide visitors toward the center of the space. From this optimal viewing point, the architecture of the wall paintings appears most legible—and also most impressive (Fig. 21).[60] Alongside other architectural details in the palace, the floor tiles enabled Bernardino to direct what observers saw, and how they saw it. The painted architecture was designed to make the room appear more expansive than it actually

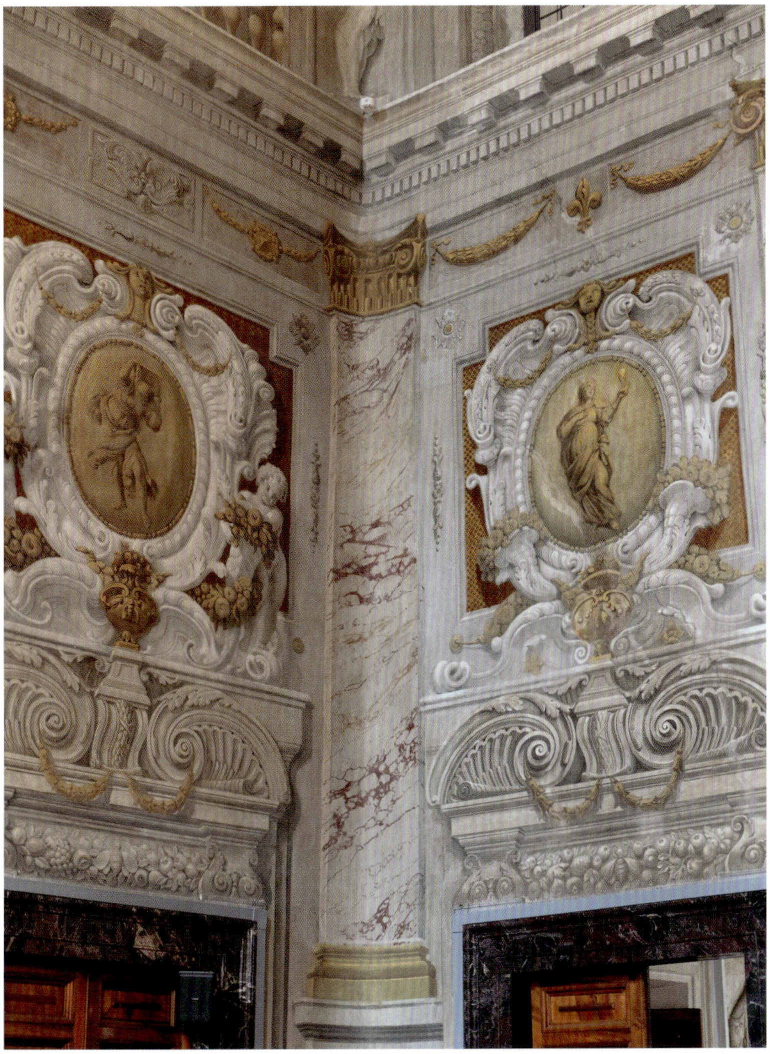

Fig. 22. Angelo Michele Colonna and Agostino Mitelli, painted marble column in the Palazzo Spada *Salone*, between 1634 and 1635.

was, the paintings conditioning observers' visual judgment by making scale difficult to comprehend, even before they reached the view of the *prospettiva* in the *Sala delle Udienze*. Visitors would either be impressed, thinking the rooms were big, or they would exercise their discernment, and in the process, be impressed by their host as the kind of sophisticate who understands the experience of visual attention and interpretation. Apparently valuing this deceptive expansion of scale, Bernardino had earlier, when he served as apostolic legate in Bologna, commissioned perspectival frescoes for his residence there that similarly magnified the apparent size of the room. Virgilio writes that the frescoes in Bologna made the space "appear twice as big" (*lo fece comparire il doppio*).[61]

The frescoes in the Palazzo Spada *Salone* abound with deceptions. A representation of a marble column in one corner of the room, although painted onto a concave surface, appears convex when observed from the center of the space (Fig. 22). Near the top of the wall to the

Fig. 23. Angelo Michele Colonna and Agostino Mitelli, fictive window and real window in the Palazzo Spada *Salone*, between 1634 and 1635.

Fig. 24. Angelo Michele Colonna and Agostino Mitelli, cornice in the Palazzo Spada *Salone*, between 1634 and 1635.

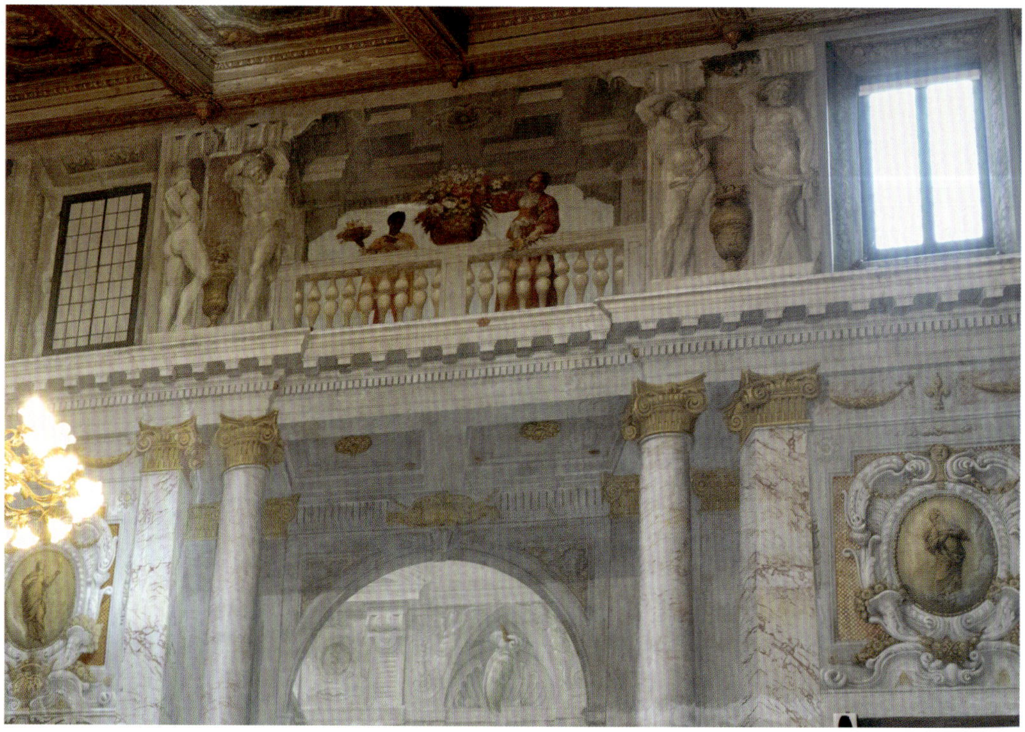

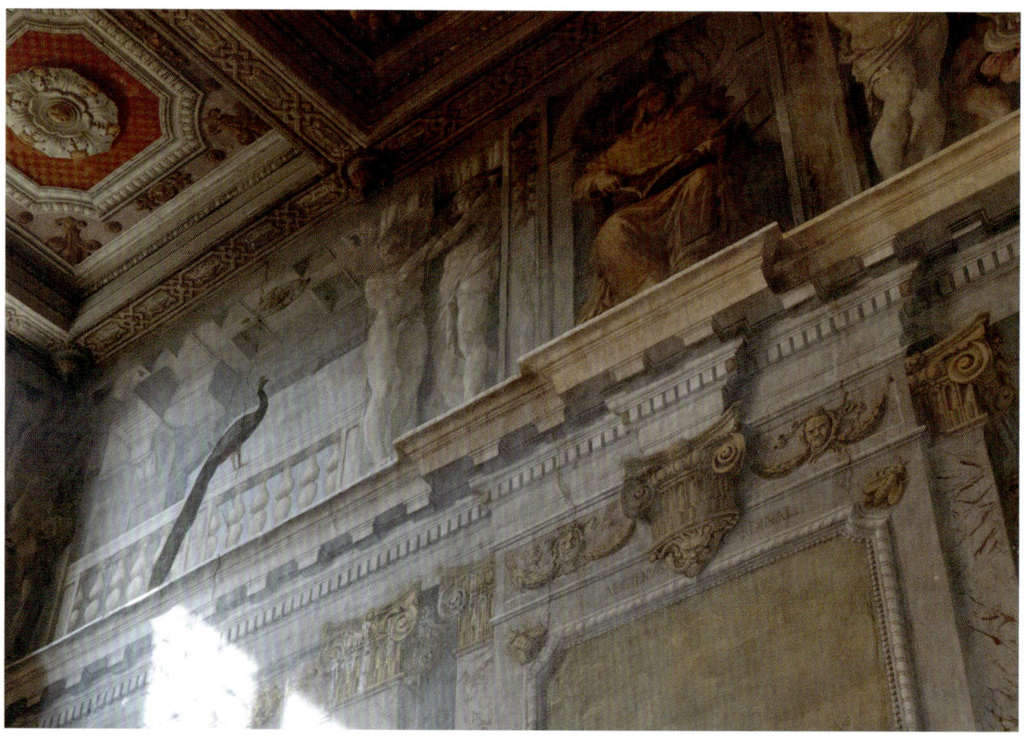

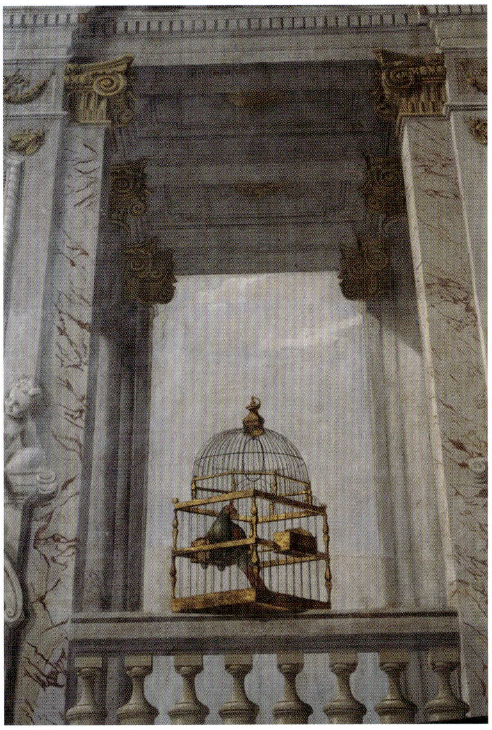

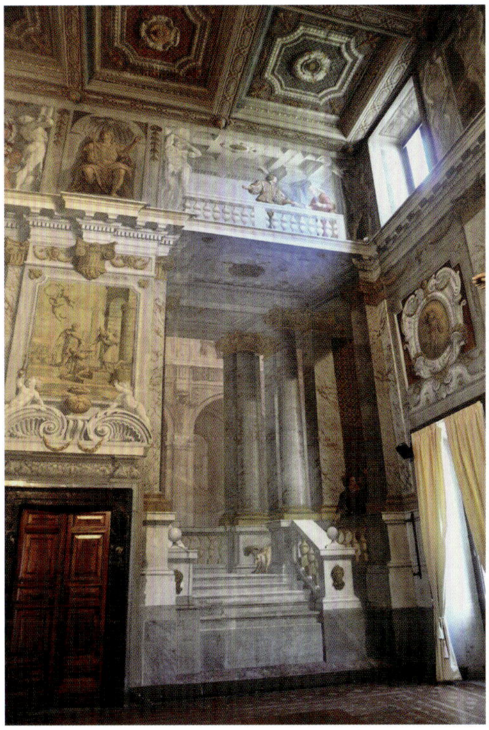

Fig. 25. Angelo Michele Colonna and Agostino Mitelli, bird in the Palazzo Spada *Salone*, between 1634 and 1635.

Fig. 26. Angelo Michele Colonna and Agostino Mitelli, fictive flight of stairs in the Palazzo Spada *Salone*, between 1634 and 1635.

right of this illusionistic column, observers discover a fictive window that mirrors an actual window on the opposite corner of the same wall (Fig. 23). Whereas the cymatium, or the uppermost molding at the top of the cornice that encircles the room, projects out from the walls in three dimensions, the elements below, like the corona and the adorning dentils, are painted (Fig. 24); this combination of actual relief with painted details makes the painted portions of the cornice particularly convincing. On a wall below the cornice, a bird (possibly a turaco species or a parrot) is shown in a cage that extends beyond the edge of a parapet in such a precarious manner that viewers might ask themselves if the cage will tumble into their space from that of the fresco (Fig. 25).[62] On the opposite wall, a fictive flight of stairs leads to an ambulatory and, as the space recedes, arches beyond an outdoor area come into view (Fig. 26). Above the ambulatory, two figures stand on a balcony, engrossed in discussion. Viewing the images today, it is hard to imagine that such fictive architecture would have been as convincing to observers as Virgilio claimed. Yet even if most seventeenth-century observers did not believe in the physical reality of the painted structures, such illusionistic frescoes may have nevertheless conveyed the risk of confusing appearance for reality and, in this way, may have stimulated the attention, refined the eyes and self-awareness of viewers, and offered compelling conversation topics that led them to appreciate Bernardino's discernment and sophistication as a patron.

Certain details in the Palazzo Spada frescoes—showing acts of visual attention and observation—would have amplified these effects of stimulating attention and intensifying awareness. In a scene from the wall described above, for instance, two men stand over a balustrade: the

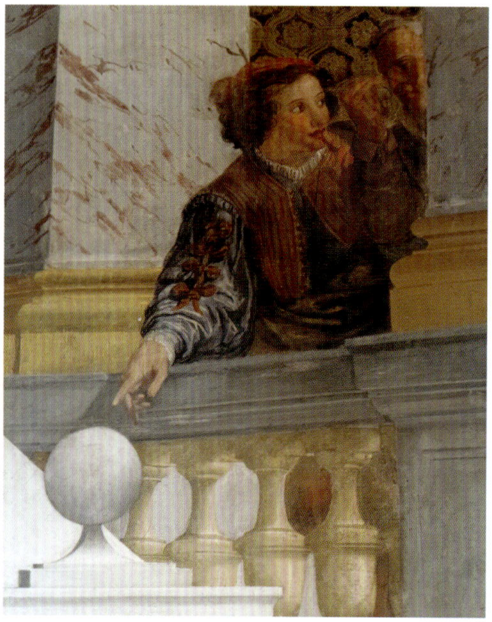

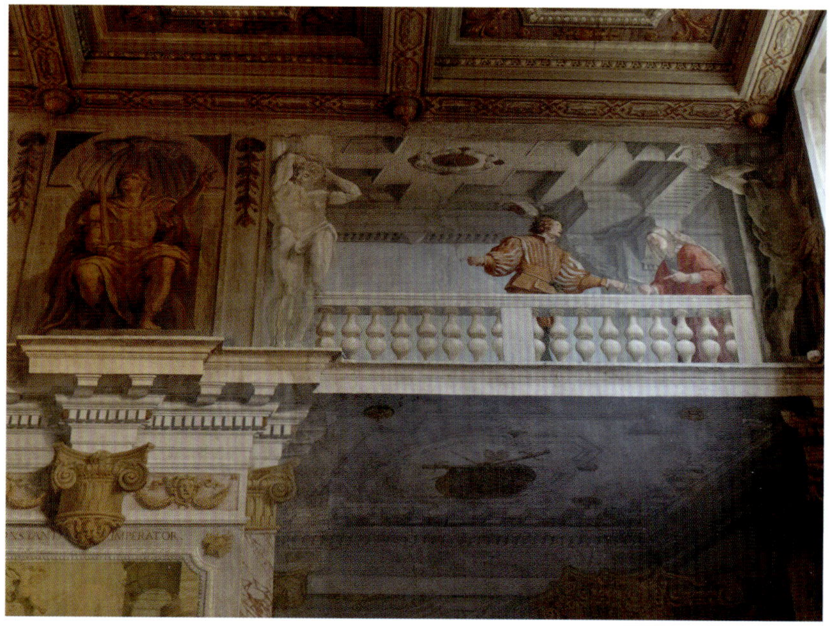

Fig. 27. Angelo Michele Colonna and Agostino Mitelli, scene of showing and observing in the Palazzo Spada *Salone*, between 1634 and 1635.

Fig. 28. Angelo Michele Colonna and Agostino Mitelli, pointing figures in the Palazzo Spada *Salone*, between 1634 and 1635.

younger man points to draw his companion's attention to a particular detail (Fig. 27); the older figure raises a pair of glasses to his eyes that he may distinguish what his interlocutor wishes him to see. Other figures distributed throughout the fictive architecture also point, urging the painting's interlocutors and the fresco's observers to pay attention to particular details: an African servant who assists a woman making a bouquet points to flowers; both figures in the ambulatory mentioned above point while engrossed in discussion; a man peering out from behind a column points upward; and a Swiss guard speaking to a trumpet player points to the right (see Fig. 23; Fig. 28, 29, and 30). A man peering through a telescope as he leans forward over a parapet proffers yet another instance of careful viewing (Fig. 31). These scenes of observant care would have brought the theme of attention to the fore—and, by picturing interlocutors in conversations about looking, would have exemplified the civil conversation that elites imagined might emerge in relation to moments of aesthetic appreciation.

Bernardino commissioned illusionistic painting of architectural elements not only within palaces but also on external walls, as recorded by the fictive windows and other details painted on the exterior surfaces of Palazzo Spada's garden side (Fig. 32). Bills from the artist Giovanni Battista Magni of Modena (c. 1592–1674) help us reconstruct the walls as they appeared in the seventeenth century. One from 1638 reads: "30 scudi for having painted the facade of the garden with columns, capitals, bases, and pillars made in ashlar and the decoration on the windows from the top of the said facade with a frame that runs around flanked by pillars made in ashlars, and for arranging the linear decoration on the top of the facade."[63] In another bill from 1641, Magni asks for further compensation, including twenty scudi for having painted a perspective with a balustrade in front of it in the palace's little garden.[64] In bills from 1652, 1653, and 1655, Bernardino is charged for Magni's rendering of a range of more illusionistic architecture details—presumably executed on the building's exterior—including, for instance, a fake iron

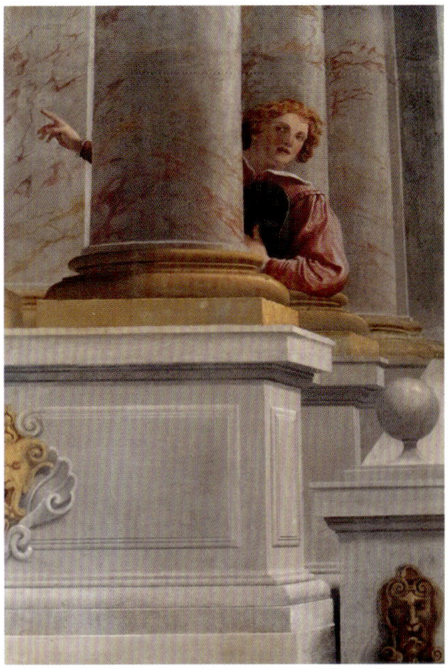

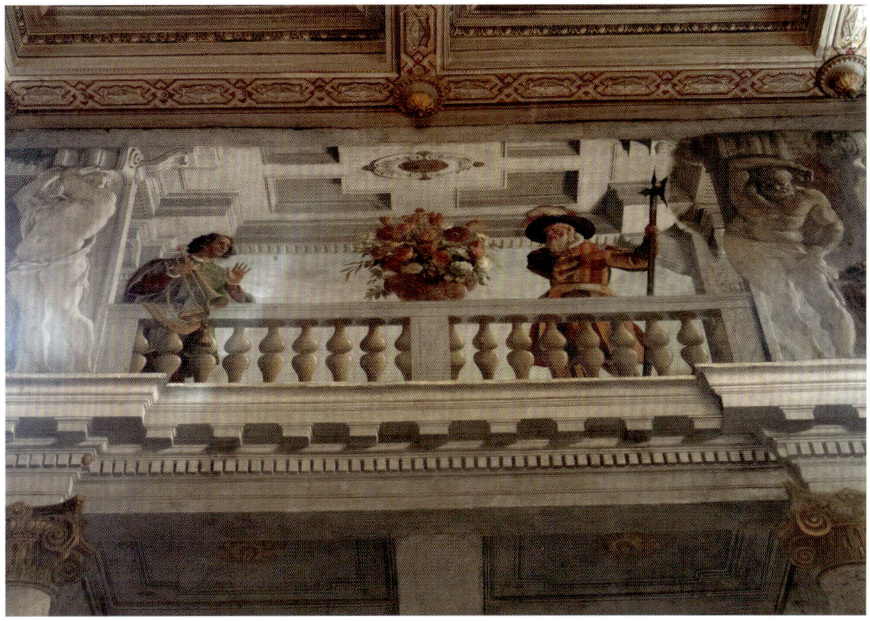

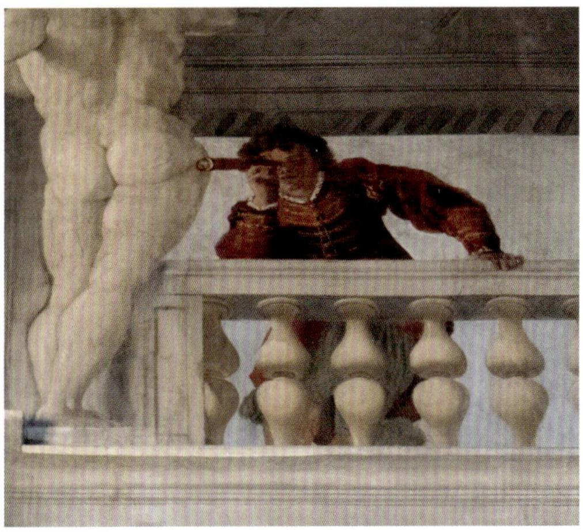

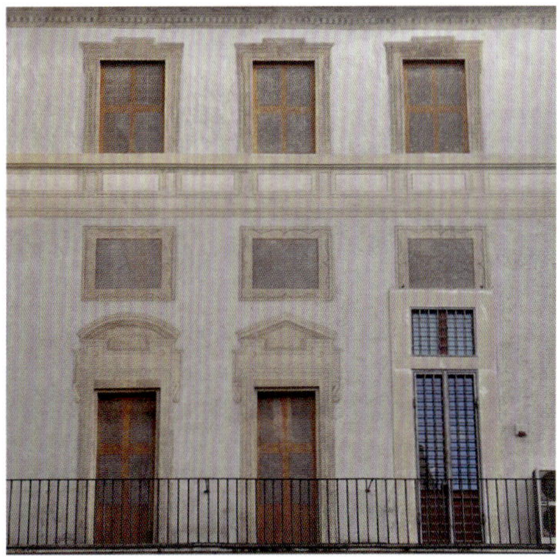

Fig. 29. Angelo Michele Colonna and Agostino Mitelli, a man peering out from behind a column who points upward in the Palazzo Spada *Salone*, between 1634 and 1635.

Fig. 30. Angelo Michele Colonna and Agostino Mitelli, a pointing Swiss guard in the Palazzo Spada *Salone*, between 1634 and 1635.

Fig. 31. Angelo Michele Colonna and Agostino Mitelli, a man peering through a telescope in the Palazzo Spada *Salone*, between 1634 and 1635.

Fig. 32. Giovanni Battista Magni of Modena, fictive windows and other illusionistic architecture painted on the garden side of Palazzo Spada, 1653.

balustrade.[65] These and other deceptive visual representations integrated around the garden, such as painted columns and an entablature, contextualize the Spada *prospettiva* (Fig. 33; and see Fig. 12).

Unlike the false balconies and windows featured on interior and exterior surfaces, the *prospettiva* could in fact be verified by touch—or walked through by a visitor. And yet its primary architectural referent, Virgilio tells us, was theatrical: the construction resembled an ephemeral piece of architecture by Borromini for the set of a Forty Hours' Devotion (also known as a *Quarant'Ore* or *Quarantore*) that Bernardino had observed.[66] In this liturgical ceremony, understood to commemorate the forty hours between the death of Christ and the Resurrection, the Blessed Sacrament is exposed over forty hours to praying clergy and laymen.

Fig. 33. Giovanni Battista Magni of Modena, painted columns and an entablature to the left of the perspectival colonnade in Palazzo Spada, 1638.

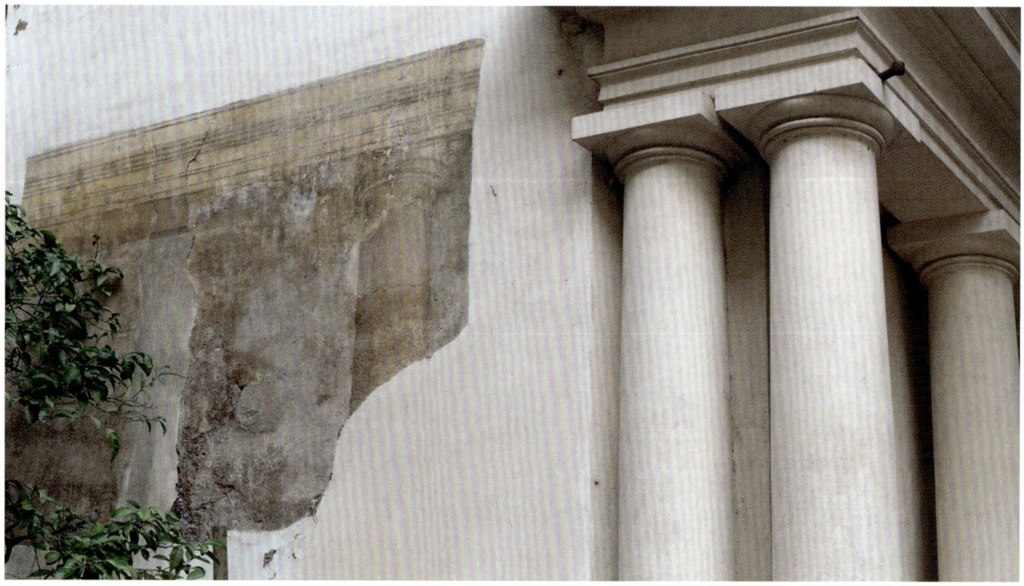

In seventeenth-century Rome, the ritual evolved into a spectacular theatrical happening. Forty Hours' Devotions featured vibrant processions, sermons, music, and temporary chapel adornments including machines and architectural edifices that emphasized and exalted the Eucharist. Since these sets were ephemeral, it is perhaps unsurprising that so few details seem to have survived of the apparatus Virgilio attributed to Borromini.[67] But like the Palazzo Spada's colonnade and its illusionistic frescoes, these ecclesiastic structures were created to confuse observers about the nature of the space they were in, and to invite the stupefaction that accompanied the denial of immediate perception. These constructions were typically lit by manifold oil lamps and candles, with the Eucharist positioned centrally, such that it appeared to cause the radiance surrounding it and draw the attention of the faithful.[68] Yes, the architecture was deceptive, but its deceptions were intended to enhance the viewer's affective response to the mystery of the sacrament of the Eucharist.

An ephemeral apparatus of the period may have employed an accelerated perspective to position the Eucharist above an intimidating and deceptively extensive architecture: the manipulation of perspective elevated the Blessed Sacrament literally within the space and also metaphorically in the minds of worshippers. A surviving etching and printed pamphlet describe this impressive apparatus, which was created for a Forty Hours' Devotion in the Jesuit Church of the Jesu (the diarist Giacinto Giglio, 1594–1671, claims that the event appeared less spectacular in real life than in the print) (Fig. 34).[69] The apparatus was designed under the direction of architect Carlo Rainaldi (1611–1691) and with the participation of painters Giovanni Maria Mariani and Domenico Rainaldi (1619–1698) for a Forty Hours' Devotion in February of the Jubilee year of 1650. In the etching, on the floor of the Temple of Solomon, a crowd of worshippers kneels in adoration; Solomon, clad in princely garb, also kneels in profound respect and reverence. A priest offers a sacrifice, and supplicants advance toward the altar, offering sacrificial animals. Christ is denoted through the Blessed Sacrament, which

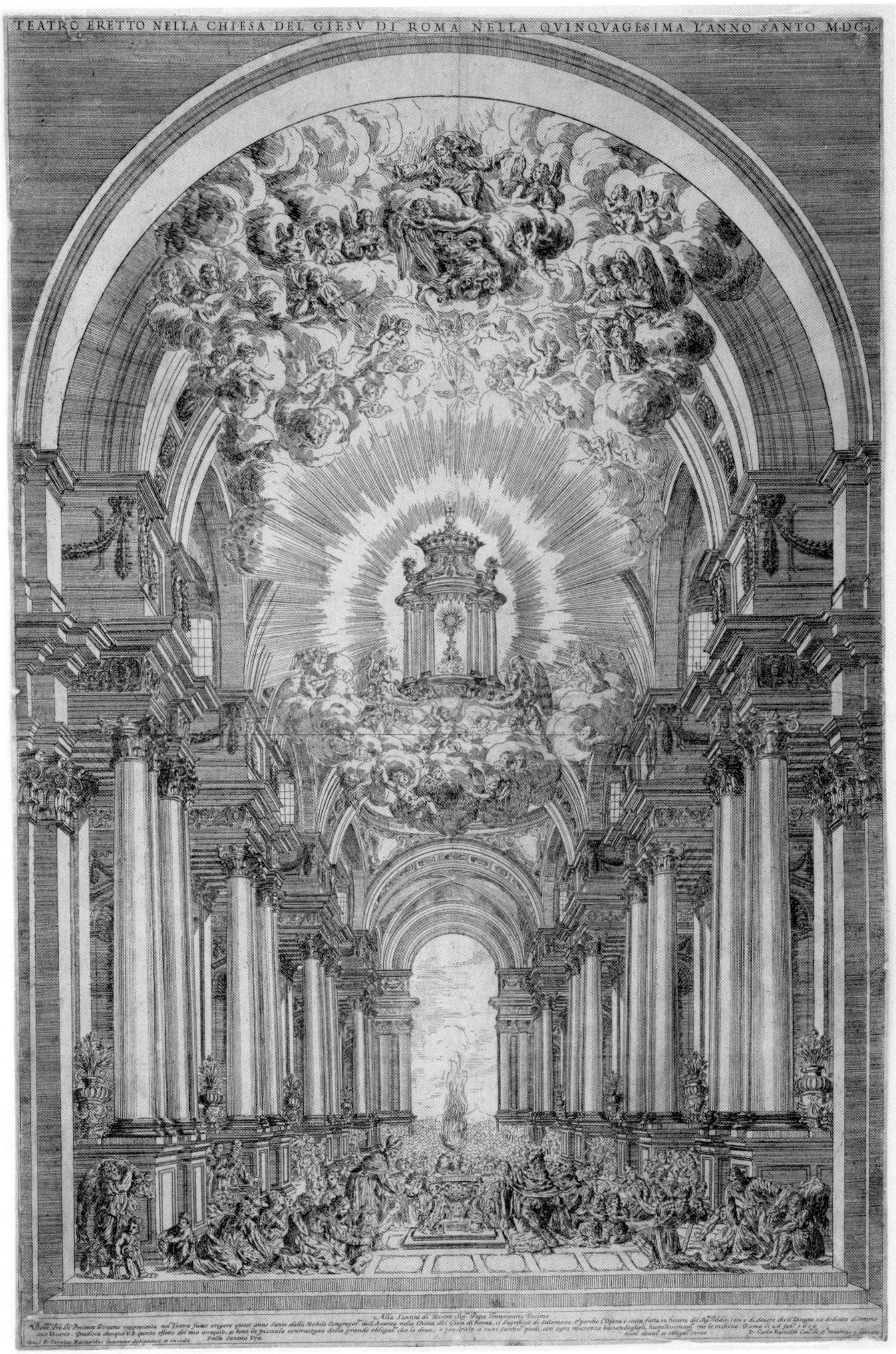

Fig. 34. Carlo Rainaldi (designer), ephemeral apparatus in the Jesuit church of the Gesù, 1650. Etching printed on paper, 27.56 × 18.03 in (70 × 45.8 cm). Museo di Roma, Gabinetto Comunale delle Stampe [inv. 86].

Fig. 35. Andrea Palladio, Teatro Olimpico, Vicenza, 1580–1585.

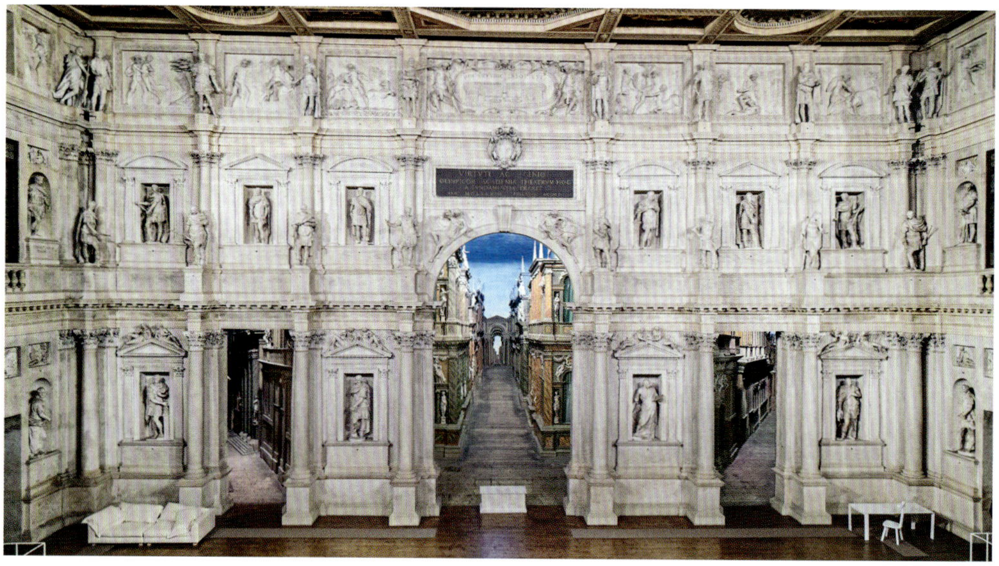

appears in glory, displayed on the image's central vertical axis in a tabernacle that angels and cherubim hold aloft. It is impossible to ascertain from the print whether the apparatus's paired columns, denoting the inside of the Temple of Jerusalem, made use of accelerated perspective, even if Giglio refers to "statues of perspective" (*Statue di prospettiva*) in his diary.[70] However, it is clear that, like the Spada *prospettiva*, these columns would have led observers' eyes to the receding distance.[71] Above, the heavens seem to have broken through the church's roof: God the Father, arms outstretched, leans forward on a cloud to supervise the action, while angels and cherubim populate the space around him. A little lower, the dove of the Holy Spirit swoops down toward the Sacrament. The pamphlet claims the architecture resembled a white alabaster, with gold accents on the vault's decorations and coffers, and on the tips of the angels' feathered wings. The tabernacle with the Sacrament seemed as if made of lapis lazuli. As Bernardino chooses to build in his palace a structure that alludes to religious theater, we can read into his perspectival deformation a wink at his expectation that his guests can draw from their knowledge of theatrical sets a discerning recognition that visual perception might be manipulated to studied effect.

The deformations that emerged in the seventeenth century as a feature of elite architecture and collecting thus carried multiple potential purposes and meanings: if spiritual affect was not always in play, most of these designs, by thwarting architectural and visual expectations and taking apart the regularity of classical designs, did invite viewers to engage intellectually with the project of seeing. Although some stylistic precedents to the Spada deformation date back to the fifteenth and sixteenth centuries, these (primarily theatrical) structures' challenges to architectural norms were limited in the centuries before baroque innovations emerged. I will not designate these early modern precedents "deformations," because they were not explicitly associated with the deformations believed to be inherent to all sublunary creations (as the *prospettiva* was in Bernardino's poem), and more simply because I have not found examples of period observers describing them with the term "deformation" or its variants. That being said,

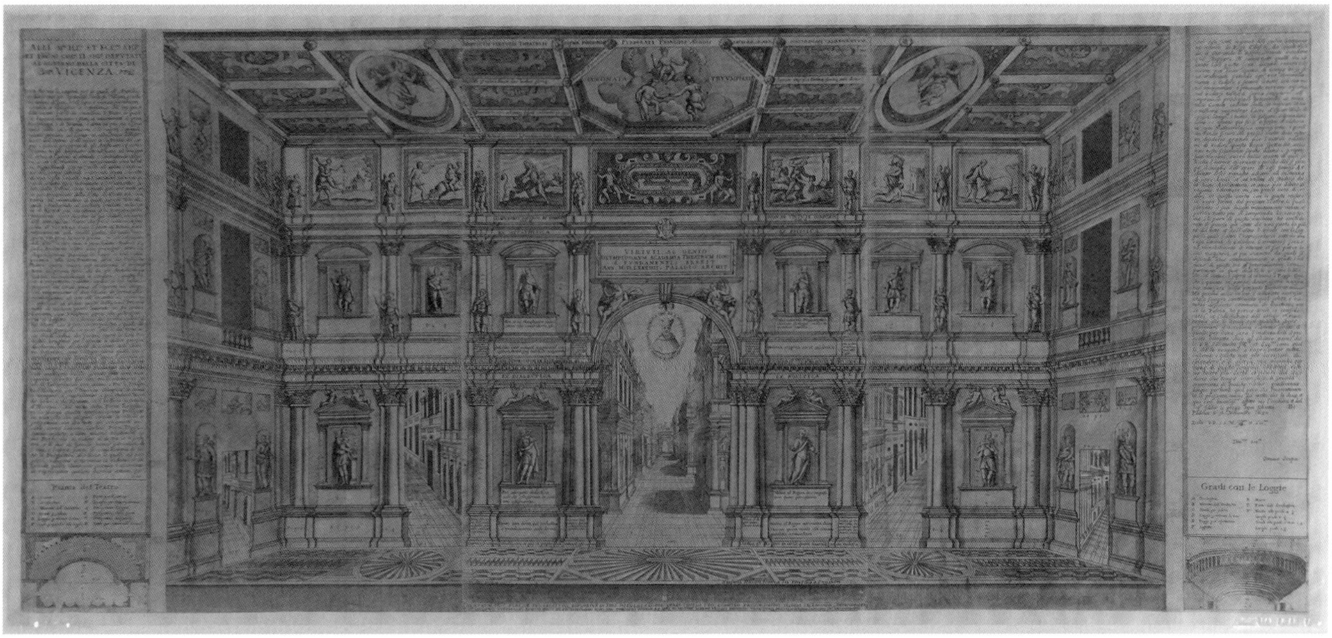

Fig. 36. Iseppo Scolari and Ottavio Revesi Bruti, proscenium and stage of Teatro Olimpico, 1620. Engraving printed on paper, 20.5 × 43.3 in (52 × 110 cm). Vicenza, Pinacoteca Civica di Palazzo Chiericati [inv. B. 2086].

some of the architectural and sculptural examples that I discuss below make use of accelerated perspective and thus particularly resemble the Spada *prospettiva*; insofar as these cultural productions undermine expectations around how classical colonnades function, future scholars may well choose to identify them with or as deformations.

For if the Spada *prospettiva* may have drawn inspiration from certain ephemeral religious structures of the seventeenth century, the colonnade (and perhaps those religious-theatrical structures as well) clearly also followed precedents from a secular tradition. Among the best known of these is the Teatro Olimpico (1580–1585) in Vicenza, designed by Andrea Palladio (1508–1580) and his student Vincenzo Scamozzi (1548–1616), who was brought in after Palladio's death to finish the perspectival scenery that continues behind the openings in the *scenae frons* (Fig. 35).[72] The white *scenae frons* stands in sharp contrast with the colors of Scamozzi's perspectival streets, modeled on Vicenza.[73] Its arched central passageway is both broader and taller than the other two doors, which are surmounted by flat entablatures. Whereas the shorter doors lead to single roads, the central arch opens onto three different streets, although it is impossible for audience members to observe all five streets simultaneously. Because the stage inclines upward and the proportions of the structures in the background diminish rapidly, audience members are deceived into believing that the streets are much longer than they really are, much as viewers of the Spada *prospettiva* are led to mistake the length of the colonnade. When the Teatro Olimpico first opened in 1558, Sophocles's (c. 496–406 BCE) *Oedipus the King*, a play based on discovery, was staged in a novel translation. The choice of play echoes the architecture's own stress on the interplay between vision and deception. When planning the Spada *prospettiva*, Bernardino and his architects could have discussed such printed images of the Teatro Olimpico's proscenium and stage as the engraving by Iseppo Scolari after Ottavio Revesi Bruti (1585–1648) that dates to 1620 (Fig. 36).[74]

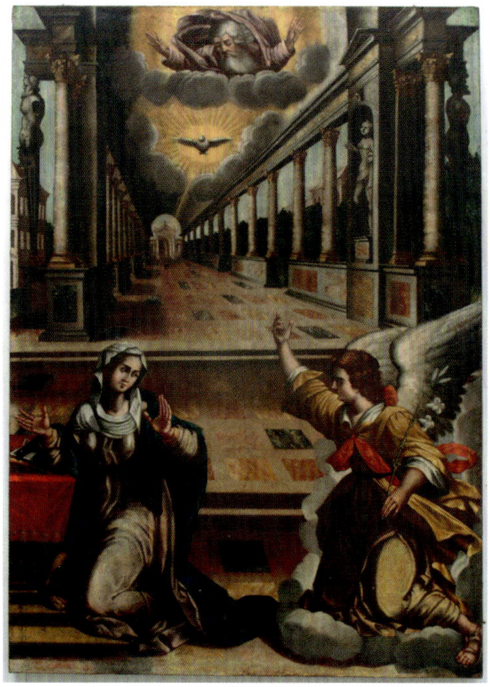

Fig. 37. Fra Giovanni Maria da Bitonto, *Annunciation*, 1631. Oil on canvas, 104.1 × 72.4 in (264.5 × 184 cm). Museo Diocesano Bitonto.

Of course, by the middle of the seventeenth century, Bernardino would not have had to look directly back at the Teatro Olimpico: in the first half of the seventeenth century, similar accelerated perspectives were appearing in painted contexts and also in three-dimensional forms. Even before he came to Rome around 1639, Bitonto had painted an *Annunciation* (1631) with the archangel Gabriel and the Virgin Mary in the foreground, before an extended Corinthian colonnade; a barrel vault in perspective above and behind the central figures is interrupted by God the Father and the Holy Spirit, appearing amid clouds and light (Fig. 37). Alongside the gridded marble floor, the colonnade conveys a regular recession into space. Within Bernardino's circle, again before the Spada *prospettiva* was completed, Bitonto had also collaborated with Bernardino's brother Virgilio on a sculpted perspectival colonnade. That distorted perspectival colonnade, constructed under Virgilio's supervision, was part of a tabernacle for the High Altar of San Paolo Maggiore in Bologna.[75] Bitonto created a wooden model for the tabernacle in 1647; both the model and the finished work, executed the following year in gilt bronze with columns made of agate, were, of course, much smaller than the Spada *prospettiva*, but they share its rapid diminution in the scale of recurring architectural components. In 1649, the tabernacle was placed on the High Altar of San Paolo Maggiore, where visitors can still view it from some 19.7 feet (6 meters) away, when standing behind the balustrade and iron door that lies before it, or from some 62.3 feet (19 meters) away, when positioned where the first pew lies today (Figs. 38, 39, and 40).[76] If the *prospettiva* was meant to be observed from a distance by a viewer positioned in the main courtyard, in the *Sala dei Rilievi*, or in the *Sala delle Udienze*, the tabernacle, positioned on the High Altar behind a balustrade and out of worshippers' reach, was similarly meant to be admired from afar. Even though Virgilio had invested

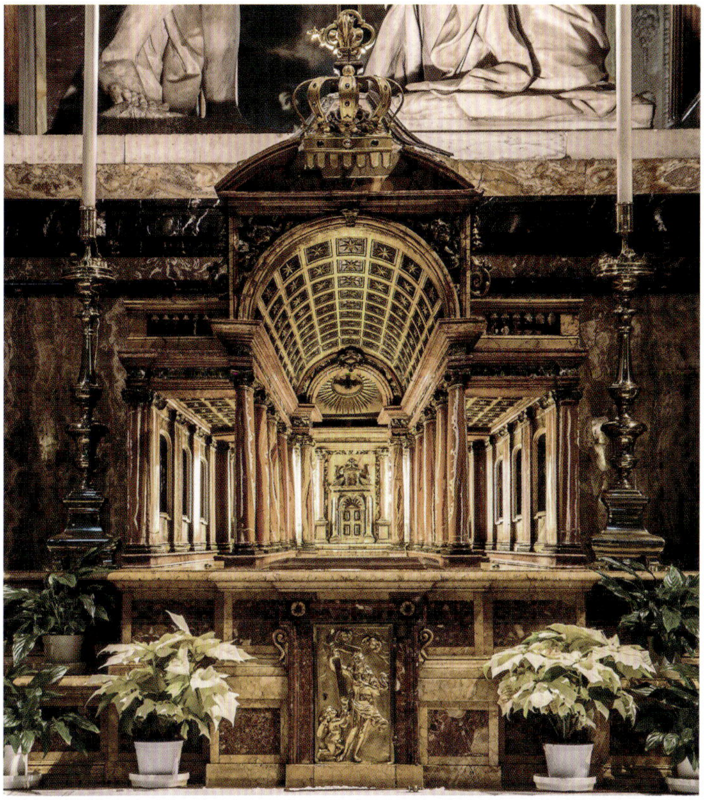

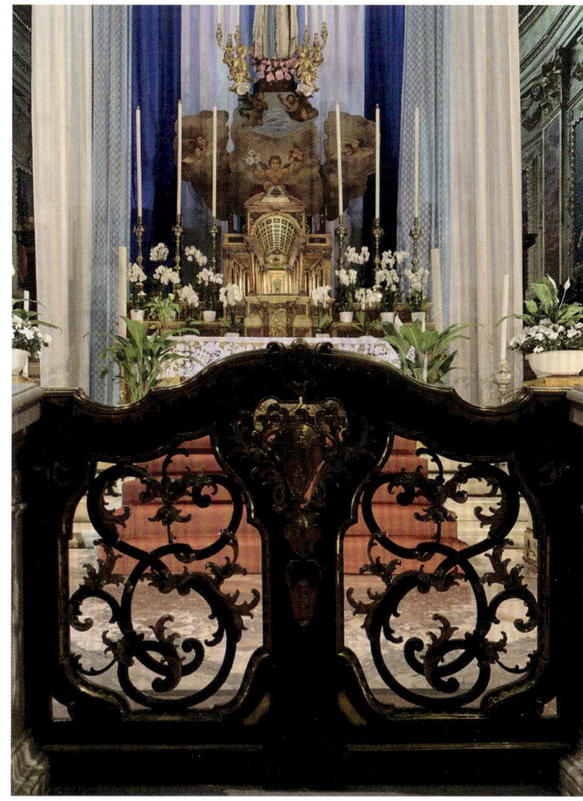

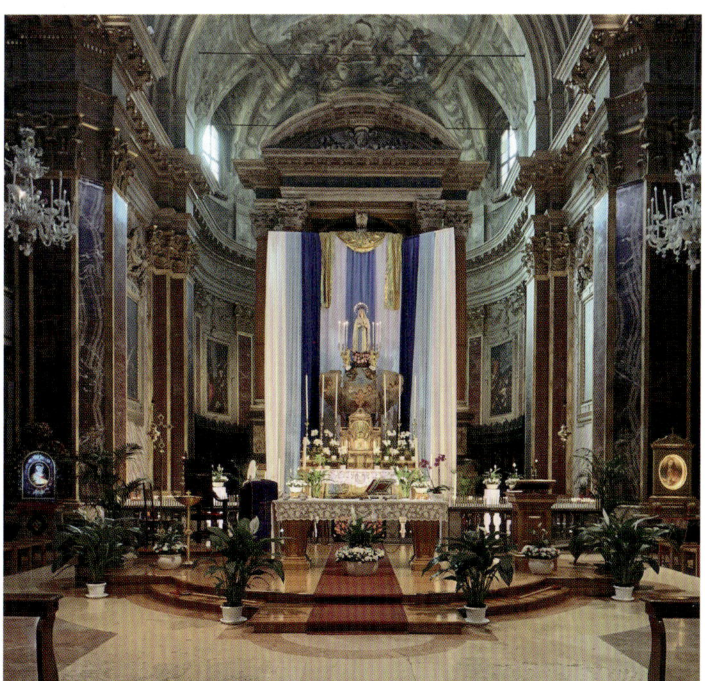

ABOVE LEFT

Fig. 38. Fra Giovanni Maria da Bitonto, tabernacle, c. 1649. High Altar of San Paolo Maggiore, Bologna.

LEFT

Fig. 40. Fra Giovanni Maria da Bitonto, view of tabernacle in San Paolo Maggiore, Bologna.

ABOVE

Fig. 39. Fra Giovanni Maria da Bitonto, view of tabernacle in San Paolo Maggiore, Bologna.

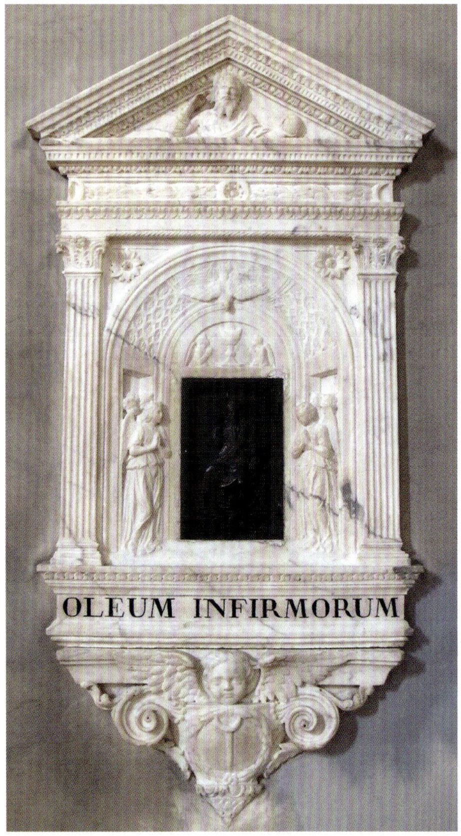

Fig. 41. Bernardo Rosselino, tabernacle, 1449. Ospedale di S. Maria Nuova, Florence.

in his own perspectival colonnade, he was critical of the cost of his brother's *prospettiva*: "It turned out as a beautiful artifact but its costs greatly exceeded its beauty."[77] Virgilio did not think his brother was getting good value for all the money he spent.

Like the Spada *prospettiva*, the tabernacle features a vault decorated with rectangular coffering. As the space recedes and the vault ends, the dove of the Holy Spirit, encircled by rays of light, appears to fly down; the dove occupies an analogous position in Bitonto's early annunciation (see Fig. 37). The tabernacle's illusionism is constructed through a technique similar to that of the Spada colonnade: a sloped floor and vault create the impression that the colonnade is longer than it really is. Both the Spada colonnade and the tabernacle amplify natural perspective.

Tabernacles functioned to hold and safeguard the consecrated Eucharist, relics, or miraculous images, as well as to display these holy entities and to exalt them to votaries. Accelerated perspectives became a popular trope in Italian Renaissance tabernacles to capture and sustain the attention of worshippers, as Paul Davies has observed.[78] Bitonto's collaboration with Virgilio belongs to an artistic tradition developed as early as the fifteenth century by artists including Mino da Fiesole (1429–1484), Desiderio da Settignano (1428–1464), Isaia da Pisa (1410–c. 1464), and Giuliano (1432–1490) and Benedetto da Maiano (1442–1497), as well as Bernardo Rosselino (1409–1464), whose 1449 tabernacle in the Ospedale di S. Maria Nuova in

Florence seems to be the oldest surviving exemplar of this type (Fig. 41).[79] These perspectival structures used accelerated perspective and converging lines to focus attention on the object of veneration, whose believed reality was contrary to human perception.[80] Just as a cat or the statue at the far end of the Spada *prospettiva* appears larger than it really is to observers unaware that the heavy freestanding Doric columns decrease in actual size as they recede in space, tabernacles that used accelerated perspective would have exaggerated the perceived size of objects of veneration that were positioned at the apparent end of the recessed space.[81] Davies has stressed, moreover, that the converging lines of perspectival tabernacles often led observers to imagine a hidden focal point within the tabernacles that may well have coincided with the placement of the holy objects behind the tabernacles' locked doors.[82] This imagined convergence further concentrated the worshipper's attention on the object of veneration and reinforced belief in the miraculous nature and sacramental power of that object. If the illusory perspective made the host loom literally larger and invited an affective response, these tabernacles also appealed to the intellect of the discerning viewer, as a tangible reminder of the gap between human perception and divine reality. The tabernacles' distortions asked the faithful to focus on and even long for the inaccessible venerated object, and would thus have supported veneration.

The seventeenth-century artists who created tabernacles and ephemeral structures for Forty Hours' Devotions used perspective much as it was used in the painted altarpieces of the Renaissance, from Masaccio's (1401–c. 1428) *Holy Trinity with the Virgin and St. John* (c. 1425) to Fra Angelico's (1394–1455) *San Marco Altarpiece* (c. 1438–1443). In these paintings, as is often noted, illusionism would have helped votaries pay attention to particular holy figures, even if the central mystery of the sacrament of the Eucharist was understood differently when Masaccio and Fra Angelico painted their frescoes than two centuries later, as Bernardino was commissioning his *prospettiva*. Transubstantiation as a core Catholic doctrine inspired a vigorous deployment of art well before the rise of anamorphic and perspectival optical devices.

Accelerated perspective enhanced devotion in large-scale architectural settings as well.[83] The tabernacles' use of accelerated perspective calls to mind the work of Renaissance architects like Donato Bramante (c. 1444–1514), who exploited this technique to enhance devotion in such buildings as Santa Maria presso San Satiro (c. 1482) in Milan, his earliest extensive exercise with the system (Figs. 42 and 43).[84] Even though worshippers can physically walk through Bramante's barrel-vaulted nave, the church's chancel is a fictional, illusionistic space executed in high-relief sculpture; although a viewer who considers the nave from a central vantage point, as seen in Figure 42, might perceive the chancel to be a fully realized structure, in fact it is too shallow to be traversed physically.[85] The converging lines of the accelerated perspective vault in S. Maria presso San Satiro would have led votaries to contemplate the focal point: the painting of the Madonna, which was considered a miracle-producing work.[86] The accelerated perspectives of Santa Maria presso San Satiro and of the tabernacles designed by Bitonto and others offer important formal precedents for the Spada *prospettiva*. In these works, designed to excite awe and religious emotion, perspectival distortion functions as a persuasive device that makes something important look more important, enabling the Church to concentrate attention and reinforce belief in the mystery of the Eucharist. The Spada deformation borrows some of the same techniques to situate the observer not before the altar but before the clever discernment of the householder and the architects he patronized.

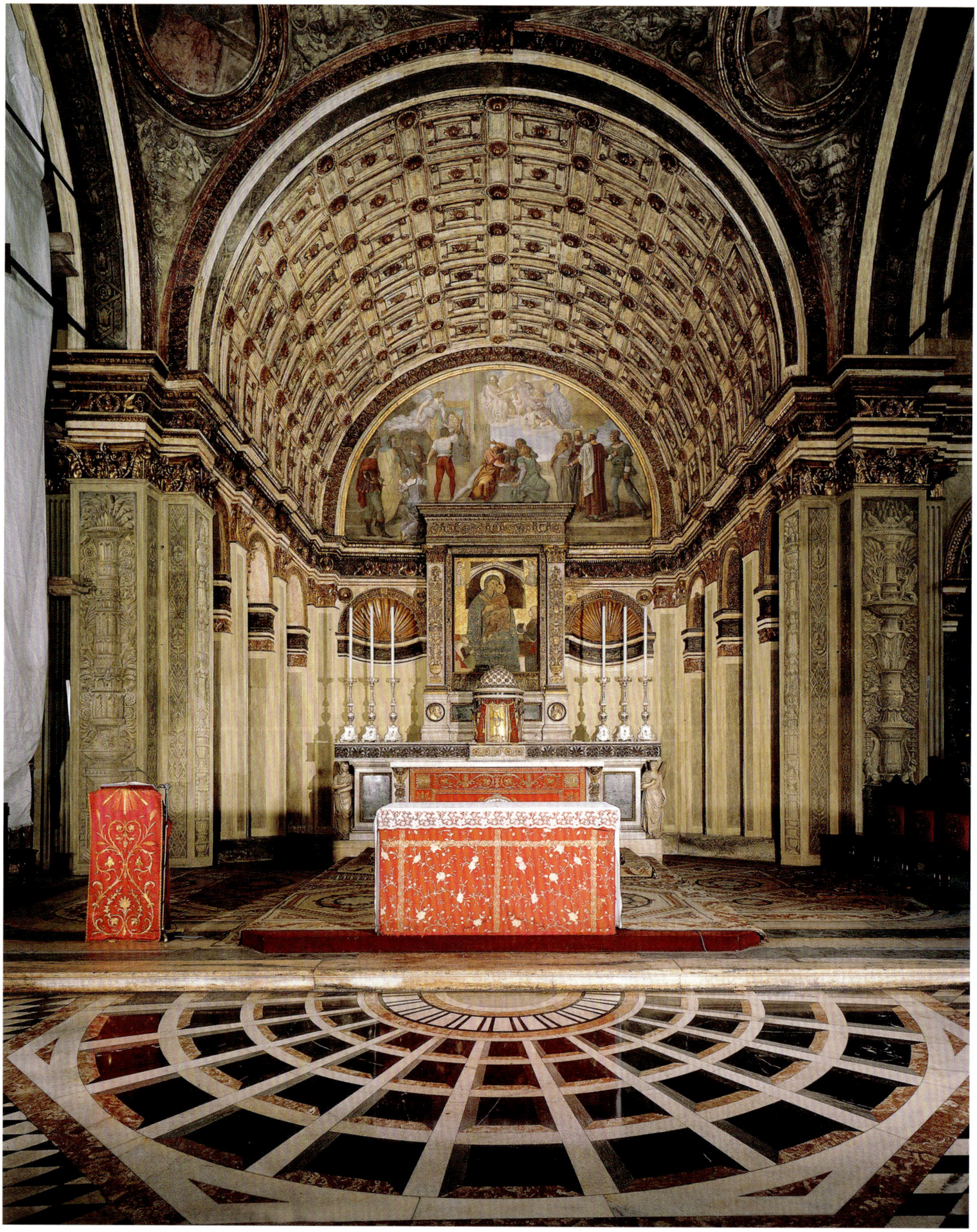

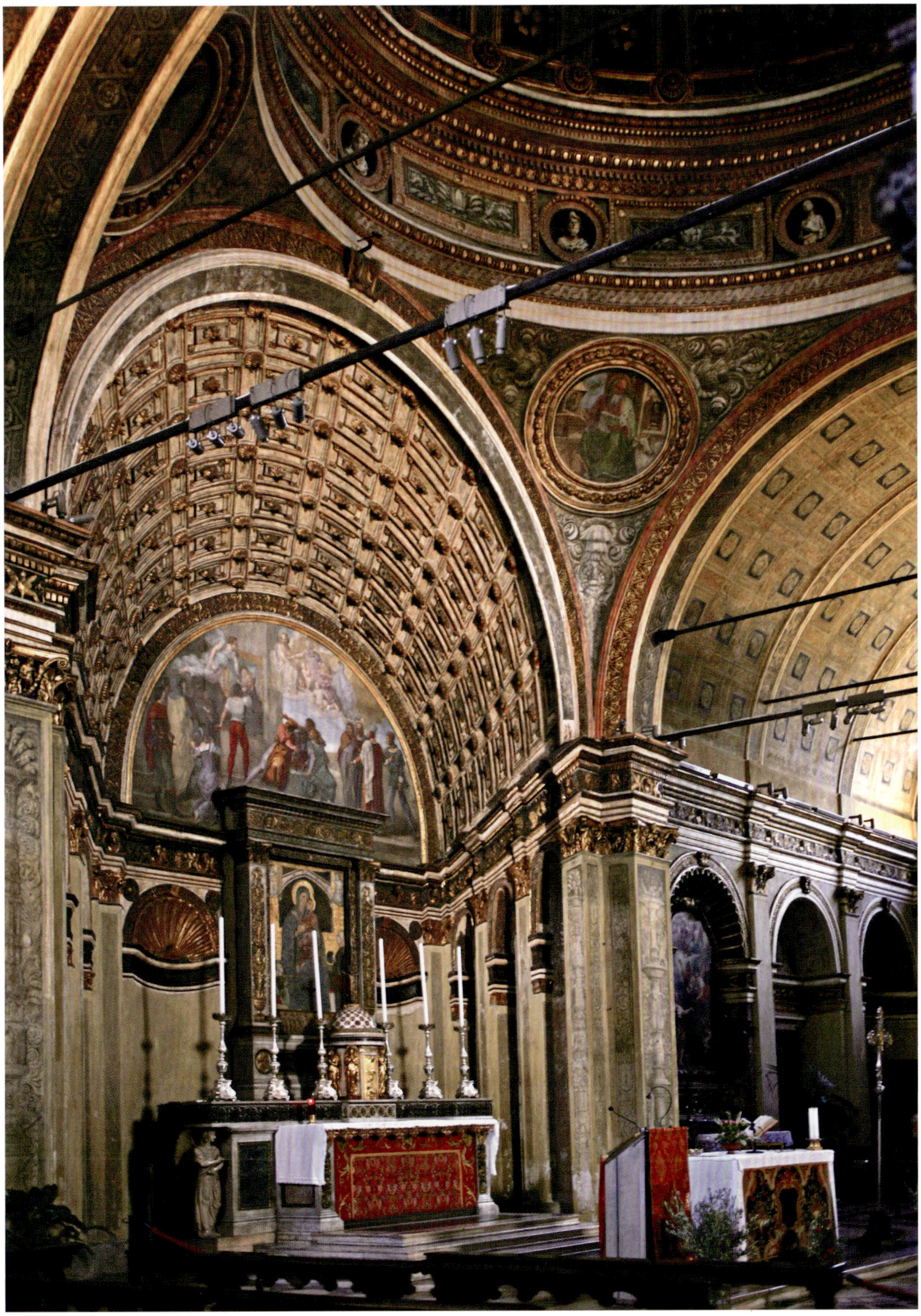

OPPOSITE

Fig. 42. Donato Bramante, Santa Maria presso San Satiro (interior), c. 1482. Milan.

LEFT

Fig. 43. Santa Maria presso San Satiro (interior), c. 1482. Milan.

Fig. 44. Gian Lorenzo Bernini, Scala Regia, 1663–1666. Vatican Palace, Vatican State.

Fig. 45. Gian Lorenzo Bernini, illusionistic coffered barrel vault in low relief in marble on a lateral wall of the Cornaro chapel, between 1647 and 1652. Santa Maria della Vittoria, Rome.

Fig. 46. Gian Lorenzo Bernini, illusionistic coffered barrel vault in low relief in marble on a lateral wall of the Cornaro chapel, between 1647 and 1652. Santa Maria della Vittoria, Rome.

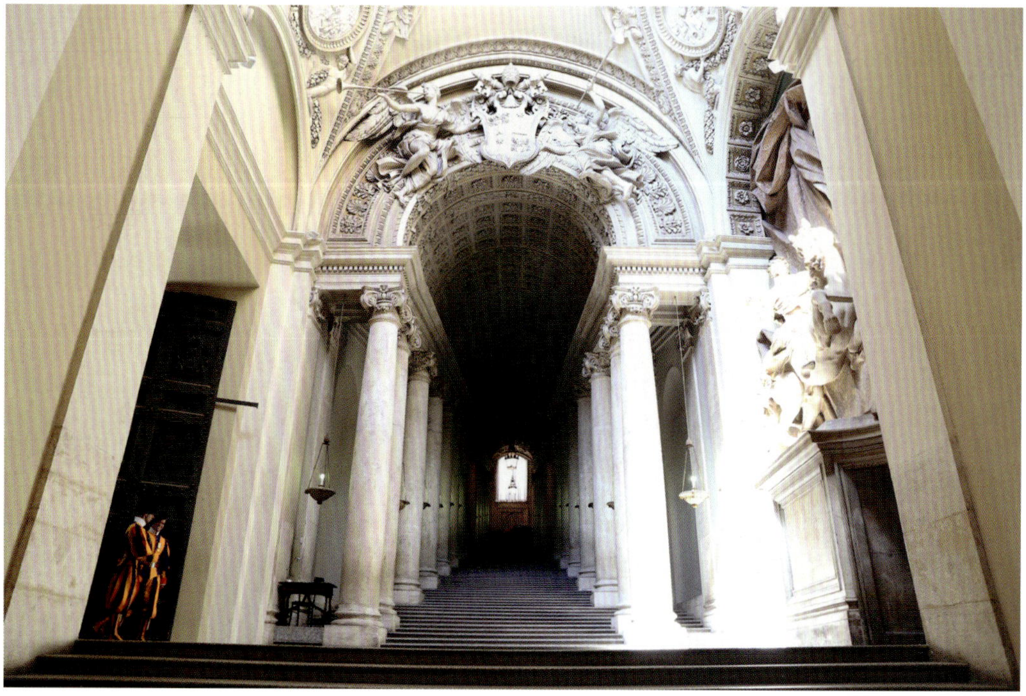

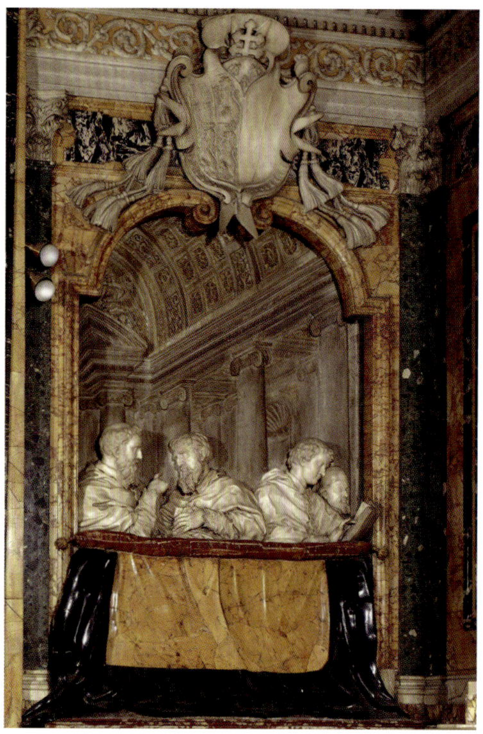

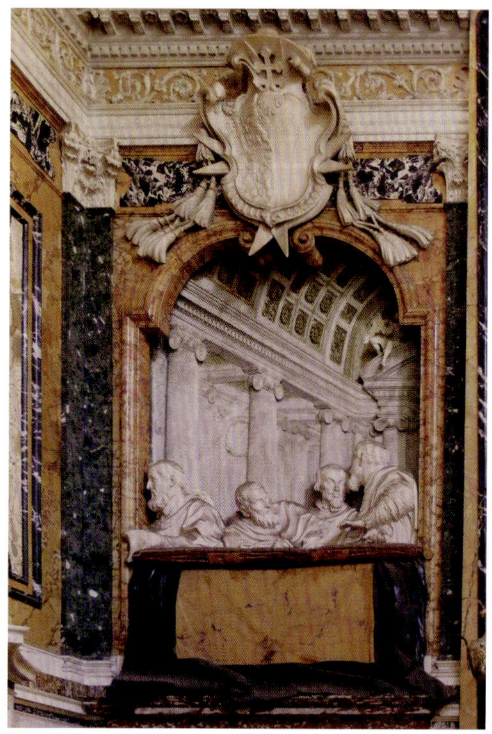

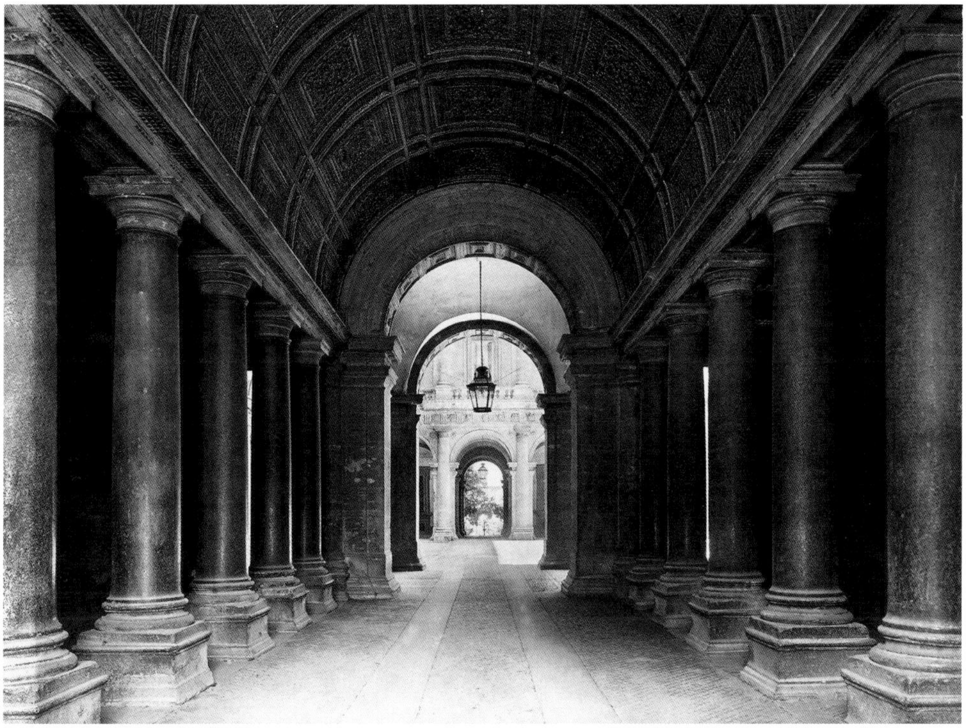

Fig. 47. Antonio da Sangallo the Younger, three-aisle atrium, 1516. Palazzo Farnese, Rome.

Bernini's *Scala Regia* of 1663–1666, completed over a decade after the Spada *prospettiva*, would further reveal how the Catholic Church used perspective—and often modified the ideals and conditions of Italian classicism—not only to stimulate votaries' faith but also to exalt the Church's moral and theological vantage point. The *Scala Regia*, the ceremonial flight of steps that functioned as the primary entrance to the Vatican palace and the main link between St. Peter's and the palace, is another prominent example of Roman perspectival illusionistic architecture (Fig. 44). Its columns are meant to appear identical in size when considered from below, but a visitor who ascends the staircase discovers that, in fact, they decrease in scale; the illusion makes the staircase appear longer than it really is. Bernini likely drew on Bitonto and Borromini's project in Palazzo Spada when designing his staircase about a decade later, although he had already sculpted illusionistic coffered barrel vaults in low relief in marble on the lateral walls of the Cornaro chapel in Santa Maria della Vittoria in Rome, between 1647 and 1652 (Figs. 45 and 46).[87] In that setting, the vaults extend behind portraits of Cornaro family members, who appear to kneel as they consider the miracle occurring before them. The architecture's illusionary effects convert the chapel into a crossing with a transept that extends to the left and right sides.

When observed from the *piano nobile* audience room in the palace, or the main courtyard, or the *Sala dei Rilievi*, the Palazzo Spada's *prospettiva* also recalls the un-deformed three-aisle atrium designed in 1516 by Antonio da Sangallo the Younger (1484–1546) that welcomes visitors to the nearby Palazzo Farnese (Fig. 47). From these vantage points, observers might be deceived into believing that the *prospettiva*, like the Palazzo Farnese atrium, presents columns

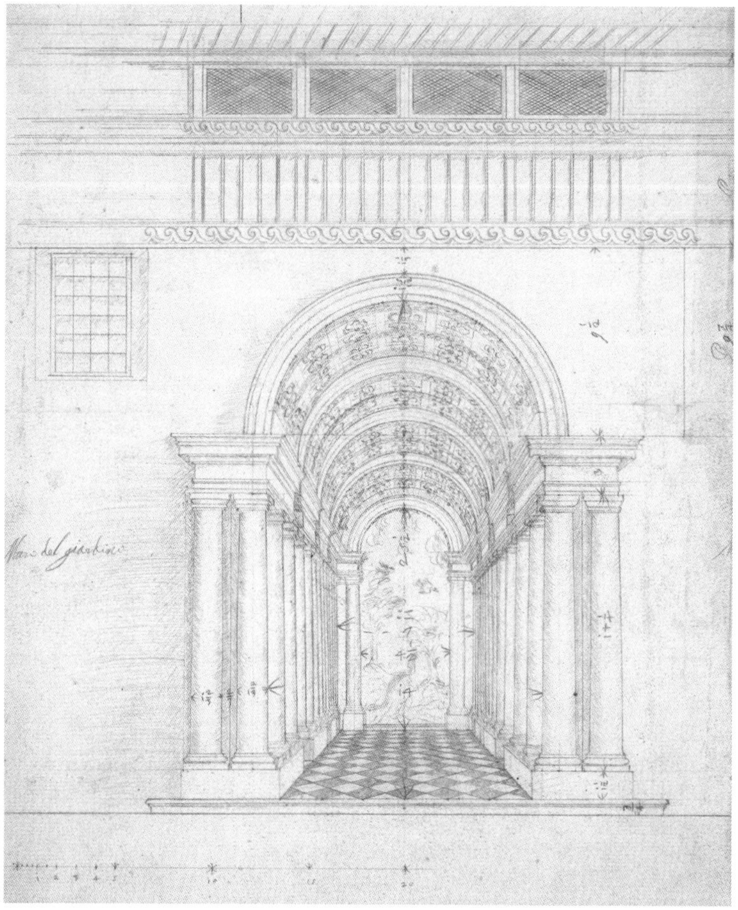

Fig. 48. Francesco Righi, preliminary elevation drawing, 1652/53. Graphite drawing on paper, 14.6 × 11.6 in (37 × 29.5 cm). The Albertina Museum, Vienna [AZRom1156].

of equal height placed at regular intervals. The Farnese atrium leads to a palace courtyard—the viewer's attention is centered on the archway leading to this further space—while the *prospettiva* leads to a shallower space centered on a statue that was not added until the second half of the nineteenth century.[88] A bill from 1653 suggests that Magni painted scenes that would have once appeared near where the statue stands today—apparently, an illusionistic painting on the wall at the far end of the *prospettiva*.[89] In the preliminary elevation drawing known as "Albertina 1156," a representation of a painting (featuring two trees, a stream, some birds) at the rear of the *prospettiva* may provide insight into what Magni's landscape would have looked like in the seventeenth century (Fig. 48).

As for most visitors, like that discerning guest who was led through the spaces of the Palazzo Spada, the deception before the Palazzo Spada colonnade dissolves quickly, if not immediately; the *prospettiva*'s true appeal lies beyond its power to deceive, in its status as an invitation for analysis and the civil conversation that defined and reinforced social categories: the *prospettiva* asks observers to discuss how the deception is created and to consider the philosophical implications of that deception. Just as Bernardino sent his moral epigram on the colonnade in a letter to Gibbes, it is tempting to speculate that he would have presented his philosophical poem on the structure to guests to liven discussions in his palace.[90] My claim regarding this work's power to incite discussion relates to the theory of "operational aesthetics" that Neil Harris develops to describe the attraction of the hoaxes created by P. T. Barnum (1810–1891). According to Harris, Barnum saw that truth versus falsehood mattered less to spectators than the satisfactions of investigating whether and how a fraud had been perpetuated; similarly, Bernardino anticipated that his visitors would be comparatively indifferent to the fact that they had been tricked, but very interested in considering how that deceit had been effected.[91]

Palazzo Spada's *prospettiva*, a deformation that violates the form and the logic of classical colonnades, provides an answer to the apparent contradiction between Bernardino's investment in "grand things under the heavens" and his view that such splendor is "nothing except specters." Deformation allows a display of magnificence and wealth to become something other than itself, that is, something other than an established and quickly recognizable signifier of affluence. That the social choreography of the palazzo ultimately focused visitors' attention on a deformation—rather than some other merely decorative expensive object—bound host and guest in a mutual display of visual discernment that carried significant social value. Archival notes document Bernardino's desire to have his nephews at hand, so that he could school them with questions and equivocal answers that would incite them to think.[92] In a similar vein, the palace's deformation was designed to confuse and thus to cue attention and discernment as well as to incite conversation. Perhaps Virgilio was wrong, and Bernardino did ultimately get his money's worth.

If Bernardino anticipated that visitors would have connected his colonnade with the atrium at the Farnese palace, he may have hoped that they would realize that he was doing something similar, but also cleverer (see Figs. 12 and 47). This comparison would have allowed him to claim the importance of his own discernment and the occasion for discernment he offered his guests, an importance above and beyond that of a regular colonnade, which would impress primarily as a display of wealth and magnificence. That Bernardino, a cardinal, drew on church practices—particularly those designed to exalt the Host and enhance the experiences of the faithful—is also significant. We recall that the converging lines of perspectival tabernacles led beholders to envision a covered focal point within a tabernacle, corresponding to the implied location of the holy objects. One wonders if in some sense Bernardino hoped to imply, metaphorically, that something important, perhaps here his genius and munificence as a patron, was lying just beyond the perspective.

Bernardino's Sundial and Periscope

Bernardino's *prospettiva* was not the only artistic feature of the palace whose effects depended on how visitors moved through space. When we juxtapose the colonnade with the monumental sundial painted in a 72.2-foot-long gallery of the palace, we recognize that these wondrous visual displays operate in a similar fashion, prolonging the visitor's journey from visual perception to intellectual discernment (Fig. 49). If the deformation choreographs a movement from disorientation to understanding, the sundial shapes a related visual experience in which guests move from a state of confusion or amazement, when initially confronted by the sundial's hyper-informativity, to discernment. That intellectual journey, via civil conversation, transpires as host and visitors physically move through the corridor, deciphering its meanings as they go.

Fig. 49. Giovanni Battista Magni of Modena and Emanuel Maignan, ceiling sundial, 1644. Palazzo Spada, Rome.

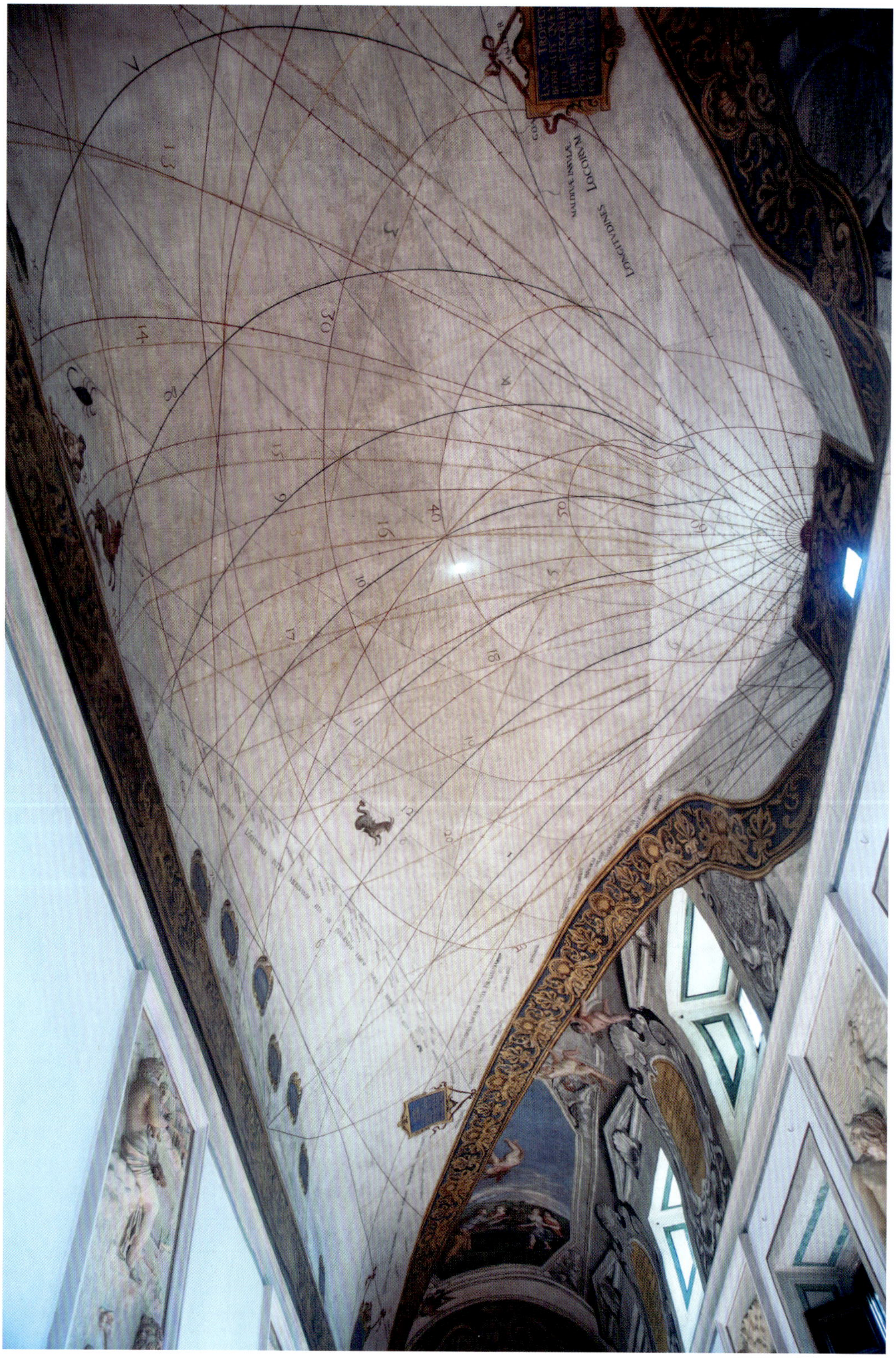

In elite circles in seventeenth-century Italy, an interest in visual deceptions often accompanied a fascination with empiricism and an enthusiasm for deploying visual representations to disseminate knowledge. Though empiricism asked observers to trust what they saw, an object like the sundial conjoins deception with empiricism (as, conversely, given how extensively those who created such deformations as the *prospettiva* relied on optics and mathematics, deformations themselves conjoined deception with the study of natural science). Although the sundial is, of course, a different sort of object than the *prospettiva*, I take the deformation as a starting point for thinking about how the sundial operated: alongside other purposes, both functioned as courtly conversation pieces that demanded specific spatial performances linked to social power. As argued earlier, the *prospettiva* asked, perhaps demanded, that an interested party decipher it before the patron, who already knew how the deception was generated; the sundial, whose content is similarly nonintuitive, likewise demanded an audience capable of accepting the challenge laid out by its maker. For a brief period, it would even have been possible for visitors to view and discuss the sundial and colonnade simultaneously: a hole was carved into the sundial gallery in 1655; though it would be patched over not long after, for a time the opening offered an additional view onto the perspectival colonnade from the space with the sundial.[93]

Magni, the same artist who painted much of the palace's other illusionistic architecture, also painted the complex ceiling sundial, between 1644 and 1645. He produced the sundial in collaboration with the Minim friar Emanuel Maignan, a talented mathematician and turner whom the French philosopher and lexicographer Pierre Bayle (1647–1706) characterizes as "one of the greatest philosophers of the seventeenth century."[94] Maignan had created another large-scale sundial in the Minim convent of Trinità dei Monti in 1637 (see Fig. 70).

Virgilio explains that Bernardino commissioned the Palazzo Spado sundial from an "excellent mathematician," who used reflected sunlight to tell time: "In the vault of a gallery that was already there he commissioned a clock from an excellent mathematician. With this clock, you can look at the time in a covered place where no ray of sunshine arrives except by reflection, and that ray shows the time in place of a gnomon."[95] In other words, Magni executed the sundial based on Maignan's design. Lionello Neppi, Giulia Martina Weston, and Ulrike Feist have discussed the Palazzo Spada sundial and its accompanying verses (inscribed next to the corridor's mezzanine windows) to illustrate the patron's relationship to scientific curiosities as instruments of self-promotion.[96] My interpretation encompasses this theme of self-promotion by arguing that the sundial, like the *prospettiva*, stages a movement, in both spatial and mental terms, from confusion to a greater clarity that guests in turn associate with the host who has enabled their physical and intellectual journeys.

The Palazzo Spada sundial is a reflecting, or catoptric, dial. It employs a small circular mirror on a windowsill to bounce rays of sunlight or moonlight onto the arched ceiling of the corridor.[97] The ceiling is painted with a tangle of overlapping, curving lines, representing mathematical projections of the celestial sphere. The lines indicate the angles of altitude and azimuth of the Sun or moon; their declination and position in the zodiac; the limits of the astrological houses; and the time of day found by the Sun according to the systems of common astronomical hours, Italian hours, Babylonian hours, and unequal hours.[98] Planetary hours can be determined with the assistance of a table painted on a nearby wall.[99] Since the gallery is oriented from northeast to southwest, with the wall carrying the aperture and mirror facing

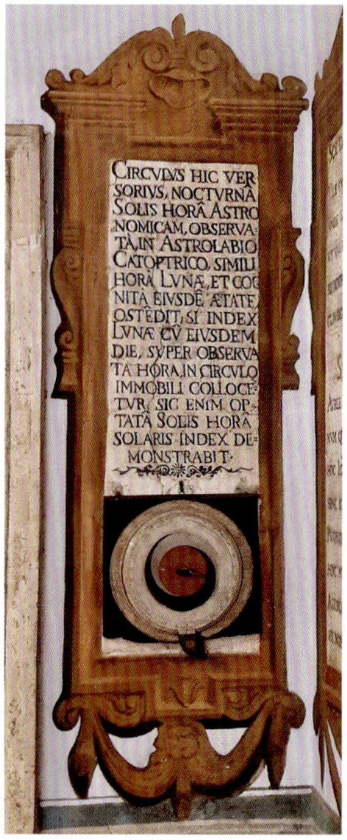

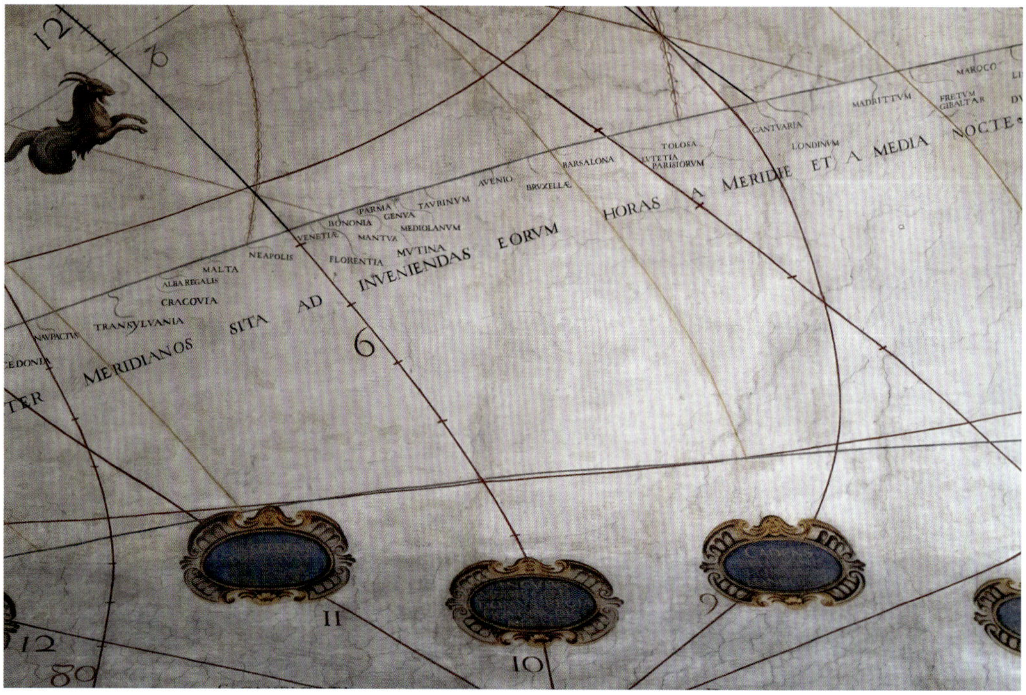

Fig. 50. Instructions for using the sundial as a moon dial, 1644. Palazzo Spada, Rome.

Fig. 51. Ceiling sundial (detail), 1644. Palazzo Spada, Rome.

almost directly southeast, the sundial works only during the first half of the day, showing hours from 5 a.m. to 1 p.m. However, the sundial can also be activated by moonlight during the second and third weeks of the lunar month (from the first quarter to the third quarter phases, and especially close to the full moon) when the moon is sufficiently bright. A second table on a gallery wall gives instructions for using the sundial as a moon dial to ascertain the solar time at night; a volvelle aids with the conversion of time by moonlight into solar time (Fig. 50).[100] The names of some fifty cities are arranged on the sundial according to their longitudes, indicating how to calculate the time in Rome when it is noon at those locations (Fig. 51).

Because the sundial displays a barrage of interesting information, rather than only presenting recognizable signifiers of wealth, it too transforms a display of magnificence and wealth into a topic for discussion. In the end, of course, it is a luxury item, as is the perspectival colonnade, and among the luxuries the sundial displays is the sheer wealth of information it offers. Was all its information necessary? Certainly not. In theory, Bernardino could have used the canonical, unequal-hour system to know when to recite the divine office, or the Italian hours to meet up with a local merchant, or the planetary hours to schedule when to take medicine on the advice of his physician, or the solar declination lines to know the month and season, and so forth. But in practice, the point of all this astronomical and cosmographical information was to dazzle visitors. The instrument would have encouraged discussions of its intricacies, and it would have impressed visitors with the mathematical sophistication of designer and patron, their host's ability to disentangle the sundial's many overlapping lines, and Bernardino's munificence as a patron of horology and astronomy. That the Palazzo Spada gallery was a reflection sundial—bringing time indoors—made it even more wondrous.

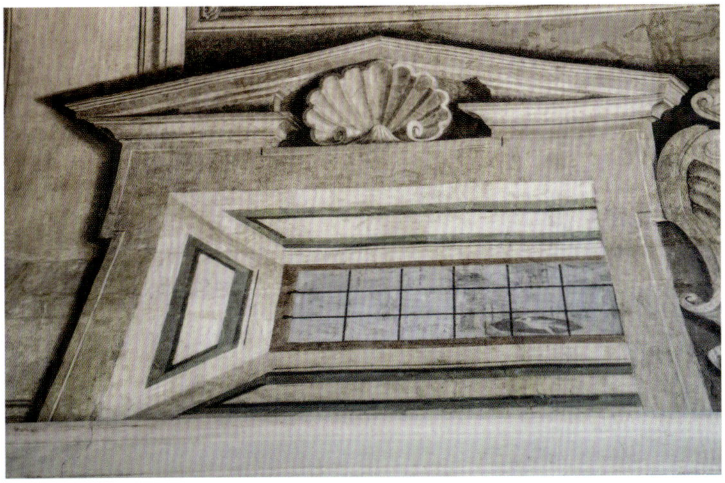

Fig. 52. Ceiling sundial (detail of fictive window), 1644. Palazzo Spada, Rome.

If the display of knowledge and technical prowess constituted the sundial's primary wonder, Magni's execution of the design used illusionistic effects to shape further the viewer's experience. In a 1644 bill reviewed by Borromini, Magni charges Bernardino for whitewashing the vault of the gallery with white lead and then painting texts, the twelve zodiac signs, a frieze, and "seven life-size putti who seem to sustain the satin cloth in which the sundial is painted."[101] By placing the hyper-informative sundial onto a fictive cloth, Magni encourages a playful mistrust of visual experience: the flexible textile could at any moment start to billow in a manner that belies—at least visually—the stable fixity on which the sundial relies. Other fictive details include a painted window with a simulated sculptural relief of a shell (Fig. 52). Yet if the sundial offers a whimsical experience rather than a sober empiricism, the incorporated deceptions remind observers that misleading appearances are everywhere (and thus that their discernment might be required at any time). And these feints do not ultimately overwhelm the tremendous mathematical and astronomical achievement of the dial itself.

In its totality, the project exemplifies in multiple ways how Catholic Reformation leaders patronized scientific achievements: The sundial evinces the cardinal's support for Maignan's work and the cardinal's commitment to visualizing horological and astronomical knowledge. But it does not, in the end, display any new advancements in horology or astronomy.[102] Many sundials in the second half of the sixteenth century already presented such copious amounts of information. Instead, the sundial's primary function was arguably social, identifying its patron as a source of metaphorical light, or, more abstractly, illumination. In his book *Perspectiva horaria* (Perspective pertaining to hours, 1648), Maignan studies sundials from both a practical and theoretical perspective.[103] But its dedication to Bernardino, who sponsored its publication, notably flatters the cardinal as a figurative Sun:

> You illuminate not only Rome, the head of the world, but also the whole world with its head, by means of the imparted rays of your praise and true glory: because you favor even our Order of Minims, beyond many other religious families, with this prerogative of your illustrious Protection; therefore, it is (the case) that with its free and

> native genius, my Perspective looks back at you with its ray of triple light, and flees to you as if to its Sun.[104]

Just as the Sun makes it possible for observers to see, Bernardino, through his patronage of illusionistic and empiricist works in Palazzo Spada—and in his role, since 1642, as cardinal protector of the Minims—sharpens the attention and discrimination of those around him. (Virgilio owned a copy of the book; one wonders if it was given to him by his brother to sharpen his discernment.)[105]

Elsewhere, Bernardino is celebrated for his perspicacity—as specifically demonstrated by his choice to subsidize Maignan's work. A biography of Maignan by his former student Jean Saguens (d. 1718), also a Minim monk, loads superlatives on Bernardino for having commissioned the sundial and then for funding the publication of the *Perspectiva horaria*, specifically highlighting his discernment:

> His Eminence Cardinal Spada, the most loving Protector of our whole Order, stood out among the fairest judges of matters pertaining to the arts and doctrines, by his title and by his combination of dignity, nobility, expertise, authority, and the greatest experience in discerning, administering, and blessing in ecclesiastical matters. After obtaining from Maignan a famous artifact of this type—which he took care to be drawn in the most splendid portico of his most spacious buildings and adorned with smaller paintings added to the finished picture—he wished to repay such magnificence by making it of benefit to all, and so, underwriting the costs with an extravagant generosity, he accelerated the publication of Maignan's *Perspectiva horaria*.[106]

Maignan particularly stresses the importance of scale in realizing his design for the sundial. He explains that when he was preparing a visualization of his sundial for Bernardino, his collaborating engraver opted to reproduce only some of its lines, because the plate was too constricted and he wanted to avoid overwhelming viewers: "He [i.e., the engraver] was not even able to capture all the lines of the astrolabe [i.e., a reflecting sundial with astrolabe-like functions] at the same time on this narrow bronze tablet; for which reason, in order to avoid confusing the multitude, particular lines were engraved, the rest were omitted, by far the greater part, in number."[107] In contrast, with a length of more than 72 feet (21.9 meters), the palace gallery succeeded in accommodating all the empirical details conveyed by the sundial's manifold overlapping lines. The colossal architectural space offered a different phenomenological experience, one that was immersive and more suited to accommodating the information than Maignan's and the engraver's printed page, a tactile, handheld object that measures only 17.25 by 10.6 inches (44 cm × 29 cm).

The contemporary function of the sundial as a challenge to the discernment of viewers—even a problem to be solved, analogous to the disorienting *prospettiva*—is well documented. In thesis booklets authored by the Jesuit mathematics professor Pierre Bourdin (1595–1653), developed to guide students in disputations at the collège de Clermont in Paris, students are asked to explain how sundials work. Although the Society of Jesus strictly prescribed subject matter in oral examinations—called disputations—in philosophy, disputations in mathematics

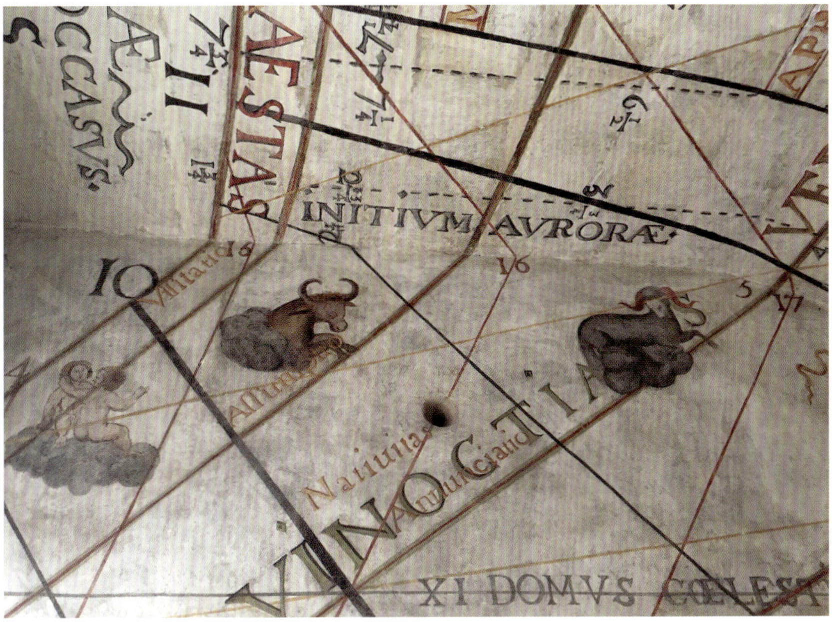

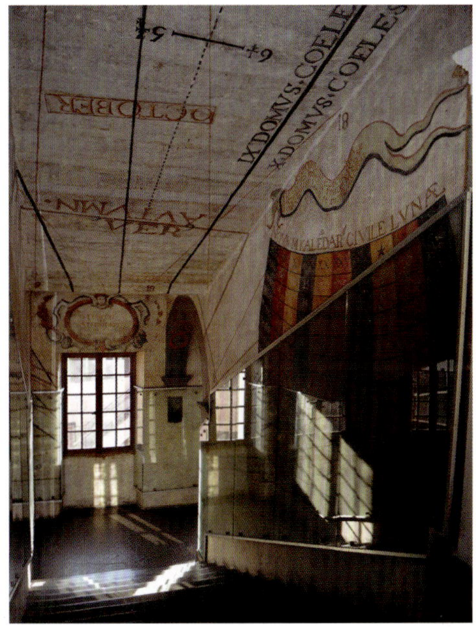

were less closely controlled.[108] Regulations for teachers stated only that every month, or every other month, the mathematics professor "should have one of the students solve a famous mathematical problem in the presence of a large gathering of students of philosophy and theology; and that, if he wished, it should be discussed afterwards."[109] Bourdin's thesis booklets resemble others of the period: thesis booklets were typically quarto-size volumes consisting of two to twenty pages of material for discussion. Their smaller format and more extensive page counts distinguish these booklets from early modern thesis prints, large-format broadsides that present a set of theses or propositions to be discussed at disputations.[110] A thesis booklet by Bourdin defended around 1639 by student Yves Henri presents the following problem: "By what artifice a sundial [*horologium solare*] must be drawn on the paneled ceiling of the Courtyard, so that it indicates the hours not with shadow, but with a narrow bright orb, which runs over the ceiling."[111] Such a "narrow bright orb" is visible on the ceiling of the Palazzo Spada gallery, a circular mark near the center of Figure 49. A prompt in another of Bourdin's thesis booklets, defended the previous year by Pierre de Cornouaille and Jacques Manchon, also demands expertise in horology: "To show various clocks, at one time, in order to indicate what hour it is, anywhere in the world, at another time, in order to measure time; what sort they are, whose artifice depends upon light, either the shadow of the Sun or the moon, and, indeed, with a candle, if you like."[112]

Some Jesuit institutions, furthering their educational aims, also featured galleries with complex ceiling sundials akin to the one in Palazzo Spada. When the Jesuit scholar and polymath Athanasius Kircher (1602–80) taught at the Jesuit College of Avignon, he created a sundial in the college's Tour de la Motte.[113] Another sundial was painted in 1673 in the Jesuit Royal college in Grenoble, now known as the Lycée Stendhal, by Jean Bonfa (1638–1724) (Figs. 53 and 54).[114] Bernardino may have learned to interpret sundials in the Jesuit institutions he attended in Italy, and then imparted this knowledge to his nephews alongside their other instructors.

Fig. 53. Jean Bonfa, sundial, 1673. Jesuit Royal college (now the Lycée Stendhal), Grenoble.

Fig. 54. Jean Bonfa, sundial, 1673. Jesuit Royal college (now the Lycée Stendhal), Grenoble.

The text of a set of epigrams, adjacent to the mezzanine windows in Bernardino's corridor with the sundial, demonstrates the period's interest in naturalism and empiricism and also flatters Bernardino with some generous imagery around his power to offer illumination to his guests.[115] The epigrams, authored by the Minim friar Guillaume du Planté, include one positing that although Magni painted the sundial's sky, the instrument features not a representation of light, but rather the Sun itself:

> In the imitation of nature a hand here attempted to depict the sky;
> but to paint the Sun it could not do.
> However, the Sun did not want to be left out of the work,
> so it paints itself and its path with its own light.[116]

Du Planté's lines, in celebrating the sundial's capacity to capture the thing itself—the Sun—rather than a representation of it, illustrate how the instrument registers the period affinity for naturalism and empiricism.

And in a further inscription, du Planté reconciles the wondrous and the deceptive, bringing natural science together with the figurative language that presents Bernardino as illuminator: here he suggests that reflecting the Sun's light to mark time on the sundial creates the illusion that Bernardino himself controls the sun:

> If there is anything new in the world, anything old in the city,
> you will see a noble exemplar of it in this palace.
> When you see various wonders in other houses, you will say,
> O what riches of the Orbs and the Urbs!
> Here, where the Sun itself shall be seen for many hours,
> and trace its path with shining lights,
> go and say, Who has been able to bring the Sun itself down from heaven?
> O what a prodigy is this of genius and of art![117]

These words present the collection of Palazzo Spada as an exceptional cabinet of wonders, one that includes a deformation, situated within a city that is, in turn, replete with wondrous collections.

That Bernardino imagined the gallery's social function much as he imagined the function of the *prospettiva*—here, guests entered the sundial gallery, walking into a marvelous scientific instrument, much as they entered into the *prospettiva,* empirically assessing the actual built environment against their initial visual perceptions—is suggested by a foldout print published in Maignan's *Perspectiva horaria*. The print shows four figures in the palace's gallery surrounded by panel paintings on the vertical walls, looking up at the sundial above and down the corridor (Fig. 55). Maignan printed this image at a scale larger than the rest of his book, and thus accentuated the magnificence of his device, even if he had acknowledged that the printed page was too small to accommodate the sundial's manifold lines. Acts of looking and conversing consume the attentive men (who perhaps include Bernardino, Virgilio, and Maignan): as though taking a cue from the many pointing figures in the *Salone*, which is adjacent to the sundial

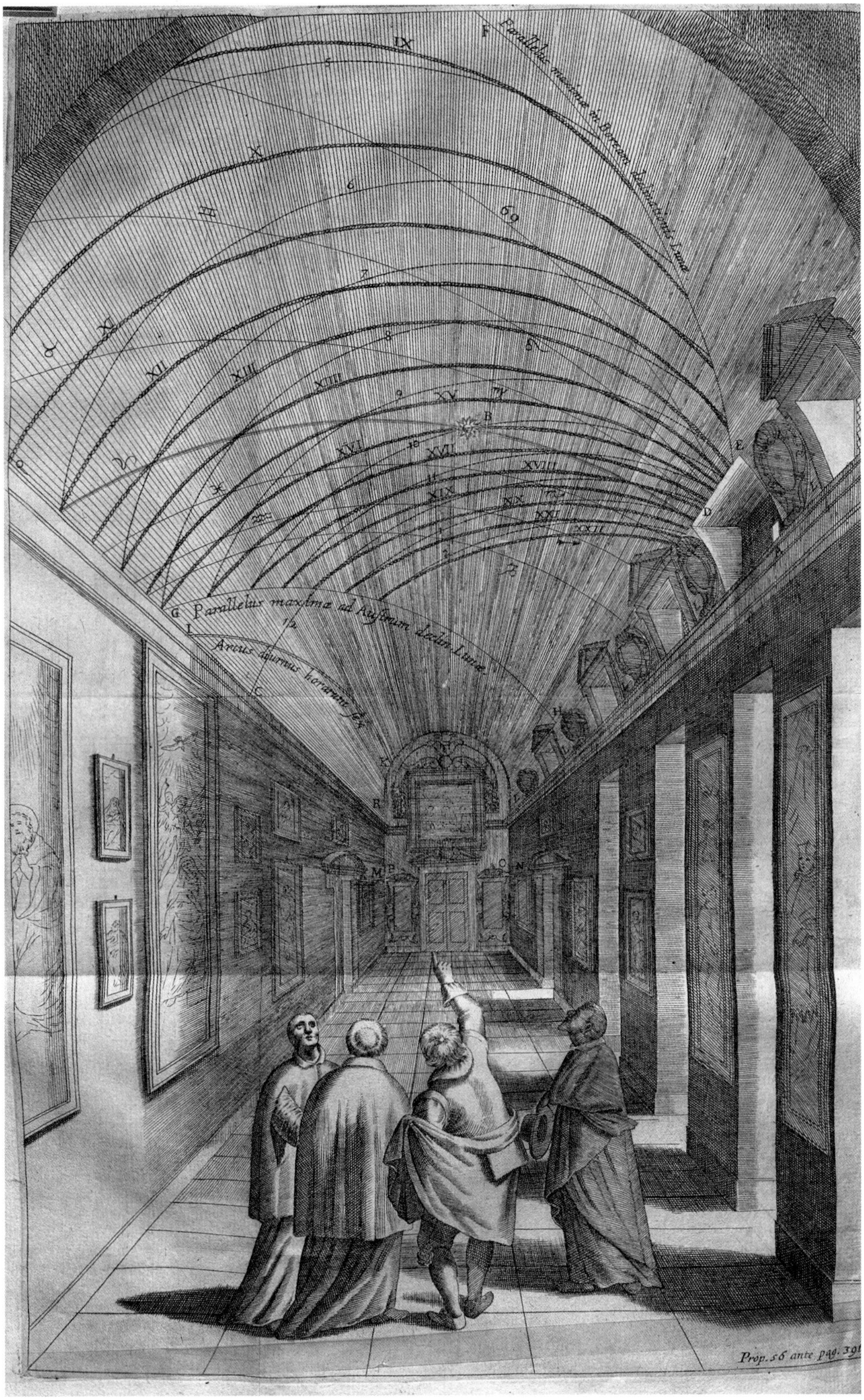

Fig. 55. Unidentified artist, four figures inside the sundial gallery from Emanuel Maignan's *Perspectiva horaria*, 1648, between pp. 390 and 391. Engraving printed on paper, 17.32 × 11.42 in (44 × 29 cm). Harvard University, Houghton Library, Cambridge, MA.

corridor, one gestures to a detail in the sundial, two gaze at the detail in question, and a fourth figure, with his back to the viewer, stares down the gallery.[118]

The sundial was praised in publications by the Minim Jean-François Niceron and the Jesuit Kircher, who stresses the relationship between ingenuity and wonder: "Not so long ago, the most Eminent Cardinal Spada arranged for a reflecting astrolabe of this kind [i.e., a reflecting sundial with astrolabe-like functions] to be delineated in his most magnificent palace; which indeed, as it was a rather ingenious work for the town, seized everyone with the greatest admiration."[119] Kircher's remark suggests that the sundial drew into Bernardino's palace elite viewers seeking to learn, in conversation with each other and perhaps their host, about this complex and wondrous apparatus. That Kircher mentions his own version of this instrument in Avignon after praising Maignan's sundial hints at future competition: Kircher would later, presumably after the publication of Maignan's *Perspectiva horaria* in 1648, accuse the Minim friar of plagiarizing his work on catoptrics.[120] (Maignan reportedly dismissed the accusation.) The disagreement implies that Kircher at least believed his work and Maignan's to be genuinely innovative in the context of seventeenth-century Roman catoptrics.

Another Bernardino commission further underscored the pervasive social orientation of his changes to Palazzo Spada. Bernardino ordered from Maignan a periscope "to watch the passers-by while remaining in the room with the windows shut."[121] This device, which cost ten scudi, was positioned in the center of the palace's facade, and would have allowed the cardinal and his nephews to expand their attention from the world within the palace's walls to the activities beyond it.[122] In a similar vein, Kircher installed a speaking tube in the Collegio Romano that linked his museum to his private apartment; a custodian used the tube to report to Kircher's bedroom who had arrived at the museum.[123] The speaking tube and Bernardino's periscope assisted with navigating Rome's complex architectural etiquette—which, as we have already seen, strictly regulated where a host must greet a particular visitor or bid them adieu, with lower-ranking guests expected to advance significantly further through the *Salone* and anterooms than higher-ranking guests, perhaps even into the audience room, before being greeted by the host.[124]

✦

Exposure to the Palazzo Spada colonnade's apparent enlargement of objects positioned toward its far end offered visitors to Bernadino's residence a visual challenge to be resolved in conversations with their host. Bernardino capitalized on social conventions that would move his guests through the spaces of the palace along predetermined routes, filling the walls along the way with architectural illusions and painted figures who pointed to occasions for observation. Guests arrived, then, in his audience room (or in the *Sala dei Rilievi*, also optimally placed opposite the colonnade) primed to deploy their visual discernment on the sight Bernardino presented to them. If each deformation or illusion involved a journey from confusion to clarity, the total journey from the entrance to the *Sala delle Udienze* offered an expanded intellectual journey through the process of visual perception and discernment. Visitors admired and analyzed the deformation in performances that enabled Bernardino to demonstrate his own discernment and cleverness as a patron.

In designing his palace's deformation in collaboration with his architects, Bernardino had drawn on rhetorical techniques found in ecclesiastical works—particularly accelerated perspective, deployed in tabernacles and the ephemeral architecture for a Forty Hours' Devotion, to enrich worshippers' devotional experience of sacred objects. As the ecclesiastical works may have reminded some worshippers of the limits of human perception, the colonnade may well have invited similar musings. In transposing the techniques of ecclesiastical works to his own residence, the church official also transposes their rhetorical power, here to reinforce a social message regarding his family's place in Roman society. Ultimately, Bernardino chose to spend money not on a regular colonnade but on a deformation that could function as a conversation piece and persuade his guests that he was a discerning patron and the mastermind of his palace.

I have suggested that we think of the *prospettiva* as enabling Bernardino to exercise social power by leading guests to move from experiential disorientation to insight. When we juxtapose that function against the sundial's implied expectation that guests would be led physically and intellectually through its intricate content, we understand that the hyper-informative, wondrous scientific instrument operated, in some ways, like the deformation. Both entities immersed observers in temporary states of confusion and then encouraged them toward discernment.

It is tempting to speculate that Bernardino's investment in his palace reflected concern for his nephews' and relatives' education. To his mind, the younger members of his family should learn to observe with attention, to apprehend meanings, to differentiate, and to characterize forms in entities ranging from architecture to astronomical phenomena, that they might ultimately discuss their observations with discernment in civil conversation with other learned observers. As the patron of "grand things under the heavens," Bernardino promoted his family and the Catholic Church. In designing his palace and amassing his collection, Bernardino not only engaged in a pleasurable pastime and created stimulating conversation pieces but also transmuted knowledge into social power.

As cardinal protector of the Minim order, Bernardino also supported the development of further deformations in their convent in Rome, artworks that were similarly conducive to visual attention, discernment, and discussion. Like Palazzo Spada, the Minim convent of Trinità dei Monti in Rome, where I will direct my attention in the next chapter, helped to propagate practices of diagnostic looking that bridged the spheres of the visual arts and science.

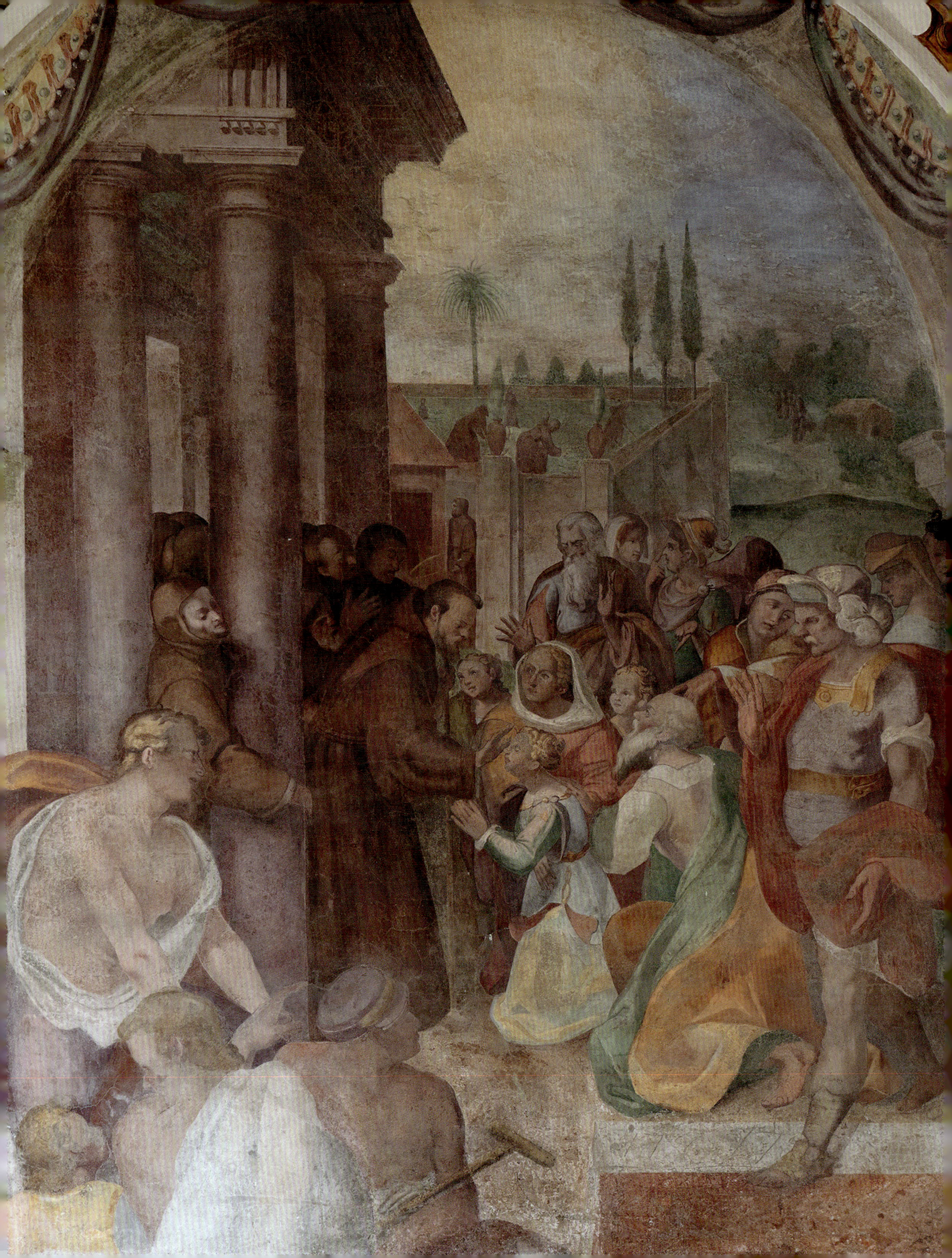

2 MINIM DEFORMATION AND REVELATION

WHILE THE LAST CHAPTER considered a key patron's motivations for building a deformation, this chapter and the two following ask why producers wanted to make artworks that expanded the space between perception and understanding. Why did seventeenth-century Catholic religious communities—members of the Minim order in this chapter, and members of Society of Jesus in the next—craft works of art that played with perspective and deployed deformation? If Bernardino was drawn to deformation because it could transform wealth into kinesthetic exploration and discerning conversation, Minim and Jesuit practitioners had other goals in mind: they valued deformations because deformations resonated with how divine revelation worked. They made art that in subject matter and process communicates a sense that human life is characterized by chaos and obscurity but also communicates that God can intervene into this chaos and blindness to reveal divine truth.

This chapter and the next two center on a species of deformation known today as anamorphosis. In its most classic form, an anamorphosis presents a confused view: an image that appears contorted when viewed straight on resolves into clarity when observed from a precise vantage point or via reflection from a specialized mirror.[1] In dislocating the viewer from one vantage point to another, anamorphoses could produce a kind of vertigo in observers, such that experiencing this image type gestured toward the transformative experience of a revelation of God's truth.[2] An anamorphosis that uses a mirror to achieve the reformation of the image is called a specular anamorphosis. Other anamorphoses, which do not use mirrors, invited observers to position themselves at a particular angle for optimal viewing. This latter variety of anamorphosis shows one scene when the image is considered from one vantage point and a different representation—in some anamorphoses a coherent one, in others an incoherent one—when the observer looks from another angle. A famous anamorphosis features in *The Ambassadors* (1533), by Hans Holbein the Younger (1497/98–1543) (Fig. 56). When observed

Detail of fig. 62

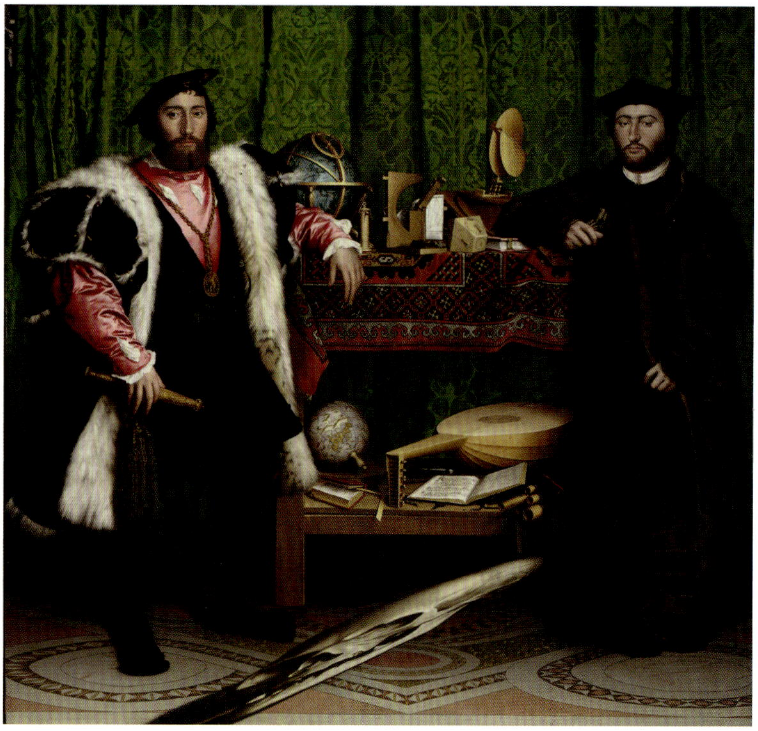

Fig. 56. Hans Holbein the Younger, *The Ambassadors*, 1533. Oil on oak, 81.4 in × 82.48 in (207 × 209.5 cm). The National Gallery, London.

frontally, a distorted skull occupies the painting's bottom region; if the image is examined from its extreme right edge, the skull appears more legible, although somewhat disfigured even from this perspective.[3]

The term *anamorphosis* comes from the Greek *ana* ("again") and *morphe* ("form"). Although the word was used by the Jesuit Gaspar Schott (1608–1666) in 1657, it was not typically deployed by the other figures introduced in this study, who instead often used variants of *deformation* in reference to this image type.[4] As well as referring to *deformations,* Niceron deploys the term *perspective curieuse* ("curious perspective") and he uses the phrase *magie artificielle* ("artificial magic") to denote optical images that play with form. When Niceron discusses the etymology of *magie,* he points to the prophetic nature that he and other Catholic Reformation thinkers ascribed to anamorphoses, a topic I will examine in depth in chapter 4: "The word *Magus* is neither Greek, nor Latin, but Persian; ... it signifies in that language the same thing as the word *Prophet* in Hebrew."[5]

Minims and Jesuits led seventeenth-century artists in creating the species of deformation known today as anamorphosis—and in examining and explaining its optical significance—because they read the anamorphic juxtapositions of unclarity and clarity as resonating with the workings of divine revelation.[6] Catholic teaching held that divine truth was revealed in an epiphany from the confusion of worldly chaos, and Minims and Jesuits in particular were attracted to anamorphosis, in both theory and practice, as an expression of this epiphany: to reform an anamorphic deformation rendered the resolved image fraught and important, giving it the character of a revelation. Minims and Jesuits were drawn to the genre for its juxtapositions of unclear and clear visual states—the tension inherent in an anamorphic image that can be resolved only when the viewer changes position or uses a mirror. These images hinged on the contrast between confusion and uncertainty, on the one hand, and clarity and apparent certainty, on the other, and they demanded that observers confront that contrast. They use them much as they use other visual-perception-centering subject matter, to foreground the contrast between human limitations and unlimited divine knowledge.

In this chapter, I suggest that Minims' attachment to the anamorphic conjunction of confusion and clarity echoed particular concerns of their founder, as expressed in an iconography that centers the saint's involvement in miracles pertaining to vision. Minims' interest in anamorphoses, like their interest in optics and the natural sciences more generally, emerges in part from this iconographic tradition, where God repeatedly intervenes to deliver or restore vision, with these varied engagements with vision enacting in some way the central project of seeking to approach God's revealed truth. In addition, the Minims saw in the workings of anamorphoses an analogy to the rigorous study they promoted of theology, mathematics, and natural philosophy. They were also drawn to how these optical tricks could reinforce in viewers' minds particular "revealed truths"—whether matters of belief or matters of historical argument (such as the primacy of the Catholic Church). This chapter and the next will stress that the images centered in the reformed anamorphosis—like the objects centered in the tabernacles with accelerated perspectives examined in chapter 1—are chosen with great intention, and chapter 4 will look still more closely at the subject matters these religious artists chose to represent within that visual system.

Anamorphic Confusion

Leonardo da Vinci (1452–1519) created the earliest known European anamorphic renderings in his Codex Atlanticus; they display the warped face of a child and an elongated eye (Fig. 57).[7] By the late sixteenth century, Galileo would mention anamorphoses (though he does not use the term) when critiquing the epic poem *Gerusalemme liberata* (*The Liberation of Jerusalem*, 1581) by Torquato Tasso (1544–1595).[8] Galileo disliked Tasso's use of allegory in the poem, inviting interpretation of every aspect as an allusion to something else—like, he argued, images that "show a human figure when looked at sideways and from a uniquely determined point of view but, when observed frontally as we naturally and normally do with other pictures, display nothing but a confused and inordinate welter of lines and colors from which we can make out, if we try hard, semblances of rivers, bare beaches, clouds, or strange chimerical shapes."[9] As Erwin Panofsky argued in a classic essay, Galileo's aesthetic judgments were colored by a bias that rewarded simple forms and shunned complication and a lack of proper proportion between corresponding things.[10] Panofsky ultimately contends that this aesthetic partiality made Galileo unable to discern the elliptical form of planetary orbits, since he saw the ellipse as a deformed circle.[11] The bias Panofsky identified in Galileo's thinking both about Tasso and about astronomy suggests the broader aesthetic implications of classical rhetoric that I discussed in this book's introduction: that clarity was an ultimate good. As we seek to understand the interpretive value that certain religious communities placed on anamorphoses in this period, two of Galileo's observations are particularly relevant: first, that looking at an anamorphosis is inherently confusing, and second, that interpreting an allegory and interpreting an anamorphosis

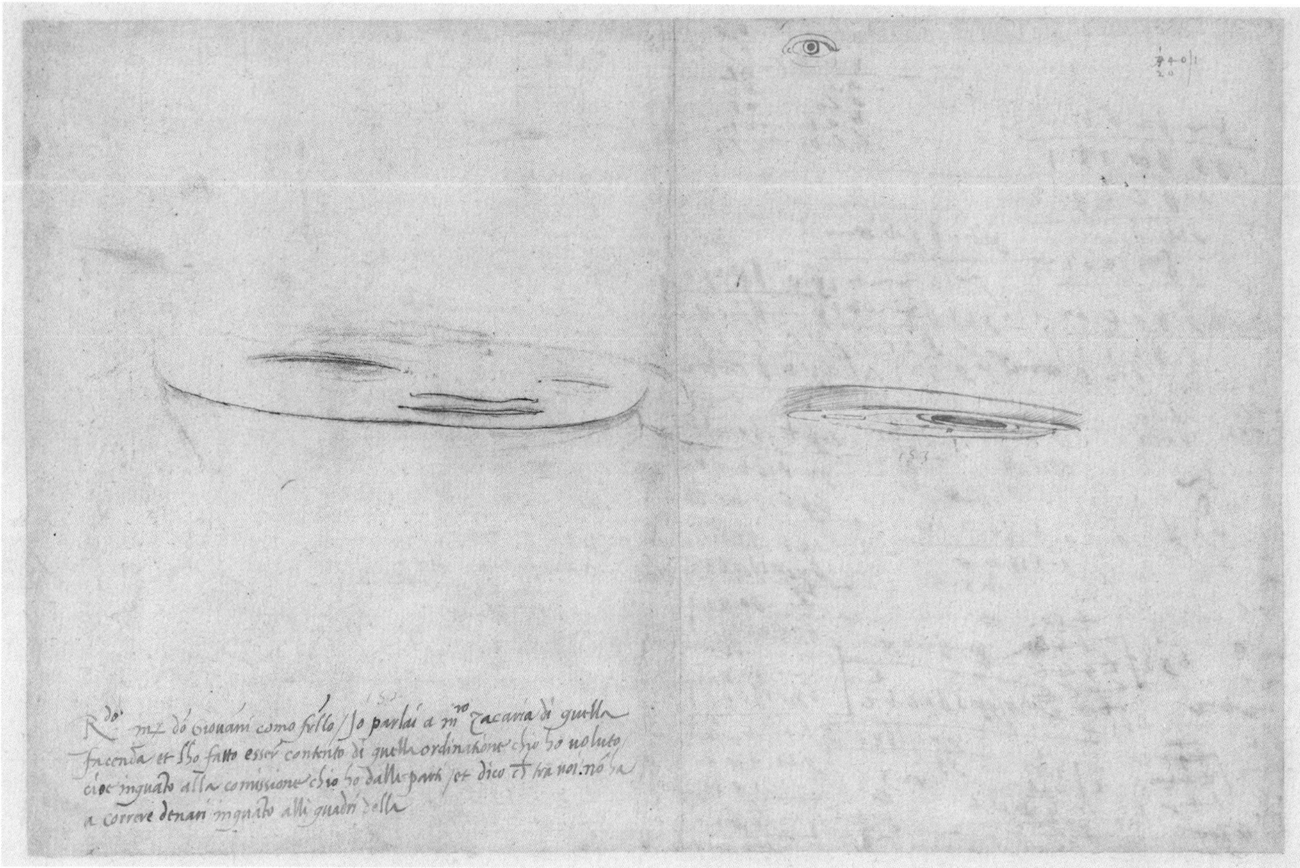

Fig. 57. Leonardo da Vinci, *Codex Atlanticus*, fol. 98r (previously 35va and 152–53), c. 1478–1518. Pen, brown ink, and traces of black pencil on paper, 10.47 × 15.75 in (26.6 × 40 cm). Biblioteca Ambrosiana, Milan.

are closely allied processes. I will return in chapter 4 to the affinity that Galileo propounds between the interpretation of an allegory and the interpretation of an anamorphosis, as this affinity is essential to contemporaries' understanding of the objects I am discussing.

Remarks from John Locke (1632–1704), in his *Essay Concerning Human Understanding* (1689), may help us further parse how early modern observers understood the "confusion" of an anamorphosis.[12] Locke explains one prevalent condition that gives rise to the production of confused ideas: "The particulars that make up any *Idea* … are so *jumbled together*, that it is not easily discernable, whether it more belongs to the Name that is given it, than to any other."[13] To clarify his analysis of this default further, he introduces anamorphoses:

> There is nothing properer to make us conceive this Confusion, than a sort of Pictures … wherein the Colours, as they are laid by the Pencil on the Table it self, mark out very odd and unusual Figures, and have no discernable order in their Position. This Draught, thus made up of parts, wherein no Symmetry nor Order appears, is, in it self, no more a confused Thing, than the Picture of a cloudy Sky; wherein though there be as little order of Colours, or Figures to be found, yet no body thinks it a confused Picture.[14]

Locke explains that the confusion one experiences in looking at an unreformed anamorphosis arises not from its asymmetry or lack of apparent order, but from the apparent distance between the image and its label or explanation. (A picture of a cloudy sky, he points out, is similarly lacking in symmetry and order—yet it is not confusing.) He writes that it is not the image that confuses, but "the applying it to some Name, to which it does no more discernibly belong, than to some other."[15] He gives examples: "When it is said to be the Picture of a Man, or *Cæsar*, then any one with reason counts it confused: because it is not discernible, in that state, to belong more to the name Man, or *Cæsar*, than to the name Baboon, or *Pompey*: which are supposed to stand for different *Ideas*, from those signified by Man, or *Cæsar*."[16] Locke's position, then, is that when we use language to describe an image (or idea), and the label and image (or idea) do not clearly relate, then the image (or idea) becomes confusing. When the entity in the lower portion of Holbein's *Ambassadors* is named a "skull," then the image confuses observers who consider the image straight-on rather than from its extreme right edge (see Fig. 56). What Locke describes around language is an alienation between the thing the anamorphosis will become when it is resolved (through a shift in vantage point or with the assistance of a mirror) and the representation of that thing prior to this reformation. Through language, observers become conscious that the deformed image departs from a clear imitation of reality, and as a result they grow confused.

One could take issue with Locke's argument here, at least in relation to Holbein's *Ambassadors*. Perhaps an observer from the frontal viewpoint would consider the deformation at the bottom of the canvas confusing even without hearing the term "skull" attached to it.[17] As Mathew Stuart has clarified, Locke could forestall this objection by explaining that the deformation occurs within a naturalistic environment that entices observers to receive the deformation as a familiar type of object. Attempting to categorize the entity involves attempting to observe it as subsumable under an idea connected with a name. Locke might explain that observers are confused by the deformation in *The Ambassadors* because it defeats their efforts to categorize it as a recognizable thing. A similar objection could be made regarding anamorphoses that are not presented within naturalistic images, such as those in Leonardo's Codex Atlanticus: these images confuse even before the words "child's face" or "eye" are applied to them (see Fig. 57). Locke might reply that observers consider these deformations in relation to other artworks produced in a context (early modern Europe) where representational and mimetic subject matter was considered a primary concern of art-making. In the contemporary visual context, observers' working expectations for artworks lead them into unsuccessful efforts to categorize an anamorphosis as a recognizable thing.

Locke describes how confusion disappears when a cylindrical anamorphosis is resolved and an observer can label the represented entity by its name:

> But when a cylindrical Mirrour, placed right, hath reduced those irregular Lines on the Table, into their due order and proportion, then the Confusion ceases, and the Eye presently sees, that it is a Man, or *Cæsar*; *i.e.* that it belongs to those Names; and that it is sufficiently distinguishable from a Baboon, or *Pompey*; *i.e.* from the *Ideas* signified by those Names. Just thus it is with our *Ideas*, which are, as it were, the Pictures of Things. No one of these mental Draughts, however the parts are put together, can

> be called confused, (for they are plainly discernible as they are,) [*sic*] till it be ranked under some ordinary Name, to which it cannot be discerned to belong, any more than it does to some other Name, of an allowed different signification.[18]

If the particulars of idea or image are jumbled, disrupting the expected relationship between word and image, then we also experience the idea or image as confused. Jumbledness, or, in Locke's words, being "*jumbled together,*" is not synonymous with confusion, but it is partly responsible for confusion. And when ideas or images are especially jumbled, Locke posits that they are more difficult to classify under names. In Locke's view, then, the gap between a deformed anamorphic form and the thing it represents is essentially linguistic—a space between an observed entity and its label. It is this gap between form and label that Minims and Jesuits hoped to take advantage of as they theorized around anamorphoses and their didactic and persuasive uses.

On Vision and Revelation

The Minims are a religious order of friars founded by Saint Francis of Paola (1416–1507) around 1435; the order flourished in the seventeenth century.[19] Minims held that votaries could learn, by observing actual and virtual images, to better recognize God's revealed truth.[20] In a secular setting, elites were expected to exhibit discernment in their interpretation of objects (including those that might delay their journey from visual perception to understanding); in a religious context, anamorphoses and the natural sciences both belong to the work of connecting to God's revealed truth. Although the order's position on the visual arts was never prescribed in an official mandate, its members' fascination with optics and religion builds on the longstanding tradition of portraying the biography of Saint Francis, where God repeatedly intervenes to deliver or restore vision.[21] Francis suffered from an ocular tumor, which was cured when he entered a Franciscan convent (he later abandoned it to found his own order); he also interceded in several miracles where sight was concerned.[22]

The Minim attention to optical miracles reflected the belief that sight is both practically important (if you look at images, your faith will be reinforced) and useful as a metaphor for comprehending. Saint Francis's optical miracles are visualized in lives of the saint, including in some of the lunettes painted in the last decades of the sixteenth century in the cloister of the Minim convent of Trinità dei Monti in Rome.[23] The convent and church, founded on behalf of the Minim order by Charles VIII of France (1470–1498), became an important center for the French community in seventeenth-century Rome.[24] In 1632, the painter Claude Lorrain (c. 1604/5–1682) produced a landscape with a representation of the convent and church in the middle distance on the left (Fig. 58).[25] Both convent and church, with its twin towers and twin steps, are readily identified today from Lorrain's painting, although the Spanish Steps, which now usher visitors up and down the hill that lies before the buildings, were not completed until the 1720s.

Some of these images represent the saint delivering sight to those born without it. One lunette represents Saint Francis providing eyes and a mouth to a child born without these organs (Fig. 59). A Latin inscription below, attributed to the French humanist Marc Antoine Muret (1526–1585), informs readers that the saint's smeared spit caused the child's eyes to emerge and mouth to be opened. An engraving by Adriaen Lommelin (1637–c. 1673) from the

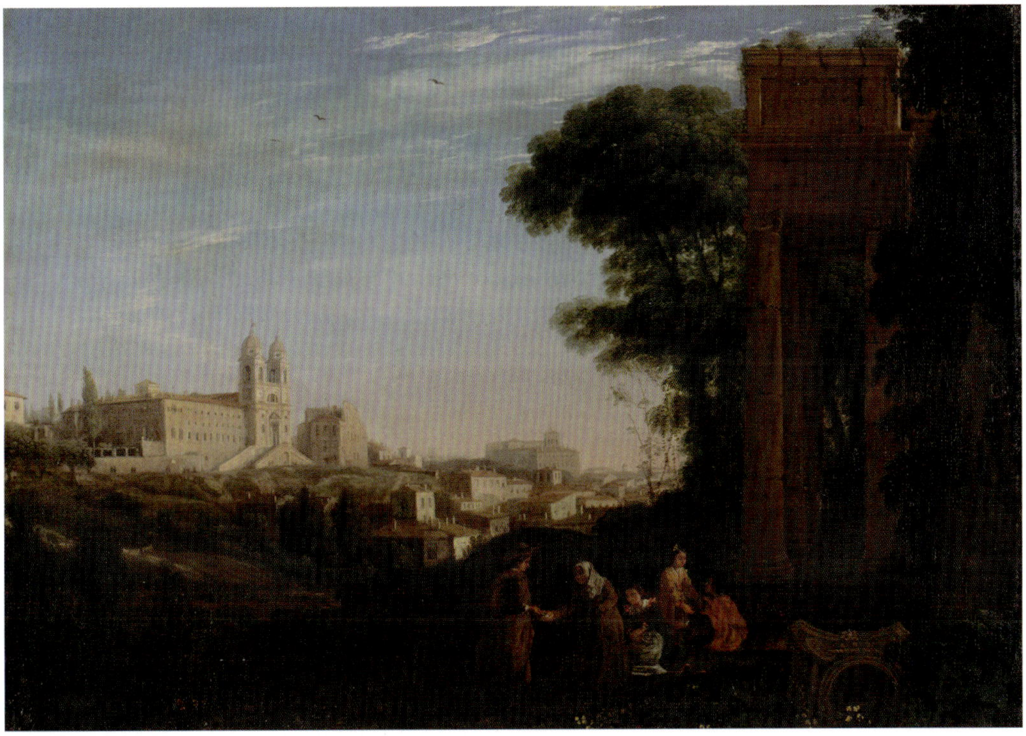

Fig. 58. Claude Lorrain, *A View in Rome*, 1632. Oil on canvas, 23.74 × 33.07 in (60.3 × 84 cm). National Gallery, London.

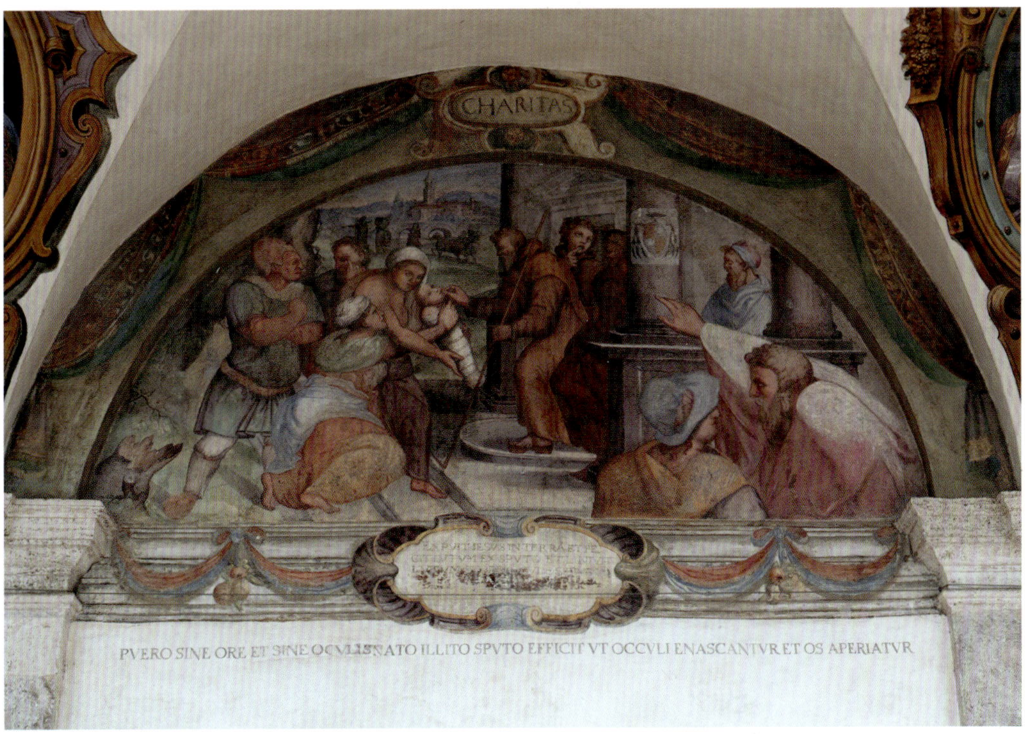

Fig. 59. Unidentified artist, Saint Francis providing eyes and a mouth to a child born without these organs, c. 1580. Fresco. Cloister, convent of Trinità dei Monti, Rome.

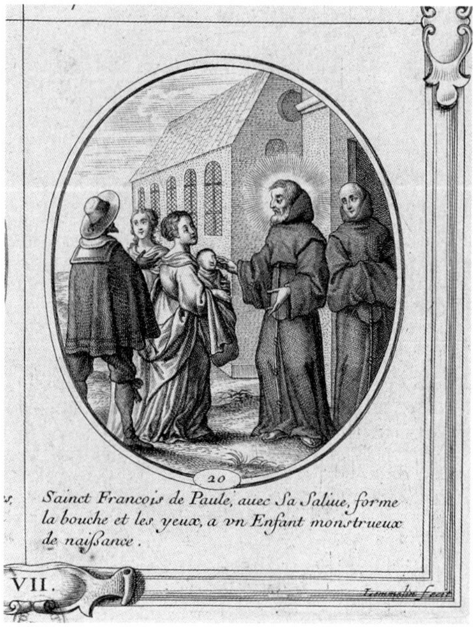

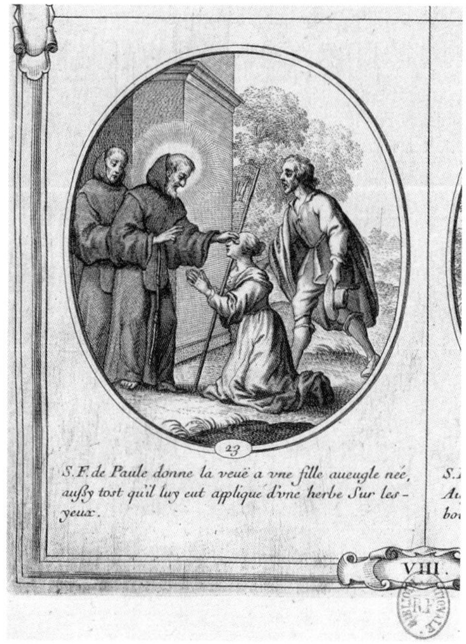

Fig. 60. Adriaen Lommelin (engraver), Saint Francis of Paola providing eyes and a mouth to a child born without these organs from Antoine Dondé's *Les figures*, 1664, plate 7. Engraving printed on paper, 4.33 × 3.14 in (11 × 8 cm). BnF, Paris [Reserve 4-RD-83].

Fig. 61. Adriaen Lommelin (engraver), Saint Francis of Paola curing a blind woman from Antoine Dondé's *Les figures*, 1664, plate 8. Engraving printed on paper, 4.33 × 3.14 in (11 × 8 cm). BnF, Paris [Reserve 4-RD-83].

Minim friar Antoine Dondé's life of Saint Francis of Paola also depicts this miracle (Fig. 60).[26] The underlying inscription reads: "Saint Francis of Paola, with his saliva, forms the mouth and the eyes of an infant, monstrous from birth."[27] The infant is pictured just as he experiences the miracle, with his left eye and mouth partially formed and his right eye yet to be created. Another of Lommelin's scenes portrays the saint curing a blind woman who kneels before him (Fig. 61). Annotated, "Saint Francis of Paola gives sight to a girl, who was born blind, as soon as he applies an herb to her eyes," the print again visualizes the miracle in progress.[28] This scene of healing is also included in a large lunette in the Trinità dei Monti cloister that has been attributed to Pasquale Cati (c. 1550–c. 1620) and Giorgio Picchi (c. 1555–1605) (Fig. 62). Saint Francis's hand is on the verge of touching the girl's face such that one senses the anticipation felt by the figures around her. This scene may have been especially inspiring to the friars in the convent who ran the apothecary and cultivated medicinal herbs. A third engraving, by Nicolas de Poilly (1626–1686), from Dondé's life of Saint Francis of Paola, further hints at limits on human visual capacity, while also suggesting that insight may be granted to the faithful. The image portrays the praying saint as invisible "to fifty armed men that the King of Naples had sent to take him" (Fig. 63).[29] Although the print's viewers are granted the ability to see the saint kneeling before an altar, the soldiers scattered around him in the composition are unaware of his presence. By portraying miracles in progress, these artists coax observers to meditate on Saint Francis's actions seemingly as they take place. The outsize role of vision in Saint Francis's biography may well have helped to inspire the Minim fascination with optical illusions. Moreover, the iconography of his life suggests that God, here following on the intercession of Saint Francis, can grant to the faithful an ability to see (both physically and metaphorically) that is denied to those who would persecute the church.

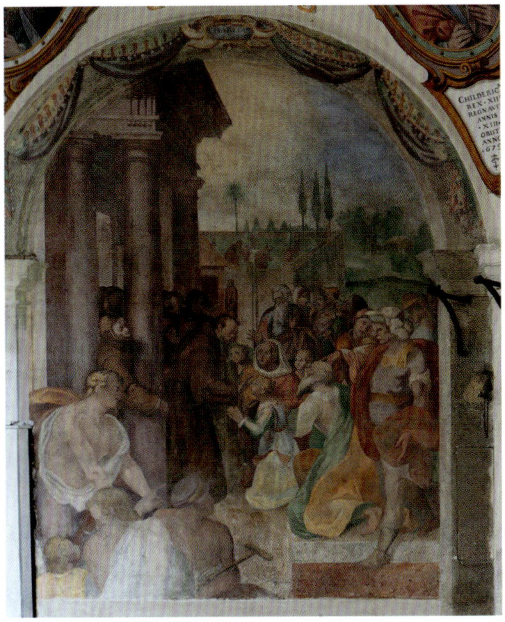

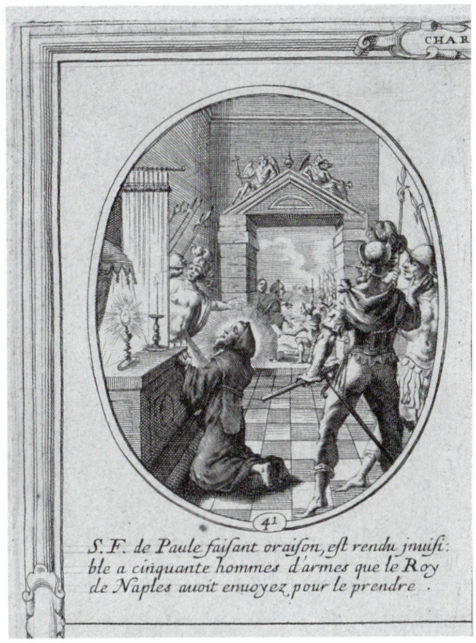

Fig. 62. Attributed to Pasquale Cati and Giorgio Picchi, Saint Francis curing a blind woman, c. 1580. Fresco. Cloister, convent of Trinità dei Monti, Rome.

Fig. 63. Nicolas de Poilly (engraver), Saint Francis of Paola as invisible from Antoine Dondé's *Les figures*, 1664, plate 13. Engraving printed on paper, 4.33 × 3.14 in (11 × 8 cm). BnF, Paris [Reserve 4-RD-83].

Dondé describes the volume's engravings of visual miracles as instruments of persuasion:

> Each of these images is a silent preacher, who exhorts us to virtue, since they show us nothing other than the example or the reward of the same virtue: if someone is persuaded in considering them, I will have achieved the aim of my work, which is nothing other than the glory of God in the exaltation and imitation of Saint Francis of Paola.[30]

The idea that pictures are Bibles for the poor reverberates behind Dondé's phrase "a silent preacher."[31] Dondé's account of imagery as a powerful vehicle for contemplation, expression, and persuasion, moreover, coincides with practices well established in late sixteenth- and seventeenth-century print culture.[32] Beyond exhibiting the history of the Minim order's founder—and his notable expertise in ocular matters—these images came to be valued by his seventeenth-century followers as opportunities for veneration or for meditating on the eye-opening power of God.

In lives of Saint Francis, the images that portray optical miracles are typically naturalistic in nature. But anamorphic representations of the saint also survive in books and paintings, offering a more challenging relationship between the saint and the exercise of perception, one that delays recognition and understanding. Insofar as anamorphosis offers deformation and reformation—appearing or disappearing depending on viewpoint or other tools—any anamorphosis encourages attention to the reality that how we observe impacts how and what we see. In the naturalistic images, Saint Francis restored or generated vision—and thus visual perception—in a series of miracles. But anamorphic images of the saint refined beholders' understanding, revealing to them the importance of observational conditions, such as the direction from which one looks, a topic that I will explore further in chapter 5. From one

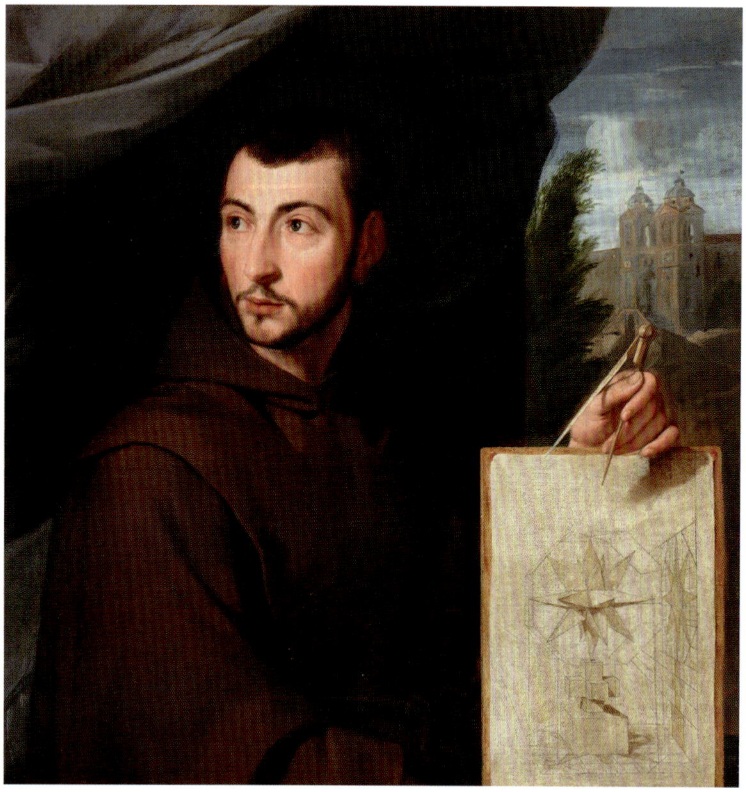

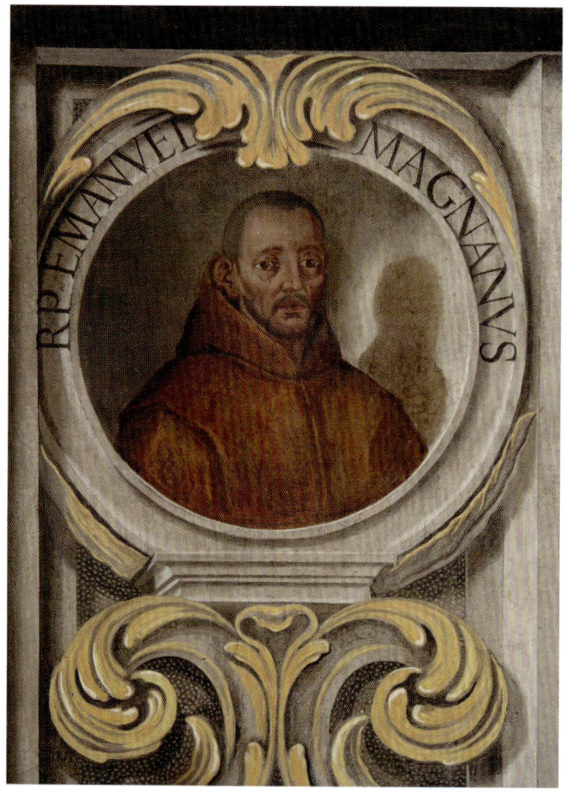

ABOVE

Fig. 64. Unidentified artist, *Portrait of Jean-François Niceron*, seventeenth century. Oil painting on canvas, 29.1 × 28.9 in (74 × 73.3 cm). Private collection.

Fig. 65. Unidentified artist, *Portrait of Maignan*, 1693. Fresco. Library, convent of Trinità dei Monti, Rome.

OPPOSITE

Fig. 66. Jean-François Niceron, *Saint Francis of Paola*, c. 1635. Cylindrical anamorphic oil painting on canvas, 26.18 × 19.69 in (66.5 × 50 cm). Galleria Nazionale d'Arte Antica, Palazzo Barberini, Rome [inv. n. 1955].

perspective, an apparently deformed visual object might have elicited intense perceptual confusion and uncertainty, yet from another angle, when appropriately reformed, the image revealed the saint. Furthermore, since these anamorphic images often picture Saint Francis in prayer and undergoing a private, meditative experience, they offer a further layer of reflection: contemplating the reformed image of the saint at prayer may have communicated a lesson around God's ultimate control over visual perception, understanding, and revelation. Like other images of the saint, these works would have also encouraged exaltation of Saint Francis of Paola.

The two protagonists in the story of Minim anamorphosis wanted to use the images to reinforce a message around humans' lack of perception as compared to divine wisdom. Both were theorists and practitioners, and together they were largely responsible for establishing how the mechanisms and logistics of anamorphoses fit within the Minim worldview. One was the mathematician and perspectival theorist Jean-François Niceron; the other was Emanuel Maignan, the mathematician, theologian, philosopher, and turner (Figs. 64 and 65).[33] Maignan had designed the sundial in Palazzo Spada discussed in chapter 1, and both scholars had taught Minim students in Rome: Niceron in mathematics and Maignan in philosophy and theology.

Around 1635, when Niceron resided in the Place Royale convent in Paris, he is said to have produced an anamorphic oil painting featuring Saint Francis of Paola that could be reformed in a cylindrical mirror (Fig. 66).[34] An examination of the painting without a mirror, as in Figure 66, reveals Saint Francis's body curved and elongated to a nearly unrecognizable degree. His

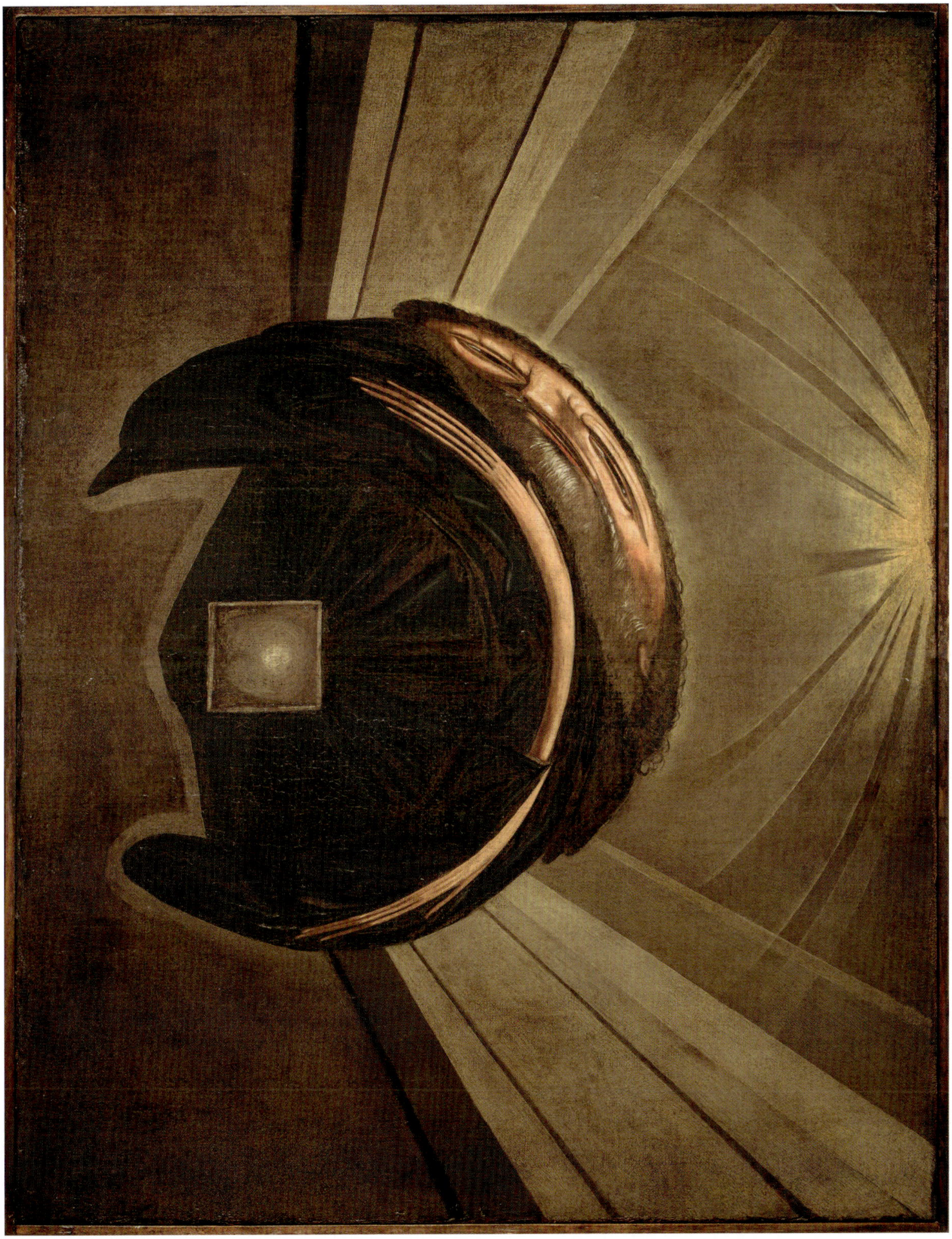

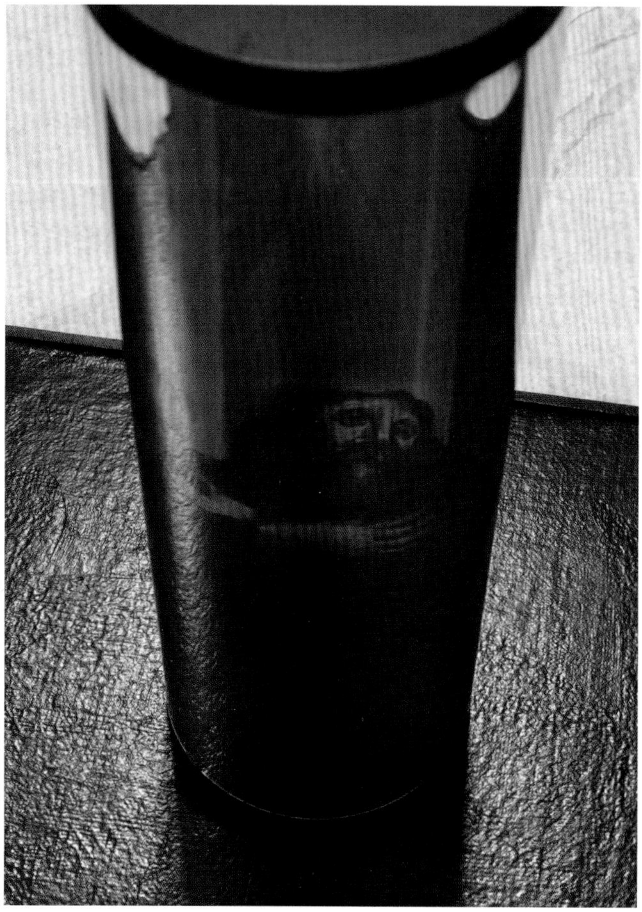

Fig. 67. *Saint Francis of Paola*, c. 1635. Oil on canvas. Galleria Nazionale d'Arte Antica, Palazzo Barberini, Rome [inv. n. 1955].

hands, forearms, and face project above the dark brown hues of his habit. The structure of the space is incoherent, although from a source that seems to originate beyond the upper border a spotlight is cast on the saint. The deformed portrayal of Saint Francis was meant to be observed on a horizontal surface; the resulting vantage point—the viewer likely bending over a painting on a table—would have contributed further to his apparent deformation.[35] The photograph in Figure 67 demonstrates how Niceron's distended image is rectified when it is reflected in a cylindrical mirror: the saint crosses his arms over his chest in prayer, a light from a now apparently heavenly source shining down on him from above, gesturing us toward the source of true illumination and vision. I invite the reader to curve the reflective sheet included in the back pocket of this book into a cylinder and to reform the deformation reproduced in Figure 66 (or other cylindrical anamorphic images printed throughout this text).

Niceron also designed a foldout engraving, visualizing a method for producing the kind of specular anamorphosis of the saint he achieved in the painted canvas, and included that image in his study of optical marvels, *La perspective curieuse* of 1638 (Fig. 68).[36] An inscription notes the location and measurements of the cylindrical mirror: "The little circle FGHI is the width of the cylinder[;] And the big circle KLMNO &c represents its base."[37] As the anno-

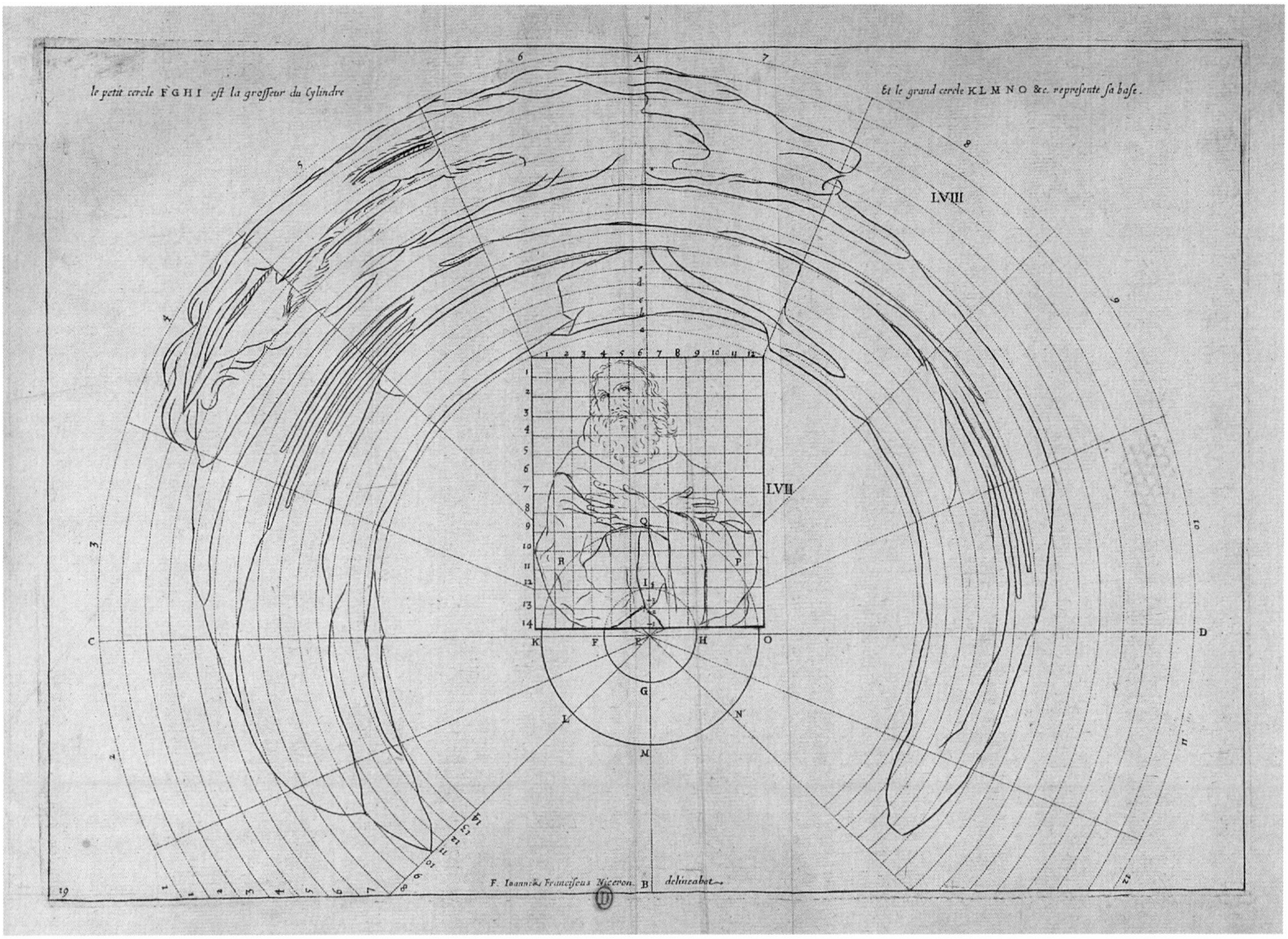

tation makes clear, the engraving is drawn to scale. The outline of Saint Francis's distorted likeness is integrated into a broad arc. His rectified image, inscribed with a grid, occupies the print's central area, where the specular cylinder would stand. In this image, as in the anamorphic panel produced a few years earlier, the saint's hands are crossed at his chest; with upturned eyes and his head leaning to the side, he appears to be in rapture. Observers are to imagine the specular cylinder reflecting the properly proportioned image of Saint Francis in ecstasy. Presented to observers, then, is an image of divine revelation to the saint, with the mechanical operation of the anamorphosis functionally echoing the movement from human chaos to clarity that was understood to represent humans' relationship to God and his creation. Beyond this potential religious benefit, Niceron trains users of his book to gain familiarity with specular anamorphoses—and encourages them to try creating anamorphoses of their own—by explaining how they are made. The foldout engraving in *La perspective curieuse* is no larger than two standard pages in the book. Presumably, Niceron opted for a foldout print so that observers could copy the precise measurements of the specular cylinder's base with ease and

Fig. 68. Jean-François Niceron (designer), method for producing a specular anamorphosis from Niceron's *La perspective curieuse*, 1638, table 19. Engraving printed on paper, 10.35 × 14.49 in (26.3 × 36.8 cm). Bibliothèque de l'Institut national d'histoire de l'art, collections Jacques Doucet, 4 RES 876.

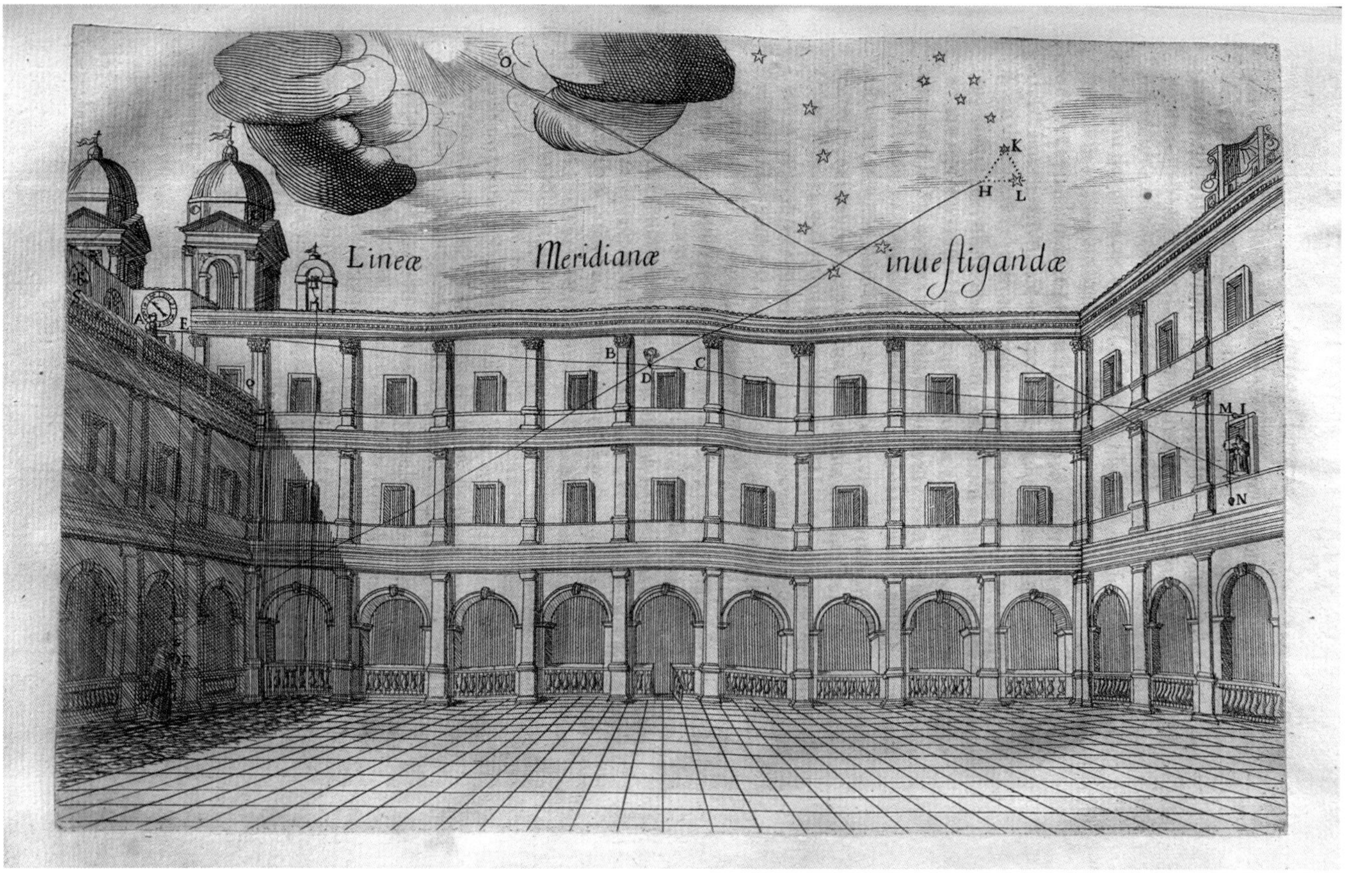

Fig. 69. Unidentified artist, Trinità dei Monti courtyard from Emanuel Maignan's *Perspectiva horaria*, 1648, between pp. 138 and 139. Engraving printed on paper, 7.13 × 11.35 in (18.1 × 26.3 cm). Harvard University, Houghton Library, Cambridge, MA.

without the impediment of the book's gutter. The design of Niceron's book facilitates readers' acts of making.

Guiding the Faithful Toward God's Truth

That the Minims saw potential for the successive experiences of anamorphic deformation and reformation to communicate a lesson to audiences around God's control over visual perception and revelation is further manifest in the multiple large-scale anamorphic frescoes that were painted on the corridor walls of the convent of Trinità dei Monti in Rome and in the Place Royale convent in Paris. The Parisian frescoes have not survived, but in Rome, in the building's first-floor corridors, the frescoes created by the Minims Niceron and Maignan are visible today. In his *Perspectiva horaria*, Maignan includes a representation of the Santissima Trinità courtyard with the church's twin towers portrayed in the upper left corner (Fig. 69). The exterior of the corridor with Maignan's anamorphosis faces the engraving's observer; Niceron's anamorphic fresco stands on the opposite side of the courtyard (that is, on the observer's side) on the northern arm and the eastern corridor of the convent's first floor. The corridor on the right side of the engraving connects to the galleries with the anamorphic wall frescoes and features a giant sundial also painted by Maignan (see Fig. 70).

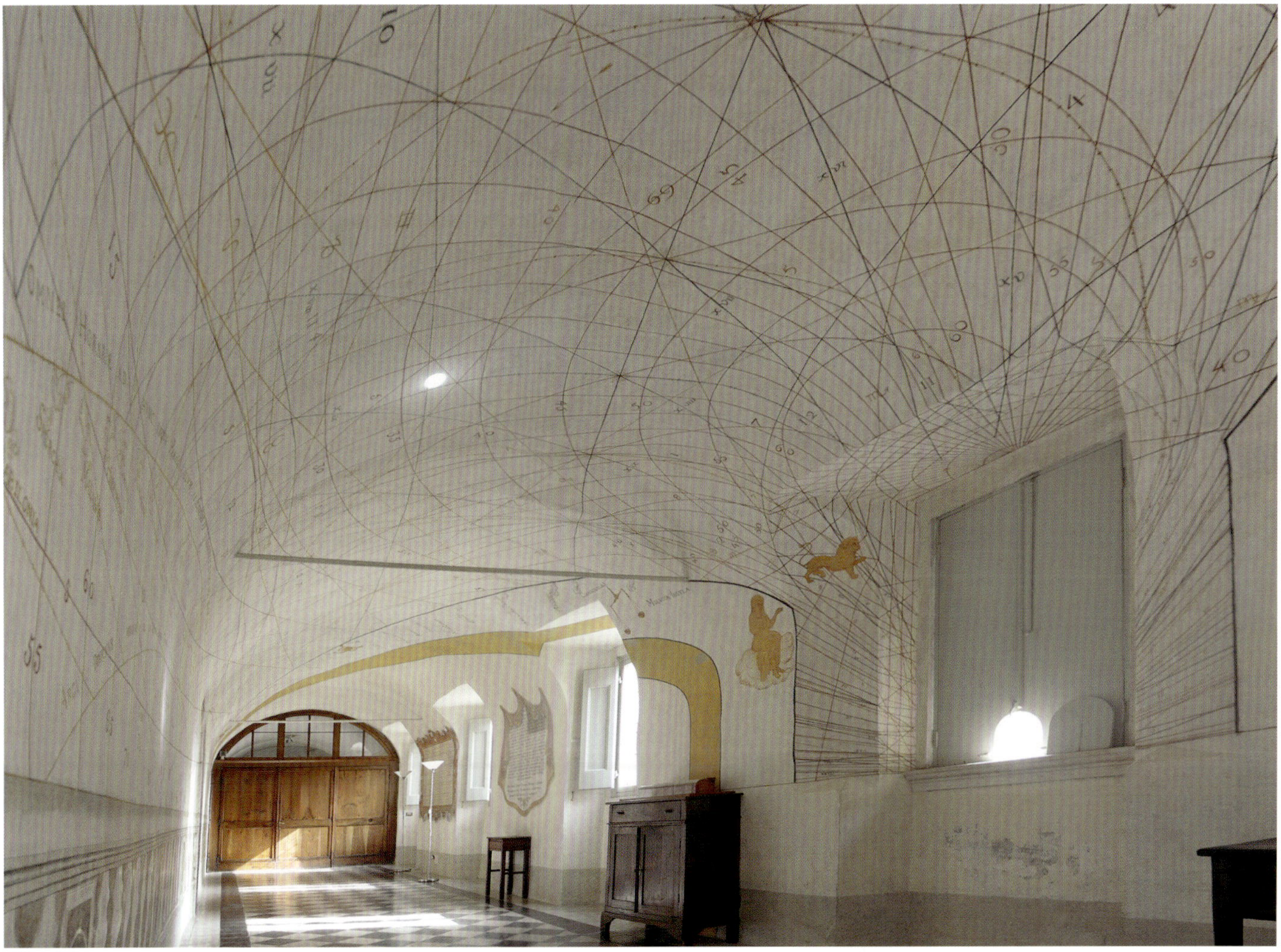

Fig. 70. Emanuel Maignan, sundial, 1637. Fresco. Corridor, convent of Trinità dei Monti, Rome.

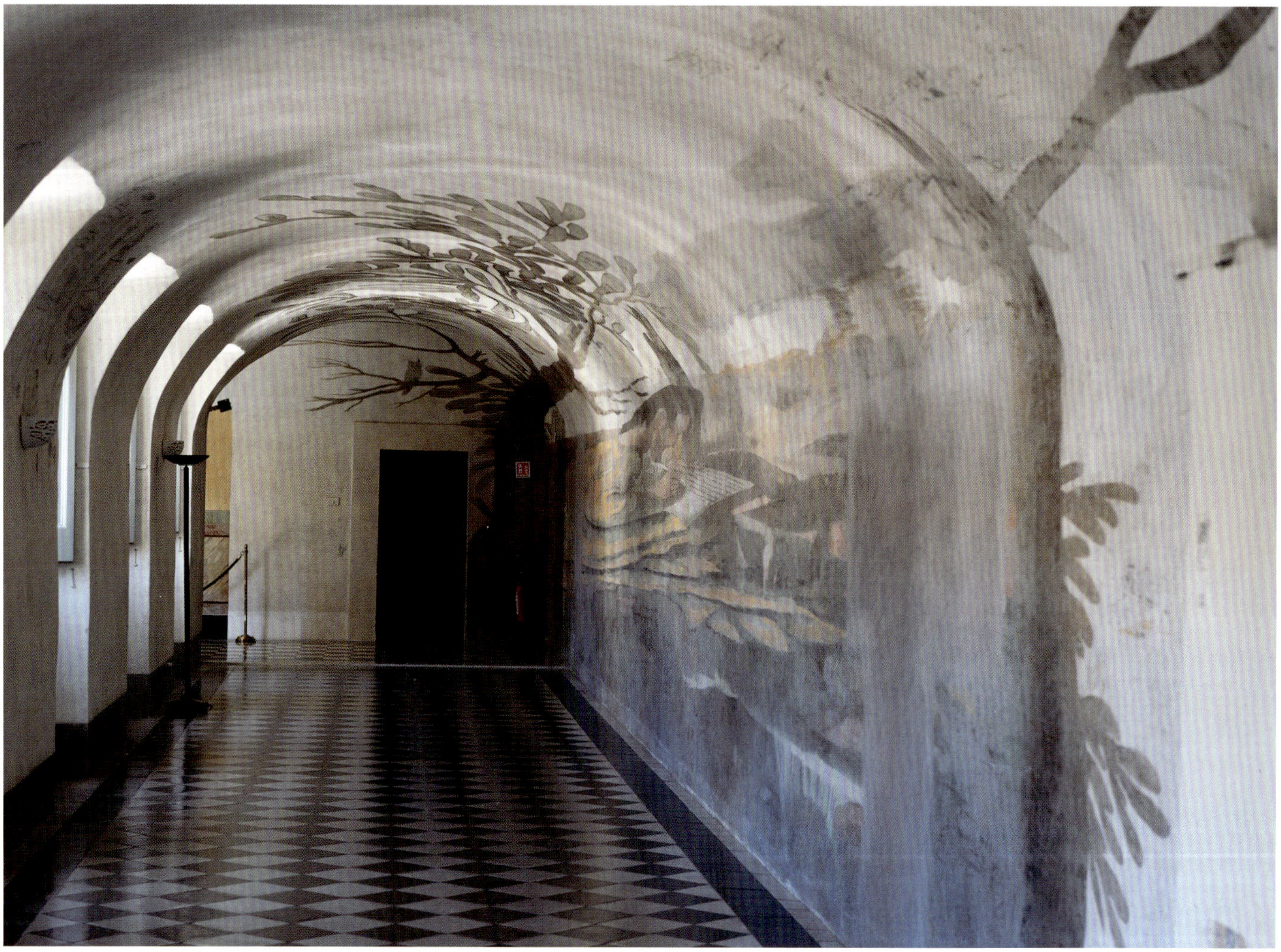

Fig. 71. Jean-François Niceron, *Saint John the Evangelist and Landscape*, 1639–1640. Anamorphic fresco *a secco*, 8 ft 6.36 in × 57 ft 6.5 in (2.60 × 17.54 m). Corridor, convent of Trinità dei Monti, Rome.

Niceron produced his anamorphic fresco in Santissima Trinità between 1639 and early 1640 (Fig. 71 and Gatefold 1). This work was only uncovered, beneath thick layers of paint and plaster, in the course of restoration work that began in 2005 and as a result has received little consideration in English-language art historical literature.[38] Because it was painted *a secco*, its pigments did not become a part of the wall, as in the *buon fresco* technique; the painting is thus in poor condition, which makes it difficult to interpret.[39] Moreover, although this fresco, like Maignan's, was originally painted in grisaille, colors were added to it at the beginning of the nineteenth century.[40] Despite its poor condition and chromatic alterations, the fresco's impressive dimensions allow it to command the attention of those who pass through the space. And in the seventeenth century, its anamorphosis would have reminded beholders (then nearly all male, since women generally were not permitted to enter the convent) that the vantage point determines what is seen.[41] The composition grows out of an enormous olive tree nestled into the northeastern corner of the hallway: an owl looks down from one of its branches, and

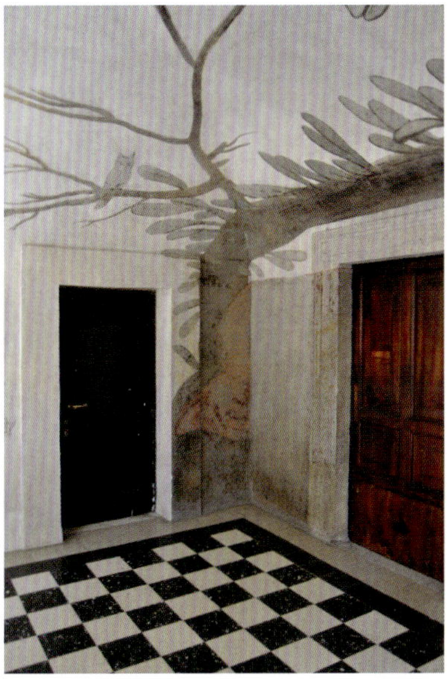

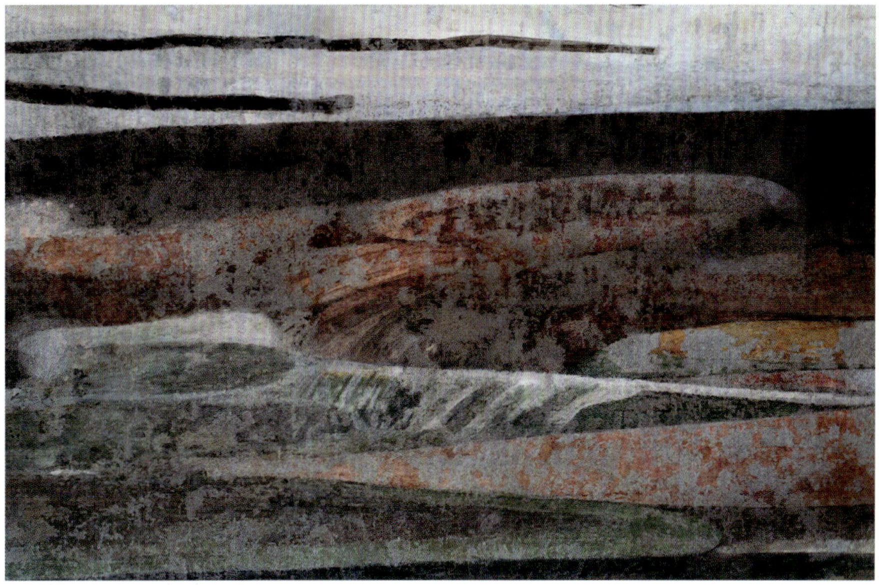

observers walking clockwise around the courtyard move parallel to the tree's outstretched branches (Fig. 72). A landscape of craggy cliffs, pathways, and waterfalls unfolds below (see Gatefold 1). Only as beholders arrive at the floor's southeastern corner does the natural scenery turn illegible as the fresco reforms instead into an image of Saint John of Patmos writing (see Fig. 71). Observers recognize a profile view of the larger-than-life-size bearded saint, clad in fabrics that today appear yellow; he leans forward, fully absorbed in bringing pen to paper. From this end of the hall, observers perceive Saint John holding a quill pen, though the same detail, when viewed head-on, looks to be a waterfall (Fig. 73; and see Fig. 71). Eliciting deceptive perceptual states in the mode of all anamorphoses, the image invites confusion and uncertainty as it oscillates between the gigantic Saint John and the miniature vignettes within the landscape.

Fig. 72. *Saint John the Evangelist and Landscape* (detail), 1639–1640. Corridor, convent of Trinità dei Monti, Rome.

Fig. 73. *Saint John the Evangelist and Landscape* (detail), 1639–1640. Corridor, convent of Trinità dei Monti, Rome.

Niceron's fresco in Trinità dei Monti illustrates a genre of anamorphosis different from that presented in Figures 66 and 67: the reformation of the large fresco relies on the observer's motion through space (rather as one arrives at a fuller comprehension of the *prospettiva* at Palazzo Spada as one moves through the space of the colonnade); to reform the smaller anamorphic panel from an incoherent image, the observer may remain stationary but change the image by placing a cylindrical mirror. The Trinità dei Monti fresco is experienced as a transition from one relatively coherent image to an incoherent image to a second coherent image; the portrait panel is seen first as incoherent, then resolves into coherence.

Maignan painted his momentous anamorphic fresco of Saint Francis—which still spans the first-floor western corridor above the Trinità dei Monti convent's courtyard—in 1642 (Fig. 74 and Gatefold 2).[42] Like Niceron's anamorphic portrait of Saint John, Maignan's fresco was painted *a secco*, and it is also in poor condition, although it is in better shape than Niceron's anamorphosis.[43] The work, like the painting of Saint John, invites observers to pay attention as they walk through the space—and to notice how a change in vantage point shifts appearances. In the process, the fresco reminds observers that they can take both literal and metaphorical

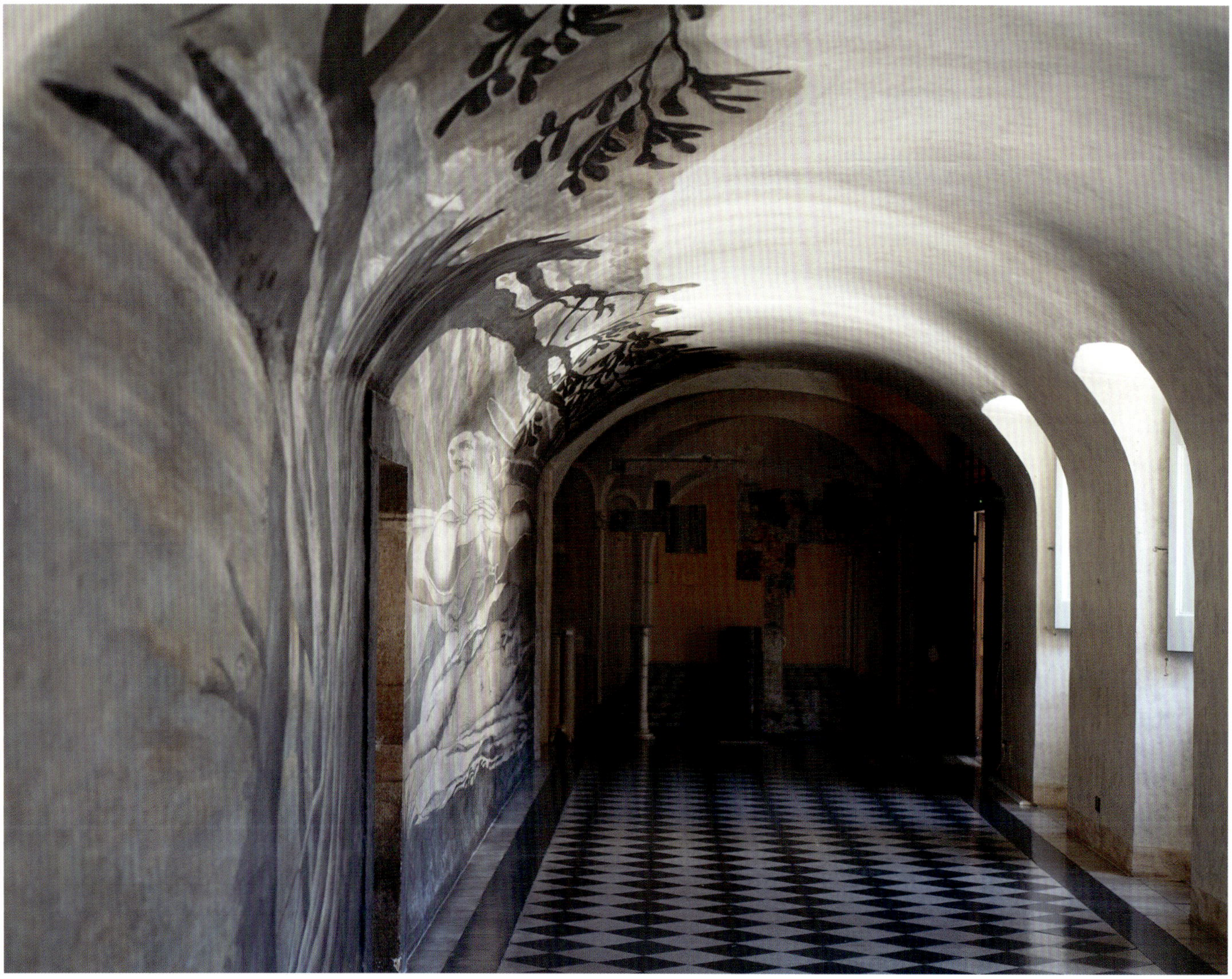

Fig. 74. Emanuel Maignan, *Saint Francis of Paola and Landscape*, 1642. Anamorphic fresco *a secco*, 7 ft 2.5 in × 61 ft 8 in (2.20 × 18.80 m). Corridor, convent of Trinità dei Monti, Rome.

steps to "see better," to improve their understanding. Indirectly, it further communicates the mediating role of the Church and its leaders, here Minim friars, in guiding the faithful toward truth. When approaching the composition from the left, beholders find another oversized bearded saint beneath an olive tree that echoes the tree painted by Niceron on the opposite side of the courtyard in 1639–1640 (see Figs. 74 and 71). Saint Francis kneels with his hands clasped in prayer and his eyes rolled heavenward to indicate his communication with the sacred.[44] In both Niceron's and Maignan's Trinità dei Monti frescoes, the trees outline the top and left sections of the composition and remain legible as observers traverse the space, whereas the representations in the areas below them seem to change. In Maignan's fresco, as observers traverse the corridor, moving north along the painted wall, the likeness of the saint grows illegible, though the olive tree remains coherent. In this fresco, as the viewer's vantage point

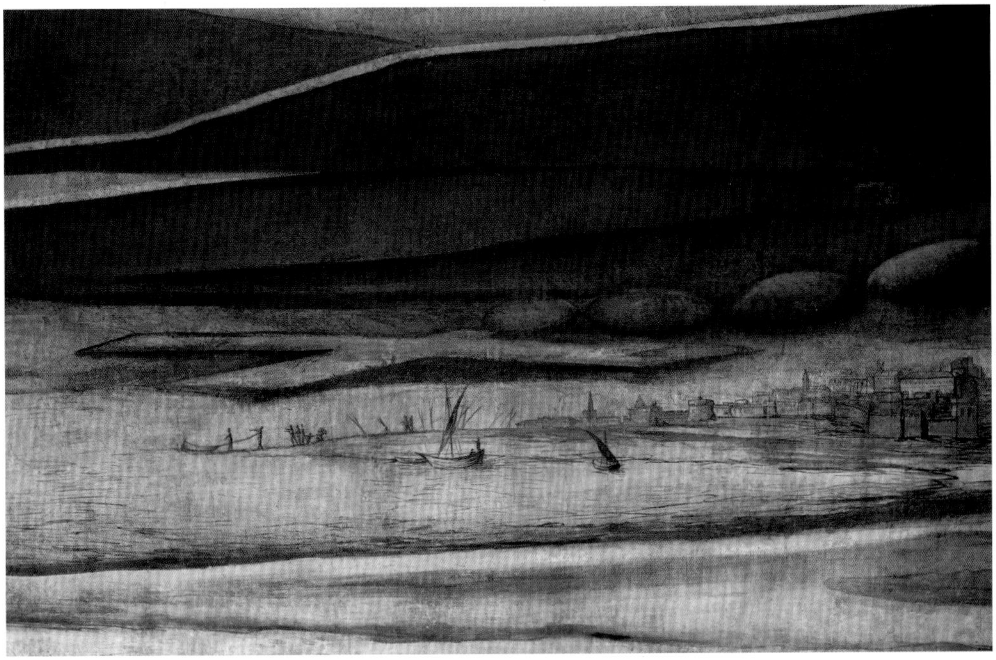

Fig. 75. *Saint Francis of Paola and Landscape* (detail), 1642. Corridor, convent of Trinità dei Monti, Rome.

changes, the saint seems replaced by a bird's-eye perspective on a rural and maritime scene, featuring tiny sailing and fishing boats, villages, bell towers, fortified bases, roads connecting hills, bridges, and a myriad of small figures (see Gatefold 2). Not all elements from the representation of Saint Francis transform into landscape elements, even when observed head-on. Notably, the saint's cross and prayer beads remain fully visible, so that as observers move in front of the fresco to the northern side of the corridor, they must negotiate between one image and the other (Fig. 75). If from a northern vantage point, the fresco does not turn into a fully controllable image, from the far southern end of the corridor, the portrait of Saint Francis appears more unified, as the landscape details are too small to intrude. Presumably, this difference in levels of clarity can be related in part as well to how the landscape was molded after what the figure needed and not the other way around, given that the former is more malleable than the latter. This fresco would have immersed Minims and visitors who walked along the corridor and initially approached the composition from the left in successive states of perceptual confusion and then comparative clarity and apparent certainty, as they observed at first a clear image that seemed to deform into chaos just as another, more coherent (though not perfectly coherent) one came into view. Looking at the fresco induces the vertigo of anamorphosis.

The Jesuit Schott describes the painting as he saw it during frequent visits to the Minims' convent in Rome:

> I saw most often, and I showed to others, in Rome, on the Pincian Hill, in the famous convent of the Minims ... the effigy of Saint Francis of Paola, stretched out on the longest wall of the walkway ... made by the most learned man of the stated order, Emanuel Maignan, with that art, so that, from an individual, determined place, established

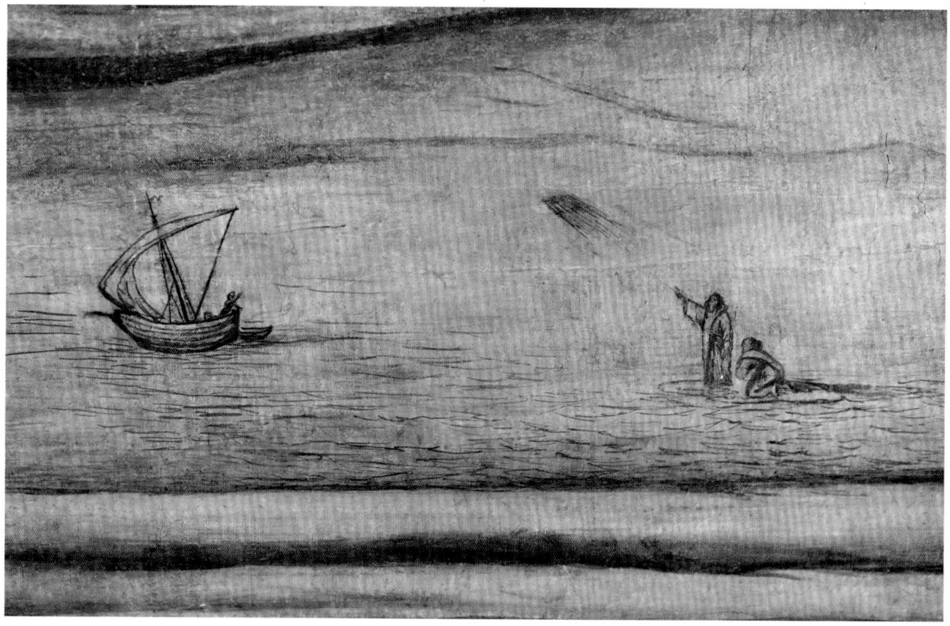

Fig. 76. *Saint Francis of Paola and Landscape* (detail), 1642. Corridor, convent of Trinità dei Monti, Rome.

> at the farthest point of the walkway, it [i.e., the effigy of Saint Francis] would appear; beyond that place, indeed, and especially when examined from the opposite place, it represented the leaves of trees, men, animals, and other similar things.[45]

Schott's remarks specifically point to the importance of viewpoint when observing the fresco, and to enthusiasm for his fresco and the Roman convent outside the Minim order. Far from condemning the image for a lack of clarity, he praises the detailed composition:

> He [i.e., Maignan] took care that there be painted, with smaller figures among the more pronounced features of the deformed image, more distinguished reminders of the virtues and works of Saint Francis of Paola, so that someone looking from the opposite point would be detained by this pleasing fallacy and not suspect that in that place there was anything else hidden by this art.[46]

For Schott, Maignan's anamorphosis was an instrument of trickery, secrecy, and pleasure that detained observers, captured their attention, and held them in place, with a noticeable emphasis on delay—not only highlighting that viewpoint conditions observational experience but hinting that a person more informed about optical marvels might be more likely to continue or to look further, rather than be detained by the "pleasing fallacy." Schott also focused on how the deformation encourages sustained contemplation of the pious works of Saint Francis.

Schott's discussion of the fresco drew on Maignan's own words in the *Perspectiva horaria*, a passage where the latter characterizes the figures representing the works of Saint Francis as "pleasant fallacies" (*iucunda fallacia*) that detain—he, too, uses the verb *detinere* (to hold)—observers' attention on particular images:

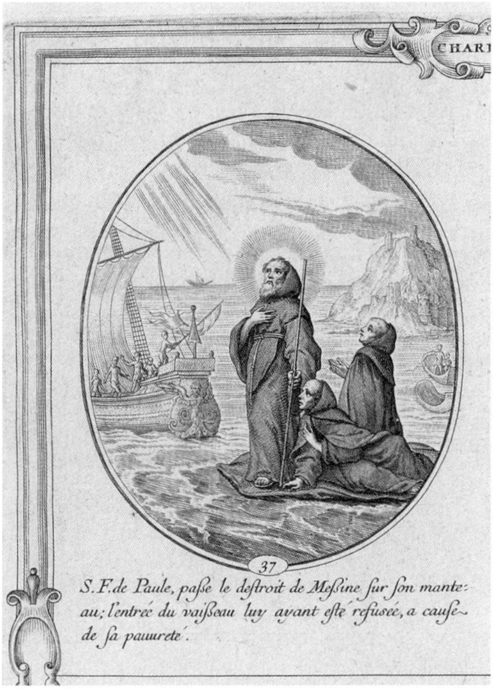

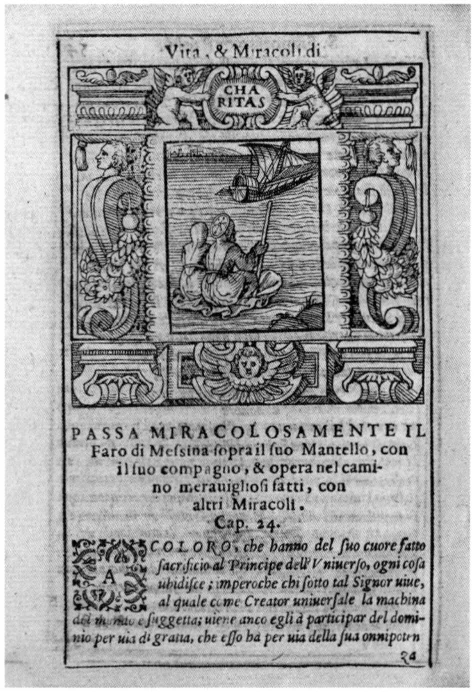

Fig. 77. Adriaen Lommelin (engraver), Saint Francis of Paola crossing the Strait of Messina on his cloak with companions from Antoine Dondé's *Les figures*, 1664, plate 12. Engraving printed on paper, 4.33 × 3.14 in (11 × 8 cm). BnF, Paris [Reserve 4-RD-83].

Fig. 78. Unidentified artist, image of Saint Francis crossing the Strait of Messina on his cloak with a companion from Paolo Regio's *La miracolosa*, 1581, 54ᵛ. Printed on paper, dimensions of the page are 6.34 × 4.41 in (16.1 × 11.2 cm). Bayerische Staatsbibliothek München [V.ss. 355 g].

> In these little mountains, roads, buildings, and so on, tiny little figures are painted with a light pencil, which are in view and seen openly: indeed, these pleasant fallacies detain the viewer, so that he does not doubt that there is anything else there which the art is hiding. In that likeness of our most Sacred Francis, which I deformed, I saw to it that the more illustrious records of the virtues and works of the Father were painted with tiny figures of this type.[47]

Maignan emphasizes that his fresco presents the marks of God through Saint Francis's actions. In one diminutive scene, for instance, Saint Francis crosses the Strait of Messina on his cloak with a companion (Fig. 76). This episode features in an engraving by Lommelin from Dondé's life of Saint Francis of Paola and in an anonymous woodcut in Paolo Regio's (1545–1607) life of the saint (Figs. 77 and 78).[48] Dondé's explanatory inscription reads: "Saint Francis of Paola, passes the Strait of Messina on his coat; entry onto the vessel having been refused to him, because of his poverty."[49] In Maignan's fresco, Saint Francis is presumably the more active, standing figure; the ship that denied passage to him and his companion sails on the left (see Fig. 76). The diminutive scale of this and other scenes induces observers to approach the painting closely and to take turns, one at a time, in attending to and contemplating the miracles before them. In asking the viewer of a large piece of art (a wall fresco) to stop and look closely at it—to "read" the wall as they would a book—the fresco signals to the observer that they should be prepared to engage in the interpretive work that is associated with spiritual reading. As images in illustrated lives of Saint Francis focused, organized, and deepened friars' personal meditations on the saint's life and thus enabled them to see God's works and revealed truth more clearly, so too might have Maignan's representation of miraculous occurrences in his anamorphic fresco in Rome.

In the passage cited above, Maignan, like Schott, foregrounds the concealment enacted within his work. He is confident that the scenes will engross observers, such that they remain unaware of the image of the saint. And he is likely correct that most observers, when approaching the fresco from the right, would "not doubt" the art, or suspect that it is "hiding" another image. Even if some elements—like the beads and the cross—indicate to attentive visitors that something else is in fact going on, they may not realize that another larger representation is "hidden" within the fresco. Maignan makes it clear that he wants beholders to experience the condition of being led into error through sight: "All of this work is composed in order to deceive vision."[50] A visitor who experienced the corridor's deceptions—deceived insofar as they would see the same image one way and then another—might increase their visual dexterity, perhaps might come to understand its mechanics by reading Maignan's *Perspectiva horaria*. In addition, the repeated experience of deformations reformed conveyed a message pertaining to God's command over vision and revelation.

If one possible interpretation of these large fresco anamorphoses is that they reinforced the Church's role in leading the faithful to the proper vantage point for understanding God's revealed truth, their subject matter might also reinforce the primacy of the Catholic Church's claim to that guiding role, as is apparent in the anamorphic paintings described in an unrealized proposal that Virgilio Spada and Maignan developed together for a villa for Cardinal Camillo Pamphilj (1622–1666).[51] The plan was probably initiated after Camillo was made cardinal in 1644, with the villa meant to be located on the slopes of the Gianicolo in Rome outside the Porta S. Pancrazio.[52] In a document elaborating the project, Maignan writes:

> On the east and west sides of the same room, there will be painted using the art of perspective an effigy of, for example, Saint Peter on one side, and on the other, one of Our Lord, or others as might please His Eminence, but with such skill that, to those in the room, they will not be apparent as will some other image to the liking of His Eminence. Instead, the effigies will appear to those who stand outside the hall in the northern loggia as they look through a small hole made for this purpose.[53]

Maignan's proposal reveals how the mechanics of anamorphoses called attention to particular religious messaging from precise vantage points. Saint Peter was the first bishop of Rome, whom popes claimed to be succeeding and who was given his role by Christ. Catholic theologians foregrounded Saint Peter to tie the Catholic Church directly to Christ.[54] The crystallized image of Christ and Saint Peter would have been particularly appealing to Catholic Reformation religious figures: the "truth" here being revealed to insiders was one that is narratively central to the Roman Catholic tradition. Similarly, in revealing a Saint Francis of Paola in communication with the sacred, Maignan was revealing another "truth" central to his order. In chapter 4, we will analyze in greater detail the "truth" that Niceron's expansive Roman anamorphosis exposes.

Additional play with illusions and with what I have described as a process of reforming that transmits a message concerning God's power over vision and revelation is manifest in Maignan's

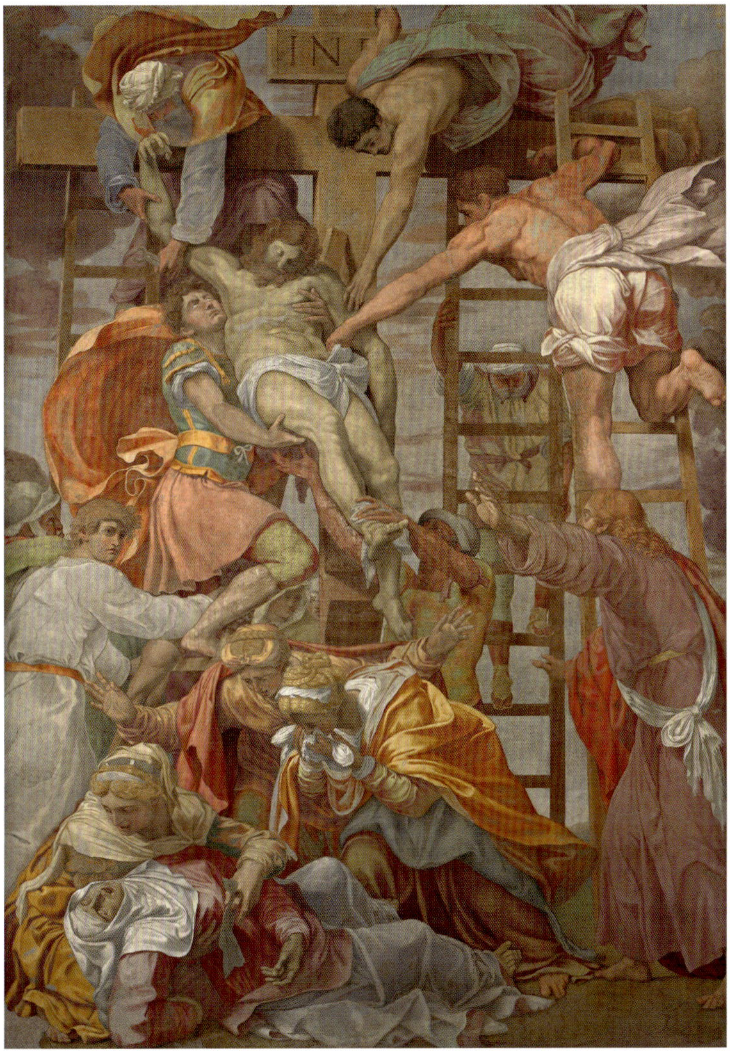

Fig. 79. Daniele da Volterra, *Descent from the Cross*, c. 1545. Fresco, 132.75 × 91 in (337 × 231 cm). Church, Trinità dei Monti, Rome.

suggestion that a variety of different figures could be painted on a wall of the villa's southern loggia so as to appear as a single figure when examined through an eyepiece (the multifocal beveled lens).[55] Maignan may have conceived of this instrument based on conversations with Niceron, who describes such a device in *La perspective curieuse* (1638), which he had had apparently produced and kept in the library of the Minim convent in the Place Royale in Paris.[56]

Niceron's *Perspective curieuse*, of course, offers further evidence of how Minim visual culture was defined by visual manipulation—deformations and other optical reformations and trickery—not least to reinforce teachings around miraculous mysteries that were not immediately accessible to human reason. In that text Niceron discusses a curious optical effect in Daniele da Volterra's (1509–1566) *Descent from the Cross* (c. 1545)—which he would have known from an original in Trinità dei Monti and a copy in the Place Royale church in Paris (Fig. 79).[57] Describing the motion of Christ's body as it is being taken down from the cross,

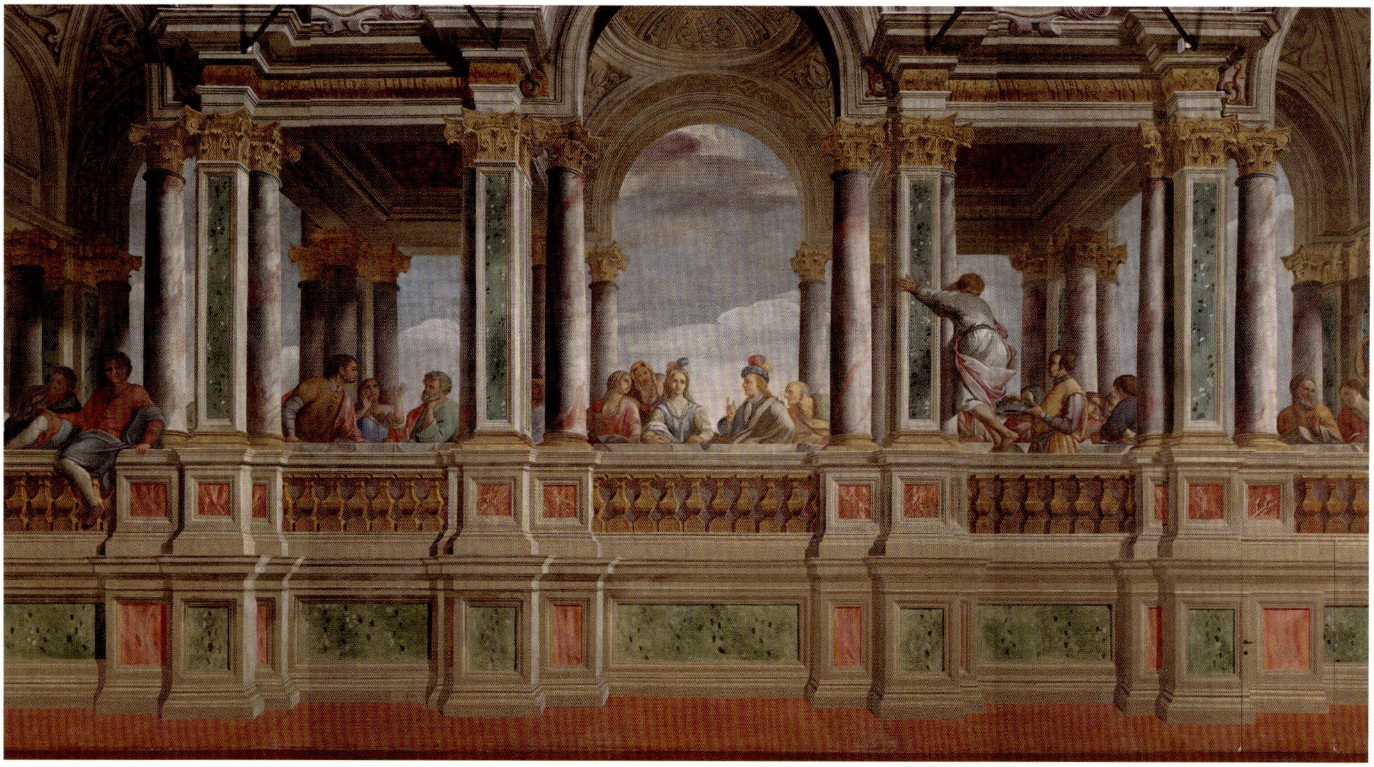

ABOVE

Fig. 80. Andrea Pozzo, *Wedding Feast at Cana* (detail), c. 1694. Frescoes painted *a secco*. Refectory, convent of Trinità dei Monti, Rome.

OPPOSITE

Fig. 81. *Wedding Feast at Cana* (detail), c. 1694. Refectory, convent of Trinità dei Monti, Rome.

Fig. 82. *Wedding Feast at Cana* (detail), c. 1694. Refectory, convent of Trinità dei Monti, Rome.

Niceron observes that the figure seems visually to rise—perhaps hinting at the Resurrection—as the viewer moves from left to right:

> A fresco in a chapel ... showing a *Descent from the Cross* in which Christ is so depicted that viewed from the left, He seems to be lying down and leaning across the picture, with His right foot thrust out toward the left; but viewed from the other side, His whole body appears almost vertical, much more foreshortened, and the foot which seemed to protrude on the left side appears to advance toward the right.[58]

Visual illusion in another set of late seventeenth-century frescoes centers the wedding at Cana, a moment when something that is disappointing to the senses and inappropriate for a wedding because it was not celebratory enough (water) is miraculously reformed into something that satisfies (wine of the best quality). The Italian Jesuit painter and art theoretician Andrea Pozzo is said to have produced some of the frescoes painted *a secco*, around 1694, on the walls and vault of the convent of Santissima Trinità dei Monti's refectory (Figs. 80, 81, and 82).[59] While the attribution to Pozzo is debated, it is unquestionable that these paintings, which depict the wedding feast at Cana within a fictive architectural setting, also foreground optical illusion and trickery. Painted details repeatedly undermine the boundary between the imaginary world of the frescoes and the space of the beholder: One wedding guest, dressed in red, blue, and white, leans back toward a column while dangling his left leg over a balustrade,

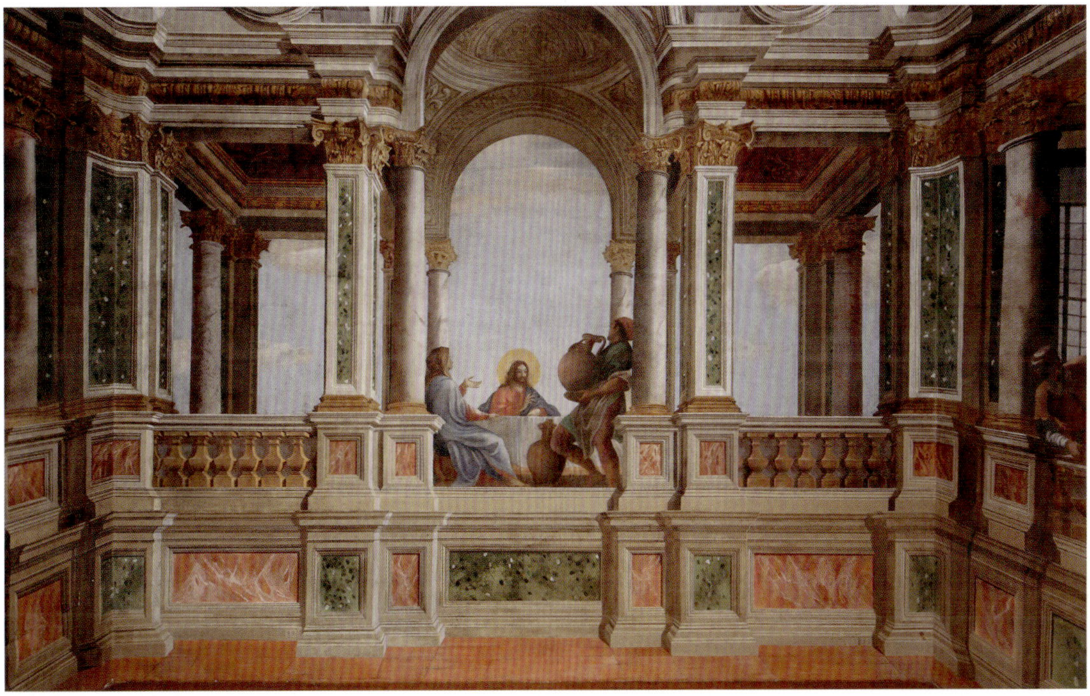

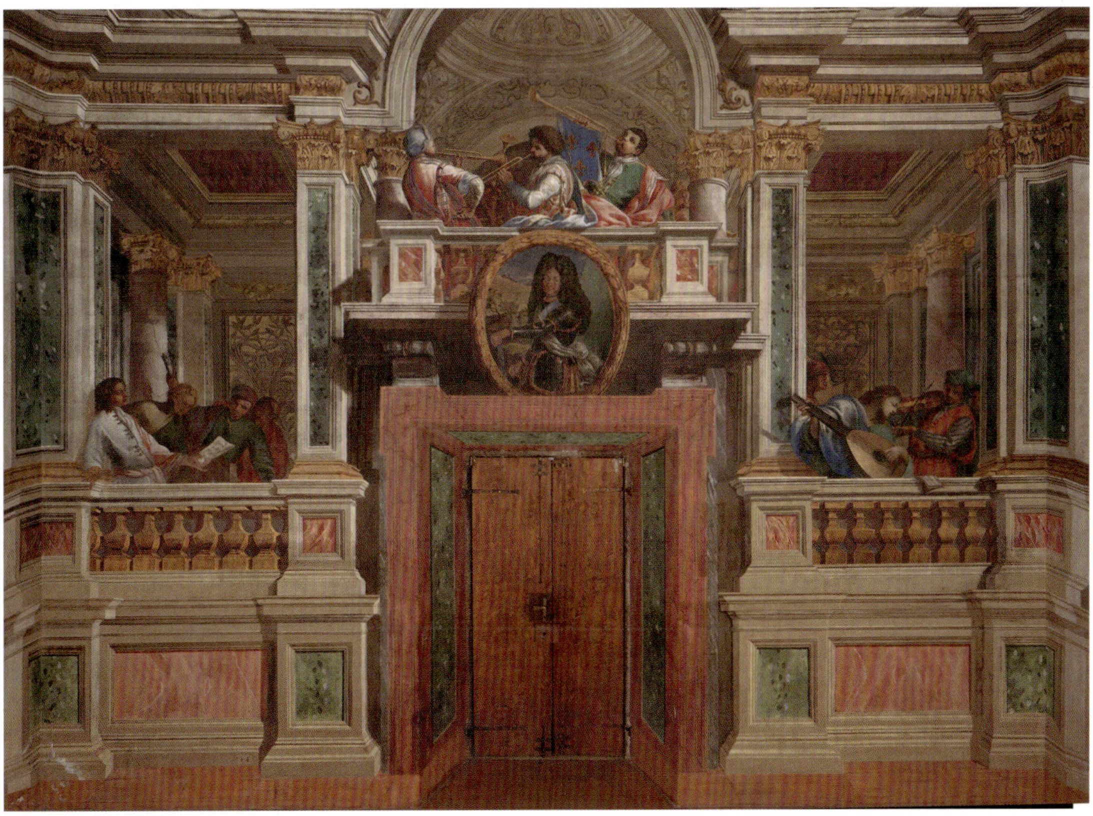

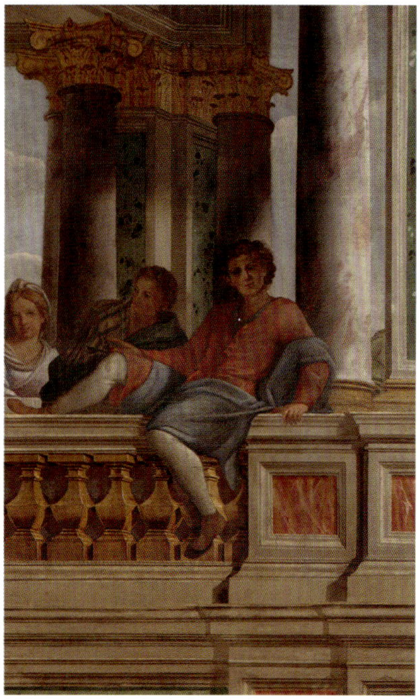
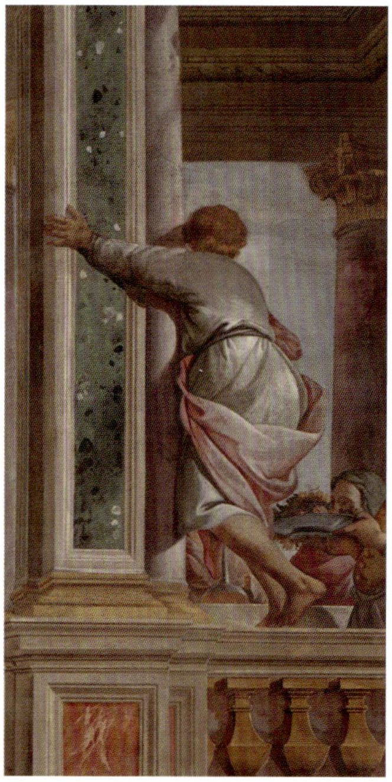
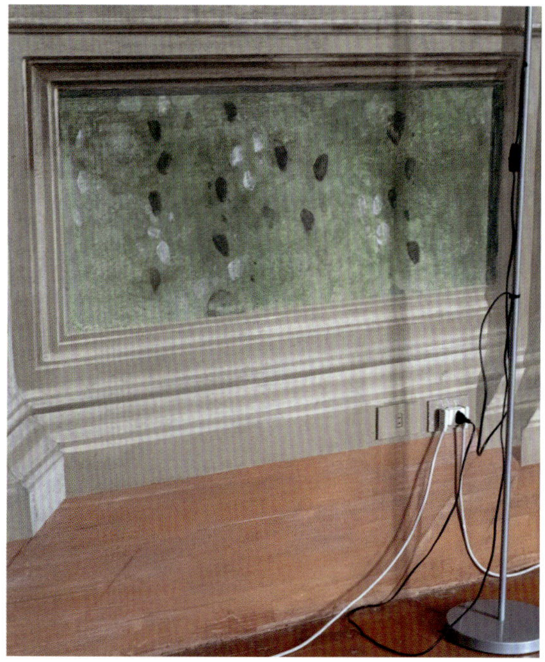

Fig. 83. *Wedding Feast at Cana* (detail), c. 1694. Refectory, convent of Trinità dei Monti, Rome.

Fig. 84. *Wedding Feast at Cana* (detail), c. 1694. Refectory, convent of Trinità dei Monti, Rome.

Fig. 85. *Wedding Feast at Cana* (detail), c. 1694. Refectory, convent of Trinità dei Monti, Rome.

apparently on the verge of slipping into the observer's space (Fig. 83). Another guest to the right, dressed in white and pink, stands on the balustrade and clutches a pilaster with one hand (Fig. 84); he too seems in danger of falling out of the painting and into the beholder's realm. The room's pavement, moreover, seems to extend into the space of the fresco, causing the refectory to appear more expansive than it really is (Fig. 85). As in an anamorphic image, certain details in the architecture, such as columns and pilasters, appear distorted and deformed resolving only when they are viewed from a predetermined point centered on the representation of Jesus (Fig. 86).

Elsewhere, and perhaps especially in a later period, illusionism in the art produced under Minims' patronage would echo a growing interest in the inevitable passing of time (a message that ultimately underscored nearly all religious messaging): Charles-Louis Clérisseau's (1721–1820) *Stanza delle rovine* (*Ruins Room*, c. 1767) for the Santissima Trinità dei Monti convent gives the impression that one has entered a room in a state of decay, even as the images distributed throughout the convent challenge observers to reflect on deception (Fig. 87).[60] This space was commissioned by the Minim Thomas LeSeur (1703–1770), a celebrated theologian, mathematician, and botanist who coauthored a commentated edition of the *Philosophiae Naturalis Principia Mathematica* (*Mathematical Principles of Natural Philosophy*) of Sir Isaac Newton (1642–1727). Responding to his patron's interests in physics and botany, Clérisseau integrated Newton's *Principia* into a niche in his fresco as well as numerous plants and scientific instruments (Figs. 88 and 89).[61] In addition, outside the convent in the Piazza di Spagna, known as the Piazza della Trinità until the end of the seventeenth century, artists were often commissioned to stage lavish festivals to honor important events, such as a military success or the birth of an heir to the throne.[62] Thus in 1662, to celebrate the birth of the son of Louis XIV

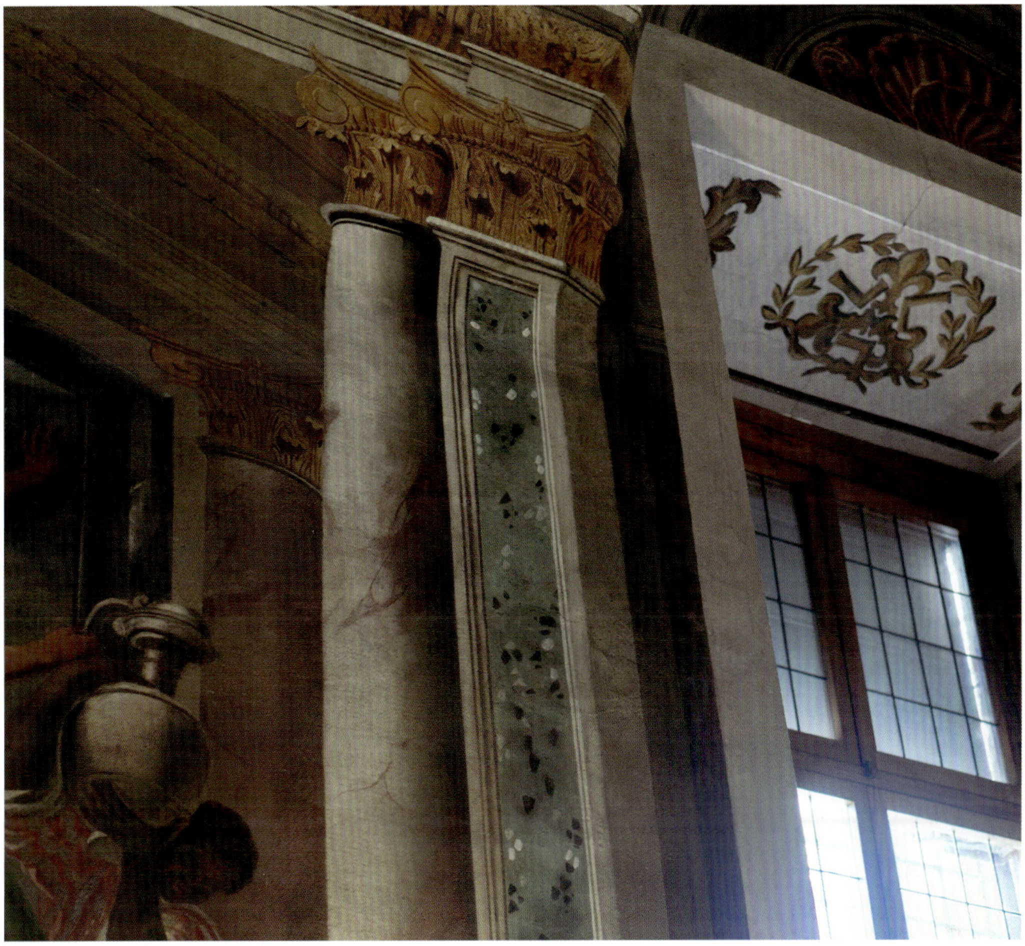

Fig. 86. *Wedding Feast at Cana* (detail of a deformation), c. 1694. Refectory, convent of Trinità dei Monti, Rome.

the year before, Bernini collaborated with Johann Paul Schor (1615–1674) to create an elaborate illusionistic apparatus on the hill in front of Trinità dei Monti.[63] The structure is recorded in an etching by Dominique Barrière that shows an enormous, erupting volcano emerging from the hillside under the church (Fig. 90).[64] Above, partially obscuring the church facade and surrounded by smoke, appears a gigantic silver dolphin (the Dauphin), representing the eldest son of the king of France. Still higher, the French crown is grasped by the papier-mâché figures of Hymen and Peace. Below, Discord falls into the fiery volcano. In the opening of the two bell towers, a large capital "L" was placed on one side and an "M" on the other, in reference to Louis XIV and his spouse Maria Theresa of Spain (1638–1683). At night, a firework display began. Music, food, and wine were also provided.

If these later uses of illusionistic visual art in and around Trinità dei Monti were focused on marking or reminding viewers of the passage of time in a historical sense, the earlier uses seem more directly focused on the time that elapses between the moment of seeing something with one's limited human perception and the moment of understanding. Yet all, in some sense, highlight the limits of human perception.

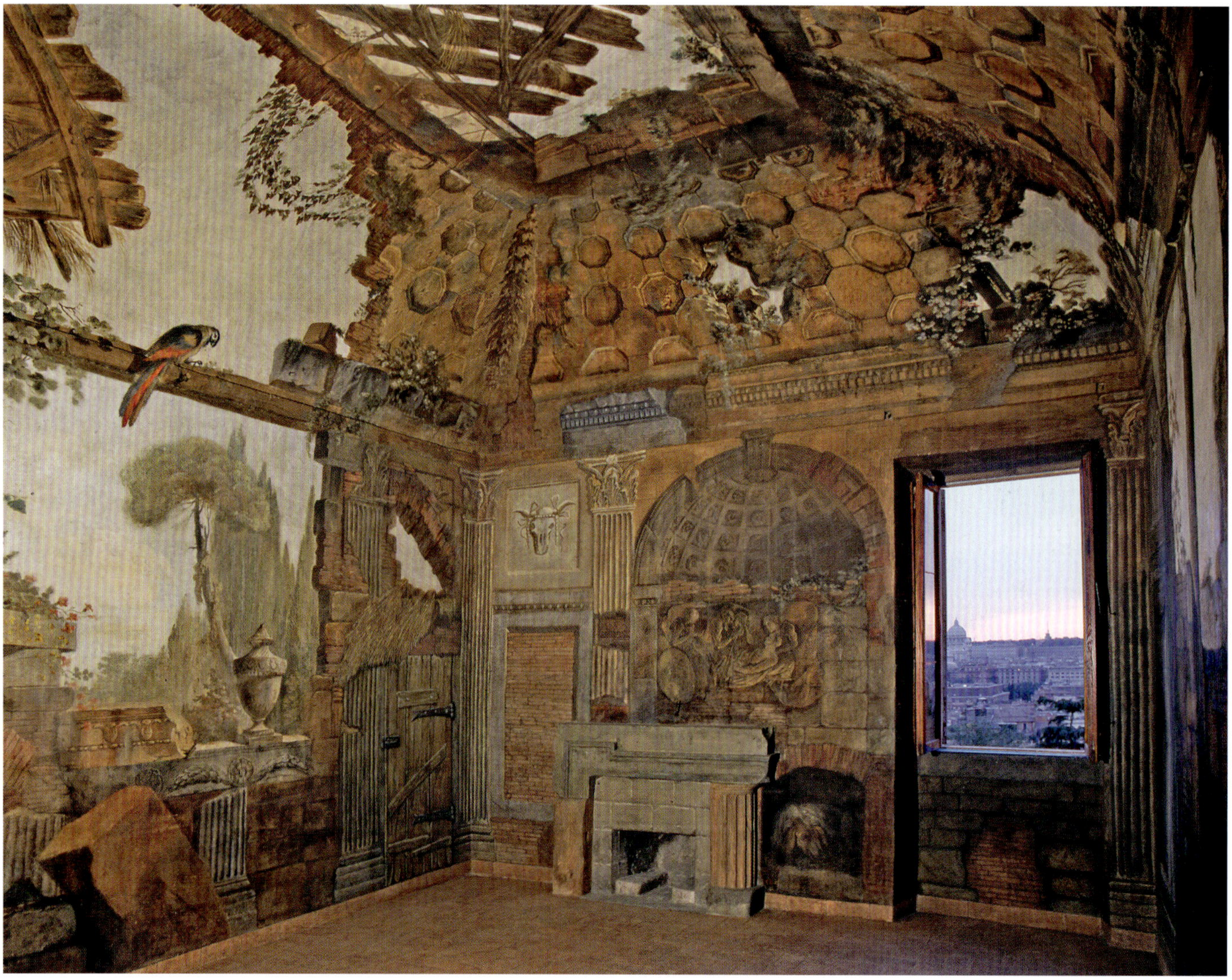

Fig. 87. Charles-Louis Clérisseau, *Stanza delle rovine*, c. 1767. Fresco. Dormitory, convent of Trinità dei Monti, Rome.

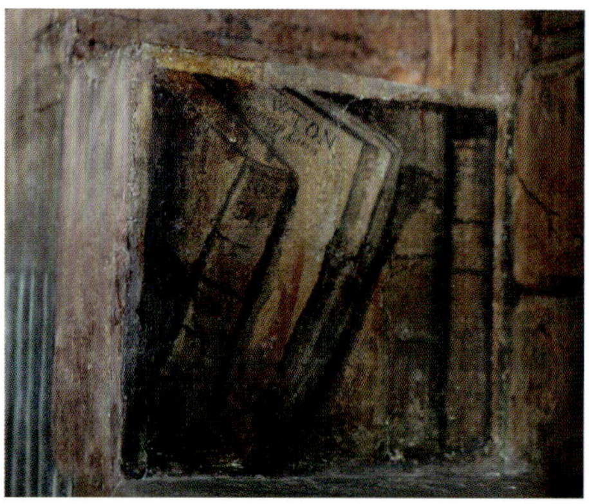

Fig. 88. Charles-Louis Clérisseau, *Stanza delle rovine* (detail), c. 1767. Fresco. Dormitory, convent of Trinità dei Monti, Rome.

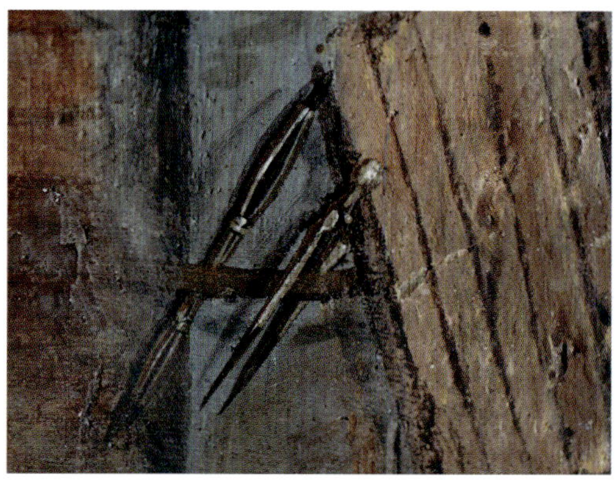

Fig. 89. Charles-Louis Clérisseau, *Stanza delle rovine* (detail), c. 1767. Fresco. Dormitory, convent of Trinità dei Monti, Rome.

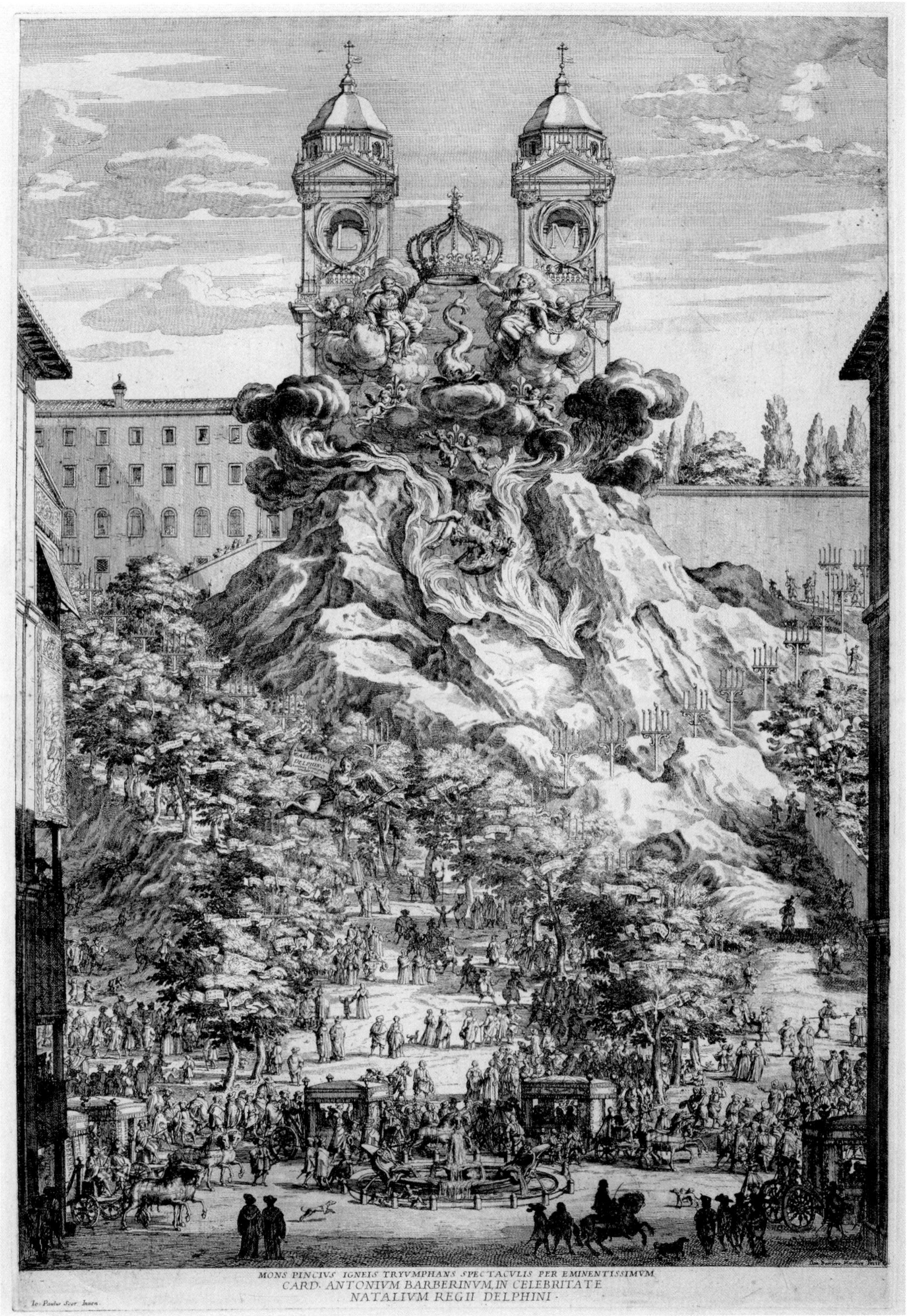

Fig. 90. Dominique Barrière (etcher), illusionistic apparatus on the hill in front of Trinità dei Monti, 1662. Etching printed on paper, 36.34 × 25.31 in (92.3 × 64.3 cm). Museo di Roma.

Using Science to See God's Creation

The Minims who were interested in exploring the operations of optics—how and why the eye sees what it sees and what tricks one might play to deform or prolong that process—were also interested in other scientific explorations, including into the operations of the universe and how time is measured through space. If their work in visual art asked humans to recognize and accept that human perception was partial and that God's miraculous workings might not always be immediately accessible to the human eye, their work in natural science looked for ways to augment human vision and observational skill—in other words, to look more closely at God's creation. Whether they were making anamorphoses or making sundials, they used the tools of optics and other natural sciences to better see—and better understand—God's created world. In the preface to his *La perspective curieuse*, Niceron stresses, moreover, that optics provides great benefits to multiple sciences from astronomy, which "calls to it for help, for precise measurement of the position and movement of the planets, by the means of the astrolabe and other instruments which follow the line of sight," to natural philosophy, which "verifies most of its experiments by its means."[65] I have mentioned the giant sundial, designed by Maignan for the Trinità dei Monti convent—it connects to the galleries with the anamorphic wall frescoes, in the corridor on the right side of the engraving in the *Perspectiva horaria* (see Figs. 69 and 70). Like the sundial Maignan designed for Bernardino Spada's palace, the sundial at the Minim convent offers a surplus of horological information. Further information would have been provided by a second sundial, discovered and partially restored in 2006–2007, which is located on the second level of the convent in the western gallery.[66] Whereas Maignan and Niceron celebrated such intellectual endeavors, other members of their order discouraged the pursuit of knowledge and study as distractions that prohibited friars from approaching the divine.[67]

Some Minims' investment in naturalism—like deformation, a mode in which information might be conveyed visually—can similarly be seen in the work of Charles Plumier. The Minim friar, who studied turning under Maignan (and also published on that craft), devoted his life to disseminating information on natural history in richly illustrated books and manifold drawings produced in France and during voyages to the French Antilles in 1687, 1689, and 1694.[68] On his death in 1704, Plumier left behind handwritten notes, sketches, drawings, watercolors, etchings, and engravings in his cell in the Place Royale convent in Paris.[69] About 6,000 drawings survive today: some 4,300 portray his botanical studies, 1,500 feature his zoological studies of animals, including fish, lizards, mollusks, and birds, and around 200 show turned objects, saddles, figures, and other entities.[70] Comments from the English naturalist Martin Lister (1639–1712), who visited Plumier in the Minim Place Royale convent in Paris in 1698 after his third trip to the Caribbean, underscore just how miraculous it is that so many of the friar's papers survived: "He was more than once Shipwrackt, and lost his *Specimens* of all things, but preserved his Papers, as having fortunately lodged them in other Vessels."[71]

Plumier's first botanical publication, *Description des plantes de l'Amérique* (*Description of the Plants of America*, 1693), appeared in a folio edition with 108 plates, a high number that points to the centrality of looking in Plumier's endeavors. A manuscript volume in the Muséum national d'Histoire naturelle in Paris features seventeen oil paintings on vellum by Plumier, some of these apparently preparatory designs for the 1693 book.[72] The plant on folio twelve of

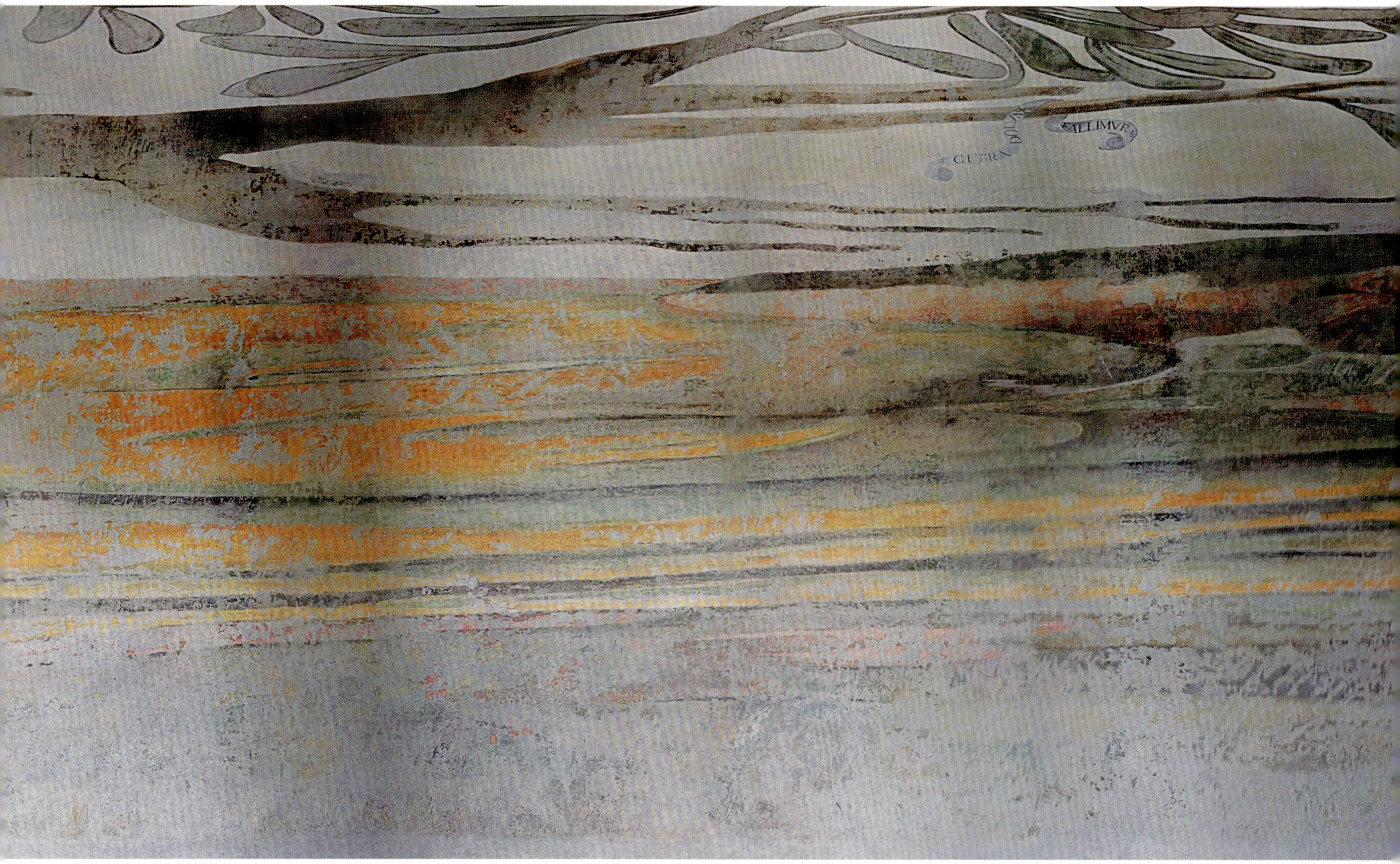

Gatefold 1. Jean-François Niceron, *Saint John the Evangelist and Landscape*, 1639–1640. Anamorphic fresco *a secco*, 8 ft 6.36 in × 57 ft 6.5 in (2.60 × 17.54 m). Corridor, convent of Trinità dei Monti, Rome.

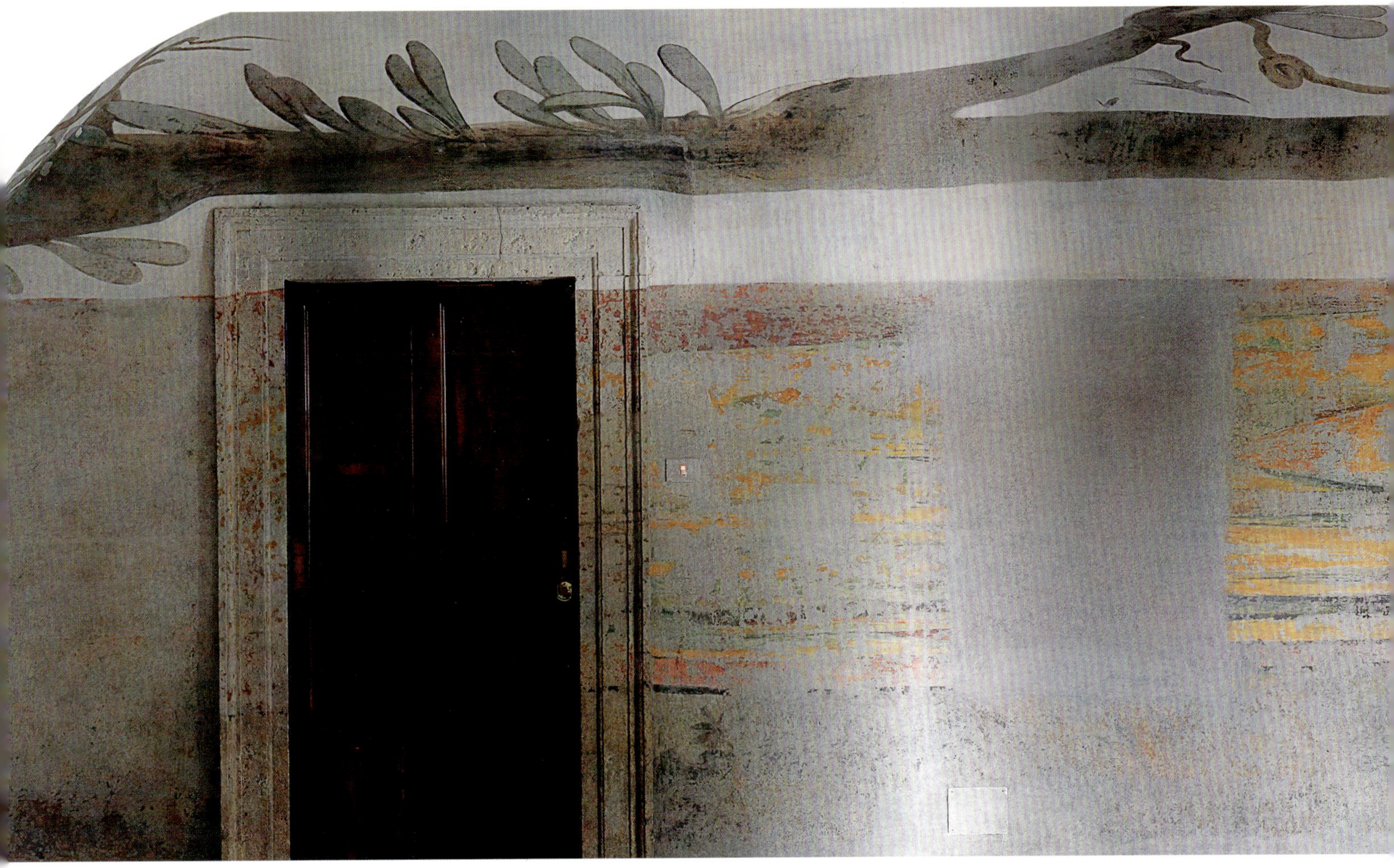

SAINT JOHN THE EVANGELIST

AND LANDSCAPE

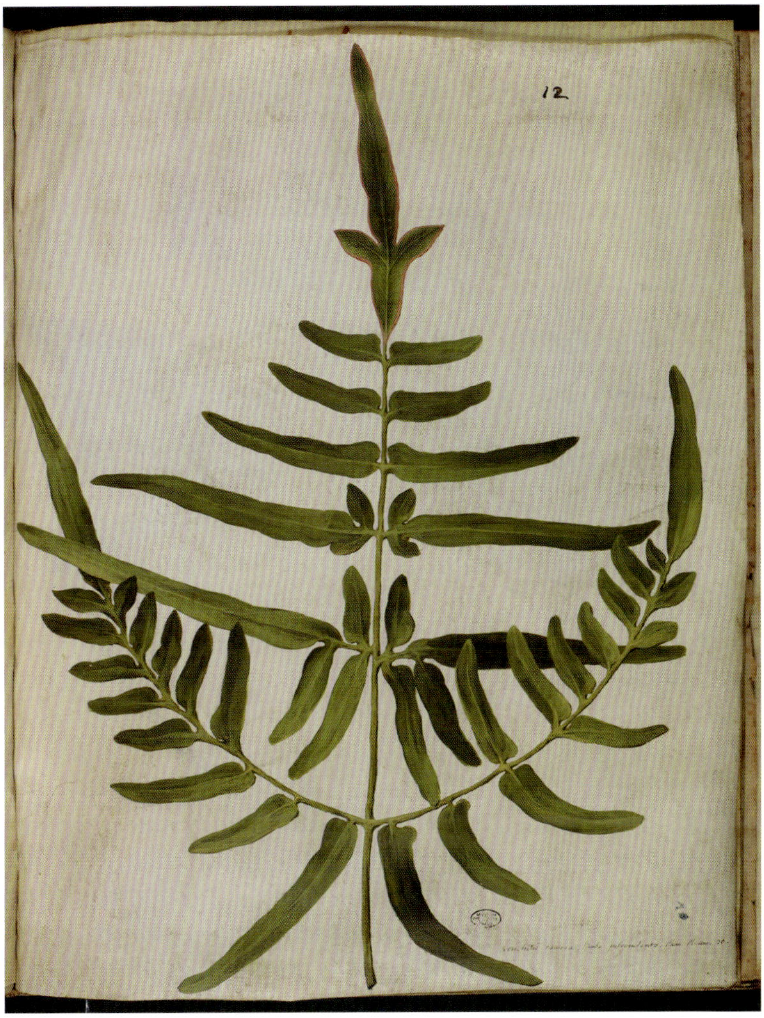

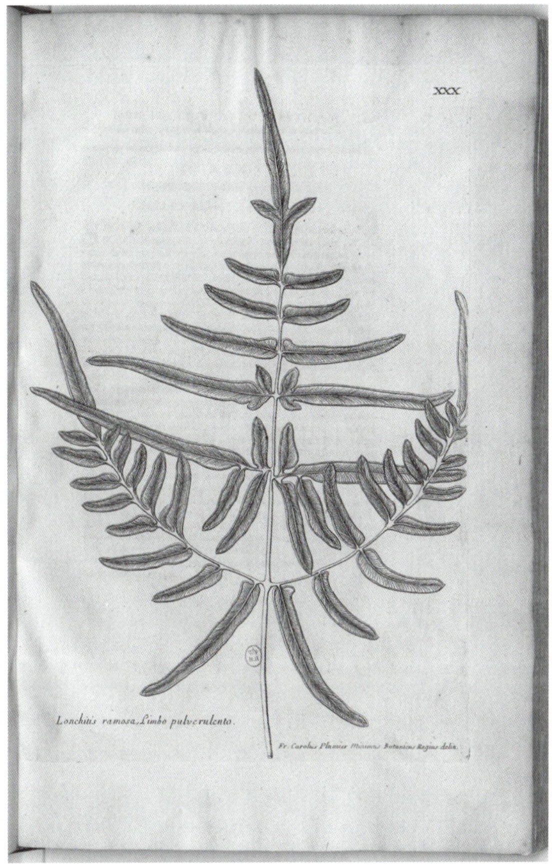

the manuscript, for instance, closely resembles that in plate thirty of the *Description des plantes*; both images represent a fern, whose leaves are superimposed in the same way, which Plumier calls the *Lonchitis ramosa, limbo pulverulento,* also known as the *Pteris mutilata* L. (Figs. 91 and 92).[73] In the colored manuscript painting, Plumier presents the most detail on the plant's highest leaf, outlined in a light brown and displaying individual veins. He carefully differentiates among darker and lighter shades of green as well, to convey the leaf's texture and to show how light and shadows might be distributed along its surface. This leaf seems to serve as an abbreviated visual record of what all the leaves looked like, in a kind of pictorial shorthand. Assuming the unidentified engraver of plate thirty in the *Description des plantes* used the manuscript painting as a guide, it shows a clear understanding of Plumier's intent: the veins of all the leaves—not just the top one—have been carefully rendered in the printed design (see Fig. 92). In an annotation on the engraving, Plumier characterizes himself as the designer and "Botanist of the King" (*Botanicus Regius*). Natural imagery abounds as well in the details of Niceron's

Fig. 91. Charles Plumier, *Lonchitis ramosa, limbo pulverulento.* Oil painting on vellum, 12.56 × 9.76 in (31.9 × 24.8 cm). Muséum national d'Histoire naturelle, Direction générale des collections, Bibliothèque centrale, Paris [ms. Res 37, fol. 12].

Fig. 92. Charles Plumier (designer), *Lonchitis ramosa, limbo pulverulento* in *Description des plantes*, 1693, plate 30. Engraving printed on paper, 16.42 × 10.67 in (41.7 × 27.1 cm). BnF, département Sciences et techniques, Paris [S-860].

Fig. 93. *Saint Francis of Paola and Landscape* (detail), 1642. Corridor, convent of Trinità dei Monti, Rome.

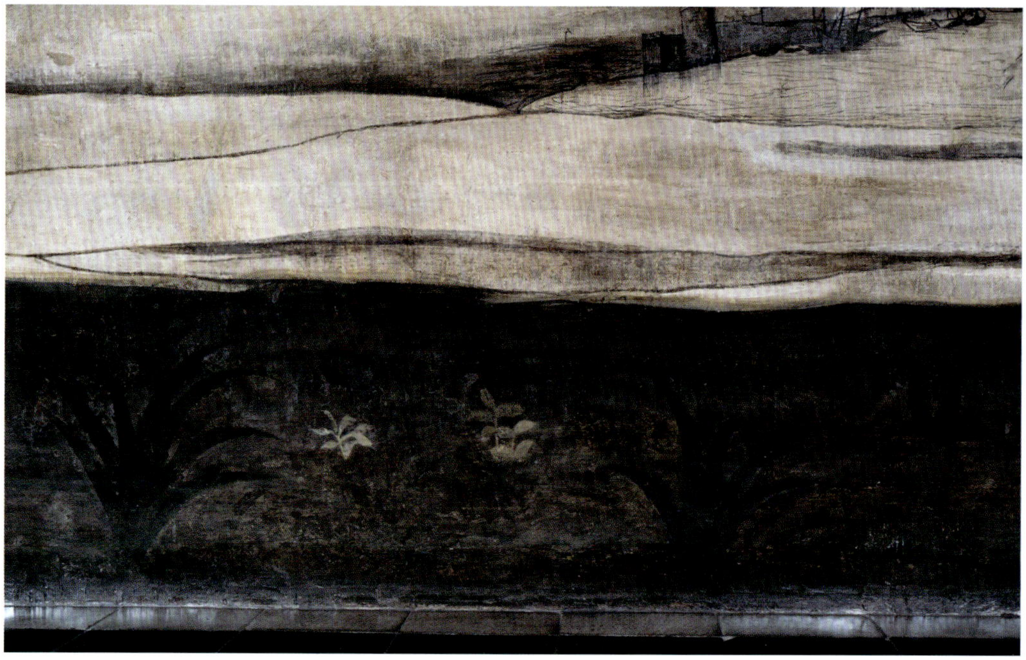

and Maignan's anamorphic wall frescoes: ferns and flowers sprout from the earth, snails and insects crawl up a tree trunk, an owl peers down from a branch, birds fly between branches or perch on a branch while a snake slithers by (Figs. 93, 94, 95, 96, and 97).

Similarly characteristic of some Minims' commitment, within the natural sciences, to augment human visual perception and consequently their understanding of the marvels of divine creation is an often-cited passage from a diary entry by the French magistrate Olivier Lefèvre d'Ormesson (1616–1686), recording a discussion with Niceron in the Place Royale monastery in Paris in 1645 around an image that conveyed information about the Moon: "Sunday, November 26 … at the Minims, where the Father Niceron shows me a map made in Brussels of a world inside the Moon, in which all the lands, mountains, islands, were named after kings and great figures from this century, and the map was very similar to that of the queen and that of Mazarin."[74] P.J.S. Whitmore offers the compelling suggestion that the map in question is the *Plenilunii lumina austriaca philippica* (*The Light of the Full Moon, for Philip the Austrian*, 1645), designed by Michel-Florent van Langren (1598–1675) (Fig. 98).[75] Niceron's presentation of selenography that displays "a world inside the Moon" reflects the sustained impact of Galileo's *Sidereus Nuncius* (*Starry Messenger*, 1610) thirty-five years after its publication. As is often noted, when Galileo looked at the surface of the Moon using a telescope—human perception augmented through the science of optics—he saw the rough surfaces of the Earth, as opposed to the perfectly smooth surface of Aristotelian-Ptolemaic cosmology. Galileo writes that the Moon's surface is "not smooth … but, on the contrary, uneven, rough, and crowded with depressions and bulges. And it is like the face of the Earth itself, which is marked here and there with chains of mountains and depths of valleys."[76] The map that Niceron showed d'Ormesson built on Galileo's telescopic findings concerning the Moon's imperfections.

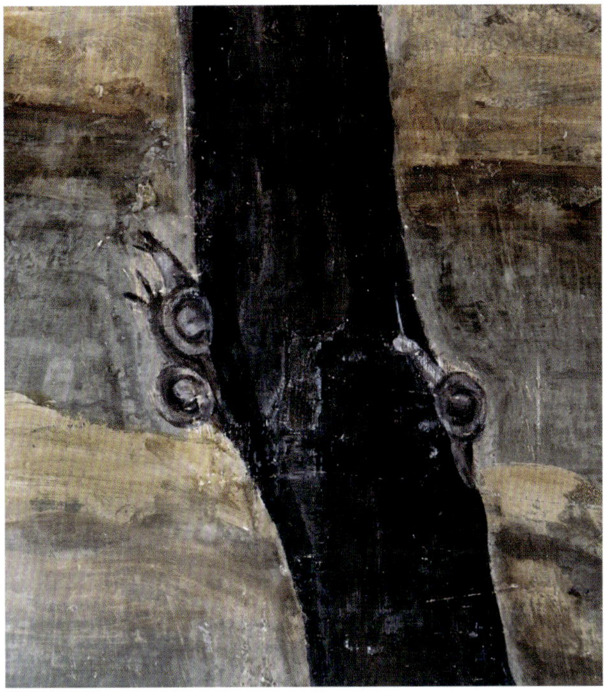

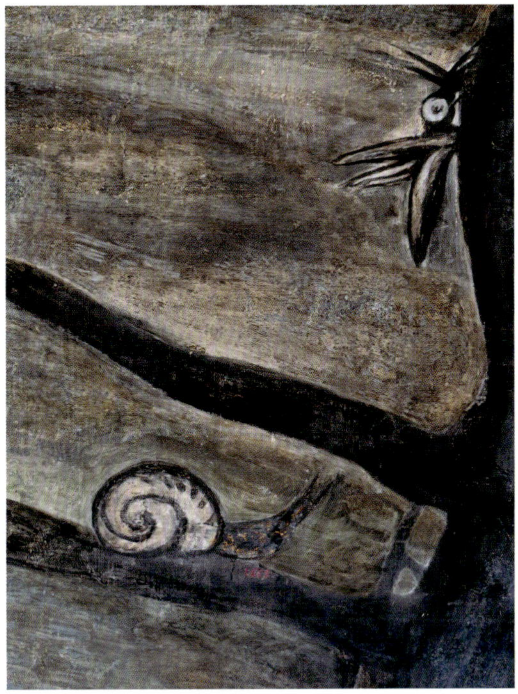

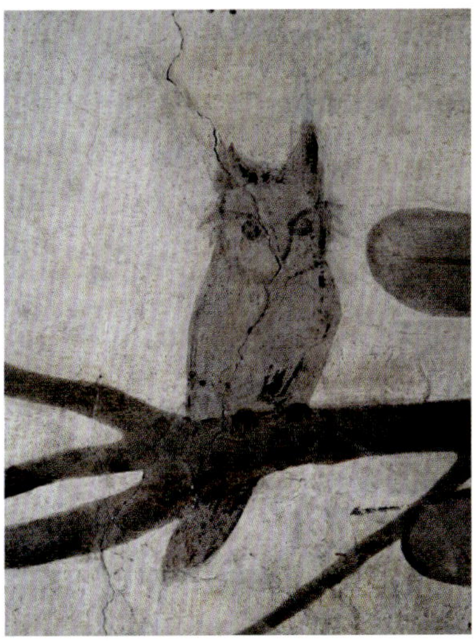

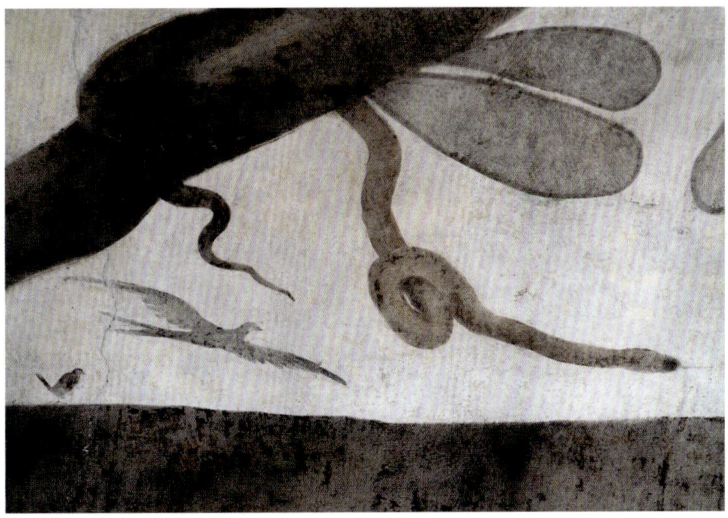

TOP LEFT

Fig. 94. *Saint Francis of Paola and Landscape* (detail), 1642. Corridor, convent of Trinità dei Monti, Rome.

TOP RIGHT

Fig. 95. *Saint Francis of Paola and Landscape* (detail), 1642. Corridor, convent of Trinità dei Monti, Rome.

BOTTOM LEFT

Fig. 96. *Saint Francis of Paola and Landscape* (detail), 1642. Corridor, convent of Trinità dei Monti, Rome.

BOTTOM RIGHT

Fig. 97. *Saint John the Evangelist and Landscape* (detail), 1639–1640. Corridor, convent of Trinità dei Monti, Rome.

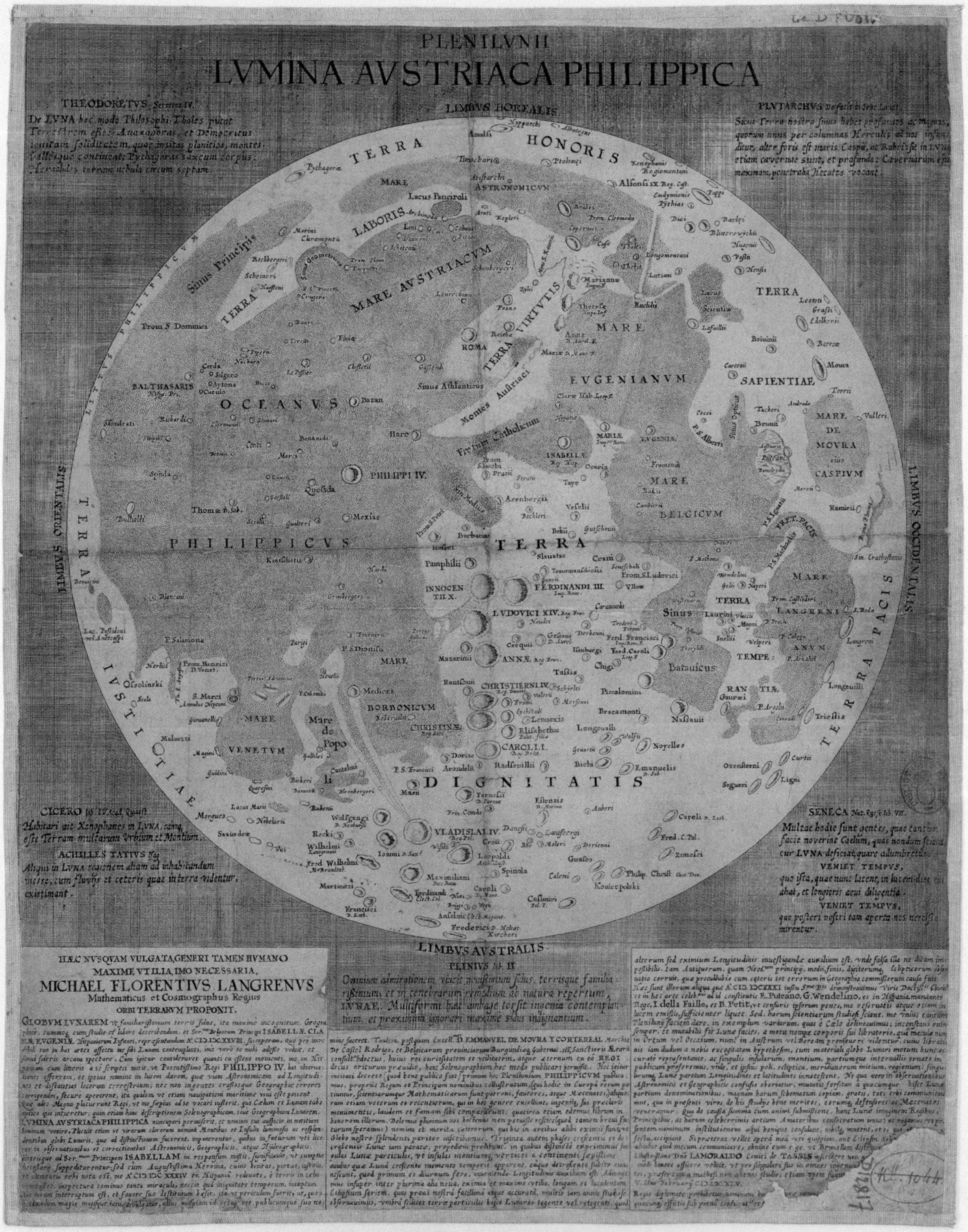
PLENILVNII
LVMINA AVSTRIACA PHILIPPICA
LIMBVS BOREALIS
TERRA HONORIS
MARE AVSTRIACVM
OCEANVS PHILIPPICVS
MARE EVGENIANVM
SAPIENTIAE
MARE BELGICVM
TERRA
DIGNITATIS
MARE VENETVM
TERRA PACIS
LIMBVS ORIENTALIS
LIMBVS OCCIDENTALIS
LIMBVS AVSTRALIS
HÆC NVSQVAM VVLGATA GENERI TAMEN HVMANO
MAXIME VTILIA, IMO NECESSARIA,
MICHAEL FLORENTIVS LANGRENVS
Mathematicus et Cosmographus Regius
ORBI TERRARVM PROPONIT.

+

The Minim visual deformations we have considered in this chapter fall into two broad categories of visual work. One category of works used anamorphoses as a metaphor for divine revelation. This mode can be understood to build upon the traditional iconography in the life of Saint Francis, the numerous naturalistic images that show him healing blind figures. While prolonging observation and inviting repeated viewing, anamorphoses excited vertiginous observational experiences that reminded the observer that God controls many aspects of humans' visual capacity, to the point that he can introduce into their perception something that others around them might say is "not really there." These works led observers to think about humans' tendency to be deceived, and our related need to rely on divine intervention. The second category of visual work we might understand as more subject-specific: deformations that focused observers' attention on particular messages, from historical arguments concerning the primacy of the Catholic Church (as expressed through Saint Peter with the keys to heaven) to matters of belief (like the Resurrection). Like Bernardino Spada, Minims valued deformations for their power to control what people saw as they moved through space along predetermined pathways—to shape their thoughts and model understandings around humans' limited capacity to comprehend visual input.

OPPOSITE

Fig. 98. Michel-Florent van Langren, *Plenilunii lumina austriaca philippica*, 1645. Engraving printed on paper, 19.29 × 14.96 in (49 × 38 cm). BnF, département des Cartes et plans, Paris [GE D-12817].

Sanatur Cæcus natus.

Ioan. ix. Anno xxxij.

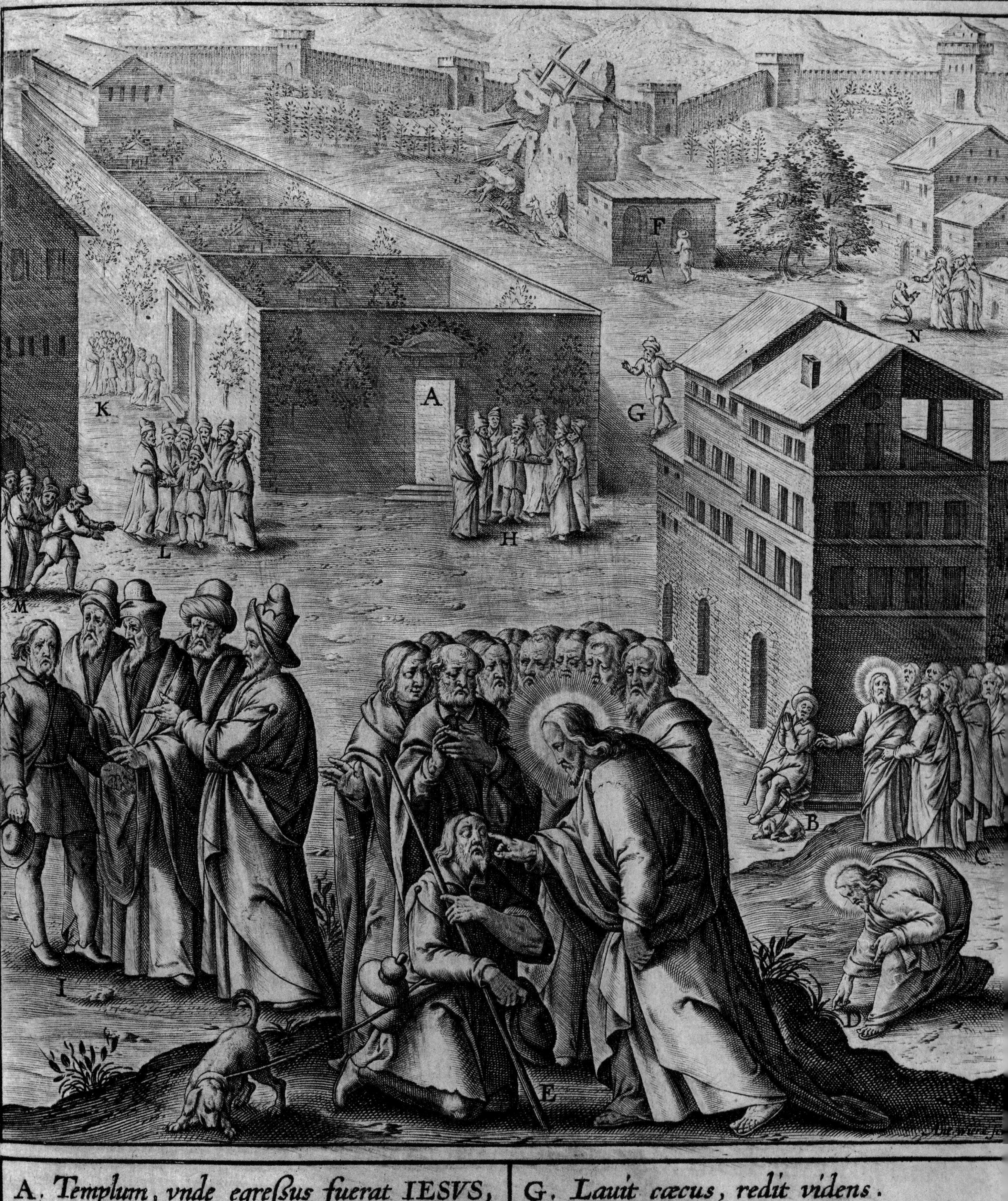

A. Templum, vnde egreſſus fuerat IESVS,

G. Lauit cæcus, redit videns.

3 JESUIT DEFORMATION AND MEDITATION

If we look at our work immediately after completing it, we are still too involved; if too long afterward, we cannot pick up the thread again.

It is like looking at pictures that are too near or too far away. There is just one indivisible point, which is the right place.

Others are too near, too far, too high, or too low. In painting the rules of perspective decide it, but how will it be decided when it comes to truth and morality?[1]

—Blaise Pascal, *Pensées*

IF THE MINIMS' visual art tradition draws observers' focus to the limits of human visual perception and to God's capacity to offer visual clarity both literally and in the more abstract sense of understanding, the Jesuits' intense interest in disciplined meditation suggests that human perception (and hence understanding) is limited not simply by human nature but also by human behavior. Jesuit visual culture should be viewed in relation to their practical guidance on meditation, which often relied on images from the life of Christ—and, beyond those, anamorphoses that played with perspective and needed to be seen from "just one indivisible point, which is the right place"—provided as meditative tools. We might recognize in Jesuits' commitment to anamorphoses that these deformations could take observers a crucial step beyond what they could accomplish with other images they deployed, including naturalistic meditative prints that invited observers on a spiritual journey through imagined spaces. Jesuits, like Minims, used anamorphosis as a metaphor for revelation (and a tool for teaching religious content), but they also deployed the genre as a means of reinforcing an image more securely in the mind—in other words, as a tool to facilitate meditation. Jesuits championed the anamorphic conjunction of confusion and clarity because these images offered discipline for the naturally

Detail of fig. 100

inattentive mind and directed it toward spiritual understanding that supported their mission "to strive especially for the progress of souls in Christian life and doctrine and the propagation of the faith."[2] Moreover, much like Minims, they saw deformations as complementing their studies in theology, mathematics, and natural philosophy.

Jesuit Spirituality

The Society of Jesus grew out of a group of companions formed in Paris in 1534, papally ratified in Rome in 1540. To understand their particular interest in deformations—images that urge observers to move their bodies through space to gain understanding—we can turn to the Jesuits' interpretation of rhetoric and their appreciation for the common early modern visualization of education as a journey. Jesuits shared Minims' interest in the ability of anamorphoses to represent a revelation of God's truth but also evinced an acute commitment to working at meditative visualizations: the Society embraced the power of verbal, visual, and virtual images and acts of image-making to clarify, strengthen, and interpret religious experiences.[3] They held that by contemplating art and architecture, and by creating and contemplating mental images, the faithful could hone their visual understanding in relation to divine reality. The early modern Jesuit model of spirituality was one of training, learning, meditation, and practice, including practice centered on movement between visual and mental prompts: members of the Society relied on visual materials among other resources to fulfill their mission of teaching people how to glimpse a divine reality and to associate higher level thinking—religious knowledge, understanding, or revelation—with the everyday.

Ignatius's treatise *Exercitia spiritualia* (*Spiritual Exercises*, 1548) presents a method for spiritual direction and meditation, to be undertaken one-on-one with a Jesuit confessor or in a retreat setting; while the text does not discuss anamorphosis, it suggests how early modern Jesuits might have understood and promulgated it as an artistic technique. The work functions as a kind of spiritual manual for religious leaders who might guide others through a month of prayer and meditation on sin, the life of Christ, the Passion, and the Resurrection; it also contains rules concerning the discernment of one's own spiritual movements, rules meant "for understanding the different movements produced in the soul."[4] Ignatian meditations, as Ignatius describes in the prelude to one exercise, ask the person engaged in following the course to imagine a religious image into a concrete physical setting:

> *The First Prelude* is a composition made by imagining the place. Here we should take notice of the following. When a contemplation or meditation is about something that can be gazed on, for example, a contemplation of Christ our Lord, who is visible, the composition will be to see in imagination the physical place where that which I want to contemplate is taking place. By physical place I mean, for instance, a temple or a mountain where Jesus Christ or our Lady happens to be, in accordance with the topic I desire to contemplate.[5]

Ignatius conceives of meditation as a pictorial practice. Just as the visual artist represents scenes, the meditator must visualize imagined ones through the method known as "composition of place."

Although there were other pathways to the meditative life, a number of Jesuit treatises devised to guide meditative prayer followed Ignatius in explicitly thematizing prayer through the creation of mental images. In a chapter from his *Via vitae aeternae* concerning the method of meditating, the Jesuit Antonius Sucquet explains the utility of images. Because an "unbridled" imagination is perceived as a spiritual risk—it's hard to "stick to pious thoughts"—Jesuits in general practice seek to help people discipline their attention (*attentio*) via visual imaging. The mandate is to picture something holy, in a specific setting. He writes:

> These [images] will be able to help everyone, since the mind is arid and dull; but it will especially help simpler folk to concentrate their attention and to meditate more usefully. [A]fter the sin of Adam [the imagination] became wholly unbridled; the unstable mind was made to wander over the whole course of the earth, and was not allowed to stick to pious thoughts. In order to restrain it, it seemed appropriate to fix it, as it were, by means of pious cogitations and by images. The blessed Ignatius approved of just this when he wrote to Nadal that he should publish the Life of Christ represented by beautiful images, for the specific convenience of the meditator. Experience teaches us how true this is; and the Holy Church, which makes considerable use of images, confirms this by its authority.[6]

Sucquet's account of a wandering attention that impedes meditation—a distracted mind that needs to be restrained and fixed—recalls the vice that medieval theologians and monastic authors, such as Bernard of Clairvaux (1090–1153), referred to as *curiositas* (the penchant to be distracted) as well as the monastic concept of *evagatio* ("aimless wandering").[7] Augustine (354–430) describes curiosity as the *concupiscentia oculorum* ("the concupiscence of the eyes") (1 John 2:16).[8] According to Sucquet, the postlapsarian mind and imagination are unfocused and confused. The wanton, aimless, and disordered minds and imaginations of humans require actual and virtual images to capture their attention, to lead them to spiritual comprehension. This concern with fixing attention clarifies why Jesuits and other members of the Catholic Reformation Church embraced images as powerful vehicles to incite observant care and then to persuade and reform confused souls: focusing on a visual stimulus disciplines the wandering attention. Anamorphoses, then, as memorable images that concentrate and direct visual attention, supported the work of meditation.

Jesuit authors furthered the process of directing attention by providing printed images for the use of meditating individuals, especially images from the life of Christ that might be studied and ultimately interiorized for meditative use. As Sucquet observes, Ignatius urged the Jesuit Jerome Nadal to "[represent] by beautiful images" the life of Christ, which Nadal would do in his meditative treatise *Adnotationes et meditationes in Evangelia* (*Annotations and Meditations on the Gospels*, 1595).[9] The first part of Nadal's celebrated book provides an extensive discussion of the Gospels, along with meditations that transmute the episodes from the life of Christ into events to be contemplated; these events can also, at times, work as metaphors for various modes of meditation. The work's second part, titled *Evangelicae historiae imagines* (*Pictures of the Gospel Stories*) and in fact published in 1593—two years before the inaugural edition of the *Adnotationes et meditationes*—displays 153 engravings that variously order, support, and augment the votary's spiritual exercises.[10] The folio-size engravings, which were mainly designed by Bernardino Passeri (c. 1540–1596) and Marten de Vos (1532–1603) and mostly cut by Jan (1549–c. 1620), Hieronymus (1553–1619), and Antonius Wierix (c. 1555/59–1604) of Antwerp, show key moments in Christ's life. They constitute the most intricate and subtle print series for which the Jesuits were patrons.

The miraculous healings of blindness that feature in lives of Christ—and in the *Adnotationes et meditationes* in particular—help contextualize this trope as it appears in the lives of Saint Francis of Paola discussed in the previous chapter. Not unlike the Minims, the Jesuits turned to narrative elements around blindness (and miraculous cures for blindness) to draw attention to the delay between necessarily limited human perception and the potential for understanding through divine revelation. If it seems the Minims were mostly focused on making people aware of their perceptive limits as opposed to God's unlimited capacity, the Jesuits were also actively interested in disciplining individuals' attention. The image for chapter 19 of the *Adnotationes et meditationes*, for example, illustrates Christ curing blind men from Jericho, as narrated in Luke 18, Matthew 20, and Mark 10 (Fig. 99).[11] In this and other prints from Nadal's work, meditative itineraries are charted onto wide-ranging land- and cityscapes annotated with capital letters that link to inscriptions below the image. By attaching details to the locations through which sacred figures travel, these and the other annotations (some of which are printed in the first part of the book) build upon the Ignatian notion of "composition made by imagining the place": the "beautiful images" of events, land- and cityscapes in Christ's life offer to the meditator models for his own visualization. The images are designed to be internalized as the physical space(s) in which the meditator should situate the memorable spiritual image. In the background of the image for chapter 19, the walled city of Jericho lies at "A"; Christ appears to the left and a bit closer to the foreground. Crowds both precede (at "B") and follow Him (at "C"). In the area nearest to the observer, Christ cures a blind man, who kneels in an expression of his faith. An annotation under the captions specifies that the print could symbolize not only the curing of the beggar prior to Christ's entry into Jericho but also another event described in the Gospels: the curing of two blind men following His departure from Jericho.

The accompanying meditation opens with an exhortation to the votary—evidence that the Jesuits believed one could actively discipline perception—to attend to the blind man, so as to gain his capacity for meditative prayer: "Come, all you whom the light of prayer and the knowledge of spiritual things have forsaken; hearken to the blind evangelical; let the eyes of your heart be opened."[12] The votary is then given the following language for how to respond to the subsequent shift from blindness to understanding: "Lord, I truly hear You, sense Your mysteries, but do not see You, am lacking celestial light; o infinite light, I beseech You. [And now] I see, good Jesus; looking upon the joyful ray of Your light, I will follow You and praise You forever."[13] The meditation and image lead the votary not just to observe and imagine the blind men regaining their sight but also to personally gain understanding of their own spiritual movement through an analogous visual progression from obscurity to clarity.

The engraving for chapter 59 similarly encourages the viewer / reader to enact the journey from their own metaphorical blindness to sight and understanding. This image portrays another miraculous healing of blindness, this one based on an episode in John 9 (Fig. 100).[14] Captions again guide the votary through the print. The letter "A" in the background by the open

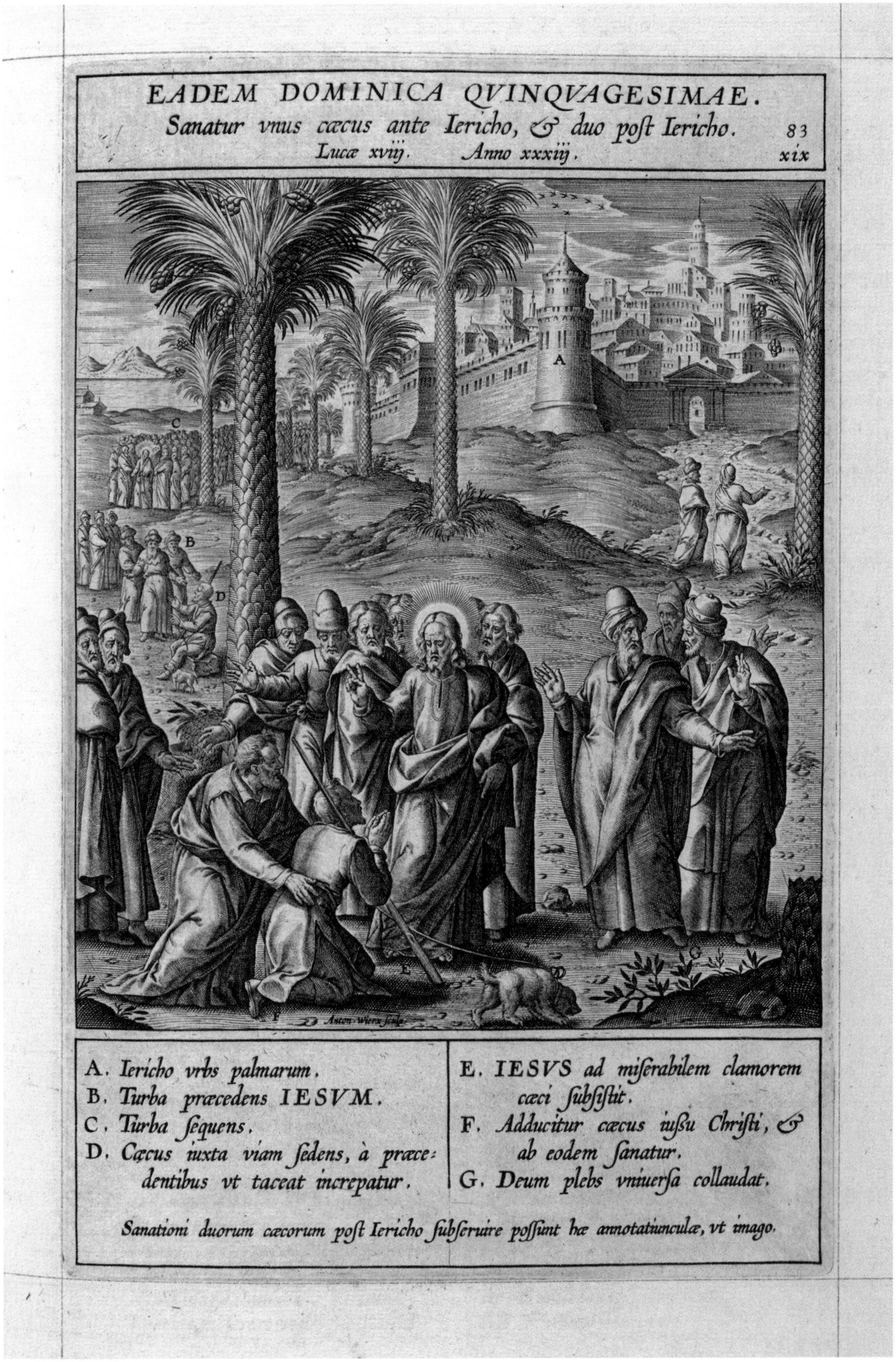

Fig. 99. Antonius Wierix after Bernardino Passeri, *A Blind Man Is Healed before Jericho, and Two Others beyond It* from Nadal's *Adnotationes et meditationes*, 1595. Engraving printed on paper, 9 × 5.5 in (23.5 × 14.9 cm). The Huntington Library, San Marino, California.

door to the temple of Jerusalem marks the first step of the meditative itinerary. Moving closer toward the foreground is the second stop at "B," where a blind supplicant sits outside a building. Behind him stands Christ; the apostles, at "C," inquire if the man's blindness was a penalty for sin. Christ corrects this misapprehension. Moving further into the foreground at "D," Christ leans forward and spits on the dirt to produce some clay. At "E" He applies the clay to the blind man's eyes and instructs him to wash in the pool of Siloam. After following His instructions at "F," the man regains his sight and returns to his neighbors at "G." At "H," his community fails to recognize him, even though the meditator contemplating the image can still see the cured man, and he is led at "I" to the Pharisees, who are divided over whether to believe his story. The next step on this circuitous meditative itinerary occurs at "K," in a moment that is difficult to decipher, as it is engraved without much detail. Annotations reveal that here the cured man appeals to his parents. The parents do not substantiate Christ's role in the miracle, and at "L" the man is called to the Pharisees again; he proclaims his faith in Jesus, and as a result, at "M," he is kicked out of the synagogue. In the last stop on the meditative itinerary, at "N," Jesus discloses Himself as the Son of God to the kneeling cured man. In stark contrast to the details near the foreground, this episode of revelation is again engraved without much detail, and therefore is difficult to discern. The obscurity in this detail of revelation is not gratuitous and does not merely signal that the event occurs in the distant background.[15] Here, too, the subject matter and form work in tandem to lead observers to reflect on the difficulty of spiritual understanding: only the man born blind has enough faith to see Christ; even the print's observer, who must strain his eyes to search for the details described in the annotations, is lacking in interpretive capacity. The obscurity may well have motivated observers to realize their own blind spots in their journeys toward faith and invited them to correct their vision, including via the exercises in meditation recommended by the Jesuits.

If the engraving above was obscure—it did not immediately satisfy the viewer by delivering clarity, as might be preferred under some understandings of classic rhetoric—we can see it as nevertheless rhetorical in being persuasive. To better understand the function of the images in the *Via vitae aeternae*, the *Veridicus Christianus*, or the *Evangelicae historiae imagines* (and in other meditative works, such as the lives of Saint Francis of Paola) we can again turn to rhetoric, particularly the works of Aristotle, whose influence still dominated Jesuit philosophy in the seventeenth century—and more specifically to the three modes of persuasion and the notion of *enargeia*, both discussed in the *Rhetoric* (1356a1–20 and 1411b24ff).[16] First, Aristotle writes, an orator persuades by character, that is, through their standing, wisdom, and discrimination, such that listeners are made to think that the speaker is credible. An image, then, could persuade by emphasizing a subject's moral and mental qualities, establishing the subject's standing as a credible source of insight. Because the prints in the *Evangelicae historiae imagines* present Christ, they would have been persuasive in this sense to early modern exercitants. Second, Aristotle argues, a public speaker could persuade by arousing the emotions of listeners: by directing emotion, the orator might move listeners into a frame of mind conducive to receiving the arguments as correct. In this vein, a lavish illusionistic wall painting could have persuaded by arousing feelings of wonder at the spectacle, which, as we have seen above in chapter 1, might, in turn, convince the viewer to value the work's patron or, perhaps, the patron's views—or, by extension, to increase respect for the Catholic Church as patron,

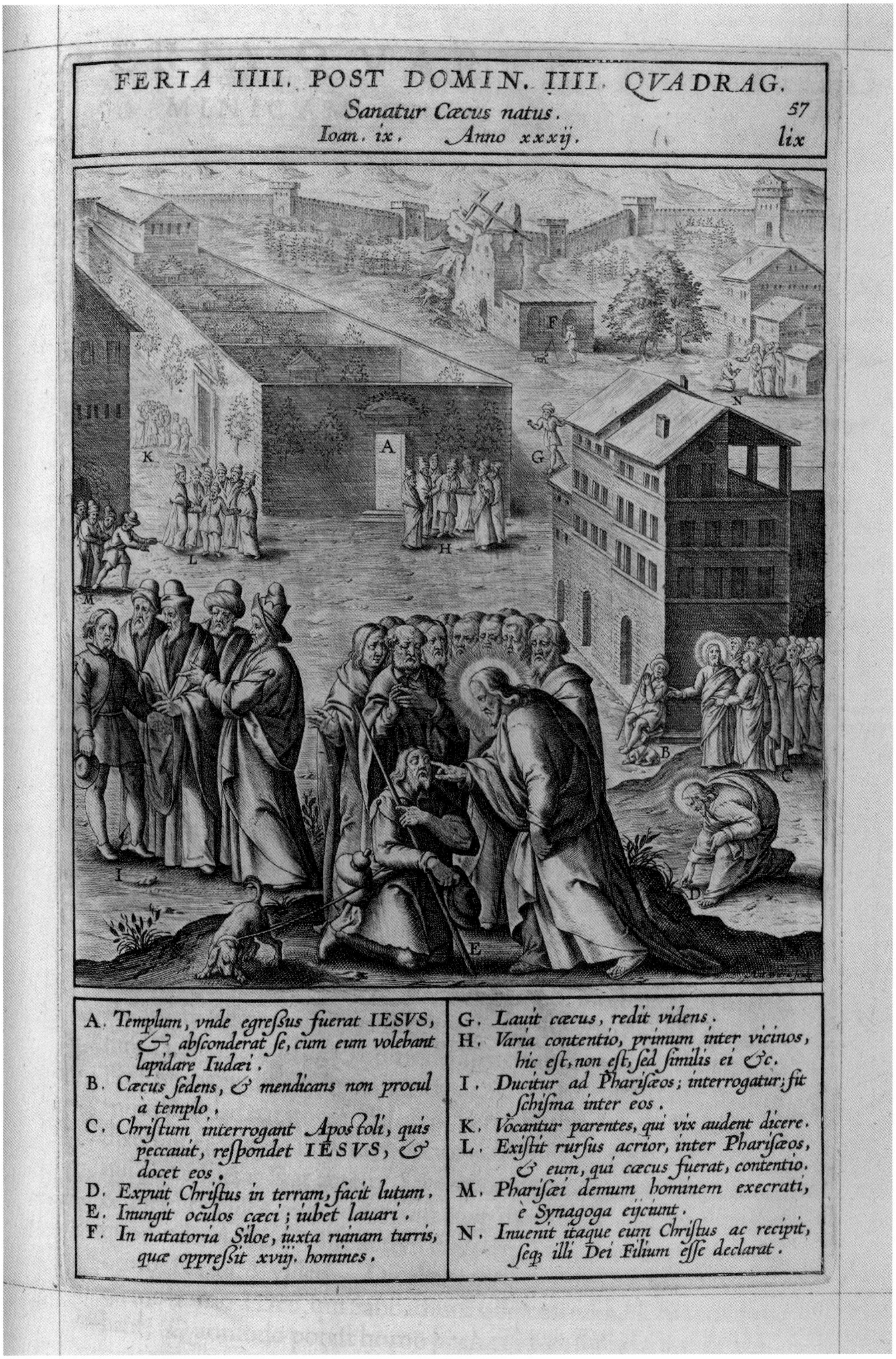

Fig. 100. Antonius Wierix after Bernardino Passeri, *Healing of the Man Born Blind* from Jerome Nadal's *Adnotationes et meditationes*, 1595. Engraving printed on paper, 9 × 5.5 in (23.3 × 14.7 cm). The Huntington Library, San Marino, California.

increase devotion toward objects of veneration, or reinforce conviction. Third, Aristotle claims, an orator could persuade through an argument that proves or appears to prove that something is true—in other words, by logic. This method of persuasion could draw on deductions or on similar cases brought forward as examples, analogies, or allusions. The Jesuit-run Seminario Romano (Roman Seminary), a boarding school for elite male students established in 1564, could be said to have deployed this third method when it adopted its emblem: a bundle of delicate laurel sprouts emerging from the soil of an enclosed garden, accompanied by the motto *Pomis sua nomina servant* ("They preserve their fame with their fruit"). The image persuades observers that just as the sprouts would develop into admirable trees, so the young seminarians would grow into honorable men.[17]

If Jesuit visual culture was rhetorical in these Aristotelian senses—persuading via its elevated subject, its emotional force, or by logic—its power ultimately lay in its vividness. The Greek term *enargeia*, from *argos* ("shining light"), denotes vividness, clearness, distinctness, or liveliness.[18] According to Aristotle, vividness is an especially convincing stylistic tactic when presenting a topic to an audience, and he recommends it in specifically visual terms: "making our hearers see things."[19] Quintilian proposes that this vividness leads audiences to feel they are present at the event the speaker discusses: "It is a great virtue to express our subject clearly and in such a way that it seems to be actually seen."[20] Notwithstanding the obscurity of certain details, the images in the *Via vitae aeternae*, the *Veridicus Christianus*, the *Evangelicae historiae imagines*, and other meditative works are persuasive in part because of their intense vividness. These sacred images are designed to steer votaries toward entities beyond themselves—that is, Christ, God the Father, and other spiritual entities.

Responding to Reformation criticisms of religious imagery, Jesuit theologians defended images as tools that could explain (to the extent that human understanding permitted) the wonder of the Incarnation—in other words, God becoming flesh in the person of Jesus Christ.[21] Moreover, insofar as the Incarnation was interpreted as an instance of godly image-production resulting in Christ, the *imago Dei* ("image of God"), it could be used to sanction the creation of more religious images in imitation of Christ. In his *Adnotationes et meditationes*, Nadal expands upon this justification of religious imagery and holds that meditative life depends on images, because in Christ God made Himself visibly manifest and able to be represented.[22]

The meditative itineraries that chart the *peregrinationes* ("pilgrimages") of sacred figures in the *Evangelicae historiae imagines* were meant to direct the prayer of young men preparing to become Jesuit priests and missionaries.[23] These students were expected to dwell on the images in the *Evangelicae historiae imagines*, internalize them, and then refer to those images in the course of their daily prayers. Jesuit missionaries carried with them the *Adnotationes et meditationes*, and the *Evangelicae historiae imagines* would become the model for two books produced in seventeenth-century China: the *Song nianzhu guicheng* (*Rules for Reciting the Rosary*, c. 1619–1623), ascribed to Gaspar Ferreira, S.J. (Fei Qigui, 1571–1649) and João da Rocha, S.J. (Luo Ruwang, 1583–1623), and the *Tianzhu jiangsheng chuxiang jingjie* (*Illustrated Explanation of the Incarnation of the Lord of Heaven*, 1637), attributed to Giulio Aleni, S.J. (Ai Rulüe, 1592–1649).[24] Spiritual itineraries delineated within sweeping, sequential landscapes also feature in other Catholic Reformation meditative treatises, broadsides, and paintings, including in the engravings Boëtius à Bolswert (1585–1633) prepared for *Via vitae aeternae*.[25]

The meditative itineraries charted in printed works illustrate the metaphorical conception, widespread in sixteenth- and seventeenth-century Europe, of life as a spiritual journey or pilgrimage. The analogy between learning and movement along a path features frequently in pedagogical spaces and documents produced throughout early modern Europe, from the organization of botanical gardens to images of the *Tabula Cebetis* (*Table of Cebes*) and philosophical thesis prints made in educational establishments.[26] In the thesis print *Typus necessitatis logicae ad alias scientias capessendas* (*Scheme of the Necessity of Logic for Grasping the Other Branches of Knowledge*, 1622; hereafter *Typus*)—engraved by Léonard Gaultier (1560/61–1635) and an anonymous engraver of lettering, designed by the Carmelite philosophy professor Jean Chéron (1596–1673), and published by Jean Messager (1572–1649)—dozens of characters travel along slippery pathways, ascend precipitous cliffs, cross rolling hills, flat plains, wooded terrains, and sail through a turbulent sea in search of wisdom, symbolized by a temple (Fig. 101).[27] The tradition of pilgrimage, furthermore, indicates strong analogies between walking and spiritual progress. It flourished in Catholic areas during the early modern period, with novel pilgrimage sites springing up, for instance, to the north of Trieste as well as in Montserrat in the sixteenth century.[28]

Walking, as Frances Gage has pointed out, was in itself understood to repair both the body and the spirit.[29] In describing the benefits of walking in the San Lorenzo cloisters, the Florentine academician Pier Francesco Giambullari (1495–1555), declared it "of no less service to the health of the body than to the recreation of the mind."[30] Part of the intellectual appeal of walking among sixteenth- and seventeenth-century thinkers lay in its popularity among ancient peripatetic philosophers.[31] Diogenes Laertius (third century CE) states that Aristotle's Lyceum featured a *peripatos* (likely a colonnade), in which he could stroll while discussing philosophical matters with students.[32] The physician Girolamo Mercuriale (1530–1606), who taught at the University of Padua between 1569 and 1587, asserted that "the plane trees of Athens in the walkway of the Academy were famous," referring to a space in ancient Athens for perambulations and learning.[33]

Jesuit meditative itineraries, like ancient and early modern theories of perambulation, resonate with the transits of observers through fixed places like those shown in the vast panoramic landscapes of Maignan's and Niceron's anamorphic wall paintings (see Gatefolds 1 and 2). Maignan's landscape, as noted in the previous chapter, features scenes—for instance, Saint Francis of Paola passing the Strait of Messina on his coat—that are also visualized in printed meditative lives of Saint Francis of Paola, and the books similarly focused and sustained the imagination and minds of votaries. If in the corridors of the Minims the observer was guided from place to place by a companion—or simply by the need to get to a destination at the end of the passage—in books, journeys followed series labels, like the alphabetic labels we have seen above, to signal to the reader where he should go next. Both associated movement—walking literal or virtual—with spiritual elevation.

Jesuit Deformations

The Jesuits, like the Minims, decorated their institutions with large-scale anamorphic wall paintings that led observers to move through a space along a predetermined path. One such fresco survives in the room of Saint Peter above the Jesuit church of the collège royal Bourbon in Aix-en-Provence (Fig. 102).[34] According to Pascal Julien, who estimates that the work was

Fig. 101. Jean Chéron (designer) and Léonard Gaultier (engraver), *Typus*, 1622. Engraving printed on paper, 29.1 × 18.5 in (74 × 47 cm). Princeton University Library, Special Collections.

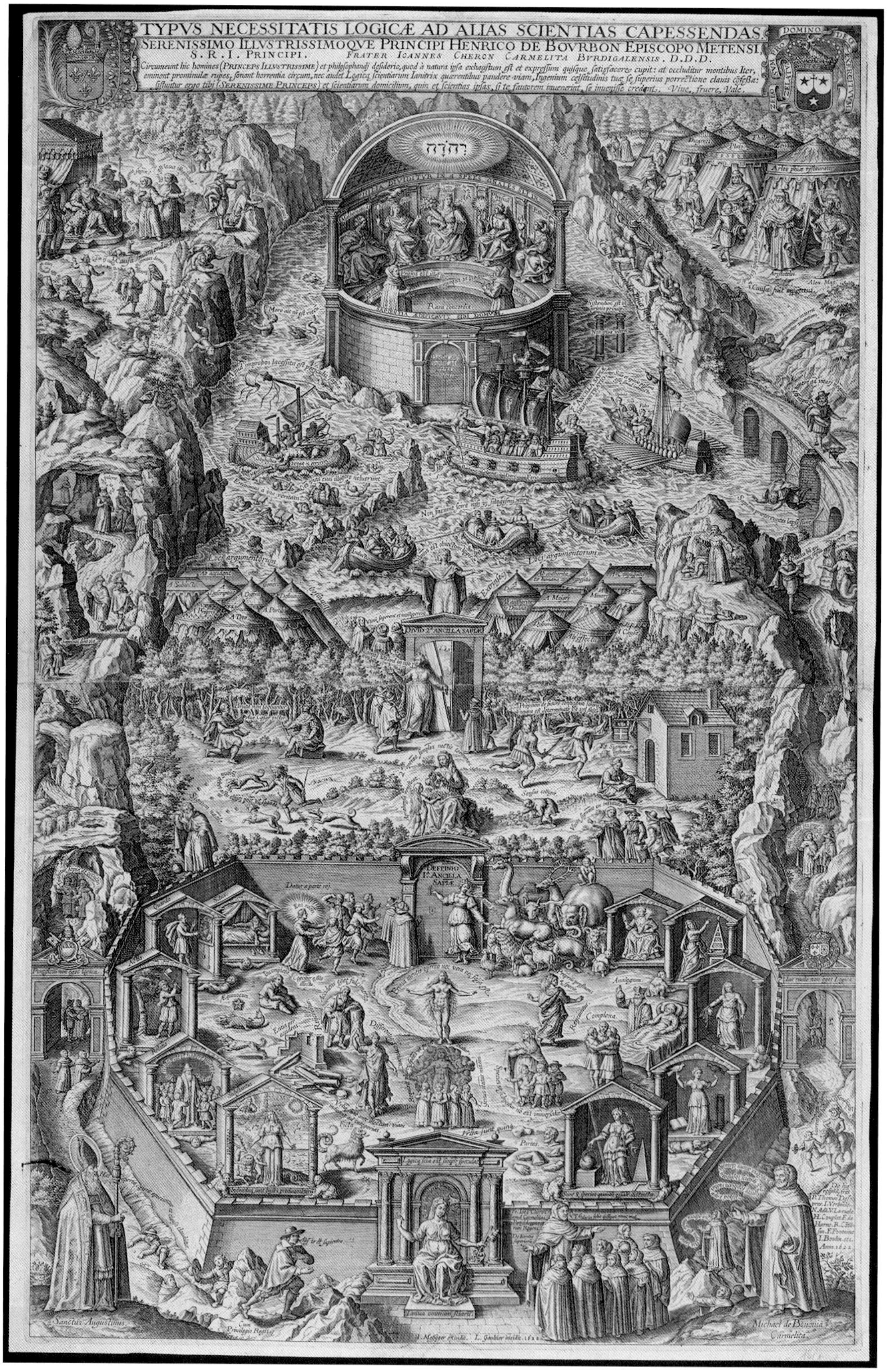

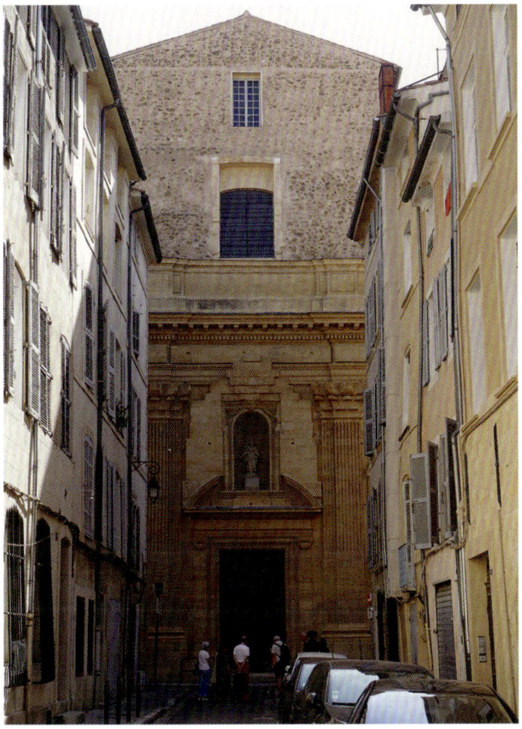

Fig. 102. Exterior view of the Jesuit church of the collège royal Bourbon, Aix-en-Provence.

produced toward the beginning of the eighteenth century, it is the sole surviving early modern anamorphic wall painting in France.[35] The image is in poor condition and is covered in graffiti. When seen straight on, it pictures a suspended piece of large fabric on which is depicted a detailed coastal landscape with houses, fortifications, mills, and boats—a landscape that resonates with Jesuit meditative itineraries in prints (Figs. 103 and 104). From an oblique view near the wall and from the left, the painting presents a kneeling and praying Saint Peter (Fig. 105). The gold and yellow keys to Heaven and Earth hang from a ribbon clasped in his praying hands. A snail crawls near the saint's knees and the rooster before him implies that the painting visualizes the saint's repentance for his denial of Christ, although this detail is difficult to read, as the image is greatly damaged here.[36] Where head-on the painting presents a town, the same detail examined from the left appears to be Saint Peter's hair and beard (see Figs. 104 and 105). Like the anamorphic frescoes of Maignan and Niceron, this painting requires observers to walk along the wall to interpret its imagery and to discover the reformed image of Saint Peter. Jesuits, like Minims (as in Maignan's proposed anamorphic wall painting for the villa for Cardinal Pamphilj), would have ascribed particular significance to the fact that the work clarifies into an image of Saint Peter, reading his role as the first head of Christ's church as supporting the Catholic claim to be the one true church.

Jesuit students were not only expected to interpret anamorphoses; they were required to gain a practical understanding of how to produce these "deformed" images. The thesis booklet we saw in chapter 1 that Pierre Bourdin developed for a disputation at the collège de Clermont, defended around 1639 by the student Yves Henri, offers the following challenge:

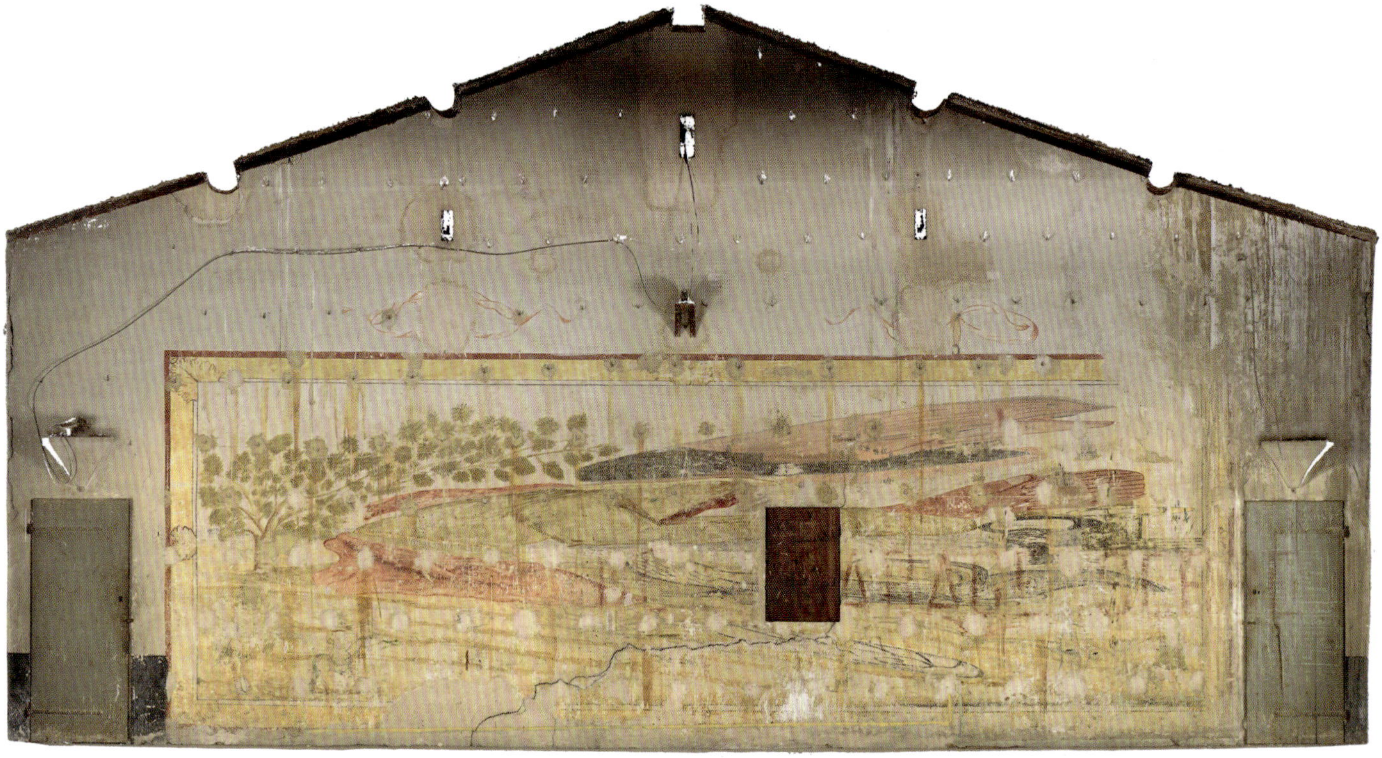

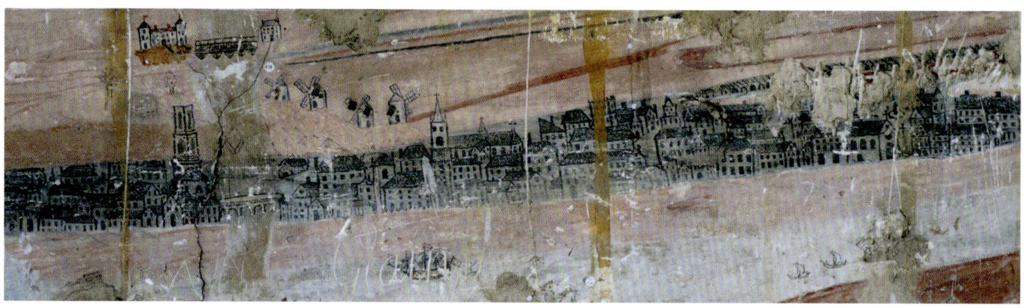

Fig. 103. Unidentified artist, *Anamorphosis Showing Coastal Landscape and Kneeling and Praying Saint Peter,* around the beginning of the eighteenth century. Fresco, 119.29 × 332.28 in (303 × 844 cm). Room of Saint Peter, Jesuit church of the collège royal Bourbon, Aix-en-Provence.

Fig. 104. *Anamorphosis Showing Coastal Landscape and Kneeling and Praying Saint Peter* (detail), around the beginning of the eighteenth century. Room of Saint Peter, Jesuit church of the collège royal Bourbon, Aix-en-Provence.

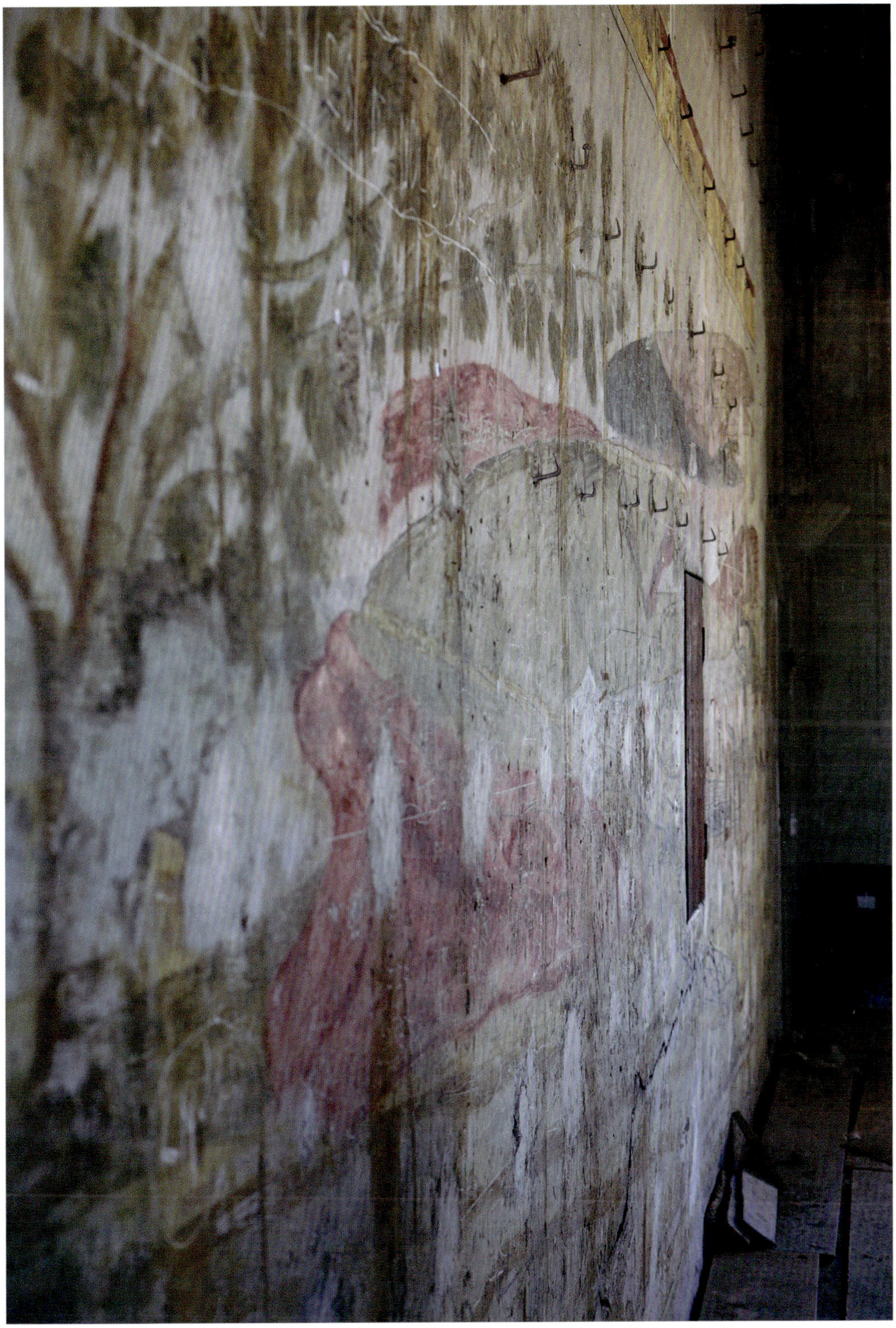

Fig. 105. Oblique view of *Anamorphosis Showing Coastal Landscape and Kneeling and Praying Saint Peter,* around the beginning of the eighteenth century. Room of Saint Peter, Jesuit church of the collège royal Bourbon, Aix-en-Provence.

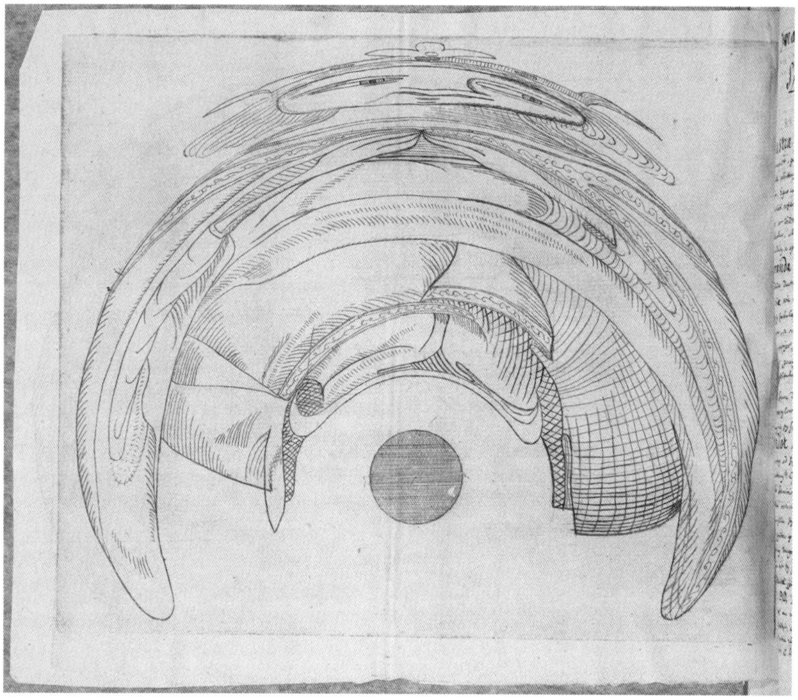

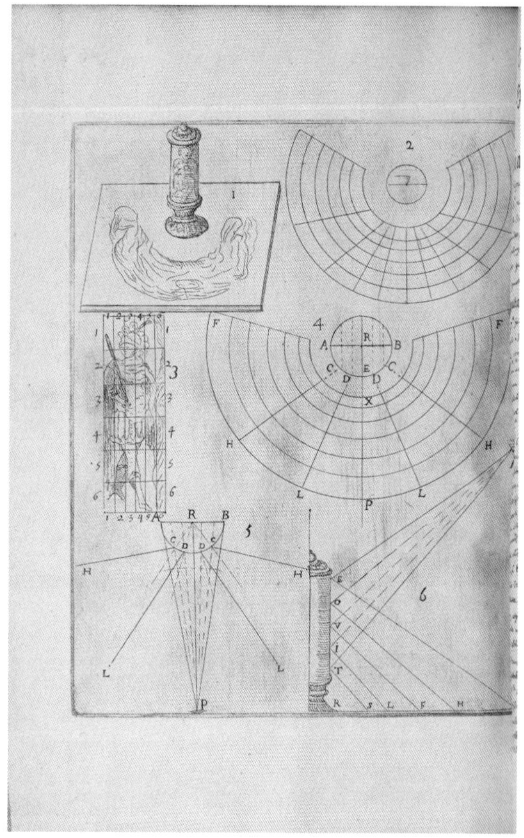

Fig. 106. Drawing of an anamorphic image that relies on a cylindrical mirror in the 1636–1638 notebook of Paul Mercier, 10.83 × 11.81 in (27.5 × 30 cm). BnF, Département des manuscrits, Paris [ms. Lat. 17862, fol. 839bis[v]].

Fig. 107. Drawings pertaining to anamorphic images that rely on cylindrical mirrors in the 1636–1638 notebook of Paul Mercier. 12.13 × 7.87 in (30.8 × 20 cm). BnF, Département des manuscrits, Paris [ms. Lat. 17862, fol. 843bis[v]].

> To explain by what artifice those images must be drawn, which are painted on the whole ceiling of the longest portico, or wall, and on the lower pavement, although they are deformed, and like monsters, because of their immoderate width, although the true height has been retained; if, nevertheless, they are viewed from a designated place, such as the entrance, they appear perfect, and represent the correct likeness of a man.[37]

Further evidence for the study of anamorphic images in Jesuit classrooms can be found in manuscript notes made during Bourdin's classes in 1636–38 by a student named Paul Mercier, who sketched anamorphic images that relied on cylindrical mirrors (Figs. 106, 107, and 108).[38]

Anamorphoses and other images and optical devices became valued tools not only in European Jesuit institutions like the collège in Aix-en-Provence or the collège de Clermont but also in overseas missions. Jesuits practiced, as Roland Barthes (1915–1980) puts it, "a radical imperialism of the image."[39] The Flemish missionary Ferdinand Verbiest, S.J. (1623–1688) describes how the Kangxi Emperor (1654–1722) admired anamorphic wall paintings by Claudio Filippo Grimaldi, S.J. (1639–1712) in Beijing:

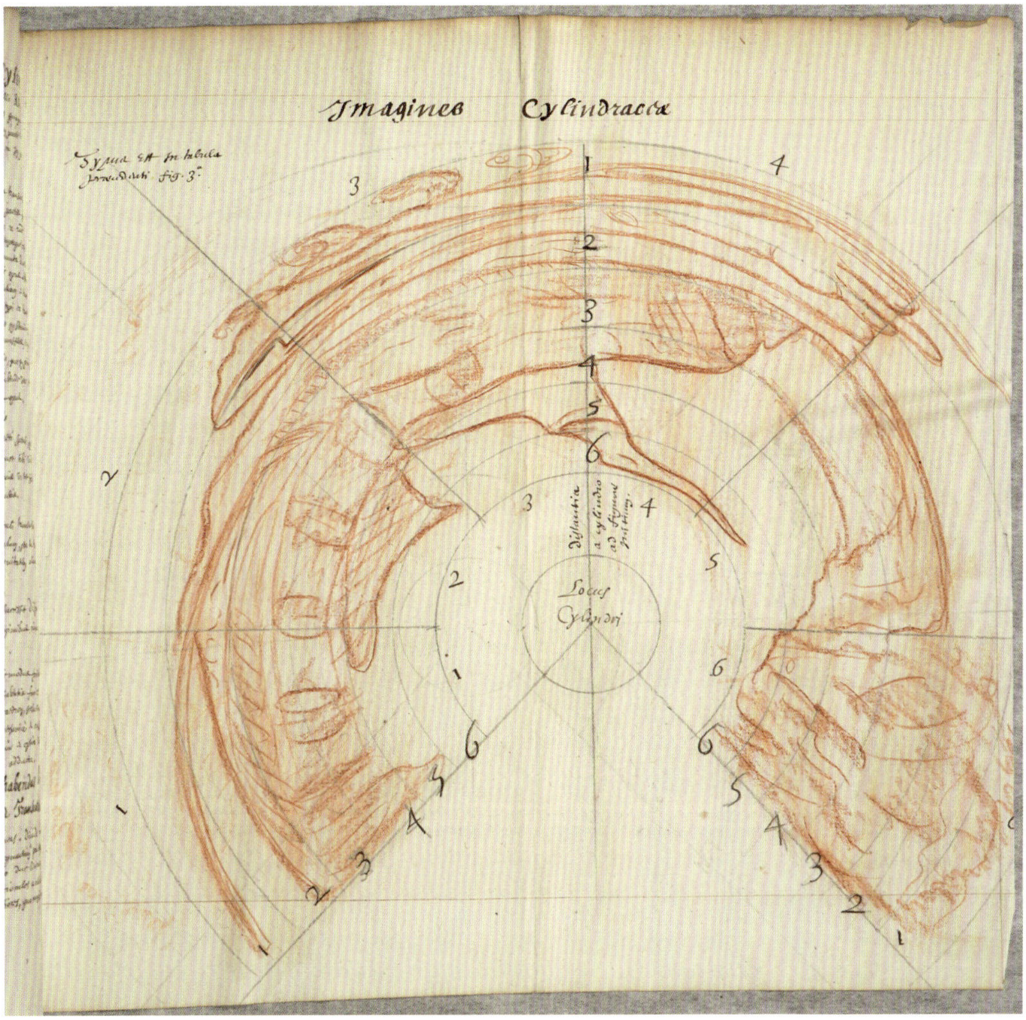

Fig. 108. Drawing of an anamorphic image that relies on a cylindrical mirror in the 1636–1638 notebook of Paul Mercier. 11.42 × 11.42 in (29 × 29 cm). BnF, Département des manuscrits, Paris [ms. Lat. 17862, fol. 845^{v}].

The garden of our residence in Beijing is enclosed by four walls, and on each of them Father Filippo Grimaldi has drawn, according to the optical laws, a picture which, in line with the length of the wall, extends over about fifty feet. When viewed from the front, it shows only mountains, woods, hunting-parties etc., but, if one looks at it from a specific angle, it conjures up a very accurately delineated picture of … a man or at least of a living being. The day that the Emperor visited our residence he looked at these pictures, one by one, long and thoroughly, and up to the present day dignitaries and the most prominent mandarins continue to frequent our residence daily, to look at these and other similar (specimens). There are none that do not admire this art, the more so since the wall-surfaces extend so far and are very irregular, and since some of them are interrupted by doors and windows so that they tear apart the pictures considerably and interrupt them, even when they are only looked at from the front.[40]

Grimaldi's reliance on "the optical laws" underscores the scientific nature of anamorphic deformation, anamorphoses' engagement with mathematics as well as the operations of the eye. The hallways that bear the Trinità dei Monti frescoes are also disrupted by doors, allowing us to better conceptualize the architectural interruptions that Verbiest refers to (see Gatefolds 1 and 2). Artists would have had to keep these voids in mind when considering where to position their representations, so that doors or windows did not disrupt the coherence of the reformed images. To enable his fresco to appear continuous when observed from the southern end of the corridor, Maignan painted the strike jambs of the doors interfering with his frescoes (see Fig. 74).

In a secondhand account, compiled by the Jesuit historian Jean-Baptiste du Halde (1674–1743)—apparently contemporary and apparently based at least partly on Verbiest's writing—we find text suggesting that Grimaldi's paintings may have featured both types of anamorphoses described above, that is, both anamorphoses that resolved when the viewer moved through space to a designated viewing location, and anamorphoses that required mirrors to resolve:

> Father Grimaldi organized another spectacle of the marvels of optics in the Jesuits' garden in Beijing, at which the emperor and his entourage were greatly amazed. He made on the four walls four human figures, each figure measured the same length as the wall, which was fifty feet. As he had perfectly followed the rules of optics, one could see head-on only mountains, forests, hunts, and other things of this nature. But from a certain point you could perceive the figure of a well-made, well-proportioned man.
>
> The emperor honored the house of the Jesuits with his presence and considered these figures for a long time and with admiration. The important and principal Mandarins who came there in crowds were equally surprised. What struck them most was to see such regular and exact figures drawn on very irregular walls that were broken by several doors and windows.
>
> It would take too long to report on all the figures drawn confusedly and that one could only see distinctly by standing at a certain point or by using conical, cylindrical, or pyramidal mirrors to rectify the vision, and on the many other marvels of optics that Father Grimaldi presented to the finest minds in China, drawing equally their surprise and their admiration.[41]

We see, then, in du Halde's description, not only the anamorphoses that depend on motion through space (requiring that the viewer stand at "a certain point"), but the use of varying kinds of mirrors "to rectify the vision." Neither Verbiest nor du Halde remarks on the identities of the figures portrayed in Grimaldi's paintings; undoubtedly, the subjects were designed to inspire meditation, and the landscapes, like the prints in the *Evangelicae historiae imagines*, set forth meditative itineraries. Verbiest and du Halde focus instead on how the representations within the paintings oscillate between potential meanings as well as on the reactions of observers. Whether these accounts accurately report Chinese observers' responses remains an open question. Not in doubt is that they evince Jesuit expectations that such illusionistic imagery—spectacles of "the marvels of optics" (*des merveilles de l'Optique*)—would elicit wonder and, in the process, could heighten interest in Catholicism.

Alfred Gell's classic study "The Technology of Enchantment and the Enchantment of Technology" helps to clarify how stupefying, wondrous, and curious Catholic Reformation images and architectures might have focused attention on Catholicism.[42] Gell defines the enchantment of technology as "the power that technical processes have of casting a spell over us so that we see the real world in an enchanted form."[43] Unexpected structures like San Carlo alle Quattro Fontane or deformed images like the fresco in the collège royal Bourbon in Aix-en-Provence that resisted straightforward intellectual understanding of how they were produced would have "enchanted" observers and fixed their attention, causing the works to appear as reflections of a transcendent power, while also, of course, inducing close looking. The sense of "casting a spell" recognizes that some seventeenth-century beholders, though knowing the works were created by humans, would have perceived them as the product of exceptional, and perhaps divinely inspired, creative ability. Gell explains: "In reconstructing the processes which brought the work of art into existence, [the spectator] is obliged to posit a creative agency which transcends his own and, hovering in the background, the power of the collectivity on whose behalf the artist exercised his technical mastery."[44] For the Jesuit illusionistic imagery produced in China and for most of the other examples discussed in this book, that collectivity is the Catholic Church.

Illusionistic imagery was seen as crucial to advancing the messages of the Church—and its authority as a mediating figure able to decode truths that might be hard for the laity to interpret.[45] This authority-supporting function was so valued that one Flemish missionary reportedly threatened to berate a colleague for insufficient zeal in the production of wondrous images. According to the Jesuit Gilles Estrix (1624–1694), the Flemish missionary François de Rougement, S.J. (1624–1676) challenged his fellow Flemish missionary Philippe Couplet, S.J. (1623–1693) to create perspectival images, believing they would draw potential converts: "If Your Reverence ... does not decorate the portico of his house in Shanghai by means of perspective paintings and other curiosities, as I have done here, I will accuse him of deficient zeal."[46]

Beyond their power as curiosities or wonders—attracting visitors to a church-controlled site (and potentially, as Fra Juan remarked of San Carlo above, encouraging return visits)—we have seen deformations' power to shape how observers moved through the spaces defined by deformed images or structures. A similar shaping of movement can be seen again in relation to some of the most striking Jesuit-produced deformations: feigned domes. Here immediately come to mind the magnificent ceiling paintings in Sant'Ignazio by the Jesuit Andrea Pozzo, who, as noted above, is also credited with the refectory of Trinità dei Monti.[47] Pozzo's faux dome, completed in 1685, represents a drum on a painted canvas in the church's transept (Fig. 109).[48] The anamorphic drum is most persuasive if observed from a central area of the nave antecedent to the crossing marked by a marble disk in the floor (Fig. 110). From other vantage points far removed from this marble disk the dome is distorted to such an extent that the Italian abbot, art historian, collector, and economist Lione Pascoli (1674–1744) characterized its appearance as "deformed and falling" (*sformato, e cadente*).[49] Given that Sant'Ignazio was the church of the Collegio Romano, the Jesuits' most important learning institution, we might read the sophistication of Pozzo's work as offering an occasion for observers to apply their knowledge of how such visual effects were produced.

When confronted with Pozzo's dome in Sant'Ignazio, from all but one general viewpoint

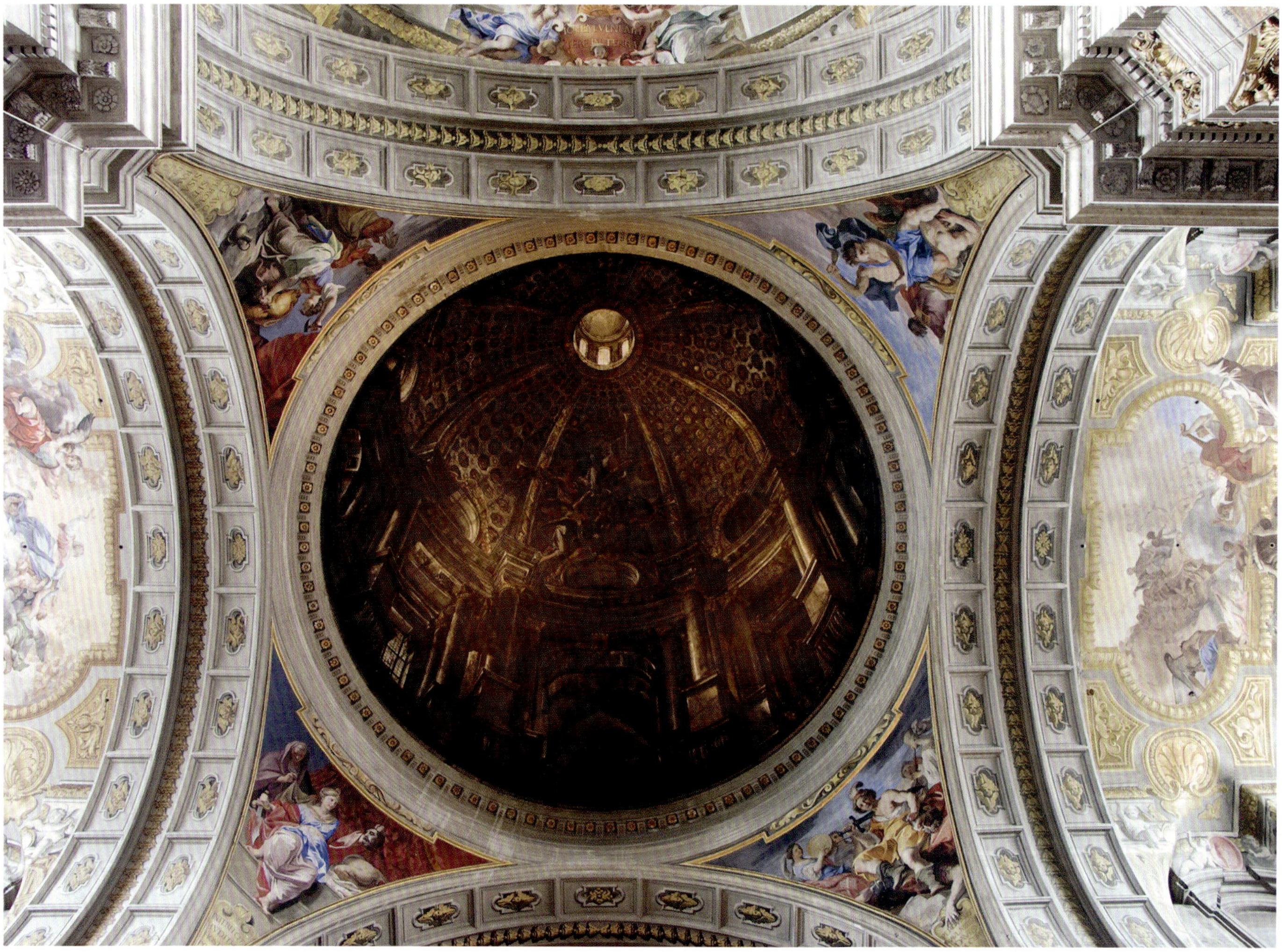

Fig. 109. Andrea Pozzo, dome fresco in Sant'Ignazio in Rome, completed in 1685.

Fig. 110. Disc in the floor of the nave of Sant'Ignazio.

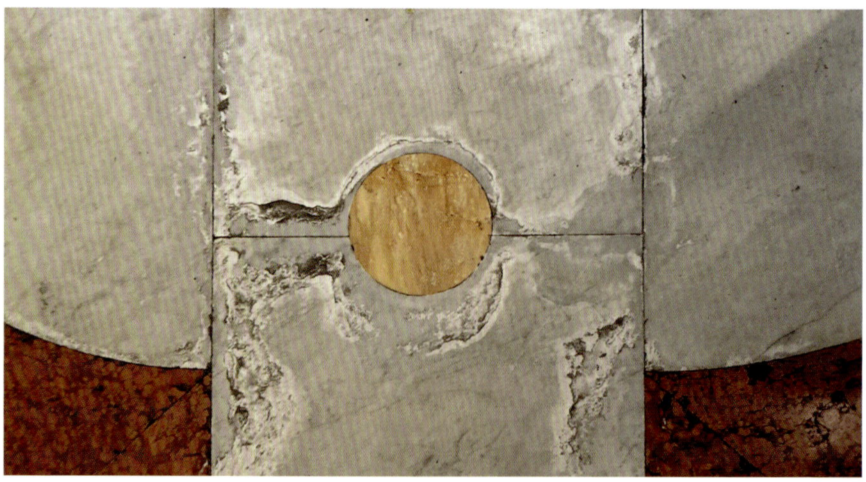

(marked by the marble disk but extending a little beyond it), observers will be somewhat confused. A similar experience holds for other anamorphoses: Maignan writes of how they resolve into a "neat, orderly" image only "from a certain point."[50] And we might see this feature as decidedly didactic. Jesuits, like the rest of the Catholic Reformation Church, held that laypeople were unable to discover truth and attain salvation on their own. They needed God's grace as well as the Church as mediator between the faithful and God on the path to securing salvation. If Lutherans contended that Jesus's message could be understood by every human being directly, the illusionistic imagery of early modern Catholicism can be read as presenting an allegory of the Catholic Church's role as a mediator.[51] The illusionistic imagery points to the divine, but it also shows that one reaches a clearer understanding of the divine by following the guidance of the Church.

In emphasizing how Pozzo used deformation to control how the faithful move in Sant'-Ignazio, I build on Andrew Horn's argument that Pozzo's works actively engage observers as performers in religious theater.[52] Among the other frescoes by Pozzo that Horn considers in support of this thesis is the Corridor of Saint Ignatius (c. 1682–1686) in the Casa Professa of the Jesu in Rome. The passageway links a garden courtyard to the rooms where the saint spent his final decade before dying in 1556. The corridor frescoes juxtapose fictive architecture with images that extol the saint by visualizing stories and "miracles" from his life (Fig. 111). These frescoes were also meant to be observed from only one marked position in the corridor. Stepping away from this vantage point, angels, figures, and architecture appear distorted. Horn presents Pozzo's anamorphic corridor as another example of spiritual theater, characterizing it as "a place of performed pilgrimage."[53]

In her account of the corridor, Lyle Massey rightly stresses that Pozzo did not consider distortions to be a disadvantage, but rather a benefit in an artwork because they forced observers to move into and out of the optimal vantage point.[54] I part from Massey, however, when she draws a contrast between the images of Niceron and Maignan, on the one hand, and those of Pozzo, on the other, suggesting that deformation was not important to Niceron and Maignan because their images "were no doubt intended to be seen from peepholes or controlled viewpoints."[55] Deformation was important to all three artists, not only to the work of Pozzo. Even if Niceron and Maignan arranged for their works to be seen from specific perspectives that resolved the images of the saints (as is indicated by Maignan's reference to the "small hole" outside the hall in the northern loggia in the villa plans for Cardinal Pamphilj), they were not exclusively seen this way: observers in Trinità dei Monti needed to traverse corridors not only to travel through the convent space—the rooms arranged around the first floor of the cloister were designated as study areas—but also to appreciate the details within Niceron's and Maignan's anamorphic landscapes.[56]

In his theoretical writing, Pozzo calls attention to the importance of the singular viewpoint that leads deformations to appear reformed: "Bear in mind that such works, in order to deceive the eye easily, must have a fixed and predetermined point from which they must be viewed, so that the viewer does not perceive the deformations and distortions which the curvature and irregularity of the vault tend to cause."[57] Pozzo's nave fresco in Sant'Ignazio in Rome, finished in 1694, celebrates the Jesuits' mission to propagate Catholicism around the globe (Fig. 112).[58] It shows Saint Ignatius receiving in his heart a ray of light—the symbol of

Fig. 111. Andrea Pozzo, Corridor of Saint Ignatius in the Casa Professa of the Jesu in Rome, c. 1682–1686.

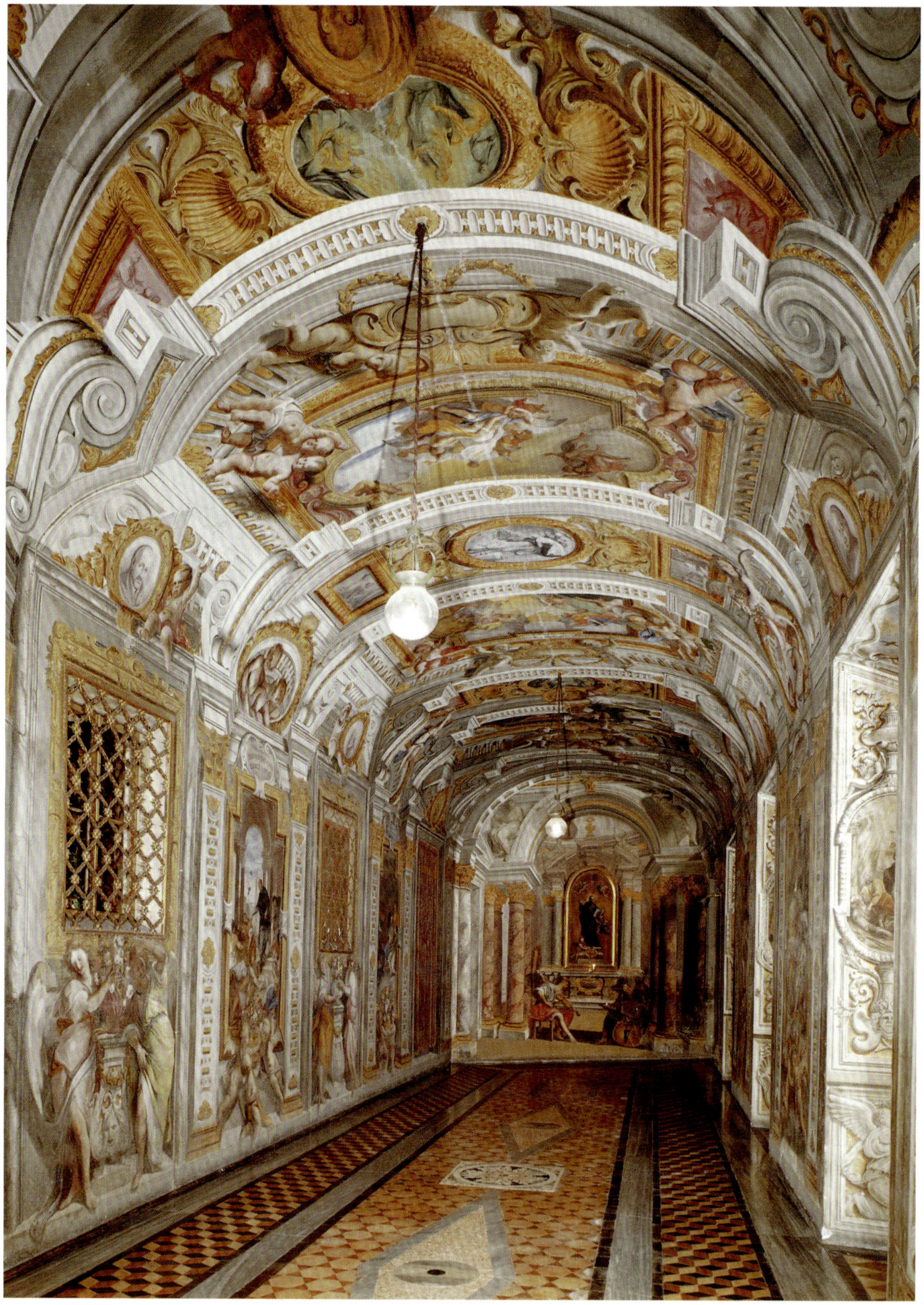

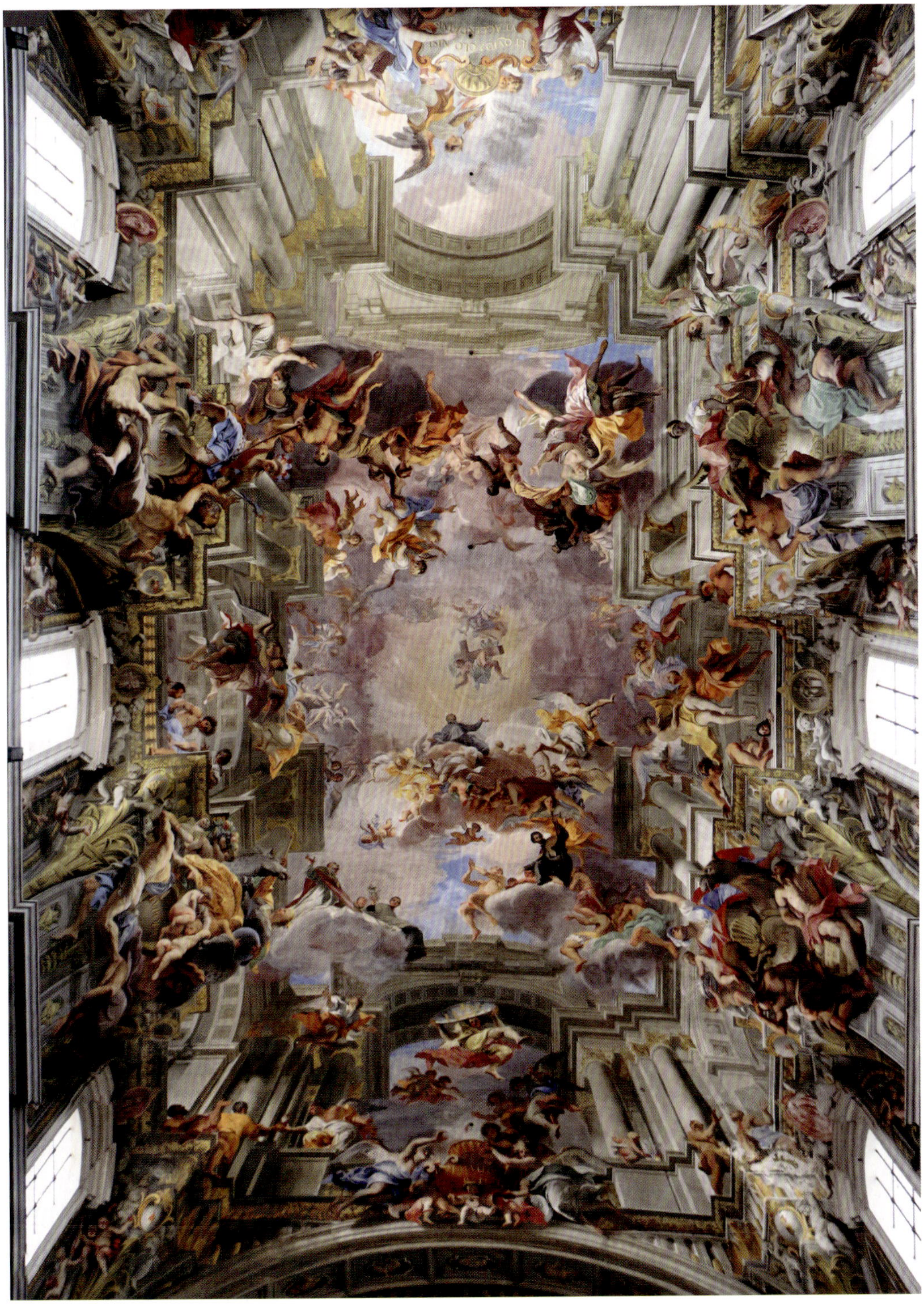

Fig. 112. Andrea Pozzo, nave fresco in Sant'Ignazio in Rome, completed in 1694.

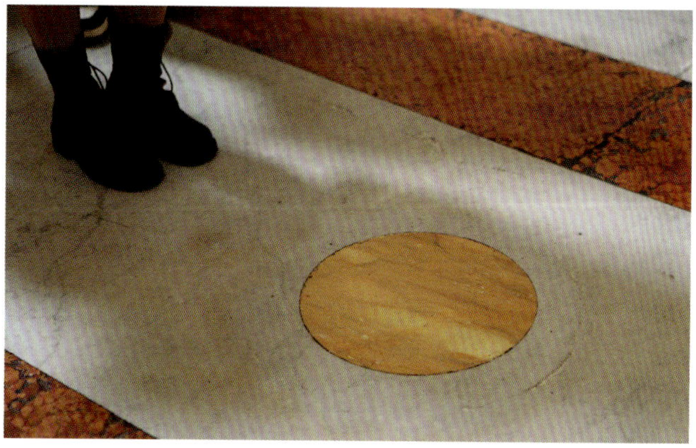

Fig. 113. Second disc in the floor of the nave of Sant'Ignazio.

divine grace—sent from Christ, that then reflects onto allegories of the four continents of Asia, Africa, America, and Europe.[59] A second disk in the nave's floor, under the fresco's Saint Ignatius, marks the "fixed and predetermined point" from which the viewer sees clearly the resolved image of Saint Ignatius's encounter with divine grace (Fig. 113). From this spot, observers may appreciate how the actual architecture and the painted architecture, which starts above the cornice and continues onto the flat ceiling, produce a persuasive visual illusion of the Apotheosis of Saint Ignatius. At the start of his text, Pozzo identifies a sharpening of spiritual skills as the goal: "Therefore, my reader, start your work happily; with the resolve always to draw all the lines of your works to the true point of the eye, which is the Divine glory."[60] The incorporation of subject matter around global propagation into the church's illusionistic imagery gestures toward the role of illusionistic art in generating awe among those who would see it and thus at its function in missionary efforts.

It is physically taxing to crane one's neck up to examine the nave fresco or the faux dome in Sant'Ignazio. One can look up at the images for only a few moments, before one must turn one's head downward again to relax. The bodily discomfort is an intentional feature of these frescoes (now undermined by the introduction, at the nave's center, of a large mirror that allows visitors to look at the frescoes without craning their necks). The frescoes as originally conceived demanded that visitors—who were not provided a mirror—move their bodies into a contorted stance, causing them to feel smaller and weaker in relation to the works they engaged above and thus eliciting their awe. Having to observe the nave or the faux dome in short spurts, moreover, adds to the apparent verisimilitude of each scene, insofar as one can never look long enough to get an overview, feel any sense of control over the frescoes, or see their illusions unravel. The painted wood ceiling of the library of the Trinità dei Monti convent, executed in 1693 by an unknown artist and showing Saint Francis of Paola in glory as well as Saint Ambrose, Saint Augustine, Saint Gregory, and Saint Jerome in its four corners, strains the bodies of observers to similar effects, as do the illusionistic architectural frescoes in the pharmacy of the Trinità dei Monti convent (Figs. 114 and 115). This required distortion of the body is not specific to religious settings and is similar to something that a visitor to the Palazzo Spada ceiling sundial would have been expected to experience, perhaps there also to reinforce feelings of humility and, in that case, their admiration for Bernardino Spada.

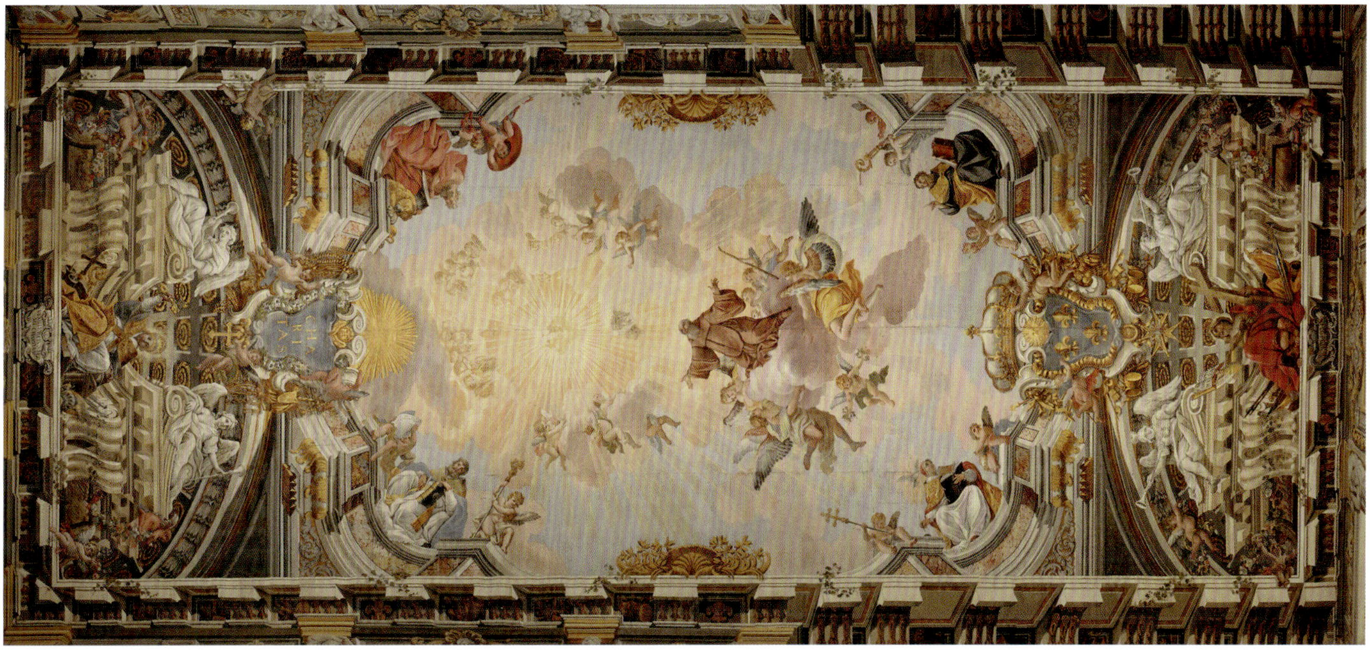

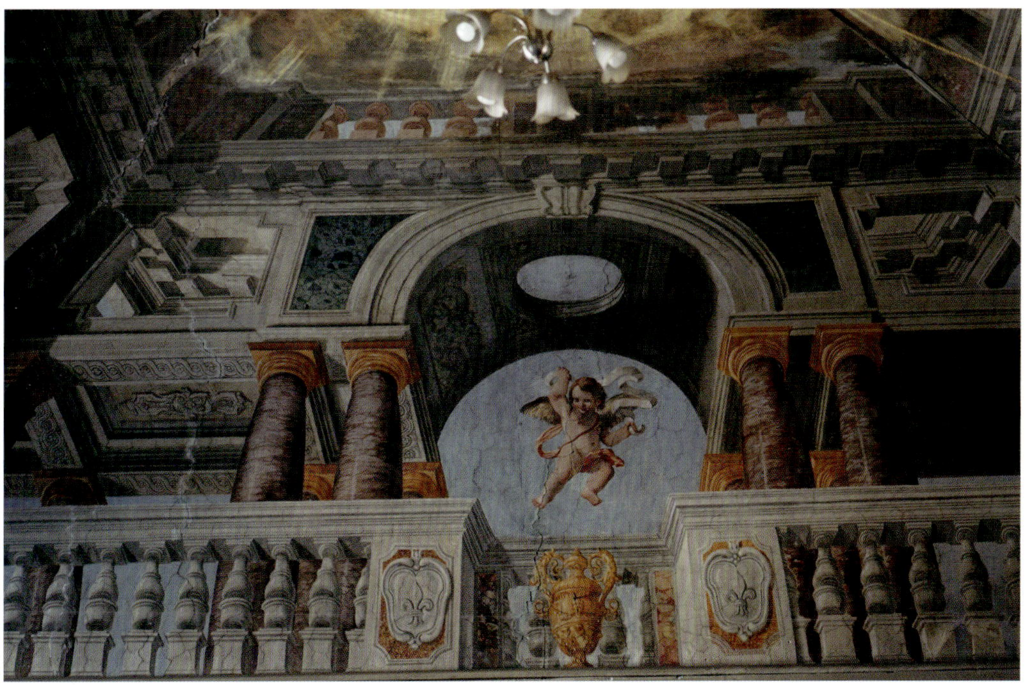

Fig. 114. Unidentified artist, painted wood ceiling, 1693. Library, convent of Trinità dei Monti, Rome.

Fig. 115. Attributed to the circle of Giovanni Francesco Romanelli (1610–1662), illusionistic fresco. Pharmacy, convent of Trinità dei Monti, Rome.

Other feigned domes integrated into Jesuit churches from Vienna to Beijing exemplify how illusionistic art manifested the Jesuit ambition to propagate Catholicism around the globe. Giovanni Gherardini (1655–c. 1729)—trained in the Bolognese workshop of Angelo Michele Colonna and Agostino Mitelli (painters of the Palazzo Spada *Salone*)—decorated the church of Our Savior, also known as the "Northern Church" (*Beitang*), in Beijing with a painted illusionistic vision and a feigned dome. The frescoes no longer survive, but an account of the paintings features in a 1704 letter from Pierre Jartoux, S.J. (1669–1720), to J. de Fontaney, S.J. (1643–1710): "The ceiling is fully painted. It is divided into three parts: the middle represents an open dome of a rich architecture, made up of marble columns which support a range of arches surmounted by a beautiful balustrade ... one sees above God the Father seated on a group of angels in the midst of clouds, and holding the world in his hand."[61] Another feigned dome was painted between 1715 and 1728 by Giuseppe Castiglione, S.J. (1688–1766), in the church of St. Joseph, the Portuguese "East Church" (*Dongtang*) in Beijing.[62] The messaging of these feigned domes, which appeared in a multitude of sites, centers around the general power of the Catholic Church. When this assertion of Church power was deployed in a missionary setting, it would have also embodied more particularly the Church's spiritual-colonial goals.

In addition to these man-made deformations, Jesuits saw potential to find anamorphic-like deformations in nature to invite individuals into the arms of the Catholic Church—or, as the missionaries would have understood their work, to lead the faithful to God. Athanasius Kircher describes a mountainside in Chile filled with emeralds, turquoises, and other precious stones that was discovered to represent the Virgin and baby Jesus if—but only if, as he stresses twice—observed from a particular vantage point.[63] Recalling an anamorphic artwork that oscillates between coherence and incoherence depending on the viewer's position, the miraculous appearance of the Virgin embedded in the Chilean mountainside emerges from confusion only to the observer who stands in a particular location. Kircher writes that the Church instituted a procession to the image every year to harness its power to convert indigenous populations—enacting a community-wide iteration of the exercise in leading an observer through space that we have previously seen to embody the Church's mediating role as guiding the faithful and interpreting spiritual meaning. Moreover, he stresses the image's providential character. He writes, "Nothing in nature can appear by way of chance and accidentally, which does not already lie behind the secret arrangement of divine providence that governs everything."[64]

The Chilean Jesuit priest and chronicler Alonso de Ovalle (1603–1651) integrates an anonymous engraving into his account of the Chilean mountainside (Fig. 116). The print shows the image of the Virgin and child with two indigenous figures praying before its left and right side. The inscription below reads: "The true image of another, seen in the cavity of a rock like a vaulted chapel, not drawn by any artisan, but fashioned by the author of nature from jasper of various colors, with a white face under black hair, in a cloak which is blue inside and red outside, and a robe woven of roses."[65] Given the emphasis on vantage point in seventeenth-century descriptions of this natural apparition, it is significant that the presumed European Catholic purchaser of the book is here offered the prime viewing point. The reader-viewer's privileged perspective seems to reflect an important phenomenon informing publications about the new world missions: that text and images describing the mission setting, or the miracles God is working in the Americas to impress the unconverted and converted Indigenous people with

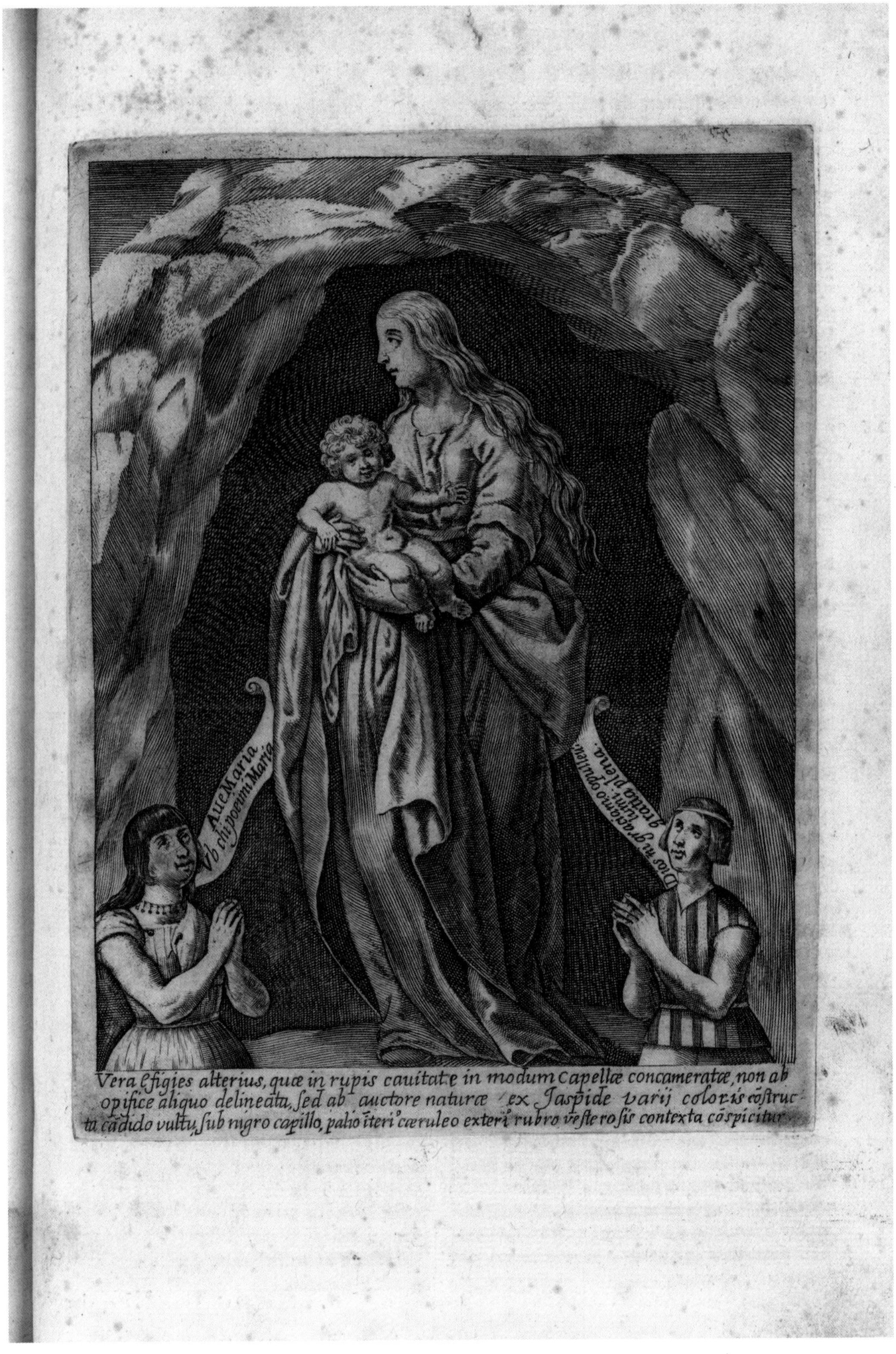

Fig. 116. Alonso de Ovalle, *Historica relacion de reyno de Chile*, 1646, following p. 392. Engraving printed on paper, 10 × 6.8 in (25.4 × 17.3 cm). John Carter Brown Library, Brown University, Providence, R.I.

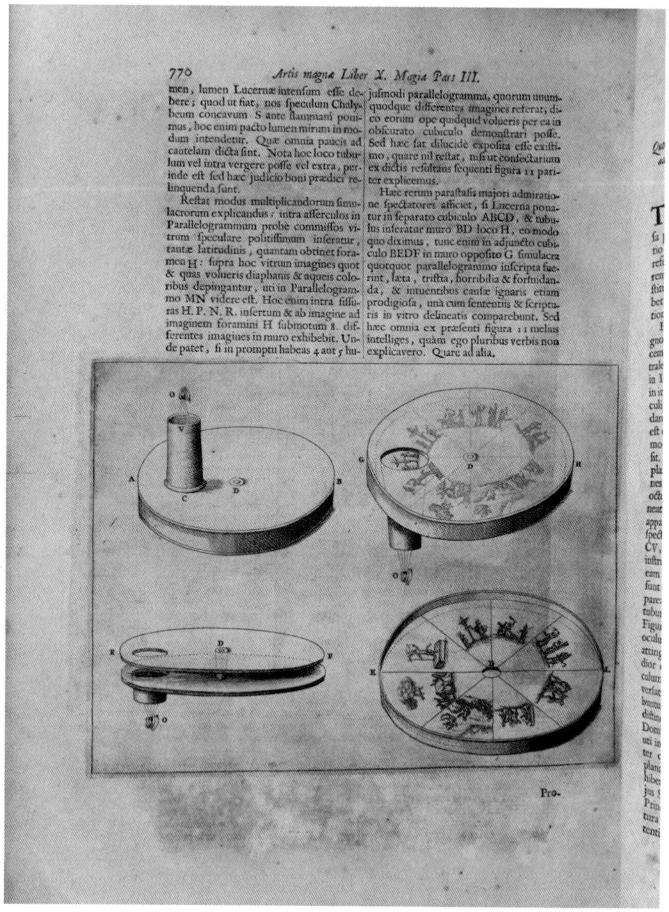
770 *Artis magnæ Liber X. Magia Pars III.*

men, lumen Lucernæ intensum esse debere; quod ut fiat, nos speculum Chalybeum concavum S ante flammam ponimus, hoc enim pacto lumen mirum in modum intendetur. Quæ omnia paucis ad cautelam dicta sint. Nota hoc loco tubulum vel intra vergere posse vel extra, perinde est sed hæc judicio boni prædici relinquenda sunt.

Restat modus multiplicandorum simulacrorum explicandus; intra asserculos in Parallelogrammum probè commissos vitrum speculare politissimum inferatur, tantæ latitudinis, quantam obtinet foramen H: supra hoc vitrum imagines quot & quas volueris diaphanis & aqueis coloribus depingantur, uti in Parallelogrammo MN videre est. Hoc enim intra fissuras H. P. N. R. insertum & ab imagine ad imaginem foramini H submotum 8. differentes imagines in muro exhibebit. Unde patet, si in promptu habeas 4 aut 5 hujusmodi parallelogramma, quorum unumquodque differentes imagines referat, dico eorum ope quidquid volueris per ea in obscurato cubiculo demonstrari posse. Sed hæc sat dilucide exposita esse existimo, quare nil restat, nisi ut consectarium ex dictis resultans sequenti figura 11 pariter explicemus.

Hæc rerum parastasis majori admiratione spectatores afficiet, si Lucerna ponatur in separato cubiculo ABCD, & tubulus inferatur muro BD loco H, eo modo quo diximus, tunc enim in adjuncto cubiculo BEDF in muro opposito G simulacra quotquot parallelogrammo inscripta fuerint, læta, tristia, horribilia & formidanda, & intuentibus causæ ignaris etiam prodigiosa, unà cum sententiis & scripturis in vitro delineatis comparebunt. Sed hæc omnia ex præsenti figura 11 melius intelliges, quàm ego pluribus verbis non explicavero. Quare ad alia.

Pro-

Fig. 117. Athanasius Kircher, *Ars magna lucis et umbrae*, 1671b, 770. The Huntington Library, San Marino, California.

His glory, would be intended, not least, to edify and reinforce the faith of the Europeans who consumed them. The Italian Jesuit Carlo Gregorio Rosignoli (1631–1707) likewise describes this Chilean natural image in terms that recall the confusion and uncertainty evoked by anamorphic paintings before one has found their resolving vantage points: "It was necessary to place oneself at a certain distance: for up close one could see nothing other than various bruises and spots without any distinction."[66]

Figuring Knowledge

If illusionistic images found in mission contexts served to display God's power to potential converts (and to those who read about the missions at home), Jesuits' search for optical tools to aid meditation led to further inventions. Kircher, who taught mathematics at the Collegio Romano, capitalized on the power of images to convey knowledge more than any other member of his Society, and authored illustrated books on topics including hieroglyphs, magnetism, mechanics, optics, light, cosmology, and Chinese visual art and botany.[67] In one publication Kircher included a picture of a "parastatic smicroscope," an instrument he deployed to represent theological images (Fig. 117).[68] The explanatory print depicts eight scenes of Christ's

SAINT FRANCIS OF PAOLA

AND LANDSCAPE

passion in the parastatic smicroscope and disembodied eyes that peer through an opening in the instrument to demonstrate how it should be deployed to enable the sequential observation of virtual images represented on a circular disk. By making it possible to examine virtual images in succession, the device served as a powerful, seemingly animated aid to spiritual meditation, assisting souls in approaching the divine. Kircher, alongside other Jesuits, including the Bolognese mathematician, astronomer, and philosopher Mario Bettini, the French mathematician Jean du Breuil (1602–1670), and Schott, took the lead in disseminating knowledge of catoptric anamorphoses and often relied on visuals to transmit this information.[69] Sketches substantiate that Kircher produced designs for illustrations in his books.[70]

Scholars have classified Kircherian images into three categories.[71] One designates knowledge gleaned directly from the senses. Kircher and other Jesuits were concerned with representing sensual knowledge as well as with illusionistic optical media, as each could be used to figure or represent divine marvels and divine creation. For Jesuits, like Minims, investigating optics—considering how the mind takes in visual information or how that process can be distorted or sustained—belonged to a broader range of scientific studies that also included astronomy and horology. A second category of Kircherian images displays plausible information that is not observed directly through the senses but that can be conjectured from experimental data. A third makes use of allegories—images that, like metaphors, use one thing to explain another. Allegories were especially valued by the Society of Jesus and by other early modern Catholic institutions for their capacity to convey complex ideas to large audiences. Allegories in frescoes (think of Pozzo's nave in Sant'Ignazio, where Asia, Africa, America, and Europe are personified) and in sculptures abound in Catholic Reformation architectures as well as in drawings and printed matter from frontispieces to illustrated thesis prints; these allegories relied on observers' erudition and challenged audiences to sharpen their powers of interpretation of geography, theology, philosophy, and other fields.[72]

To better make sense of Jesuits' appreciation for anamorphic deformations as visual puzzles (produced thanks to a knowledge of mathematics and optics) that aided the observers' spiritual growth, it is helpful to consider them beside the similarly appreciated allegories and emblems—in other words, images that visualize a motto along with an explanatory text. These visual conundrums likewise required interpretive work from an observer / reader who was conceived as mentally journeying toward spiritual growth. Jesuits produced emblems in great numbers, not least to advertise the sophistication of their teaching.[73] I have already mentioned the practical meditation guide *Veridicus Christianus*, which was only one of four emblem books that Johannes David authored in the early seventeenth century.[74] The emblem book *Pia Desideria* (*Pious Desires*, 1624) by the Flemish Jesuit Herman Hugo (1588–1629) and dedicated to Urban VIII (1568–1644) went through numerous editions, including those published in Lyon in 1625, in Antwerp in 1628, 1629, and 1632, in Milan in 1634, and in Munich and Cologne in 1635; the first edition featured prints designed by Boëtius à Bolswert.[75] This immensely popular emblem book, one of many emblem books with Christian objectives, describes paths to salvation. Emblems were also exhibited in some colleges annually or biannually to advertise the sophistication of Jesuit teaching.[76] From the 1550s onward, it became common for colleges, schools, and academies to select an emblem that was meant to express and display their aims

Fig. 118. Peter Paul Rubens and Theodor Galle, *Putti Anatomizing a Cyclops's Eye* from Franciscus Aguilonius's *Opticorum libri sex*, 1613. Engraving printed on paper, dimensions of the page are 14.17 × 9.06 in (36 × 23 cm). BnF, département Réserve des livres rares, Paris [V-1652].

Fig. 119. Peter Paul Rubens and Theodor Galle, *Demonstration* from Franciscus Aguilonius's *Opticorum libri sex*, 1613. Engraving printed on paper, dimensions of the page are 14.17 × 9.06 in (36 × 23 cm). BnF, département Réserve des livres rares, Paris [V-1652].

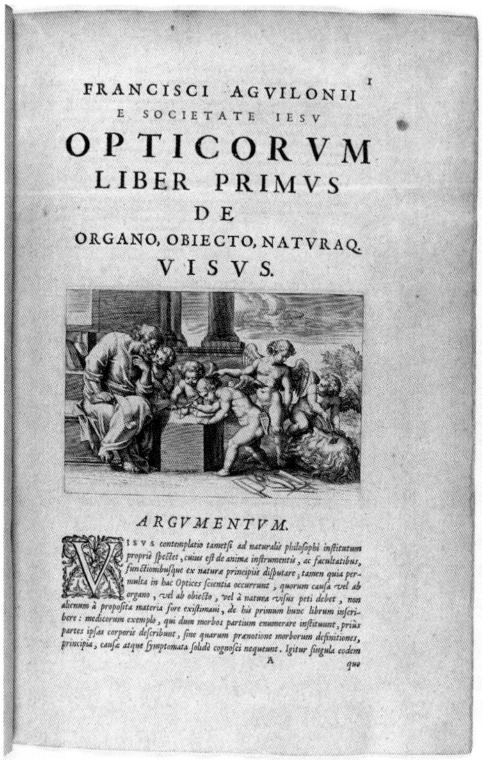

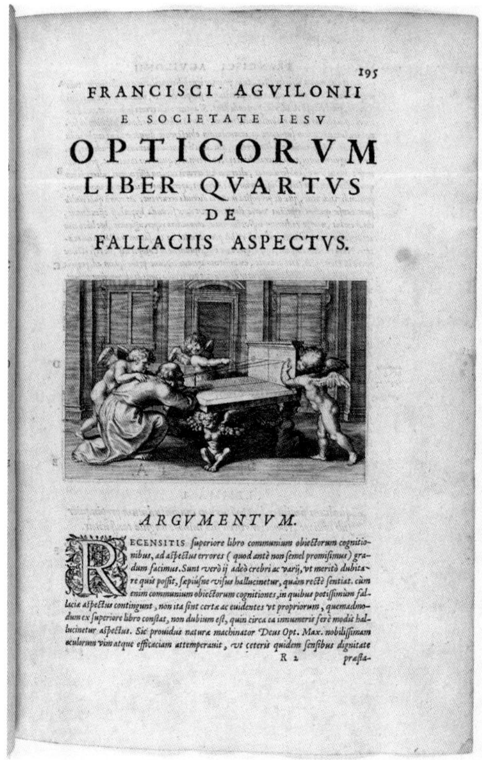

and institutional characteristics; the emblem of the Seminario Romano discussed above is an illustration of this trend.[77]

Jesuits' commitment to the visual representation of information extended to works on the mechanics of vision itself, like such illustrated textbooks on optics as the Antwerp-based Jesuit mathematician and architect Franciscus Aguilonius's (1567–1617) *Opticorum libri sex* (*Six Books on Optics*, 1613).[78] This work by the Rector of the Jesuit Maison Professe at Antwerp features a frontispiece and a vignette to open each of the six books, images designed by Peter Paul Rubens (1577–1640) and cut by Theodor Galle (1571–1633). The vignettes depict putti actively engaged in scientific experimentations that offer a persuasive demonstration of an "argument" (*argumentum*) in the text. In the first scene, putti anatomize a cyclops's eye, whose enormous scale facilitates the procedure (Fig. 118). A device in another vignette, more directly pertaining to deceptive appearances, demonstrates that an entity can seem to be in two distinct locations relative to a backdrop if it is observed first by one eye and then the other (Fig. 119). A putto guides an adult male figure to show him this optical effect. By integrating erotes into such scientific experiments, Aguilonius, Rubens, and Galle produced a playful environment of experimental study, just as the visual puzzles presented by deformations elicited pleasure.[79]

Actualized Allegories and Embodied Exegeses

Before concluding these paired chapters on Minim and Jesuit deformations, I must stress that the respective decoding projects Minims and Jesuits invited through their uses of visual deformations resonated with broader interpretive concerns in a period of intense controversy over exegetical authority. Although allegories and emblems, like some optical effects, engaged the viewer in the work of interpretation or decoding, Catholic Reformation thinkers never imagined the individual believer to be capable of approaching God's mysteries without support. Many deformations, as I have argued above, functioned as allegories for the Church's mediating role in deciphering what is difficult to see with the naked eye, especially for the laity; that institutional explanatory work ultimately and necessarily points the faithful to the divine, the transcendent God, and the source of salvation, Jesus.[80]

A well-known passage from a sermon on providence delivered on March 10, 1662, by the Jesuit-educated French bishop and theologian Jacques-Bénigne Lignel Bossuet (1627–1704) is productive to read in relation to the argument I have proposed—that "coming to understand an anamorphosis" and "coming to a proper understanding of faith though the Church's guidance" would have been recognized by educated contemporaries as analogous:

> When I consider within myself how human things are arranged in a confused, uneven, and irregular manner, I compare it often to certain paintings, which are shown usually in the libraries of the curious as a game of perspective. The first glance shows us only the shapeless lines and a mixed confusion of colors, which seems to be either the attempt of some apprentice, or the play of some child, rather than the work of a learned hand. But as soon as the one who knows the secret makes you look at them from a certain place, immediately all the uneven lines coming together in a certain way in your sight, all the confusion unravels, and you see a face with its lineaments and its proportions, where there was previously no appearance of human form. It is, it seems to me, gentlemen, a fairly natural image of the world, of its apparent confusion and its hidden correctness, that we can never notice except by looking at it from a certain point that faith in Jesus Christ helps us to discover.[81]

Discussing anamorphoses as a "game" with a hidden view, Bossuet points to a guide—"the one who knows the secret"—who can teach observers the key to examining them and deciphering their forms. In noting "how human things are arranged in a confused, uneven, and irregular manner" and referring to the "confusion" and "hidden correctness" of the world, he may allude to circumstances that persistently perplex the faithful, as when God appears to allow wicked people to receive good things, while tolerating great suffering among virtuous individuals.[82] For Bossuet, the confusing images presented initially by anamorphoses symbolize human experience of a world that seems at first morally incomprehensible, but on reexamination "from a certain point"—from a point of view, or with a mirror that reforms the initial image—is discovered as coherent through faith in Christ. Although Bossuet does not directly insert "the one who knows the secret" into his spiritual analogy, from his intense commitment to preaching and his orthodox view that Biblical exegesis was the work of the Church, not the laity, we can conclude that he saw the Church as uniquely possessing the ability to show the observer just where to position the lens or mirror.[83] Bossuet seems to trust that one who operates from faith will ultimately find that faith leads him or her to the right, in the sense of Catholic, perspective; faith is the guide, leading the viewer through doctrinal confusion to the wisdom of the Church.

In this way, Bossuet's remarks lend support to my claim that Catholic religious anamorphoses came to be understood as allegories of the Catholic Church's mediating role in pointing the faithful to the transcendent.

Bossuet's sermon further stresses the importance of finding the correct vantage point, analogizing it to replacing disorder with "wisdom":

> How great is the confusion of this painting! And does it not seem that these colors were thrown at random, only to blur the canvas or the paper, if I may speak in this manner? Perhaps you will find that what seems confusing is a hidden art; and if you know how to find the perspective from where you must look at things, all the inequalities will be rectified, and you will see only wisdom where you imagined nothing but disorder.[84]

Bossuet thus clarifies a spiritual philosophy associated with the transition elicited by anamorphoses from either perceptual confusion to clarity or from perceptual clarity to confusion to clarity.[85]

Not all early modern anamorphoses, not even all anamorphoses with religious imagery created by Catholic Reformation thinkers, were understood primarily or exclusively as allegories for the Church's mediating role. No religious allegory inheres in the cylindrical anamorphic panel in Palazzo Barberini (attributed to Niceron) that shows an amorous couple and a third figure, who pays close attention to the lovers' exploits (Fig. 120).[86] A work like this is presumably more interested in structuring social interactions through art and architectural features as we have seen in the Palazzo Spada—the patron indulges, and may invite his guests to indulge, in an elite pastime (here, perhaps, punning on a relationship between looking closely and voyeurism)—than in an unveiling of theological truths. And even as I underscore the epistemological resonances of Niceron's or Maignan's religious anamorphoses, I do not want to deny that these works too catered to monied tastes, or deny that they were appreciated for their ludic nature, even when they addressed religious themes.[87] I should further note that Protestants also created deformations—including works that were clearly intended to sway observers to the Protestant cause—that might equally enlist allegory in communicating theological messages to observers, though Protestant deformations (like Protestant allegories and emblems) are outside the purview of this book.[88]

If Catholic anamorphoses (and related illusionistic images) with religious subject matter allegorized the church's role as mediator between the faithful and the divine, these images also repositioned, in a novel format, the traditional work of exegesis. From late antiquity onward, theologians had grappled with the Bible as a difficult, sacred work that was supposed to be understood at various levels depending on the capacity and knowledge of the scholar or cleric engaged in the interpretation. Such learned scholars and theologians as Niceron, Maignan, Kircher, and Schott would have known the traditional medieval fourfold system for interpreting Scripture: the literal sense (sometimes referred to as the historical sense) pertains to episodes from the past; the allegorical level explicates something through its likeness to something else; the moral or tropological level pertains to correct conduct; and the anagogical or eschatological level has to do with the purposes of God in eternity.[89]

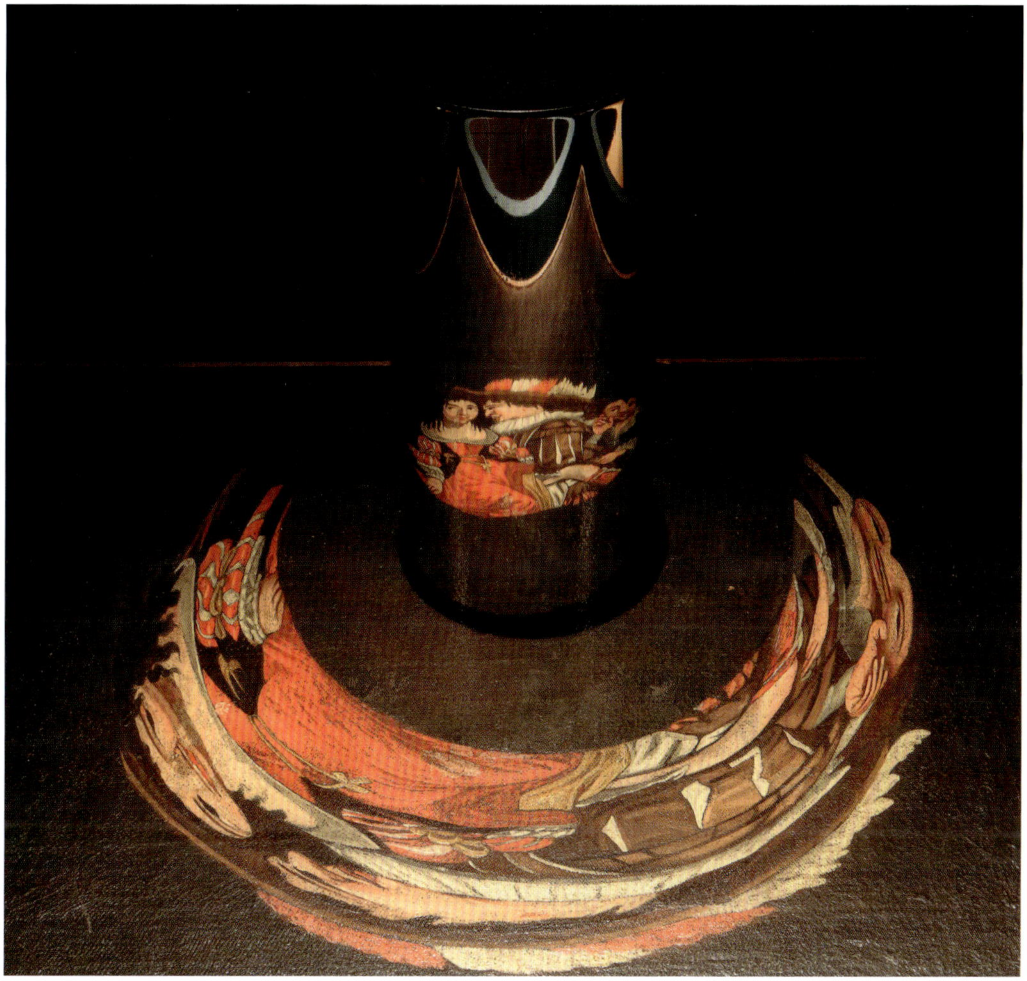

Fig. 120. Jean-François Niceron, *Scene of Married Life*, c. 1635. Cylindrical anamorphic oil painting on canvas, 19.69 × 26.18 in (50 × 66.5 cm). Galleria Nazionale d'Arte Antica, Palazzo Barberini, Rome [inv. n. 1956].

In response to intellectual developments in the fifteenth, sixteenth, and seventeenth centuries, scholars evolved new skills and tools for exegesis, and scriptural interpretation accordingly became increasingly contentious, both formally and practically. With the rise of philology, Renaissance humanists brought an improved understanding of Greek and other languages to Scripture and consequently began to question the accuracy of the Vulgate; scholars started to analyze biblical texts in relation to ancient rhetoric and dialectic and other relevant frameworks.[90] The missionary endeavors that attended the widespread colonizing efforts of Europeans meant that scholars attempting to translate theological terms and stories had to consider afresh the basic texts of Christianity. Questions of biblical interpretation became especially pressing in the sixteenth century, as Western Christendom divided into multiple factions. Protestant exegetes famously claimed that even the most impenetrable sections of the Bible, such as Genesis or Revelation, could be understood by solitary readers, aided by faith—without recourse to a skilled guide who knew its secrets. As Reformation leaders promoted the concept of *sola scriptura* ("by Scripture alone") to broad segments of society, Protestants turned away from the interpretive authority of the papacy or Catholic theologians with their fourfold division of meaning. The sheer number and variety of new printed Bibles further complicated scriptural exegesis: Theodor Dunkelgrün has noted that by 1518 printed biblical texts had appeared in multiple European vernaculars as well as in Greek, Hebrew, Aramaic, Arabic, and Ethiopic.[91] Finally, the period witnessed an increased interest in exegetical images, issued as free-standing prints or in print series and made possible by the rise of innovative publishers in Rome, Lyon, Antwerp, and other cities. Exegesis in the early modern period was never confined to written materials, and visual exegesis came increasingly to offer models for the interpretation of textual sources.[92]

We must, then, also consider the early modern Catholic enthusiasm for anamorphoses in relation to these rising concerns over interpreting Scripture. When we do so, we recognize that Catholic Reformation anamorphoses functioned much as allegory did in the theological and patristic multivalent interpretation of Scripture. The true meaning of the allegory is disguised, and the reader or viewer must use their learning to penetrate the allegory and arrive at the truth. Arriving at the true meaning of anamorphosis or related illusionistic images—whether one arrives at that meaning by changing locations, or by using a lens or mirror—is not only the work of intellect but also work done by the eyes of a person who moves through space along a predetermined path (or looks through a designated device to reform the image). Intellectual insight consists in knowing the right place to stand or how to use the provided mirror. Anamorphoses are designed to allow so-called naïfs to comprehend a truth that a scholar might determine via medieval hermeneutic schemes: visual manipulation directs the naive viewer to the same truth, by dint of what one might instead call an "actualized allegory," that is, an allegory made physically manifest through the observer's movement, or "embodied exegesis," an act of interpretation in which the body (as opposed to the mind) takes the lead in directing the interpreter to understanding. Mere seeing serves as a corollary for more intellectual modes of processing.[93] Anamorphoses seem to make the essence of theological mystery available even to observers with the most rudimentary background knowledge of theology.[94]

+

In theory, the processes of actualized allegory and embodied exegesis would have made theological exegesis more accessible to broader segments of society. In practice, however, with the exception of Sant'Ignazio, most Minim and Jesuit deformations were situated in closed, elite spaces that would not have been open to most individuals, and certainly not to most of the laity.[95] If we can presume that access to the anamorphic wall paintings in the collège royal Bourbon in Aix-en-Provence, the Santissima Trinità convent in Rome, and the Place Royale convent in Paris would have been more straightforward for members of the Society of Jesus and Minim order, then it is relevant to consider how social standing factored into entry into the Society and order. The constitution of the Society of Jesus, which generally drew its members from the upper echelons of society, cites nobility not as a necessary condition for admittance, but as a likely indicator of an applicant's suitability, with those of higher rank being deemed more suitable and those of lower rank being deemed less so.[96] Minims, by contrast, did not tend to come from aristocratic backgrounds, and their convents might be seen as less exclusive spaces for this reason and for the well-documented fact that Minim convents in Rome and Paris were open to a broad range of men of letters beyond the Minims themselves.[97] Here we might recall how the Jesuit Schott wrote of visiting the Santissima Trinità convent "most often" to admire its anamorphoses. The Place Royale convent in Paris, moreover, occupied a special role as an intellectual center. The French Minim friar and polymath Marin Mersenne (1588–1648) famously convened scholars and scientists on a weekly basis in the convent and engaged in far-reaching correspondence. Among the guests he welcomed were Pascal, the mathematician Gilles Personne de Roberval (1602–1675), the astronomer and mathematician Ismaël Boulliau (1605–1694), the philosopher, experimentalist, and observer Pierre Gassendi (1592–1655), the mathematician and collaborator of René Descartes (1596–1650) Claude Mydorge (1585–1647), the royal engineer Pierre Petit (1594–1677), and Girard Desargues. By 1639, the Palace Royale convent library had grown so esteemed that Minims made it available to learned readers; by the late seventeenth century, the collection held more than ten thousand volumes.[98]

If we revisit the Minim anamorphic corridors of the Place Royale in Paris and Santissima Trinità in Rome, thinking about allegory in its exegetical sense, then we see that novitiates from the bourgeoisie and artisan classes, as yet with limited intellectual training, would have had access to these works alongside intellectual elites. And although their floors do not seem to have had marble markers in the way of Sant'Ignazio—there was no obvious signaling on the floors of the "certain point" for viewing—the frescoed corridors, like any corridor, functioned as passageways, allowing individuals to move from one place to another. The Minim anamorphic corridors funneled students, visitors, and professors to study rooms and different parts of the convent; moving through the corridors necessarily asked that Minims and their guests travel through perceptive states from confusion to clarity that would have aided spiritual movements. In the next chapter, I will consider the theological mechanics of anamorphosis—how these actualized allegories used a kinesthetic exegesis to concretize difficult tenets.

Propositionis verba si penè singula notentur, quid, ac quibus cautionibus agendum sit indicant. Nam in faciendâ hic deformatione non speculum cylindricum, sed superficies cylindrica quæritur, in qua effigies aliqua pingatur, & planum non vbilibet, aut quomodolibet collocatum, sed horizonti parallelum volumus, vt in eo deformatio & maior, & occultior fiat. Deformatio autem à nobis facienda in propositione additum habet id aduerbij: *Ritè*, quoniam mechanicæ, aut geometricæ omnes deformationes vt ritè, ac certis regulis fiant, exemplum accipiant necesse est a naturalibus proiectionibus, ac traiectionibus luminosorum radiorum. Nos igitur emendatissimam deformationem in plano horizontali exhibituri à naturæ, quasi dicam, arte hic ordiemur.

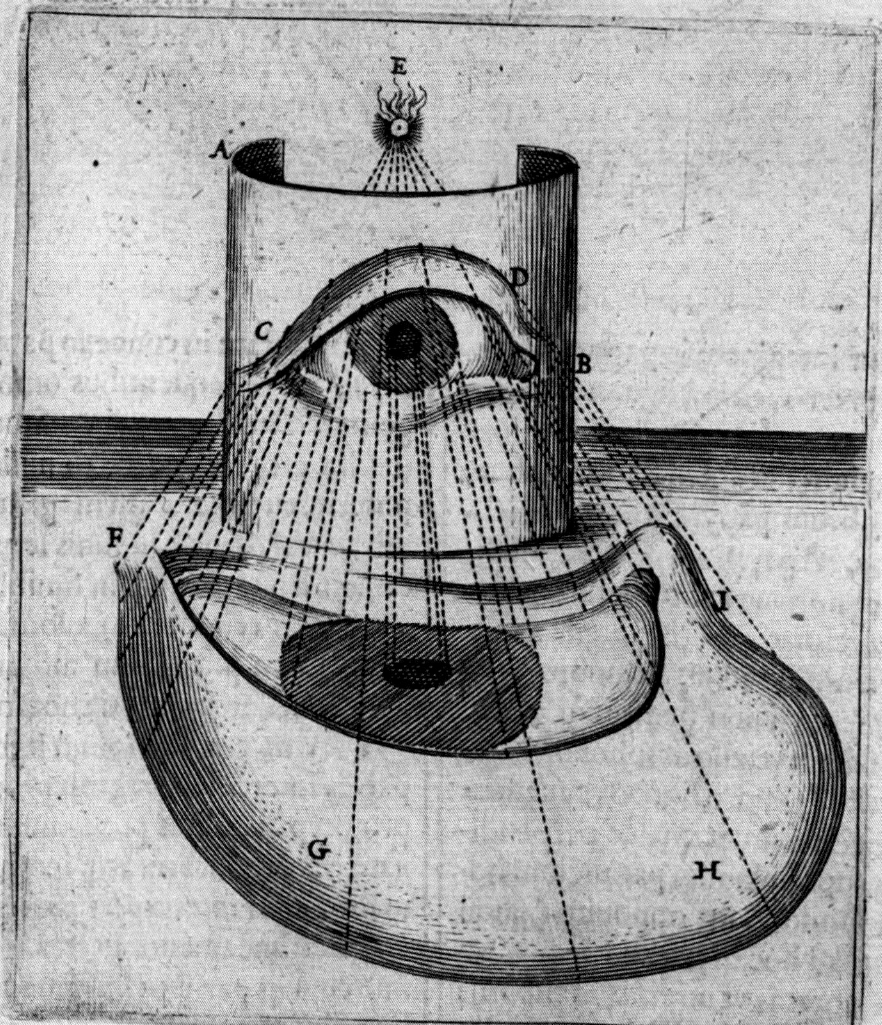

Telam, vel papyrum, vt AB, circa ligneum cylindrum incuruato; in eaq; papyrô cylindricè sic incuruatâ imaginem quampiam effinge (vt oculi CD) per quam emendatissimam ita, vt adspectanti appareat eo quamproximè modo, quo appareret emendatè picta in plana aliqua superficie. Neq; verò necessum est colores in picturam inducere, sed prima sat erunt, ac præcipua lineamenta plumbo, vel atramento signata. Ideo autem in propositione cautè positum est effigiem signandam esse in superficie cylindrica, ne quispiam erraret dum in plano picturam signaret, deinde planum picturâ imbutum incuruaret cylindricè; tunc enim effigies ritè picta in plana superficie deformaretur ab ea incuruatura cylindricâ: quæ res nostro instituto multum officeret; volumus enim, post deformatam in plano horizontali effigiem, reddere eamdem reformatam, atq; emendatam etiam in cylindrico speculo, vt mox docebo. Non apparebit autem in eo speculo emendata, nisi in cylindrica superficie, vt præcepimus, pingatur emendatè, atq; inde in planum horizontale deformata proijciatur.

4 OBSCURITY JUSTIFIED

RECOGNIZING THAT ancient rhetoric, however its principles were applied, was important to the development of early modern Italian art theory and practice—and having observed in the last chapters that Catholic practice embraced the deformations and reformations of anamorphoses as persuasive tools—it would be helpful to investigate further classical perspectives on obscurity. For if the Christian framing of human experience as a journey from obscurity and chaos toward divine revelation undoubtedly shaped early modern Catholics' thinking about anamorphoses, classical thinking around obscurity further supported contemporaries as they deliberately chose unclearness, or when they identified unclearness as desirable. In *De finibus bonorum et malorum* (2.15), for example, Cicero distinguishes between well-founded and objectionable obscurity: he holds that obscurity can be excusable if it is deliberately adopted or if it emerges from the abstruseness of the subject.[1] In the latter case, obscurity could function as a diagnostic instrument that signals the necessity of critical discourse or commentary on a text.

Obscurity in a text, as some thinkers argued, could have pedagogical value, catalyzing readers' interest and attention to engage more closely (as, we might recall, an anamorphosis could encourage a viewer to pay closer attention to the resolved image). In *De doctrina Christiana* (*On Christian Teaching*), Augustine proposes that the Bible features obscure passages to counter ennui and contempt:

> Those who fail to discover what they are looking for suffer from hunger, whereas those who do not look, because they have it in front of them, often die of boredom. In both situations the danger is lethargy.
>
> It is a wonderful and beneficial thing that the Holy Spirit organized the holy scripture so as to satisfy hunger by means of its plainer passages and remove boredom by means of its obscurer ones. Virtually nothing is unearthed from these obscurities which cannot be found quite plainly expressed somewhere else.[2]

Detail of fig. 121

Erasmus (c. 1467–1536) likewise presents a vindication of obscure allegory, insofar as it leads readers to labor for an underlying truth.[3] Erasmus's image of laboring for truth resonates with the work of observing an anamorphosis—moving in space or manipulating a mirror in the analogous work of embodied exegesis.

Writing in the late sixteenth century, Gabriele Paleotti echoes Cicero's justification of obscurity that may arise from the inherent difficulty of a subject: "Sometimes the attempt is made to express things that by nature cannot be expressed, things so recondite and difficult that there is no easy way to convey them to folk, like the operations of intelligences, the secrets of divine providence, the mysteries of predestination, and so on."[4] Although Paleotti urges artists "to refrain from such subjects as much as possible, yet assenting to them inside with true faith," this passage clarifies how Jesuits and Minims might have justified their anamorphoses vis-à-vis the rhetorical framing of sacred imagery: visual obscurity or deformation might be particularly justified for an image whose iconography pertained to "recondite and difficult" things or, as I will demonstrate, to "the secrets of divine providence [and] the mysteries of predestination."[5] It was not by chance that the providential image in the Chilean mountainside (discussed in chapter 3) featured an oscillation between confusion and clarity that, as I argued there, evoked anamorphic operations.

In chapter 1, we saw how tabernacles distorted classical colonnades through accelerated perspective to heighten the size and apparent value of the host, and in chapters 2 and 3, we saw how Jesuit and Minim anamorphoses, as a group, centered theologically or institutionally significant images that the Catholic Reformation Church wanted observers to pay close attention to, such as Saint Peter as the first head of the true Church or Saint Francis of Paola as a figure whose faith was worthy of imitation. And yet the Catholic impetus to deform and to reform seems to have been the strongest when Church figures were communicating around the central mysteries and doctrines that adherents were to accept with the most unquestioning faith, but that scholars found hardest to explain because they involved the divine reaching directly into human experience. If deformations and reformations were attractive in embodying the essential role of the Catholic Church as a mediator between the laity and God, these intentionally obscure—or justifiably obscure—images were especially attractive in embodying God's revelatory mediation into human reality. Catholic Reformation practitioners deemed visual obscurity to be actively encouraged where theological mysteries were most difficult, both because the obscurity of the mysteries justified the obscurity of the visual art and because that visual obscurity could actually be a plus, insofar as it prevented boredom in the observer. The objects in this chapter pair difficult theology with complicated anamorphic representations: through the religious and embodied exegesis of the viewer, the deformations resolve into a kind of clarity. When artists and patrons of the Catholic Reformation sought to communicate the most difficult mysteries of the Church, and they chose deformations to do this work, we see justifiable obscurity in both of its classical senses: by using deformation—anamorphoses and other optical effects—to model a mystery like the Eucharist, artists not only communicated the intrinsic difficulty of the subject matter itself, but also continued to engage observers in the progress toward understanding that we have seen in the prior chapters.

The figures in this chapter were writers and teachers of theology, as well as experts in the practical endeavor of making obscure images and objects. The Jesuit Mario Bettini and the Minims Jean-François Niceron and Emanuel Maignan all published works on mathematical curiosities, and all instructed students in their respective communities: Bettini lectured on philosophy and mathematics in Parma, while Niceron taught mathematics and Maignan philosophy and theology at Trinità dei Monti.[6] All used visual experiences to explicate theology. Their writings and their objects analyzed in this chapter were, like much of the material considered in chapters 2 and 3, produced in classical languages or held in restricted spaces that only select, often elite groups would have been able to access. Bettini's *Apiaria universae philosophiae mathematicae* (*Beehives of Universal Mathematical Philosophy*, 1642) and Maignan's *Cursus philosophicus concinnatus ex notissimis cuique principiis* (*A Philosophical Course arranged from its Best Known Principle*, 1653) and his *Sacra philosophia, sive entis supernaturalis* (*Sacred Philosophy, or of Supernatural Being*, 1672) were written in Latin and thus addressed to intellectual elites.[7] And yet, these objects and images were occasionally accessible to those from more ordinary backgrounds: Niceron's fresco in Rome (notwithstanding its Latin and Greek inscriptions) and his and the Minim friar Charles Plumier's vernacular texts could have been appreciated and consumed by those who did not belong to the intellectual and social aristocracy; although women were generally not permitted in Minim convents, the order (as noted in chapter 3) did draw members from lower social ranks. And anamorphoses like Niceron's in Trinità dei Monti invited all observers to interpret them in an embodied mode of exegesis that—unlike traditional hermeneutic schemes—did not demand great erudition from the observer / interpreter and therefore could have been meaningful to Minim students with less education.

In the works of Bettini, Niceron, and Maignan, the contrast between the perceptual confusion of deformation and the anamorphosis's resolution in clarity (and associated certainty) would help the beholder move from an obscure source requiring interpretation to a clarifying experience that facilitated an individual visual exegesis. Minim and Jesuit anamorphic and other reforming images evince their creators' awareness that highly illusionistic imagery could trigger in observers (through actualized allegories and embodied exegesis) a glimmer of insight onto such "recondite and difficult" topics (in which the divine extended into human experience) as "the secrets of divine providence, the mysteries of predestination," transubstantiation, and the mystery of the Eucharist.

Bettini and Anamorphic Revelation

In his *Apiaria*, Bettini suggests how anamorphosis can represent and even allude to an experience of prophecy. The text demonstrates how to produce an anamorphic image of an eye and how to use cylindrical mirrors to reform the image. An engraving integrated into his textual instructions for making specular anamorphoses shows a disembodied representation of an eye on a piece of fabric or paper; this image is affixed to a wooden cylinder and annotated with the letters A and B (Fig. 121).[8] The accompanying text reads: "Curve a piece of fabric or papyrus, as AB, around a wooden cylinder."[9] An image of a second eye, deformed and puddle-like and threatening to spill into the observers' space, occupies much of the print's foreground. In tandem with Bettini's text, the engraving offers instructions for representing an eye deformed such that its reflection in a column-like mirror would be coherent. Bettini recommends perforating the image affixed to the cylinder with a needle or some other stylus "as if the paper were a

OPPOSITE

Fig. 121. Unidentified artist, instructions for producing a specular anamorphosis from Mario Bettini's *Apiaria*, 1642, vol. 1, *Apiarium Quintum, Progymnasma Primum*, chap. 7, p. 7. Engraving printed on paper, measurements of the full page are 13.66 × 9.25 in (34.7 × 23.5 cm). Staats- und Stadtbibliothek, Augsburg [2 Math 11–1].

Fig. 122. Unidentified artist, specular anamorphosis from Mario Bettini's *Apiaria*, 1642, vol. 1, *Apiarium Quintum, Progymnasma Primum*, chap. 7, p. 21. Engraving printed on paper, measurements of the full page are 13.66 × 9.25 in (34.7 × 23.5 cm). Staats- und Stadtbibliothek, Augsburg [2 Math 11–1].

sieve."[10] Citing letters integrated into the prints, he continues: "Place some light before the concave, perforated paper, as in E. From E, the cast rays project, through the holes of the eye CD, which is pierced all around, onto the plane placed below, a certain interrupted and dissipated, shapeless form, as is FGHI."[11] As this passage reveals, the *Apiaria* is an instructional manual.[12]

While some engravings in the *Apiaria* show how to make specular anamorphoses, others visualize what specular anamorphoses look like once they have been created. One completed specular anamorphosis appears in a later chapter of the *Apiaria* (Fig. 122). The text clarifies the content of the integrated engraving: "An image [of a rose] erected and emended in a mirror, which the eye saw from the same point through direct sight inverted and deformed on the plane below."[13]

In guiding readers to make specular anamorphoses or at least to understand how they are constructed (though the book's explanations are occasionally difficult to follow), Bettini teaches them to observe their optical effect with greater perspicacity, to learn about deformations through acts of making, as well as through acts of looking. As will soon become apparent, Bettini is thus pairing the complex operation of the visual forms he describes with one of the thornier issues Catholicism had to grapple with: prophecy, or seeing the future. (Recall here Paleotti's "the secrets of divine providence [and] the mysteries of predestination" as examples of "recondite and difficult" theological matters.) Bettini describes his intermittent explanatory obscurity as deliberate, consistent with his project of presenting "secret" information: "Secret things must be brought forth in a semi-secretive way, and therefore elsewhere, after some things have been explained by clear examples, other things, which we will judge must only be indicated, we will deliberately gesture at abstractly, and without examples, as if under a veil."[14] With this warning Bettini not only frames anamorphoses as a secret art, but coaxes readers to be on the alert: if they observe his work acutely, they might join the select few able to decipher his book's hidden forms and, perhaps, recreate his devices.[15] His reference to a covering veil recalls a trope of spiritual exegesis as a process of unveiling ("through a glass in a dark manner").[16] Bettini's anamorphoses likewise invite a process of unveiling through the exegetical work of reforming images with cylindrical mirrors.

Bettini composed these instructions for producing anamorphoses—an eye, a rose—in the spring, probably in the year 1627. A few months later, and presumably well before the *Apiaria*'s publication in 1642, he learned that Girolamo Colonna (1604–1666) had been made a cardinal and, shortly after, archbishop of Bologna.[17] Bettini asserts that his practical explanations, centered on those images of eye (representing "foresight") and rose (the sacred purple), had predicted Colonna's elevation: "Of course, optic knowledge anticipated, in time, and pre-signified these favorable omens of my homeland."[18] Although Bettini hints that there is a relationship between anamorphoses and prophecy, he mystifies that relationship, such that it is unclear if it is in the research into the principles of optics, or in the making of an anamorphosis, or in the interpretation of the anamorphosis—or in some combination of these—that observers come into an awareness of future events. If for Bettini it was self-evident that perspectival research on optics and anamorphic images could train individuals in recognizing signs and symbols, such that they could better catch sight of divine providence at work, the link between this field of inquiry and prophecy needs explanation to modern readers. Bettini hints that the anamorphic images he has taught readers how to make also model one Church leader's gift of "pastoral foresight":

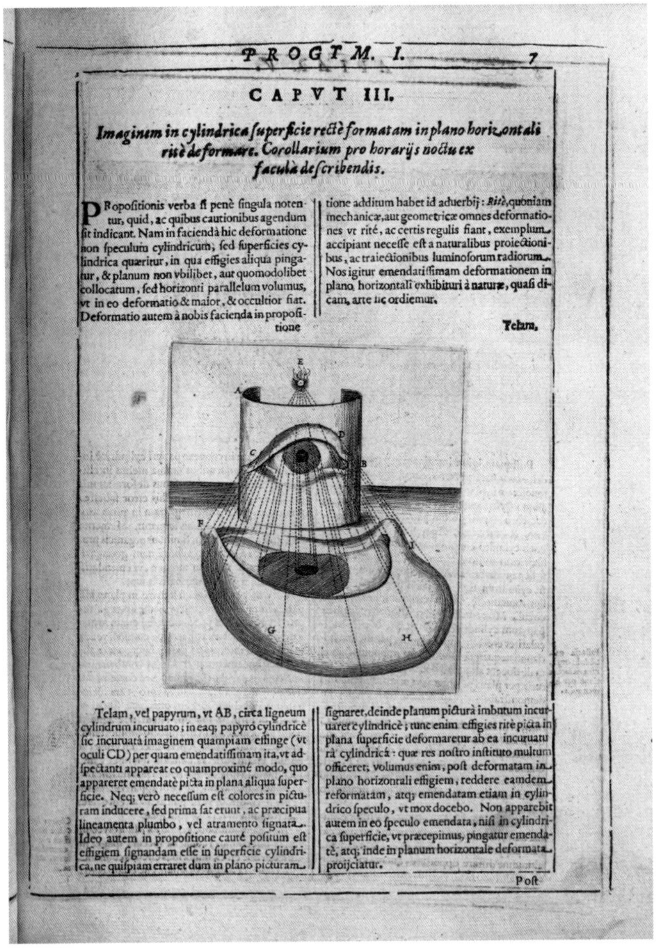

PROGYM. I. 7

CAPVT III.

Imaginem in cylindrica ſuperficie rectè formatam in plano horizontali ritè deformare. Corollarium pro horarijs noctu ex faculà deſcribendis.

Propoſitionis verba ſi penè ſingula notentur, quid, ac quibus cautionibus agendum ſit indicant. Nam in faciendà hic deformatione non ſpeculum cylindricum, ſed ſuperficies cylindrica quæritur, in qua effigies aliqua pingatur, & planum non vbilibet, aut quomodolibet collocatum, ſed horizonti parallelum volumus, vt in eo deformatio & maior, & occultior fiat. Deformatio autem à nobis facienda in propoſitione additum habet id aduerbij: *Ritè*, quoniam mechanicæ, aut geometricæ omnes deformationes vt ritè, ac certis regulis fiant, exemplum accipiant neceſſe eſt a naturalibus proiectionibus, ac traiectionibus luminoſorum radiorum. Nos igitur emendatiſſimam deformationem in plano horizontali exhibituri à naturæ, quaſi dicam, arte ſic ordiemur.

Telam, vel papyrum, vt AB, circa ligneum cylindrum incuruato; in eaq; papyrò cylindricè ſic incuruatà imaginem quampiam effinge (vt oculi CD) per quam emendatiſſimam ita, vt adſpectanti appareat eo quamproximè modo, quo appareret emendatè picta in plana aliqua ſuperficie. Neq; verò neceſſum eſt colores in picturam inducere, ſed prima ſat erunt, ac præcipua lineamenta plumbo, vel atramento ſignata. Ideo autem in propoſitione cautè poſitum eſt effigiem ſignandam eſſe in ſuperficie cylindrica, ne quiſpiam erraret dum in plano picturam ſignaret, deinde planum picturà imbutum incuruaret cylindricè; tunc enim effigies ritè picta in plana ſuperficie deformaretur ab ea incuruatura cylindricà: quæ res noſtro inſtituto multum officeret; volumus enim, poſt deformatam in plano horizontali effigiem, reddere eamdem reformatam, atq; emendatam etiam in cylindrico ſpeculo, vt mox docebo. Non apparebit autem in eo ſpeculo emendata, niſi in cylindrica ſuperficie, vt præcepimus, pingatur emendatè, atq; inde in planum horizontale deformata proijciatur.

Poſt

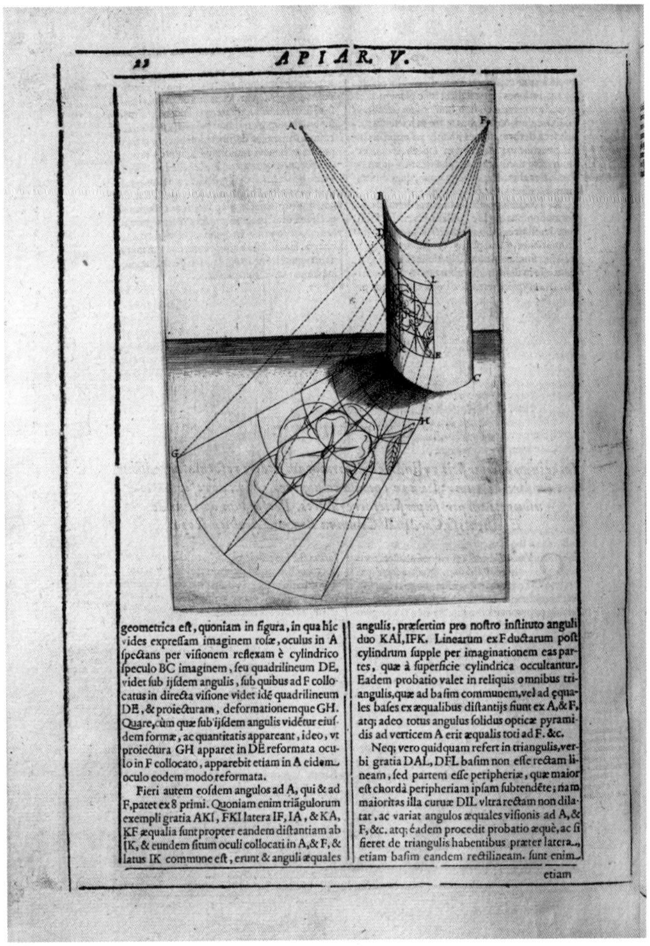

23 APIAR. V.

geometrica eſt, quoniam in figura, in qua hic vides expreſſam imaginem roſæ, oculus in A ſpectans per viſionem reflexam è cylindrico ſpeculo BC imaginem, ſeu quadrilineum DE, videt ſub ijſdem angulis, ſub quibus ad F collocatus in directa viſione videt idē quadrilineum DE, & proiecturam, deformationemque GH. Quare, cùm quæ ſub ijſdem angulis videtur eiuſdem formæ, ac quantitatis appareant, ideo, vt proiectura GH apparet in DE reformata oculo in F collocato, apparebit etiam in A eidem oculo eodem modo reformata.

Fieri autem eoſdem angulos ad A, qui & ad F, patet ex 8 primi. Quoniam enim triāgulorum exempli gratia AKI, FKI latera IF, IA, & KA, KF æqualia ſunt propter eandem diſtantiam ab IK, & eundem ſitum oculi collocati in A, & F, & latus IK commune eſt, erunt & anguli æquales angulis, præſertim pro noſtro inſtituto anguli duo KAI, IFK. Linearum ex F ductarum poſt cylindrum ſupple per imaginationem eas partes, quæ à ſuperficie cylindrica occultantur. Eadem probatio valet in reliquis omnibus triangulis, quæ ad baſim communem, vel ad æquales baſes ex æqualibus diſtantijs fiunt ex A, & F, atq; adeo totus angulus ſolidus opticæ pyramidis ad verticem A erit æqualis toti ad F. &c.

Neq; vero quidquam refert in triangulis, verbi gratia DAL, DFL baſim non eſſe rectam lineam, ſed partem eſſe peripheriæ, quæ maior eſt chordà peripheriam ipſam ſubtendēte; nam maioritas illa curuæ DIL vltra rectam non dilatat, ac variat angulos æquales viſionis ad A, & F, &c. atq; eàdem procedit probatio æquè, ac ſi fieret de triangulis habentibus præter latera etiam baſim eandem rectilineam. ſunt enim

etiam

Therefore, my reader, when you see in our work an eye in the columnar mirror, see, in your mind, a pastoral foresight in the most eminent Cardinal Colonna, and from these things, our optic reformations, know the future health of souls, and the reformed customs of the Church of Bologna (if, by chance, there should be a deformation). When you see a rose in the columnar mirror, recognize the sacred purple of our priest, and in it, the choicest rose of charity toward his people. Recognize, in the same sacred hero Colonna, as if a certain columnar mirror, in which the special lights of all these virtues appear and gleam marvelously, which virtues are sought in the sacred Prince endowed with purple, and in the best shepherd of souls. Let there be an end of the first introductory essay in this favorable optical foretelling.[19]

In a passage that evokes the *specula principum* ("mirrors of princes") genre—holding up the image of a perfect prince in which all other princes can see the ideal to which they should aspire—Bettini explains to readers that his optical devices trigger in observers an understanding of Colonna's foreknowledge, while also flattering Colonna in a form that is as elaborate as it is recherché.[20] The eye in the mirror foreshadows Colonna's eye and, in particular, his capacity to envision what must happen to ensure the salvation of his congregation and Church. The rose in the columnar mirror prefigures the color of the cardinal's vestments and Colonna's charitable actions. Colonna is also compared to the columnar mirror: as the mirror reforms a deformed eye, Colonna reforms a deformed Church; as a mirror reflects lights, Colonna reflects virtues. As Bettini and his readers undertake "optic reformations" (*nostris opticis reformationibus*)—rearranging images of distended eyes and roses in columnar mirrors—so does Cardinal Colonna reform the souls and procedures of the Bolognese Church. The columnar mirrors further allude to the name "Colonna," which means "column." This alleged similarity between visual and spiritual reformations, Bettini claims, renders his "optic knowledge" a prophetic symbol of Colonna's elevation.

Of particular interest here is Bettini's description of Colonna's "reforming" actions, offering further support for the suggestion that Minims and Jesuits conceived of anamorphoses as actualized allegories denoting the Church's mediating role in pointing the faithful to God. As experimenters in optics started with a deformed and confusing image and ended with a well-proportioned or reformed representation, Catholic Church leaders of the period began from distortions or deformations—identified, or generated, by Protestants or other reformers—and hoped to correct these in the mirror of the Catholic Reformation. We might even see the Council of Trent itself as a vast anamorphic operation.

Minim Prophecies

I proposed in chapter 2 that an early modern beholder of Niceron's and Maignan's anamorphic frescoes in Trinità dei Monti, seeing that the paintings' representations seem to shift as he moved through the corridors, may well have wondered if he was experiencing a divine revelation. The illusionism of Niceron's and Maignan's actualized allegories augments the intensity of observers' individual spiritual reaction: that illusionism invites a mode of interpretation in which the body's movement helps the mind and spirit to approach understanding. Deformations and reformations, which offered a means of modeling revelation, also offered an experience that could seem to effect revelations. These optical effects thus trained observers in visual understanding of an earthly nature that, in turn, prepared them to comprehend the mechanics of revelation and to distinguish truth in the event of divine revelation.

Given the role of anamorphoses in guiding spiritual acuity, it is suggestive that Niceron integrates the capitalized inscription "WITHOUT DECEPTION WE ARE DECEIVED" (*CITRA DOLUM FALLIMUR*) onto a banderole suspended from a branch in his fresco, with the emblematic annotation on the ubiquity of deceptions serving as a kind of thematic statement for the fresco (Fig. 123). Its text posits that to reach the truth, we must accept and move through the stage of deception and its concomitant confusion and uncertainty. And, unless we know how deceptions are produced, we will be deceived. The dictum recalls the Jesuit rhetorician, dramatist, and Marinist poet Emanuele Tesauro's (1592–1675) subsequent comments on the pedagogical pleasure of deception: "It is a secret and innate delight of the human intellect to find that it has been *playfully deceived*: the transition from illusion to disillusion is, in fact, a kind of learning by an unexpected way, and therefore most pleasing."[21]

Niceron's annotation had featured in the frontispiece of the *Perspectivae libri sex* (*Six*

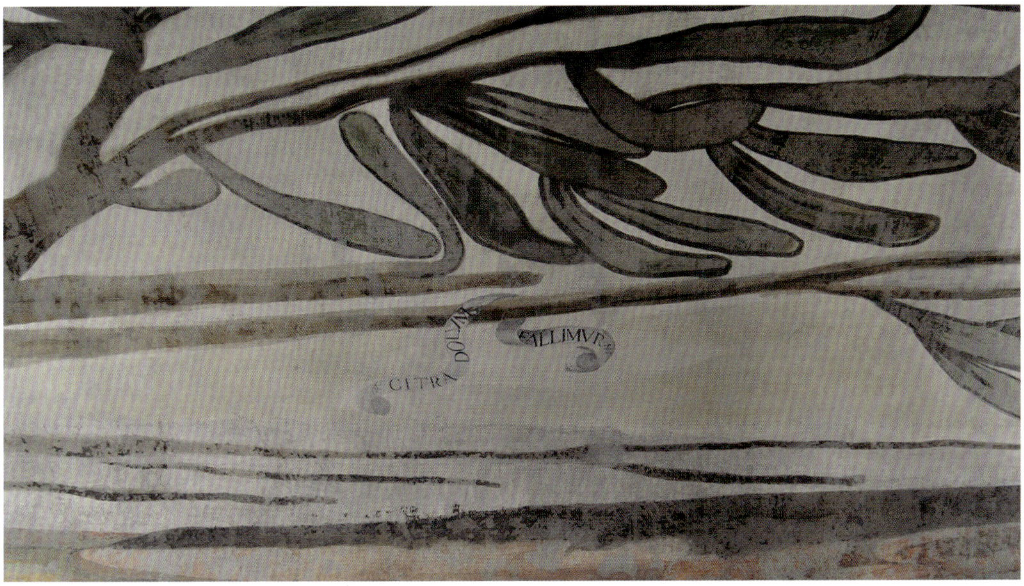

Fig. 123. *Saint John the Evangelist and Landscape* (detail), 1639–1640. Corridor, convent of Trinità dei Monti, Rome.

Books of Perspective, 1600) by the mathematician, philosopher, and astronomer Guidobaldo del Monte (1545–1607), who was the brother of Caravaggio's (1571–1610) early patron Cardinal Francesco Maria Bourbon del Monte (1549–1627) (Fig. 124).[22] In the *Perspectivae libri sex*, which foregrounds the relationship between artworks and their viewers, Guidobaldo ponders deception and vision, explaining that he will speak "about the varied appearance of visible things, and that manner of seeing which seems to deceive sight by a certain art, although the entire deceptive appearance may be done away with by means of mathematical demonstrations, which cannot lie."[23] Guidobaldo recommends that observers train themselves in mathematics, developing educated eyes so that they no longer succumb to deception. Mathematics is the tool of the truth seeker here, just as it is the chief tool of modern science. As we saw in chapter 2, Maignan likewise underlines the importance of deception in his *Perspectiva horaria*—"All of this work is composed in order to deceive vision" (*Tota huius operis, ad visum decipiendum compositi*).[24] Guidobaldo, Maignan, and Niceron draw the reader's focus to deception, and Maignan and Niceron manufacture deceptive deformations to help observers develop the discernment they need to inhabit and to work through deceptions. Much as Bossuet (see chapter 3) posits that perspective deciphered through faith can resolve the abstract bewilderment and uncertainty consequent from the world's apparent moral confusion, Guidobaldo proposes that the mathematics of perspective can counter bewilderment vis-à-vis concrete perceptual confusions and uncertainties. As Minims and Jesuits with a sophisticated interest in mathematics, Maignan, Niceron, and Bettini would have likely advocated a combination of faith and mathematical knowledge. In the preface to *La perspective curieuse*, Niceron praises perspective because it "makes us know and discern more perfectly the beautiful works of nature and of art"; elsewhere, he defines optics as "a science, which teaches to discern and to judge well the objects of sight."[25] The verb *discerner* is often grouped with the verbs *reconnaître* ("to recognize") or *connaître* ("to know")—as in this sentence from Niceron's preface.

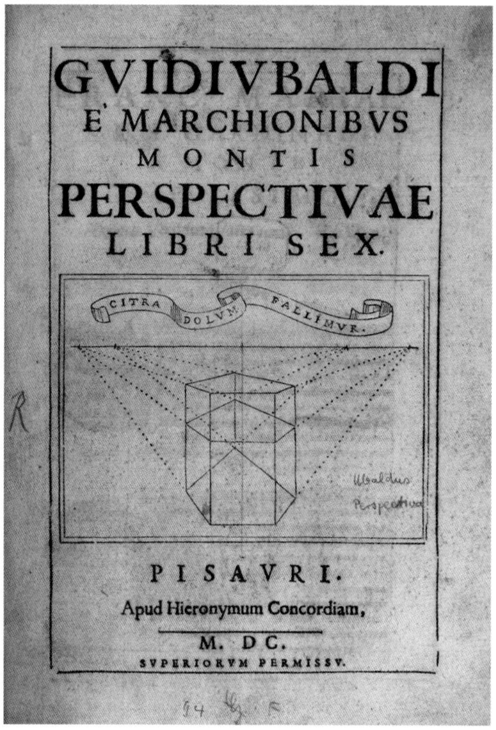
GVIDIVBALDI
E MARCHIONIBVS
MONTIS
PERSPECTIVAE
LIBRI SEX.

CITRA DOLVM FALLIMVR.

PISAVRI.
Apud Hieronymum Concordiam,
M. DC.
SVPERIORVM PERMISSV.

Fig. 124. Unidentified artist, frontispiece from Guidobaldo del Monte's *Perspectivae libri sex*, 1600. Woodcut printed on paper, dimensions of entire page are 12.1 × 8.1 in (30.7 × 20.5 cm). Bayerische Staatsbibliothek München [2 Math.a.105].

The observations on deception Niceron added to the banderole can be understood in relation to his comments throughout *La perspective curieuse* on viewing anamorphic deformations; he stresses the transition from a state of confusion and uncertainty to one of clarity, an experience echoed in Bossuet's account of anamorphoses. Observers may have gained not only perceptual clarity but also relative cognitive clarity as they came to understand the mathematics behind the ways in which they read the representations. They may have gained further cognitive clarity if, like Bossuet, they thought of the shift from perceptual confusion to clarity, through change in physical perspective, as an actualized allegory (that is, an allegory made physically manifest) for the shift from moral confusion to clarity through change in metaphysical perspective. Niceron's title for a section on producing anamorphic images captures this experience of transition: "To give the method to describe all sorts of figures ... that seem confused in appearance and from a certain point represent perfectly a proposed object" ("Donner la methode de descrire toutes sortes de figures ... qui semblent confuses en apparence & d'un certain poinct representent parfaictement un objet proposé").[26] With the phrase "from a certain point" (*d'un certain poinct*)—also deployed in passages by Bossuet and Maignan cited in chapters 2 and 3—Niceron likewise stresses the importance of viewpoint.

Niceron's *Thaumaturgus opticus* (*The Wonder-Making Optics*, 1646)—a revised Latin edition of the first two books of his *La perspective curieuse*—features an explanatory engraving illustrating the execution of a fresco of Saint John of Patmos (Fig. 125). The fresco was one of two further anamorphic wall frescoes that Niceron authored in the Place Royale convent in Paris, neither of which has survived.[27] This first fresco again represented Saint John the Evange-

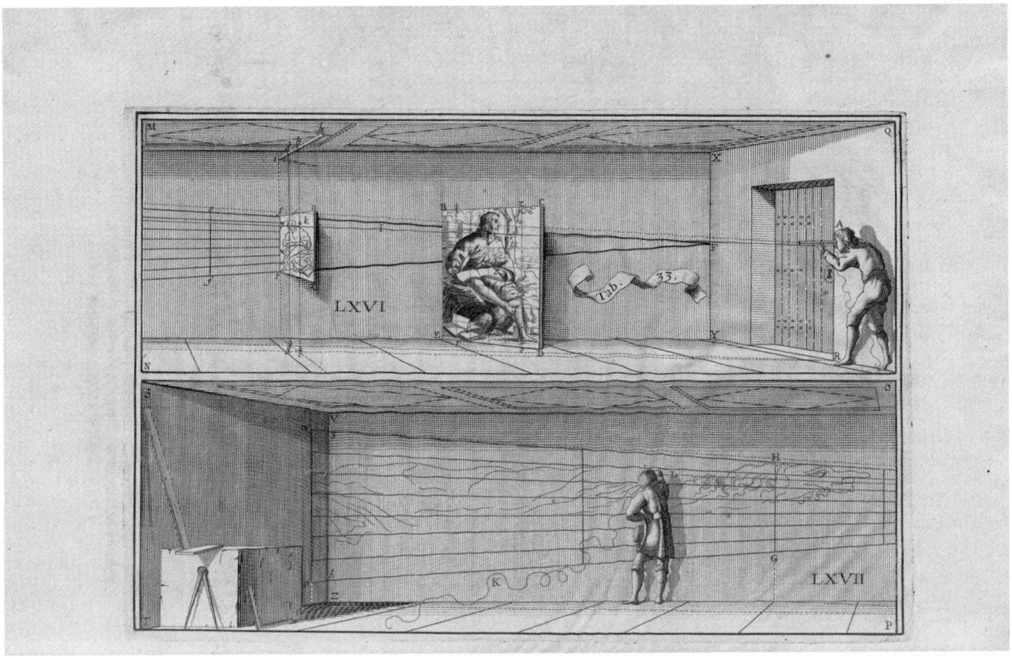

Fig. 125. Unidentified artist, table 33 from Jean-François Niceron's *Thaumaturgus opticus*, 1646. Engraving printed on paper, the dimensions of the entire page are 9.2 × 14.1 in (23.3 × 35.7 cm). Staatsbibliothek zu Berlin—Preußischer Kulturbesitz, Abteilung Handschriften und Historische Drucke [2" Mx 261–1: R].

list writing his prophetic visions on the island of Patmos; the other depicted Mary Magdalene in contemplation and was finished by Maignan after Niceron's death.[28] Niceron's discussion in *Thaumaturgus opticus* offers insight into what he might have included in the landscape of his Parisian anamorphic fresco of Saint John: "We have fashioned forests, fields, groves, temples, gardens, notable varieties of trees, plants, and flowers, caves, caverns, animals, little beasts, the curving coils of sea creatures, banks, shores, ships, fires, pools, the skiffs of fishermen, cargo ships, and triremes built for fighting."[29] As in Maignan's Roman anamorphic wall painting *Saint Francis of Paola*, these innumerable small details are likely to have detained observers and, perhaps, drawn them into meditation.

In his explanatory engraving, and throughout his *Thaumaturgus opticus*, Niceron, like Maignan and Bettini, offers readers practical guidance on how to conduct optical research themselves, so that they can gain an understanding of and familiarity with visual deceptions. Consider Niceron's discussion clarifying his procedure for coordinating the colors and shading of elements from each coextensive representation, that is, from the landscape and from the image of Saint John in the Parisian anamorphosis:

> In this way, we placed in the dark and shadowy folds of the green tunic dense forests and groves composed of many trees, in the brighter or conspicuous parts of this tunic, golden cornfields and harvests already ripe, in the white belt, a river's flow or a fountain's waters; in the white sheets of the open book, a huge lake, and in it, harbors, shores, ships, fishermen, and so on. In the head, caverns, caves, steep cliffs, rocks, edifices, indeed the ruin and downfall of the whole city of Babylon, for which we even placed the angels sounding trumpets.[30]

The two images represented within Niceron's Parisian anamorphosis of Saint John were closely related to one another, not just on a formal level but on a thematic one as well. From an oblique angle, the fresco showed Saint John on the island of Patmos writing about the Revelation; from another vantage point, the same wall painting presented the contents of the Revelation—the fall of Babylon and the trumpets that cue the apocalyptic events. Insofar as the former image foretold the latter, Niceron's anamorphosis, like Bettini's, can be seen as an actualized allegory that brings together optics and prophecy. Whereas a learned observer might see a fresco of John on the island of Patmos writing about the Revelation and immediately think about the contents of the Revelation, any observer of Niceron's painting with basic biblical knowledge can experience both John writing and the prophetic dynamics of his writing by moving in space: a viewer who cannot draw on their own knowledge of the Revelation derives almost all necessary information from the fresco itself simply by engaging in embodied exegesis, or by advancing forward and observing the image of John writing shift to make manifest that which he reveals in a process I have referred to as actualized allegory. It is the essence of a prophetic revelation that what it reveals can be seen here on Earth only obscurely ("through a glass in a dark manner"), with the implicit promise that it will be seen clearly and "face to face" in the hereafter. And hence it makes sense that Niceron takes the writing of the Apocalypse as his subject.

In the Roman anamorphic fresco as described in the *Thaumaturgus opticus*, we see that Niceron has inscribed a Revelation-related inscription on the binding of a book (Fig. 126). He comments on that inscription in the explanatory text: "I recall the projection delineated in the Roman convent, alluding in this way to the Greek words that I placed on the very book of the Evangelist that he holds before his hands, and I displayed [the words] 'A revelation of this perspective, the eyewitness of the Apocalypse' to be read by the spectators."[31] The rhetorical power of the Greek inscription "A revelation of this perspective, the eyewitness of the Apocalypse" (*tautēs optikēs apokalupsis, ho tēs apokalupseōs autoptēs*) lies in part in Niceron's wordplay.[32] He deploys a kind of pun known as an antanaclasis, which features the repetition of a word in a different sense. The word *apokalupsis* (meaning an uncovering or revealing of something) appears twice: first in the nominative and then in the genitive. The first time *apokalupsis* occurs without a definite article, but the second time it is accompanied by the definite article *tēs*, such that the reader understands the word to refer here to the biblical Revelation or the Apocalypse. In addition, the words *optikēs* ("perspective") and *autoptēs* ("eyewitness") both share the root *opt-*, which has to do with seeing. Finally, the letters *aut* appear in the first and last words of the sentence. The conjunction of all these repetitions in such a brief sentence indicates a play with sounds, suggesting that the phrase was meant to be uttered aloud, in a sort of performance that would have prompted a discussion (as in the discussions elicited by the visual art of Palazzo Spada).[33] The wordplay around variants of seeing urges a comparison of Niceron's revelation to Saint John's: as the saint presented the Revelation, Niceron presents a revelation of his perspective. The inscription suggests that the act of experiencing the fresco helps observers to understand perspective and its optical trickery and that this understanding is also related to those "secrets of divine providence [and] the mysteries of predestination." Rather than having to draw extensively on their own learning, the observer finds that the illusionism of the anamorphic representation intensifies their inner sight (and perhaps, by extension, their

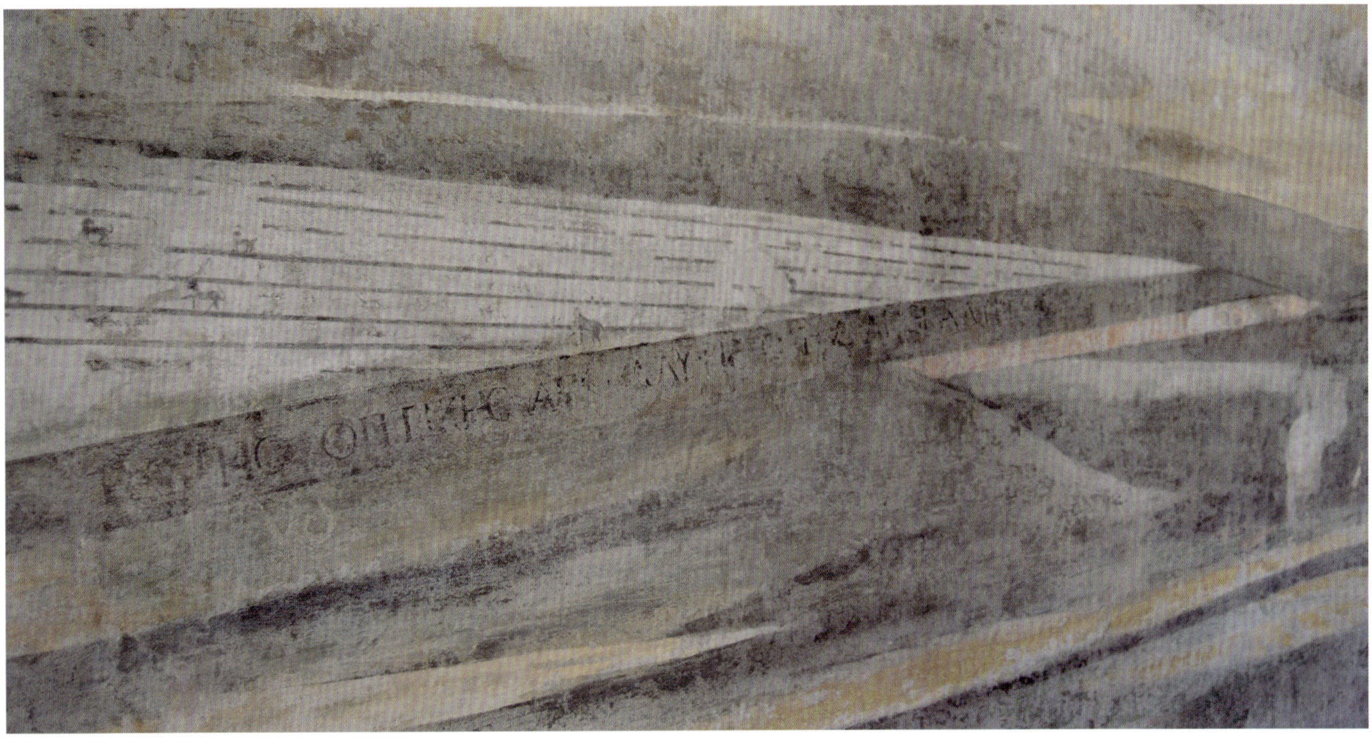

Fig. 126. *Saint John the Evangelist and Landscape* (detail), 1639–1640. Corridor, convent of Trinità dei Monti, Rome.

foresight) through the mechanics of actualized allegory—allegory is made materially apparent through the observer's movement. Niceron's frescoes and his discussion of those frescoes in the *Thaumaturgus opticus* add context to Bettini's comparison between optical research and religious revelation, his contention that optical devices could bring observers to understand at a deeper level divine prophecy. In Niceron's works an understanding of perspective, the science of optics, is conjoined with an understanding of God's divine plan. The quality of visual discernment comes to signal that of religious foresight. The future cannot be simply seen; one needs to know how to look. A Revelation of the future requires decoding.

An interest in the relationship between optics and religious revelation is not only apparent in Niceron's anamorphic wall painting; it is also manifest in his suggestion in *La perspective curieuse* to create deceptive images that represent existing prophecies. In particular, he discusses predictive images employed together with an optical device that uses a multifocal beveled lens to produce a single figure out of multiple separate figures (a device Maignan had also recommended to Cardinal Pamphilj for his villa). In his text and in tables at the end of *La perspective curieuse*, Maignan describes what makes it possible for multiple renderings of Ottoman sultans, when viewed through a perspective glass tube, to combine into a cohesive portrait of Louis XIII (Figs. 127 and 128).[34] (Both representations pertain to an existing prophecy, as we will see.) Regarding an engraving from *La perspective curieuse* reproduced in Figure 128, which shows the busts of twelve "Ottomans dressed in Turkish clothes" (*Ottomans vestus à la Turque*), Niceron writes: "This figure is more or less the copy, at least the design of a tableau that I traced and had painted two or three years ago, and which is kept presently in the library of our

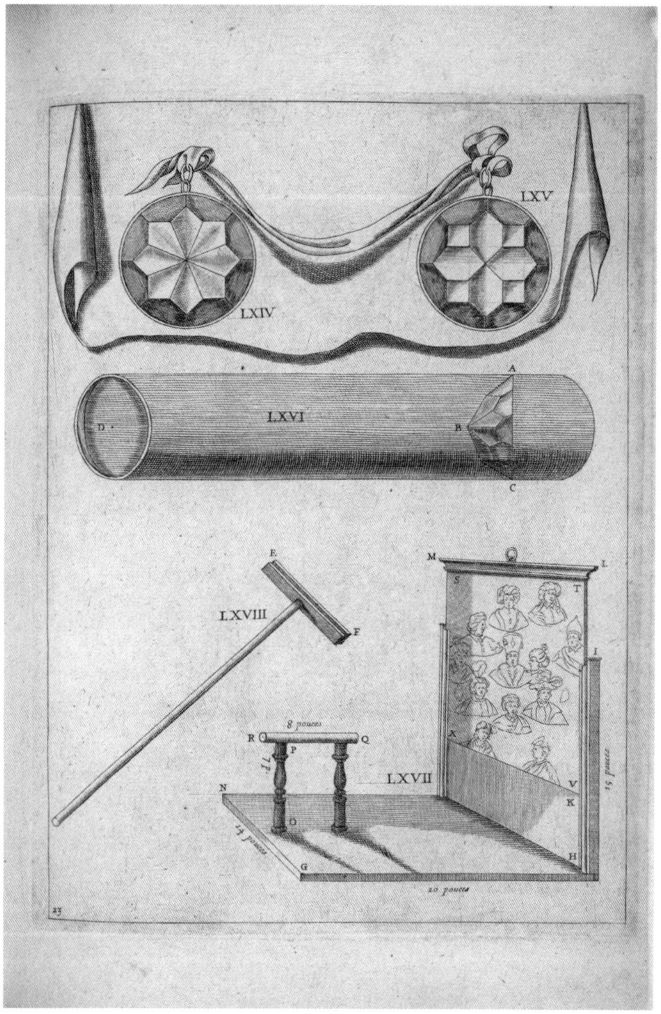

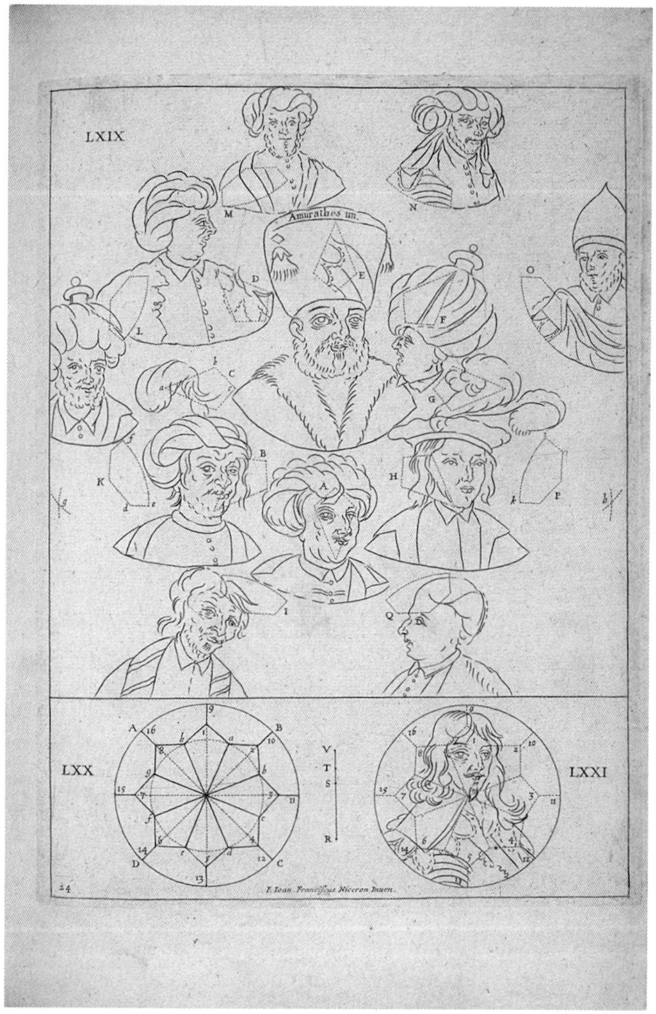

Fig. 127. Unidentified artist, plate 23 from Jean-François Niceron's *La perspective curieuse*, 1638. Engraving printed on paper, 10.67 × 7.56 in (27.1 × 19.2 cm). British library, London [L.35/4].

Fig. 128. Jean-François Niceron (designer), plate 24 from Niceron's *La perspective curieuse*, 1638. Engraving printed on paper, 10.59 × 7.56 in (26.9 × 19.2 cm). British library, London [L.35/4].

convent in the Place Royale in Paris."[35] In the frontispiece to *La perspective curieuse* engraved by Pierre Daret (c. 1604–1678), one putto looks through a glass tube at a similar rendering of sultans (Fig. 129). A second putto points to this composition with a stick: he guides his companion's observational practice and helps him to gain visual discernment, to understand how to look through the optical device and make sense of what he sees. It is tempting to imagine similar acts of guided viewing among Minims and visitors to their convents. By thwarting observers' expectations—those familiar with microscopes or telescopes would likely expect the lens to enlarge the image they had seen with the naked eye—the optical device challenges observers' assumed ability to discern formal arrangements. The device exemplifies the ways in which baroque imagery reforms forms into utterly unrecognizable shapes and in the process, also raises queries concerning discernment in acts of viewing.

Although this instrument once stored in the Place Royale convent library does not appear to have survived, a similar creation designed around 1642 under Niceron's guidance is now

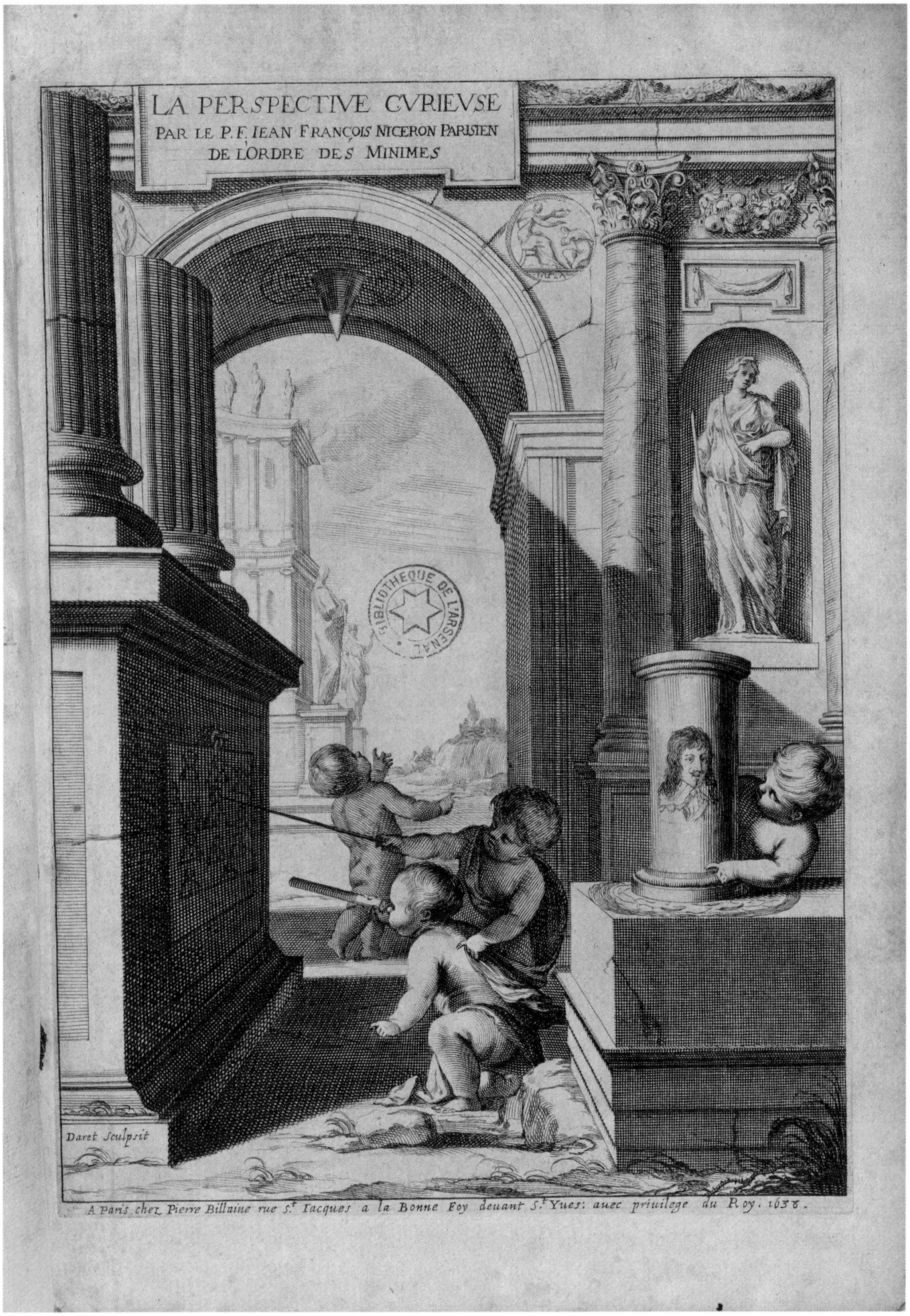

Fig. 129. Pierre Daret (engraver), frontispiece for Jean-François Niceron's *La perspective curieuse*, 1638. Engraving printed on paper, 11.22 × 7.87 in (28.5 × 20 cm). BnF, département Arsenal, Paris [FOL-S-1353 (1)].

Fig. 130. Jean-François Niceron, optical toy, 1642. Oil painting on a wooden board, fixed vertically to a second horizontal wooden board with a wooden support, 20.86 × 16.93 × 27.56 in (53 x 43 x 70 cm). Museo Galileo, Florence [Sala I, inv. 3196].

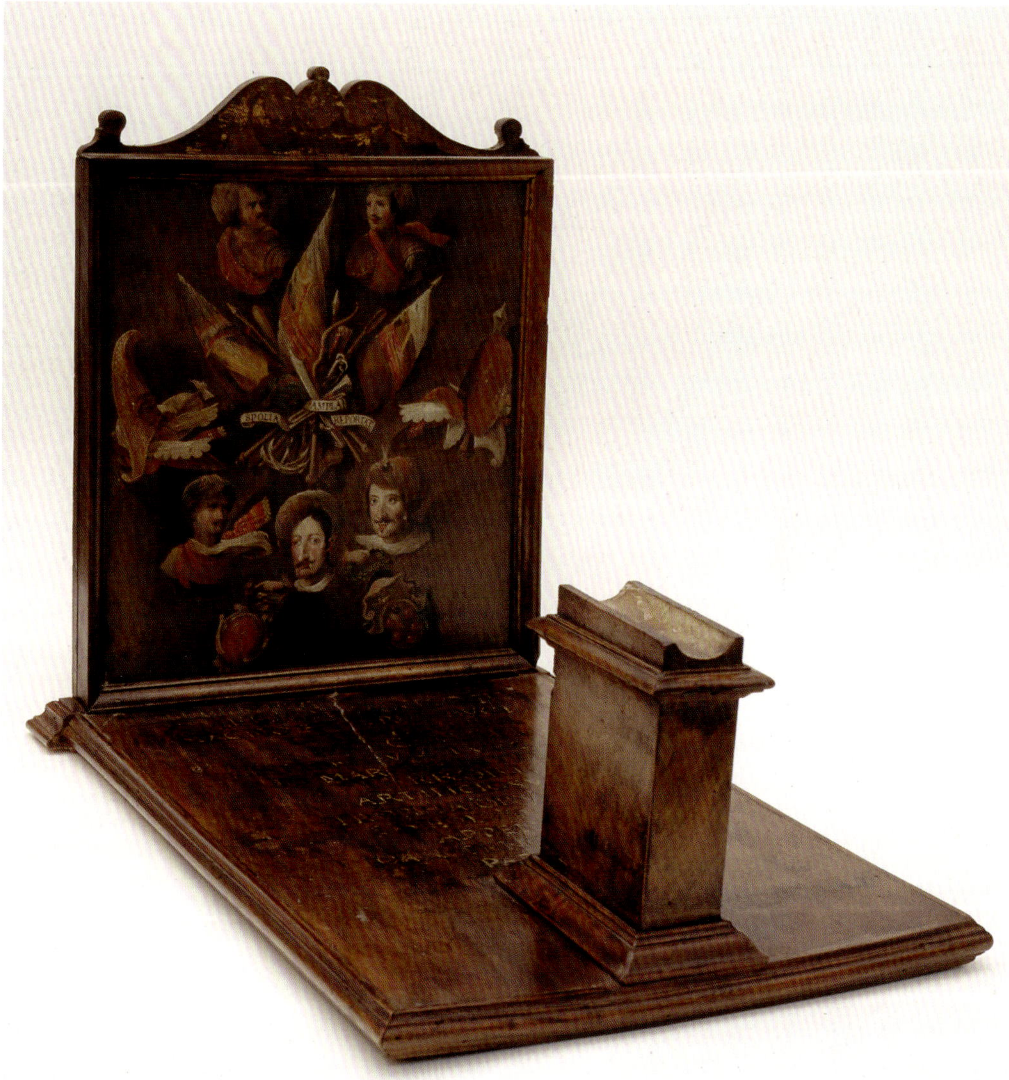

held by the Museo Galileo in Florence (Figs. 130 and 131).[36] When the Florentine painting is examined directly, it displays five sultans and the Medici coat of arms as well as various flags, shields, and spoils of war that all float above a brownish, olive-green surface (see Fig. 131). The inscription reads "It carries back ample spoils" (*Spolia ampla reportat*). The image is relatively coherent, notwithstanding the lowermost sultan's right eyelid drooping strangely alongside his wide-open left eye and the indistinctness of the background. When observed through its tube, the painting would have displayed Ferdinando II de' Medici, Grand Duke of Tuscany (1610–1670).[37]

In *La perspective curieuse*, Niceron situates his representations of sultans and Louis XIII, who was understood at the time as a reflection of God's power, in relation to a known prophecy:

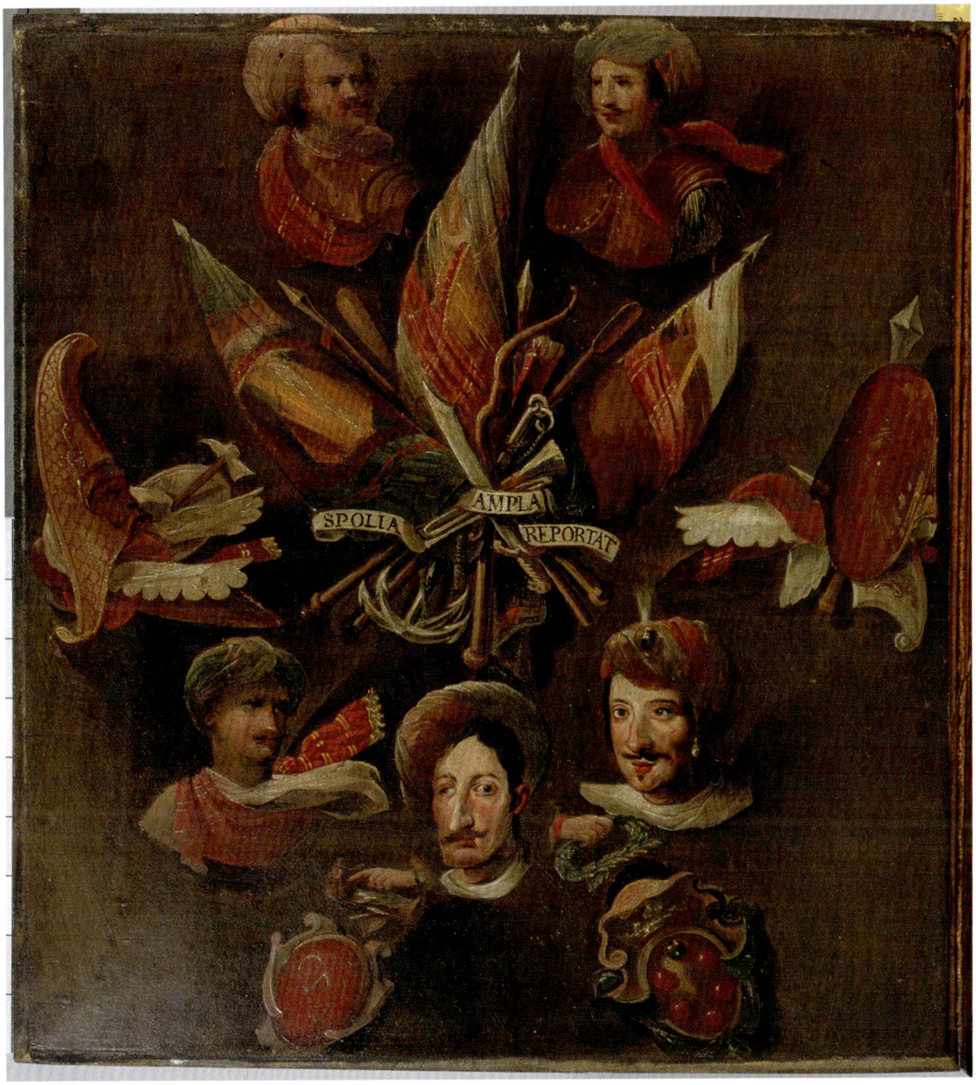

Fig. 131. Jean-François Niceron, optical toy, 1642. Oil painting on a wooden board, 20.86 × 16.93 in (53 × 43 cm). Museo Galileo, Florence [Sala I, inv. 3196].

> This design is made with regard to the prophecy ... which Muhammed formerly left to his followers. Recommending that they never offend the French monarchy, because their empire would never be ruined except by the power of one of its kings. On this wish to show that the honor of this conquest does not belong to any one other than Louis the Just [i.e., Louis XIII], we have contrived that most of the emperors in this tableau pay tribute to him, such that they contribute each some part of themselves to form his image, as if they were despoiling themselves to honor his triumph.[38]

The French scholar and orientalist Guillaume Postel (1510–1581) mentions this prophecy in a work written between 1564 and 1566.[39] He explains that the Muslim prophecy predicted that

Ottoman rule would be crushed in an attack by the so-called "Yellow Peoples" (*Ben Saphra*):[40]

> The Turks accord special authority, nearly as much as they give to their Koran, to a book of prophecies in which it is explicitly written that the prince and people of yellow color shall destroy the Turks as well as the other Ismaelites or Muhammedics, who are vulgarly called Muhammadans. From this evidence one can well believe that the Turks go to the greatest possible lengths to conceal the said prophecy from foreigners.[41]

Postel clarifies that the "Yellow Peoples" were taken to be the Gauls.[42] Niceron's device thus presents yet another instance of the science of optics and an act of reformation becoming conjoined with prophecy.

Beyond his discussion of this instrument of political propaganda, Niceron suggests further iconographies for the optical contraption that also present existing prophecies:

> One can take from the Old Testament all the figures of the same meaning and contrive that being painted and arranged on the plan according to the prescribed rules, through the telescope they only represent what is prefigured. I also had the thought several times to paint some prophets of those who spoke more expressly of the Virgin and the Incarnation, each with a flying band, on which was written the words of his Prophecy as, for example, Isaiah with these words "BEHOLD A VIRGIN SHALL CONCEIVE AND BEAR A SON" … & contrive that by the telescope we see only the Virgin with the inscription: "I SERVE THE LORD," &c.[43]

Since the Old Testament was often interpreted as prefiguring the events in the New Testament, this account again illustrates a movement from an obscure source requiring interpretation (the Old Testament) to a clarifying vision (the New Testament) through the process of actualized allegory and embodied exegesis.[44] Users of Niceron's optical device would first look at the prophets and then, a few moments later, behold the Virgin through the telescope. In this way, the temporal progress of observers would mirror the movement forward in time from the age of the prophets to that of the Virgin, if on a much smaller scale. This chronological reflection embedded in the device's use and iconography might then have contributed to the vividness of the prophecy. By means of embodied exegesis, simply by looking through a tube, observers could see and interpret events in the Old Testament as prefiguring moments in the New Testament.

The Passionflower and the Allegorical Interpretation of Nature

We have seen in prior chapters that Minim and Jesuit explorations of optics were subsumed under a broader interest in exploring the natural world, wondering at—and hoping to understand—divine creation. Jesuits and Minims turned to the practice of natural history much as they had turned to optics, because they wanted to decipher the enigma of divine providence; in this section, we will see how Catholic Reformation missionaries read a particular plant from the Americas as a naturally occurring symbol of God's being in the region, which testified to the righteousness of subsequent missionary endeavors there.

The Minim Charles Plumier writes that the study of botanical specimen provides opportunities for beholders "to cast their eyes onto the earth, to admire there those rare wonders in which his divine providence bursts forth in thousands and thousands of ways."[45] Elsewhere he declares that nature is a credit to God: "All that the earth produces blesses the name of the Lord."[46] "We are obliged to regard them," he goes on, "as so many illustrious wonders, which God has placed on the earth."[47] Plumier's mention of "illustrious wonders" and "divine providence" positions his natural history endeavors on a continuum with other Minim activities, particularly those through which friars explored "the secrets of divine providence [and] the mysteries of predestination" using visual deformations and reformations.

As Minims and Jesuits conjoined with prophecy visual deformations and reformations, they also conjoined with prophecy seemingly shifting formal manifestations of entities found in nature (as we already saw with the image in the Chilean mountainside). The book of nature was not easily opened to discernment. Nature was understood to be obscure. We might recognize the challenge of reading nature as not unlike the challenge of making sense of an anamorphic work prior to its reformation; though this is not an analogy that contemporaries described directly, they brought similar habits of reading and interpretation to each. If to understand an anamorphosis observers moved their bodies, through embodied exegesis, to understand nature they relied primarily on faith instead.

Plumier's comments about the natural world as an index of divine providence and munificence recall remarks made by the Spanish Jesuit missionary and naturalist José de Acosta (1539/40–1600) in his *Historia natural y moral de las Indias* (*Natural and Moral History of the Indies*, 1590). In this work, which also appeared in Italian (1596), French (1597), Dutch (1598), German (1601), Latin (1602), and English (1604), Acosta describes how the contemplation of nature helps man "to know and glorify the Author of them all."[48] If visiting an enormous sundial in a palace could inspire awe in relation to a patron, then contemplating the sun itself—or other elements in nature—would produce a much greater awe of God himself. The connection between natural history and theology was deeply ingrained in Jesuit thought: Ignatius, for instance, posited that the pious examination of the natural sciences could be understood as a form of prayer.[49] Acosta's book draws on his findings over the course of nearly twenty years as a missionary in Peru and Mexico as well as on his study of ancient and modern sources. Beyond paying homage to Pliny's *Natural History*, his title foreshadows an interest in not only the "natural" but also the "moral," that is, in the interrelations among nature, humanity, and the Christian worldview.

This religious exposition of nature appears in Acosta's discussion of the American plant that some early seventeenth-century observers came to call the passionflower, for visual traits they saw as revealing the Passion, that is, the sufferings of Jesus in the last days of his life:[50]

> The blossom of the ... passion flower, is believed to be a remarkable thing; they say that it has the signs of the Passion, and that the nails and the pillar and the blows can be found in it, and the crown of thorns and the wounds. There is something to be said for this belief, although in order to imagine it one needs a touch of piety to help one see it all; but much is very clearly visible.[51]

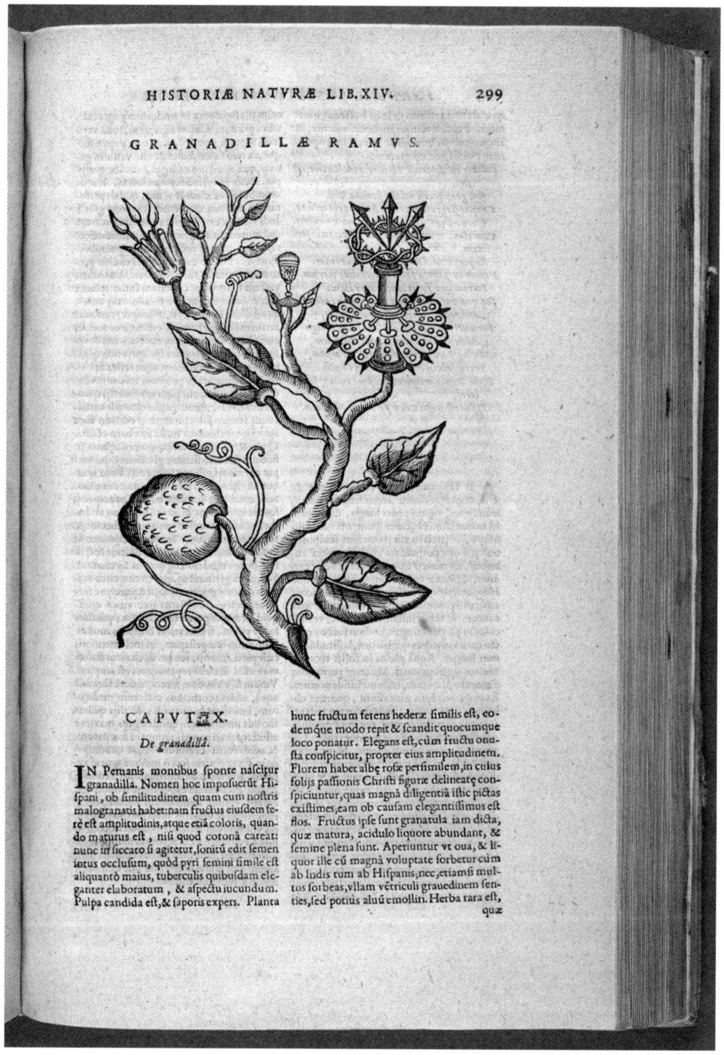

Fig. 132. Unidentified artist, passionflower from Juan Eusebio Nieremberg's *Historia naturae*, 1635, p. 299. Woodcut printed on paper, the dimensions of the page are 90.17 × 56.13 in (35.5 × 22.1 cm). The Huntington Library, San Marino, California.

Other Catholic Reformation authors, including the Jesuit Juan Eusebio Nieremberg (1595–1658), who taught at Madrid's Colegio Imperial, also analyzed the passionflower, which grew in many different species across the Americas.[52] Nieremberg integrated a woodcut of the plant with representations of the instruments of the Passion—the crown of thorns, the nails, the whips, the sponge, and the chalice—into his book published in 1635 (Fig. 132).

If when observing a Minim or Jesuit specular anamorphosis beholders needed to hold a mirror to discern the Catholic Reformation perspective, when it came to seeing the passionflower, Acosta suggested that the lens required to discern the *arma christi* was a "touch of piety." Or the empirically observed flower and the flower that "has the signs of the Passion" might be compared to the two ends of an anamorphic wall painting that leads observers to move from one relatively clear representation to confusion to another clearer representation. In drawing this analogy between the interpretation of an image (reformed by moving in space) and

Acosta's reformation of the passionflower (by looking at it through the lens of faith), I suggest that the makers of optical puzzles were using humans' creative powers in ways they understood to imitate God's *modus operandi*: looking around them at the universe that yielded its secrets slowly and reluctantly, Minims and Jesuits wondered at the divinity that created it; they created visual puzzles in imitation of God's own puzzles and in hopes of reinforcing faith in the represented mystery. Following this line of thought, some Catholic Reformation thinkers read the passionflower's imagery as an indication that God was present in America and approved of the missionary endeavors among the indigenous populations. This reading amounted to a claim of prophecy.

A shared belief that the created natural world held the secrets of its maker did not always, of course, lead to agreement over what those secrets were. Other Catholics who examined the passionflower doubted that the flower showed the *arma christi*. The Catholic apothecary John Parkinson (1567–1650), who served as principal royal botanist to King Charles I, unequivocally rejected the Christological reading in his *Paradisi in Sole Paradisus Terrestris; or, A Garden of All Sorts of Pleasant Flowers Which Our English Ayre will Permitt to Be Noursed Up* (1629): "Some superstitious Jesu-ite would faine make men beleeve, that in the flower of this plant are to be seene all the markes of our Saviours Passion; and therefore call it *Flos Passionis*; and to that end have caused figures to be drawne, and printed, with all the parts proportioned out, as thornes, nailes, speare, whippe, pillar, &c in it."[53] Parkinson presents two images of the passionflower in his book: a symbolic depiction of the flower with the instruments of the passion (to be rejected) and a naturalistic representation (to take its place).[54] Parkinson leans on the vocabulary of reforming and deforming in his acknowledgment that Jesuits will not believe his denunciation of their interpretation: "I must referre them [i.e., those who believe the Jesuits] to God, and hee knoweth the truth, and will reforme or deforme them in his time."[55] Although this passage is open to interpretation, we might infer here that Parkinson thinks that God will reform or deform the false visions of those who believe the Jesuit lies concerning the flower. According to Acosta, a "touch of piety" was needed to reform the passionflower into its Christological form. According to Parkinson, by contrast, God was needed to reform the misleading perceptions of Jesuits. This example demonstrates how Catholics could be divided in whether to adopt an allegorical reading of nature and how conflicting interpretive camps adopted the notions of deforming and reforming to opposing ends. Whether or not something was understood to be deformed depended on how one perceived.

The botanist and physician Pietro Castelli (1574–1662) also dismissed the Jesuit readings of the plant in his 1625 description of the Farnese gardens in Rome.[56] He presented the flower in two engravings, neither featuring symbolic imagery (Fig. 133).[57] Castelli's engravings and the naturalistic print in Parkinson aim to replace deceptive images with naturalistic ones. They are designed to distinguish the object of vision from the religious expectations that it had come to evoke.[58] These naturalistic images imply that the miracle is no miracle at all: it is actually a misinterpretation, a misunderstanding, and a visual deception. On the one hand, Castelli and Parkinson's version of understanding nature is opposed to the version where faith leads observers to identify a higher, spiritual truth; on the other hand, both camps spoke the same language insofar as each considered nature to be a manifestation of God's will by means of which God could be understood and adored.[59]

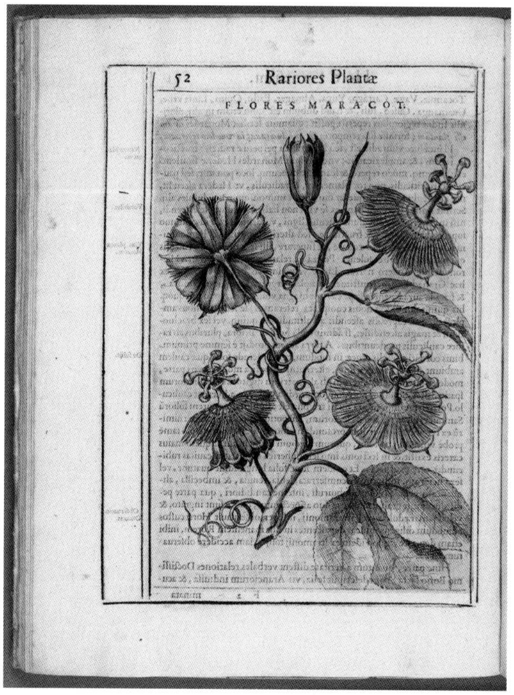

Fig. 133. Unidentified artist, passionflower from Pietro Castelli's *Exactissima descriptio*, 1625, p. 52. Etching printed on paper, the dimensions of the page are 12.2 × 8.3 in (31.1 × 21.2 cm). The Huntington Library, San Marino, California.

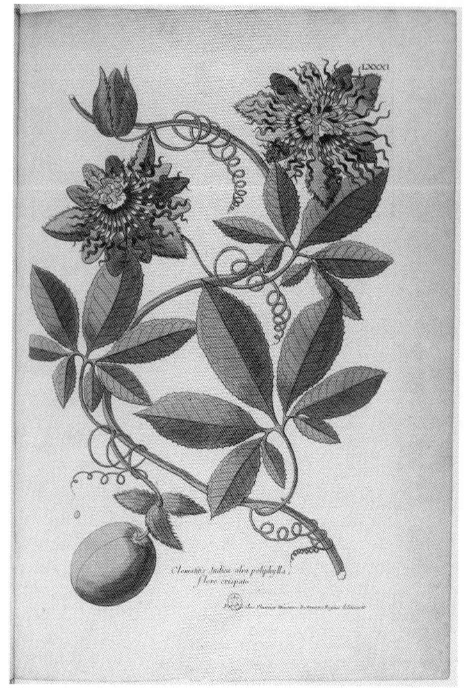

Fig. 134. Charles Plumier (designer), passionflower from Plumier's *Description des plantes*, 1693, plate 81. Colored Engraving printed on paper, 14.57 × 9.25 in (37 × 23.5 cm). BnF, département Estampes et photographie, Paris [RESERVE JD-15 (A)-PET FOL].

The naturalistic mode of representing the passionflower without Christological symbols was also adopted by members of the Roman Accademia dei Lincei and indeed by Plumier, who discusses and reproduces several different kinds of passionflowers in his *Description des plantes* (Fig. 134).[60] His naturalistic prints notwithstanding, Plumier mentions the plant's symbolic resonances in his accompanying text: "There are, however, some to which they give the more particular name of … *Fleur de la passion*, because we believe we find on the plant some marks of the Savior's passion, such as the three nails, the five wounds, the whips, and the ropes."[61] When describing the form of the passionflower reproduced in his plate 81, Plumier uses the *arma christi* to characterize the parts of the plant: "The column, the nails, and the arms where the wounds are attached, are whitish, and all speckled with red spots: the head of the nails are split in two, and the wounds are covered with a very fine yellow dust."[62] Although Plumier's prints of the passionflower do not show the *arma christi* on the flower, he still pays homage to the Christological interpretation of the specimen, reflecting a shared sympathy with the notion that nature manifests the merit of its maker.

The Eucharist and the Lens of Faith

That deformations and reformations were seen as especially valuable in cases where they could shed light on a great mystery is evident from the visual devices designed both to embody the doctrine of transubstantiation and to evoke the mystery of the Eucharist. Theological writings on the Eucharist debate one of the most discussed signs in all of Catholicism. These debates revolve partly around the exact role of perception in the doctrine of transubstantiation and as such are debates that necessarily engage with questions of representation and sense perception. The act of visually reforming emerges as a practical response to this much-discussed doctrine because of its capacity to model the miraculous transformation of the Eucharistic bread and wine.

This section considers Maignan's mode of working through the belief that the Eucharist was the body of Christ alongside his approach to anamorphoses: to explain this central mystery of the faith, he promotes instruments and operations that resemble those of anamorphosis—from relying on curved mirrors to envisioning seeing "through" appearances (and thus shifting his vision) to perceive a worldly object as a portal leading to some larger divine truth.[63] Although Maignan does not acknowledge or directly address the parallels between his approach to artworks and optical devices on the one hand and his thinking about the Eucharist on the other, similarities in his articulation of each reveal that ultimately genres like anamorphosis were not just artistic practices but also came to function as artistic expressions of theology. Moreover, analyzing Maignan's contributions to the mystery of the Eucharist clarifies how he and others might have made sense in theological terms of contemporaneous multifocal beveled lenses and their Eucharistic images.

Scholars' interest in devices that would, in visual and physical terms, embody the transformation of the Eucharistic bread and wine and evoke the mystery of the Eucharist—an interest we will see expressed below—may emerge from the enormous challenges they confronted as they tried to establish doctrine that correctly explains the nature of the transformation. Throughout this period of reformations, dramatic disagreements arose concerning the physical aspect of the consecrated elements, that is, of the Eucharistic bread and wine. These and other disagreements, broadly pitting Catholics against Lutherans, built on previous disputes that had transpired within the Catholic Church.[64] Luther articulated the doctrine of consubstantiation—following the consecration, the substances of Christ's body and blood coexisted with those of the bread and wine—developing this position against the medieval doctrine of transubstantiation—a term employed by the Fourth Lateran Council in 1215.[65] After the rediscovery of Aristotle's philosophy, the doctrine of transubstantiation had been formulated to hold that the entire substance of the bread and wine is converted into (according to Thomists) or destroyed and supplanted by (according to Scotists) the entire substance of Christ's body and blood, with solely the accidents (that is, the unessential characteristics, such as the apparent properties) remaining.[66]

Seventeenth-century conceptions of "accident" drew on Aristotle's writings as well as the third-century text *Isagoge* by the philosopher Porphyry (c. 233–c. 305). Porphyry defines accidents as items that can be present or absent from a subject without its destruction.[67] In the thesis print titled *Typus* designed by the Carmelite professor Jean Chéron (which we saw in chapter 3), accident is personified as an old man who holds himself upright by leaning on a cane and tree stump (Fig. 135). This engraving reiterates then-standard Aristotelian scholastic orthodoxies; the image features the inscription, "The accident needs support" (*Accidens eget fulcimento*). Although subjects are able to exist without some of their accidents (but not necessarily without any accidents whatsoever), it is impossible for accidents to exist without substances, in the absence of a miracle like that of the Eucharist. Chéron describes the accident as needing support because, just as the elderly man is unable to stand in the absence of support, an accident is unable to subsist in the absence of support from a substance. For example, the accident of "being short" is unable to exist in the absence of a substance characterized by shortness.

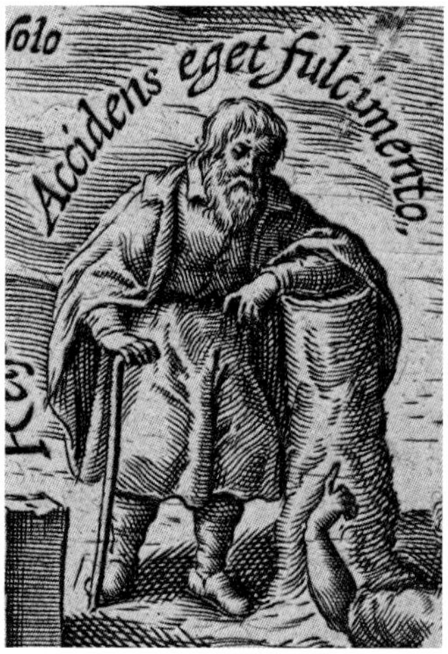

Fig. 135. Jean Chéron (designer) and Léonard Gaultier (engraver), *Typus* (detail), 1622. Princeton University Library, Special Collections.

During the thirteenth session of the Council of Trent in 1551, transubstantiation was discussed again, and more directly than during the Council of 1215. No strict explanation of how the doctrine connected to Aristotelian scholastic notions of substance and accident emerged from the Tridentine Council, but seventeenth-century Catholic scholars continued to use Aristotelian scholastic terminology to explain faith in the Eucharist. The doctrine held that once the Eucharistic wafer and wine are consecrated by a priest, they are transformed into or supplanted by Christ's body and blood; during this shift, the substances of the wafer and wine cease to exist, and the substance of Christ's body takes their place. The particular accidents of the wafer and wine (that is, their color, texture, odor, taste, and so on) remain, although not as accidents inhering in Christ's body. This conception of the Eucharist assumes what was known as the doctrine of "real accidents," according to which accidents are fully "real" insofar as they are freestanding; they can exist independent of their corresponding substances. For instance, the whiteness of the Eucharistic bread, which subsists even if the bread's substance is destroyed, is a real accident.

The Catholic hierarchy resisted any attempt to explain transubstantiation in terms other than the Aristotelian principles of substance and accident. But not all Catholic scholars agreed with the philosophical principles that others treated as foundational to explaining the doctrine. Alternative, unorthodox explanations of the miracle of the Eucharist were formulated by figures like Descartes and Maignan. These alternative explanations are worth our attention, because they demonstrate contemporaries' concerns around the relationship between sensory perception—including visual perception—and reality. In his theory of matter, Descartes held that the physical world consists of extension: "I have a complete understanding of what a body is when I think that it is merely something having extension, shape, and motion."[68] Body, in Descartes's system, is extended substance. Shape and mobility, along with size and divisibility, are modes or primary properties of extension. All other properties (taste, sharpness, solidity, etc.) are a function of these modes. For Descartes, bread's sensible appearances (*species*)—its taste, color, and smell—are not actual properties belonging to the bread; rather, they are mental events generated by material entities in motion. Descartes held, moreover, that the properties that exist in external objects like bread or wine, such as size and shape, cannot subsist without a material substance in which they remain fixed. Consequently, he rejected the notion of real accidents existing independently: "It is completely contradictory that there should be real accidents, since whatever is real can exist separately from any other subject; yet anything that can exist separately in this way is a substance, not an accident."[69] Descartes posits that if the entity is real and exists separately, then it must be a substance, as opposed to an accident.

As the French theologian and philosopher Antoine Arnauld (1612–1694) recognized, Descartes's attack on the scholastic Aristotelian doctrine of accidents and his promotion of mechanistic philosophy together threatened Catholic explanations that accidents of the Eucharistic bread and wine could persist beyond consecration:

> But what I see as likely to give the greatest offence to theologians is that according to the author's doctrines it seems that the Church's teaching concerning the sacred mysteries of the Eucharist cannot remain completely intact. We believe on faith that the substance of the bread is taken away from the bread of the Eucharist and only the accidents remain. These are extension, shape, color, smell, taste, and other qualities perceived by the senses.[70]

Given the precepts of Descartes's theory of matter, it follows that the Aristotelian scholastic doctrine underlying transubstantiation—that real accidents can be sustained independently after the transformation of the substance—becomes untenable.[71] Maignan's own description of the Eucharist, which I will turn to next, addresses the theological problem resulting from Descartes's theory of matter, that is, how observers could discern "extension, shape, color, smell, taste, and other qualities" of bread if the accidents of bread do not persist.

In his *Cursus philosophicus*, Maignan, like Descartes, rejects Aristotelian scholastic principles, which he terms "fallacious."[72] To justify observed bread-like qualities ("extension, shape, color, smell, taste") in the Eucharist without relying on the subsistence of real accidents, Maignan normalizes God's apparent deception of observers' senses (although, as I will explain below, he will insist that God is not in fact deluding those who observe the Host). Maignan gathers other instances in which Christ presented himself under novel guises, similar instances of "appearance":

> It is not a new miracle, distinct from this, through which God makes the thing appear which does not exist, namely bread which does not exist, while he causes the body of Christ to be certainly truly and really chewed under the appearance of bread, passed around under the appearance of bread, and touched and seen, and so on, under the appearance of bread; in this way, with the proportion preserved, he made himself be seen by Magdalene under the appearance of a gardener, and by the disciples going to Emmaus, under the appearance of a pilgrim.[73]

Rather than relying on Aristotelian scholastic accidents, Maignan uses another standard term of medieval Aristotelian philosophy—*species* ("appearance")—to describe what observers see in each instance, something that is not really there. Maignan emphasizes that there is nothing so exceptional about the Eucharistic miracle—about God manifesting himself as bread: it is not a "new miracle" (*novum miraculum*), though it is still a miracle, that is, Christ's sacrifice. Observers think that they perceive bread and wine when they are actually in the presence of Christ; likewise, Mary Magdalene thought she was perceiving a gardener and the disciples at Emmaus thought they were perceiving a pilgrim when they, too, were observing Christ.[74]

Maignan stresses that the pilgrim and gardener were only semblances; what was real in each instance was that Christ was present to Mary Magdalene and to the disciples in Emmaus: "It [i.e., the pilgrim or gardener] was the appearance alone: that is, although Christ was himself, he nevertheless appeared as another person, indeed, he appeared as a gardener or a traveler."[75] Maignan describes how Christ was able to disguise his presence to Mary Magdalene and the disciples in Emmaus in physical terms: "He was holding [i.e., transfixing] the eyes and, equally, the other senses of those looking on, so that they would not recognize him, hidden under this foreign appearance."[76] Maignan then compares Christ's disguise as a gardener and pilgrim to his disguise as the Eucharistic bread and wine:

> But in the same way, with the proportion preserved, it must be considered that everything which has some real existence in the Eucharist is Christ himself; but nothing real survives of the bread and the wine, that nothing real remains exactly, but only the appearance or the sight of all these things exists there, the appearance which holds the eyes and the other senses faithfully, so that they do not know that Christ is hiding himself under this appearance, as if in his proper person; but nevertheless, they in fact receive him under a foreign appearance, in the food and drink. And in this way, if you understand that Christ was able to show and exhibit the appearance of a traveler, or of a gardener, although he did not show anything entirely real of a traveler or a gardener, nothing real, I say, distinct from himself (for you will not believe that he then put on, for example, the real clothing of a traveler, which he made vanish with him soon after), you will understand well that he is equally able to exhibit the appearance of bread and wine; although, concerning bread and wine, there is nothing entirely real present which he exhibits.[77]

Avoiding the Aristotelian scholastic term "accidents," Maignan employs the term *apparentia* ("appearance") in addition to the term *species* to denote the unreal things that Christ is able to make observers see. In the presence of Christ's body, God creates the sensory appearance of bread and wine in the mind of the perceiver; the sensory appearance of bread and wine is not a "real species" outside the mind of the perceiver. Although Maignan does not directly compare his anamorphic wall frescoes to Christ's disguise as the Eucharistic bread and wine, we can reference his paintings because of a shared vocabulary of experience. With the verb *tenere*, he draws focus to the way both Christ and appearances hold or transfix the eyes and other senses of observers and capture their attention. Both Schott and Maignan also used the verb *detinere* to characterize how "pleasant fallacies" held observers as they considered the anamorphic wall painting portraying Saint Francis of Paola (which was also described in their texts as "hidden" or "hiding"). The anamorphosis fixes attention on the revealed truth, much as, in a slightly different context, the deformed accelerated perspective of a tabernacle fixes attention on the venerated Host. Christ also holds the senses of Mary Magdalene and the disciples, and the appearance of the Eucharistic bread and wine holds the senses of the faithful. Maignan would, of course, have recognized Christ's ability to hold the senses of the faithful as miraculous, while explaining deformations' capacity to hold the senses as the work of visual reasoning. Nevertheless, in that a person who walks along a corridor exhibiting an anamorphosis thinks they are observing a landscape when "really" they are also observing a praying figure, the anamorphosis can be related to Christ's disguise as a gardener or pilgrim or as the Eucharistic bread and wine: the anamorphosis is a concrete, artistic manifestation of the Minim friar's more abstract interest in humans' vulnerability to visual deception and the work of seeing "through" appearances as a recurring theme in Christianity.

Jean-Baptiste de la Grange (c. 1641–after 1680) notes the similarity between Descartes's and Maignan's disavowal of accidental forms: "Father Maignan is equally opposed to the ordinary philosophy ... he perpetually mocks the accidental forms of the peripatetic, he believes like Descartes and Gassendi that there are only spirits and bodies in the world."[78] De la Grange expresses immense disdain for Maignan: "ordinarily his opinions are the most ill-chosen of all, and the most badly delivered."[79] And his scorn is rooted in the worry that by rejecting Aristotelian scholastic accidents Maignan implies observers of the Eucharist are perpetually deceived by God:

> Because when this opinion [i.e., that there are no real accidents] is accepted, it is true to say that there is no whiteness in the Eucharist, that there is no hardness, that there is no size, nor taste, whatever we seem to see there of whiteness, to feel there of hardness, and to find there some taste; we are perpetually mistaken when we think that there is something material in the Eucharist, that is the cause of the sensations that we have of it: in the same way that those who sleep imagine they are eating and running, when in fact they are not eating or running.[80]

De la Grange is unpersuaded by Maignan, because he finds it easier to trust in the survival of freestanding "real accidents" hovering over the substance of Christ's body than to believe that what our senses give us to understand is entirely false.[81] He asserts again and again that Maignan's perspective and the Cartesian perspective are "much more difficult to believe" (*beaucoup plus difficile à croire*) than the traditional Catholic doctrine.[82] He claims that whereas the Catholic explanation of transubstantiation requires only one miracle—the survival of accidents—the opinion of the so-called new philosophers requires as many miracles as there are people who observe the sacrament: God needs to enact a miracle such that what is actually the body and blood of Christ has the appearance of bread and wine for each observer of the mass.[83] De la Grange's criticism rests on Occam's razor, the principle (ascribed to the Franciscan friar William of Ockham, c. 1285–1349) that an explanation should rely on the smallest possible number of assumptions: one miracle is more probable than many.

Before de la Grange, the Jesuit professor of theology from Lyon Théophile Raynaud (1583–1663) had been similarly appalled by Maignan's ideas concerning the Eucharist, and had conveyed his misgivings in a work that appeared in 1665.[84] His "appendix against Maignan and others" cites extensively from Maignan's *Cursus philosophicus*, including the passages discussed above on Mary Magdalene and the disciples in Emmaus.[85] Raynaud contends that it follows from Maignan's position that transubstantiation does not occur: "There is no transubstantiation, since nothing remains in common of either substance, nor can the true reasoning of the Sacrament be explained … so that the manner of the Sacramental existence does not differ from apparitions."[86] Like de la Grange, Raynaud worries that Maignan's stance implies that God deludes mankind, that he causes them to believe in "apparitions."

De la Grange's and Raynaud's disdain for Maignan was not universally shared. Schott, for one, called him, "the most learned man" of his order, as noted in chapter 2. I have already observed, moreover, that the French philosopher and writer Pierre Bayle characterizes Maignan as "one of the greatest philosophers of the seventeenth century." Bayle prefers over Descartes's version Maignan's mode of explaining the mystery of the Eucharist.[87]

When Maignan responds to Raynaud, his emphasis is on visual discernment, and he turns to the experience of illusionistic optical devices to make clear his position on the miracle of the Eucharist. In his *Sacra philosophia*, Maignan refutes Raynaud's suggestion that Maignan's theory of the Eucharist would imply God deludes mankind. In arguing that his account of the Eucharist does not imply deception by God, Maignan introduces the category of "the common people" (*vulgus*), who lack visual perspicacity, and he describes how the *vulgus* respond to reflections in a concave mirror:

> And so visible appearances, reflected from a concave mirror, particularly a parabolic mirror, represent the object in the space on the nearer side of the mirror. But this seems to be a fallacy, a phantom, an illusion, and an idol among the common people; in whose judgment nevertheless, there is no fallacy (and much less a phantom, etc.), because the same appearances [*species*] reflected from a flat mirror represent the same object, just as if it exists in the space beyond the mirror.[88]

While a reflection in a concave mirror is incomprehensible, and common or uneducated people might interpret that reflected image as a deception, Maignan explains that it is no such thing: the common observer is simply mistaken in thinking that a reflection in a concave mirror is "a fallacy, a phantom, an illusion, and an idol." He also stresses that a reflection in a flat mirror is not a deception; this judgment, we can infer, the common people take for granted, since (as Maignan explains subsequently) they have grown accustomed to flat mirrors.

Maignan next draws a distinction between the "common people," or misapprehensive viewers, and more discerning, perspicacious observers who are characterized by *sapientia* ("wisdom"): "But everyone who is wise laughs at this disparity of judgment, because it is no more a fallacy in the first case, than in the second."[89] Wise observers understand that there is no more deception in a concave mirror than in a flat mirror: neither reflection is a deception. Maignan continues: "For if in the first case the sense of a man, or of an ape, etc., is deceived, in the second case the sense, e.g., of a Molossian hound is also deceived, who, as if seeing another similar to himself, and suspecting it to be present, rages against it, barks, and leaps."[90] As common people are deceived by an incomprehensible and unusual parabolic mirror, a dog, unaccustomed to mirrors, is deceived by seeing his reflection in a flat mirror into thinking that it shows something to be present that is not present, and, as a result, responds with aggression.

Maignan emphasizes that inexpert, misapprehensive observers reach their false opinions about reflections in curved mirrors due to inexperience, which causes a lack of discernment: "The judgment of the common people concerning the fallacy of what they think about the parabolic, concave mirror, is established because they see this effect of the concave mirror more rarely."[91] By contrast, because the common people have experience with reflections in flat mirrors, they no longer read them as deceptions: "Indeed, the effect of the flat mirror ... they [i.e., the *vulgus*] do not assume for a fallacy, because they have already for a long time grown accustomed to discern similar effects every day; and so they hardly think that they are deceived."[92] Maignan's discussion of the wisdom derived from experience with optical devices, such as concave mirrors, helps to explain the fascination with developing and exhibiting such instruments and research among seventeenth-century Minim and Jesuit scholars; the devices functioned to educate observers and hone visual discernment, so that, unlike Molossian hounds or the common people, elites would not mistakenly characterize appearances as deceptive that could be fully understood through such fields as mathematics, optics, perspective, and catoptrics.

But how do these remarks on deception and the observation of reflections in flat and concave mirrors pertain to Maignan's claim that in his account of the Eucharist, God does not delude mankind? Maignan continues his explanation in visual terms, pointing out that when philosophers (that is, wise, perspicacious observers as opposed to the *vulgus*) see a reflection of the Eucharistic wafer in either a flat or a concave mirror, they know that they see its appearance as opposed to observing scholastic Aristotelian accidents:

> Since in an apparition through a mirror, whether flat or concave, with bread, according to the stated hypothesis, appearing reflected beyond the mirror or in front of it, no philosopher thinks that there are any real accidents of bread really there, beyond or in front of the mirror, but that this appearance of light happens and is consistent with the appearance emitted from bread, reflected from the mirror and taken up by the eye, which becomes the single opinion of everyone.[93]

Maignan points out that it is in no way controversial to say that the reflection of bread in a mirror does not imply the existence of real accidents of bread within that reflection. How odd it is, then, that when it comes to the Eucharist itself, scholars would look for Aristotelian scholastic accidents when there are only appearances, implying they assume bread cannot be made to appear without accidents existing as well: "It is a wonder that in the Eucharist, anyone might seek accidents in appearance alone, as if it [i.e., the appearance of bread] cannot be made here without these things, and indeed as if it can be made there."[94] Just as it is possible to produce the appearance of bread without real accidents when bread is reflected in a mirror, it is also possible, Maignan claims, in the Eucharist to produce the appearance of bread in the perceiver's mind without real accidents existing outside the perceiver's mind.

Maignan turns to divine omnipotence and discernment regarding sensory impressions to develop his argument further. He insists that God can produce appearances of bread: "In the Eucharist, in fact, an appearance, not bread, exists; but God may emit, or make, in the place of bread, appearances altogether similar to those which bread itself would emit if it were present."[95] For Maignan, an appearance of bread exists in the perceiver's mind—created there by

God—and not in the external world. God's all-powerful nature, he contends, means there is no difficulty in believing that God can produce images of bread.[96]

Maignan further investigates the value of sensory information in helping the observer discern the nature of the Eucharist. Here he contrasts the experience of the Eucharist against his earlier model of the mirror. In that instance, a reflection that initially seems to be a fallacy might be validated by the use of other senses. But in the instance of the Eucharist, the senses of touch, taste, smell, and hearing cannot help the observer to recognize that God is not deceiving the faithful:

> Because, in those appearances of mirrors which have been spoken of, whatever is a fallacy of one sense can be corrected through another; because it can certainly be recognized that the object is not really beyond the mirror, or before it at that point of the perpendicular line. However, this cannot be known about the Eucharistic bread after the consecration.[97]

Although this passage is somewhat ambiguous, I take it that Maignan holds that the common people, with experience—and the support of other senses than vision—come to realize that reflections in concave mirrors that they initially deemed to be fallacious are not deceptions. His point is that mirror illusions are overcome because observers cannot touch or smell the mirrored bread, and here there is no other sense that will resolve the illusion of the bread in the Eucharist, so (he explains) God directly warns the faithful, protecting them against deception. Maignan emphasizes that Christ's words as recalled in the sacrament—"This is my Body"—necessarily have a powerful impact, overcoming what the person seems to perceive:[98]

> In the soul of the faithful the word of Christ is more powerful in removing the occasion of an illusion, a fallacy, etc., than a thousand perceived experiences are for a common person, or even for a notable philosopher. Therefore, from these things, it is already certain that there is no occurrence of deception, delusions, apparitions, etc., from appearances in the Eucharist.[99]

Maignan believes that when, during the Mass, the priest cites Christ's assertion that the bread and wine are His body and blood, this utterance will counter any misunderstandings produced by the sensory experiences of observers, whether inexpert or discerning, even when the latter category includes "a notable philosopher."

Notwithstanding these alternative accounts, the institutional position of the seventeenth-century Catholic Church leaders on the mystery of the Eucharist remained resolutely Aristotelian. We might assume, then—despite Maignan's influence in regard to optics and the visual arts—that the devices I discuss below were crafted with the aim not only of reinforcing faith in the mystery of the Eucharist, but also of reinforcing the Church's doctrinal explanation of the transformation of bread and wine. As recently discussed by Domenico Collacciani and Sophie Roux, the Jesuits used a multifocal beveled lens together with a Eucharistic iconography that seems to insist on the accepted messaging around transubstantiation. When observed with the naked eye, these images displayed the host surrounded by angels, but when examined through the lens, they appeared to visualize Jesus Christ.[100] Jean du Breuil describes the iconographic effect: "I saw one time, a picture … in the middle of which was painted a host, like those in relation to which one says the mass, and around this host, there were many angels in a devotional posture. But when one looked through the hole of the lens, one saw nothing other than an image of the little Jesus."[101] The lens, which came to be known as the Lens of Faith (*Specillum fidei*), enabled the observer to witness an image of transubstantiation as well as reinforcing belief in the miracle of the Eucharist.[102] With a visual playfulness addressed to the vogue for encoded and symbolic images, the device served to heighten the faith of the beholder through ocular evidence that seemed to confirm the doctrine of transubstantiation and strengthen belief in the Eucharist.

This Eucharistic iconography and its accompanying lens feature in thesis booklets authored by Pierre Bourdin.[103] The same 1639 thesis booklet that we saw in earlier chapters—where collège de Clermont student Yves Henri was asked to explain how one would draw a sundial on a ceiling or an anamorphic fresco—also presents an account of the "Lens of Faith" and its accompanying iconography:[104]

> We will show the Lens of faith, that is, a picture, in which the sacrament of the sacrosanct Eucharist is depicted: imagine the perfect Host in the middle, and around it, adoring Angels, and the entire Heavenly curia erected in the splendor of the spectacle. If, indeed, you look through the lens, positioned away from the region of the Host, you will not see Angels, but the Lord of the Angels; not the Host, but he who it truly is, and it is believed that the Lord Christ, represented graphically, hides under the covering of the Host.[105]

The observer of the represented bread has not yet seen what lies behind its veil. The image examined through the "Lens of Faith" is more truthful, revealing to observers what the host is "truly" (*vere*), namely, Christ.

The upper three cartouches printed on the frontispiece to Henri's thesis booklet display drawings by an anonymous hand of a multifocal beveled lens (Fig. 136). In all three sketches, a disembodied eye peers through the lens, which is positioned in front of an image. The three representations, though difficult to decipher, appear to depict—from left to right—celestial entities, a land or seascape with a sun on the left and a flock of flying birds, and a vase with flowers. Although none of the images seems to visualize the host, the conjunction of these representations and the multifocal beveled lens attests to the popularity of this device in Jesuit pedagogy. Further drawings of the instrument that explain its mechanics feature in manuscript notes made during Bourdin's classes of 1636–1638 by a student named Paul Mercier (Figs. 137 and 138).[106]

Another engraved representation of the lens is integrated into a thesis booklet by Bourdin that was used in 1640 by a student at the collège de Clermont named Charles Potier (Fig. 139, labeled "E").[107] In this anonymous print, a disembodied eye peers through the lens at an image of a Holy Face with an aureole of fiery light that is surrounded by four cherubs. This representation seems to present the image of Christ as he would have appeared when a representation of the host was examined through the "Lens of Faith." The program of ceremonial events in Potier's thesis booklet refers to experiments with the "Lens of Faith": "On the same day, after

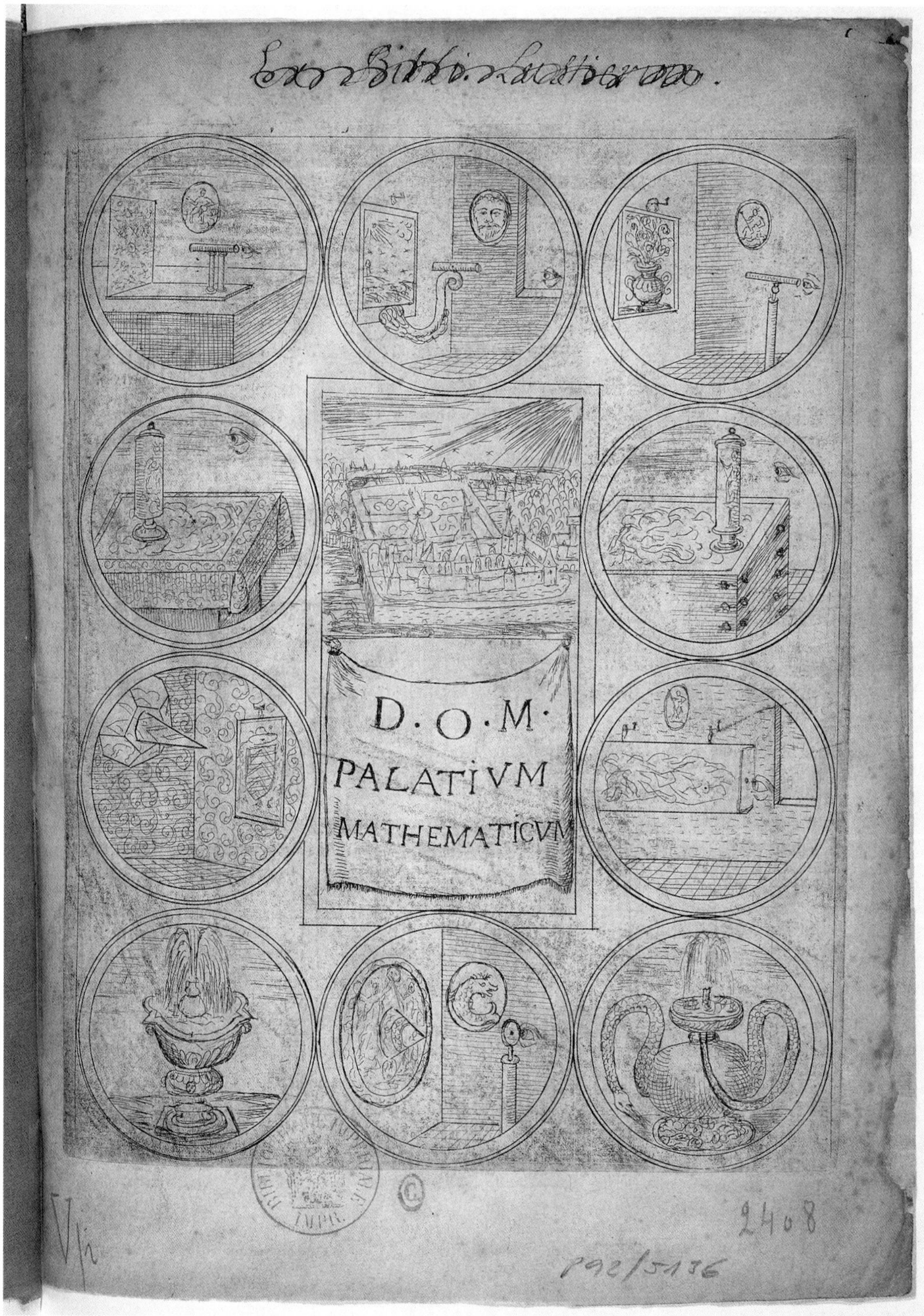

Fig. 136. Unidentified artist, frontispiece to Yves Henri's *Augustæ cœli reginæ Mariæ Virgini Deiparæ ejusque integerrimo sponso D. Josepho Palatium mathematicum D.D.C. [Dicat, dedicat, consecrat] Yvo Henri Briocensis, pro annua celebritate literaria collegii Claromontani Societatis Jesu*, 1639. Etching printed on paper, in-4to. BnF, département Littératures et art, Paris [Tolbiac -Rez-de-jardin - magasin - VP-2408].

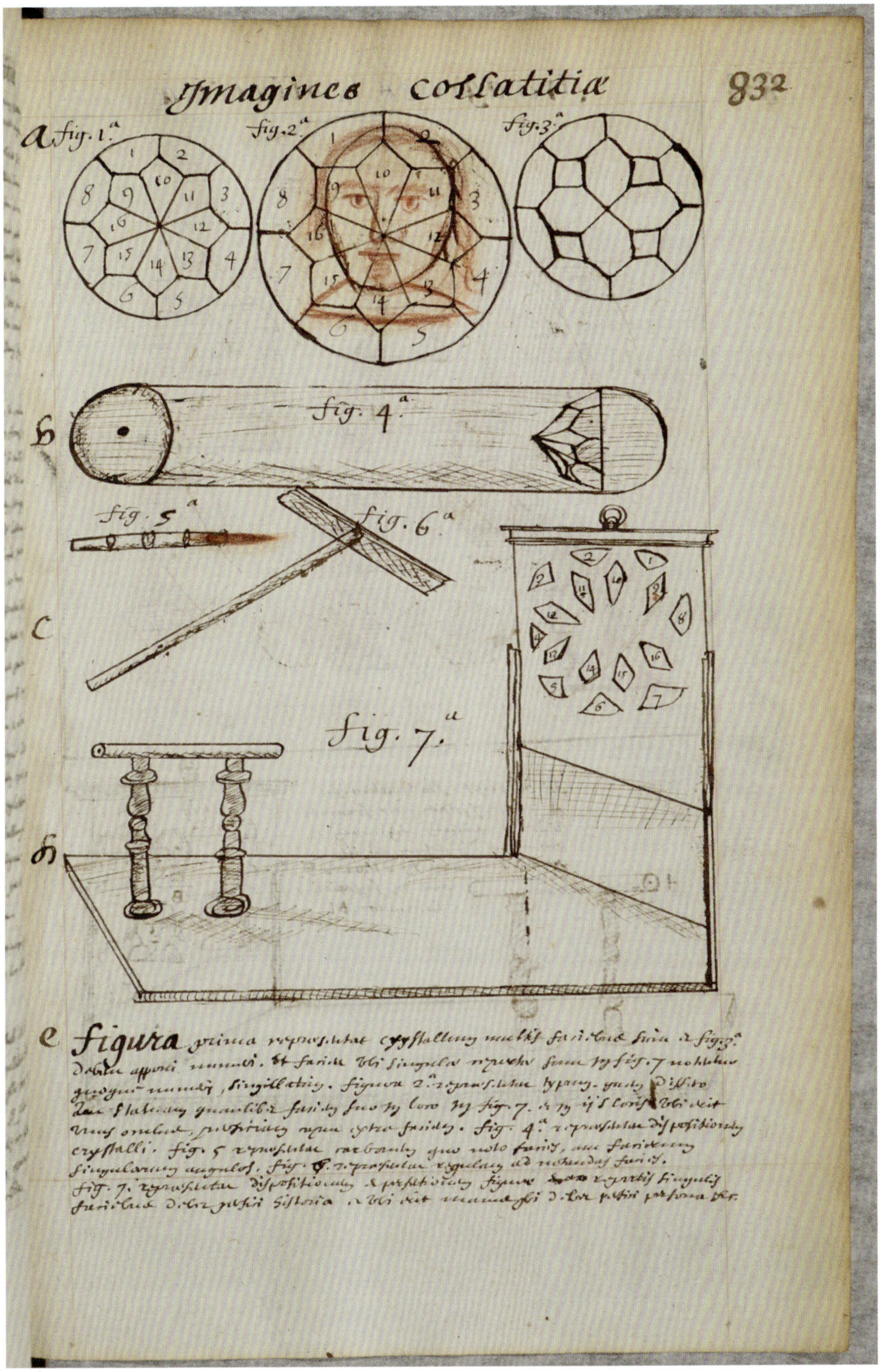

Fig. 137. Drawing of an optical contraption from the 1636–1638 notebook of Paul Mercier, 12.2 × 7.48 in (31 × 19 cm). BnF, Département des manuscrits, Paris [ms. Lat. 17862, fol. 832r].

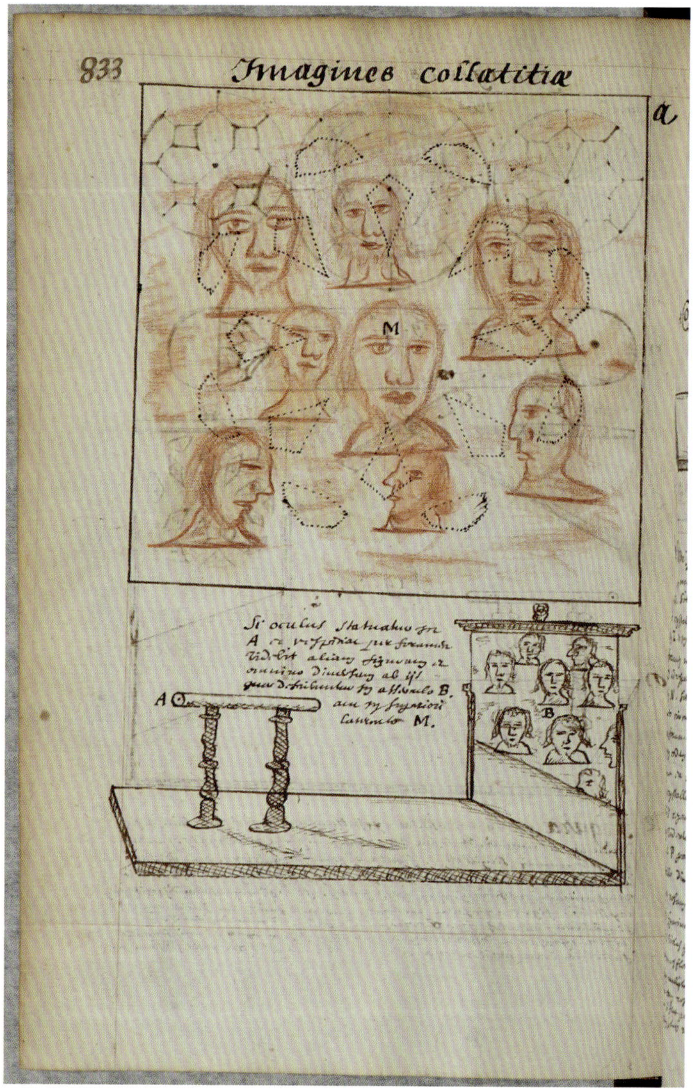

Fig. 138. Drawing of an optical contraption from the 1636–1638 notebook of Paul Mercier, 12.2 × 7.48 in (31 × 19 cm). BnF, Département des manuscrits, Paris [ms. Lat. 17862, fol. 833ᵛ].

mid-day, the contests were celebrated again, about 2 o'clock, and a demonstration of the Lens of Faith opened them, and a demonstration of similar images, which, having been described from the principles of refraction by many people, represented something new and unexpected."[108] The same engraving of the "Lens of Faith" is included in the 1661 edition of Bourdin's *Le cours de mathematique* (*The Mathematics Course*).[109]

Is there any evidence that Maignan or other Minim friars were aware of Bourdin's experiments with the "Lens of Faith"? Although I have not found a direct link to Maignan, documents suggest some connection between Bourdin and members of the Minim Place Royale convent in Paris. A volume in the Bibliothèque du Conservatoire des arts et métiers that includes, among other materials, thesis booklets by Bourdin carries a manuscript inscription on the title page of the first document: "F. Joannis francisci Niceron 1641. De conventu Minimorum Parisiensium."[110]

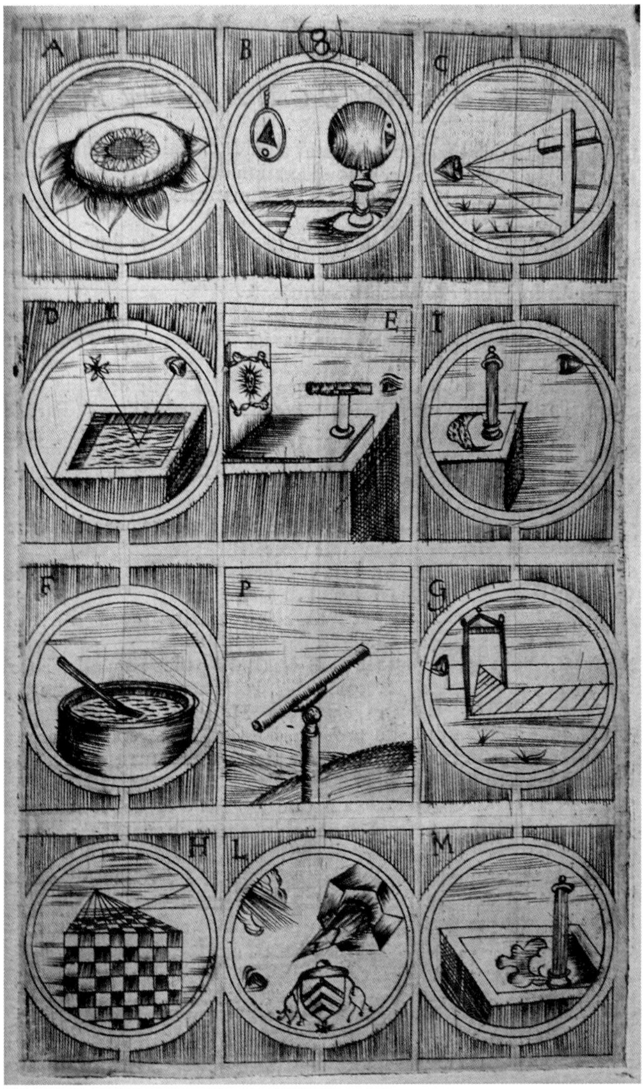

Fig. 139. Unidentified artist, image integrated into page eight of Charles Potier's *D.O.M. [Deo Optimo Maximo] Encyclopædia mathematica, ad agones panegyricos in Claromontano Parisiensis Societatis Jesu collegio. Agonista Carolus Potier Castrotheodoricus*, 1640. Engraving printed on paper, 8.58 × 6.38 in (21.8 × 16.2 cm). Bibliothèque Sainte-Geneviève, Paris [Magasin Fonds ancien 4 V 63 INV 539 (P.5)].

The volume's binding displays the arms of the Minims of the Place Royale monastery in Paris—implying that it was assembled by members of the monastery. The thesis booklet defended by Charles Potier is among those integrated into the volume.[111] It is known that Bourdin, in turn, owned Niceron's *La perspective curieuse*: in a copy of the book held by the Bibliothèque de la Sorbonne, the cover page is inscribed to Bourdin in the author's hand: "For the Reverend Father Bourdin Professor of Mathematics at the collège de Clermont of the Company of Jesuits: from the author, his very humble servant" (Fig. 140).[112] Here we should also recall the reputation of the Minim Place Royale convent as a gathering site for scholars. It seems at least possible that Niceron, when he arrived in Rome from Paris in 1639 and again in 1641, might have mentioned Bourdin's pedagogical displays with the "Lens of Faith" to Maignan and that these discussions, in turn, might have fed into Maignan's own ideas about the miracle of the Eucharist.

Fig. 140. Copy of *La perspective curieuse* with a dedication to Pierre Bourdin in Jean-François Niceron's hand. 13.8 × 9.1 in (35 × 23 cm). Bibliothèque de la Sorbonne, Paris.

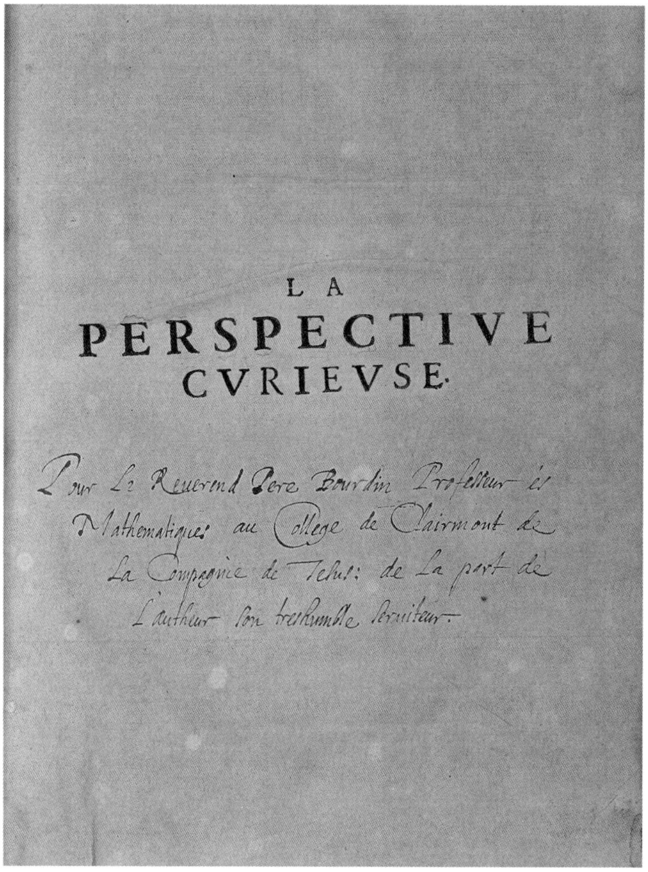

But what was really going on intellectually and theologically with such devices? Belief in divine revelations necessitates a mode of contemplating reality that takes into account both the natural and the supernatural but subsumes the former under the power of the latter. In a miracle, the continuity of the natural could always be broken by the supernatural.[113] Devices like the Lens of Faith are ingenious human contraptions that are intelligible through drawings and visual reasoning. Although these devices and their initially incomprehensible reformations bolstered the power of the Catholic Church and its leaders, in the end, they fail to demonstrate anything truly.

✦

In chapters 2 and 3, we saw that Catholic Reformation leaders created visual deformations and reformations to direct the faithful toward certain truths and to analogize these optical devices to big-picture questions concerning the messiness of human existence and the contrasting clarity of God's divine plan (with the Catholic Church as a crucial intermediary for the faithful, directing them toward the divine). In this chapter, we have seen how the interest among Catholic Reformation leaders in visual deformations and reformations intensified in relation to the faith's key theological mysteries: "the secrets of divine providence, the mysteries of predestination," and the Eucharist. In regard to the latter, visual devices were not only used as rhetorical tools to explain the mystery of the Eucharist and the doctrine of transubstantiation—a difficult doctrine to grasp—but were created specifically to manifest that mystery and doctrine, with images of the Eucharist bread resolving into images of Christ. The didactic importance of these devices is ultimately hard to measure. On the one hand, tools like the Lens of Faith remind us that during the Catholic Reformation clergy worried about errors of understanding—perhaps especially that individuals might not comprehend the doctrine of transubstantiation or believe in the mystery of the Eucharist—and consequently they drew on all the optical and rhetorical tools they could to lead people to embrace this central doctrine and mystery of the faith. On the other hand, as it is likely such devices were primarily deployed within Minim and Jesuit classrooms and convents, it is tempting to speculate that they were used primarily to preach to the already convinced (although, as we have seen, the unorthodox theories of the Minim Maignan suggested to some that he did not accept the doctrine of transubstantiation).

Anamorphoses and multifocal beveled lenses functioned as actualized allegory or allegories that made theological enigmas evident through physical elucidating motions, that is, through embodied exegesis, that captured attention and honed observers' visual discernment. The move from confusion and uncertainty to greater clarity and apparent certainty in the interaction with an anamorphic image or in glimpsing an image of Christ through the lens of faith is like the move from confusion and uncertainty to clarity and apparent certainty when we progress from faith and revelation here on Earth to the direct viewing of the divine in the Empyrean. For some, discernment in the science of seeing became coterminous with comprehension in seeing not only the present but also the future—although even with proper visual education, no one could, of course, see as fully on Earth as in the hereafter.

The illusionistic artworks and devices discussed in this chapter, from anamorphoses to lenses of faith, were entertaining and diverting. But they were also an integral component of a baroque culture that harnessed rhetorical strategies—justifiable obscurity, actualized allegory, and embodied exegesis—to bolster Catholic faith. The movement from the disorientation of experiencing a deformation to the clarity of its resolution facilitated an individual visual exegesis that could bring the observer into communion with the transcendent, albeit an imperfect communion.

The close association between the form and content of visual and spatial deformations as I have discussed them in the book's first four chapters might seem to suggest an early modern consensus that the process of visual discernment could be tightly controlled by an authority or directed to a certain outcome. Yet in the next chapter, we will discover that as more seventeenth-century architects and thinkers thought about the technical aspects of deformations, a further debate emerged around individual observers' capacity for discernment. There was no consensus around the certainty with which particular questions in moral theology could be resolved, and there was no consensus when it came to determining whether what people believed they saw corresponded to the way things were (as we have seen in the debates surrounding the passionflower). From this lack of consensus around visual discernment, there emerged further disagreement around what might correctly be labeled a deformation and around whether and how an authority like the Catholic Church should deploy visual and spatial deformations.

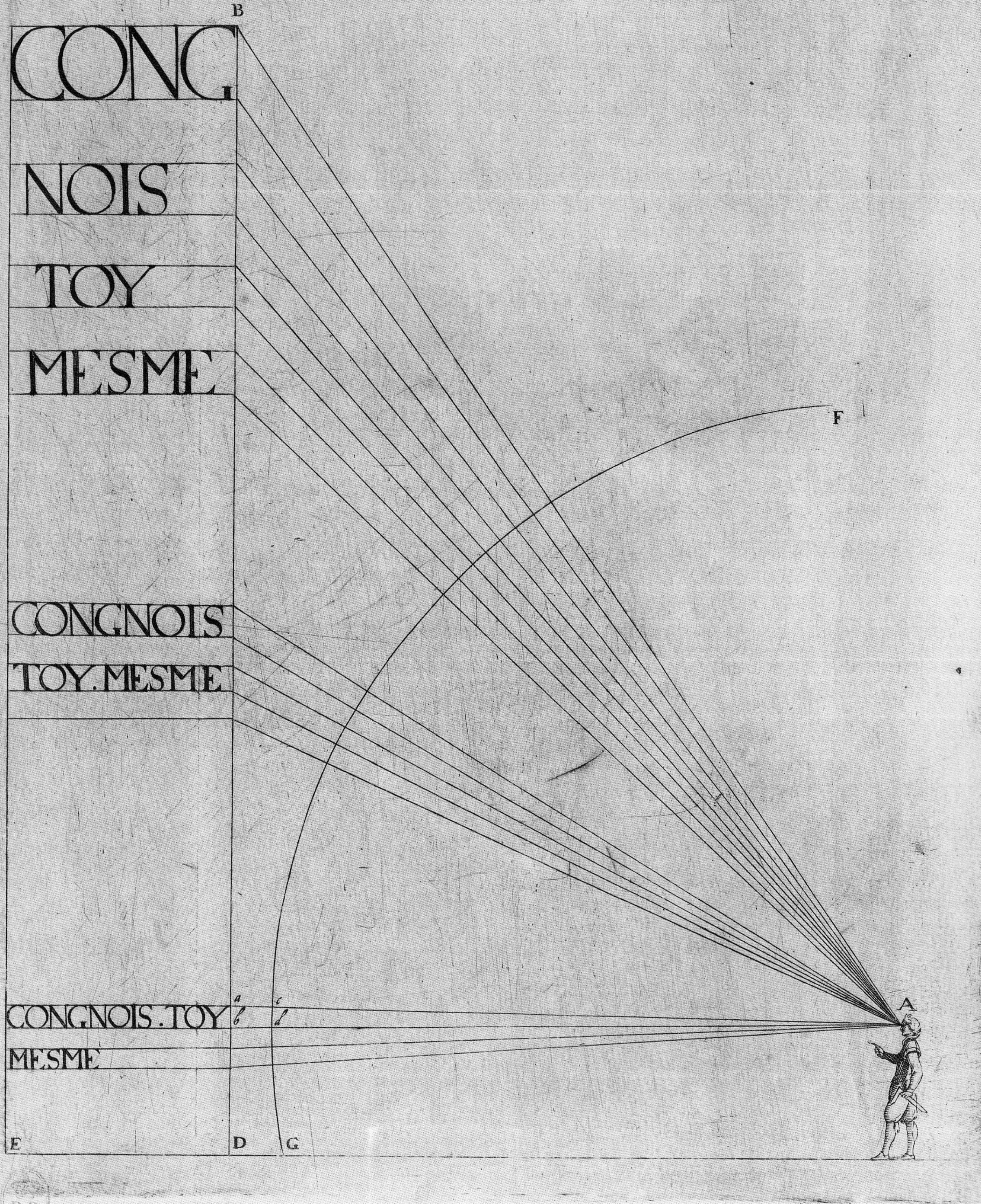
B
CONG
NOIS
TOY
MESME
F
CONGNOIS
TOY. MESME
a c
CONGNOIS . TOY b d
A
MESME
E D G

5 CARAMUEL'S ARCHITECTURAL REFORMS

THE PATRONS, practitioners, and theorists we have seen so far differed in the ends toward which they deployed deformations and other optical devices: some wanted to explain the workings of the created world and some the mysteries of the faith. Bernardino Spada may have hoped the visitors who contemplated the complexity of his sundial gallery would consider the divine glory of the creator, but there and in his *prospettiva* as well as the other formal galleries of his palace where his guests walked, he also hoped to impress them with his own social power and his sophisticated understanding of the visual and spatial. We have also seen broad agreement that deformations usefully introduced uncertainty—at least initially—and could gain rhetorical power via the journey from obscurity to clarity. But if the previous chapters have examined figures who valued the experience of deformations and their eventual progress toward clarity and resolution, this chapter focuses on a figure who sought to avoid deformation—to forestall completely the initial vertiginous experience that others found so powerful.

This chapter revolves around a single character, the Spanish theologian and polymath Juan Caramuel y Lobkowitz, who offered instructions for how to create architecture that averted the experience of deformation, advocating building "with grace, and without deformity" (*con gracia, y sin deformidad*).[1] Caramuel, who was born in Madrid but spent much of his adult life in Italy, conceived of his architectural writing as an extension of classical architectural theory; he wanted to avoid deformations for the same reasons that he promoted harmony and proportion in architectural works. Unlike the figures introduced in earlier chapters, he did not value visual and spatial deformation as a tool for reinforcing personal or institutional authority or for instructing the faithful in particular doctrines and observances. His discomfort with indulging, even momentarily, the initial experience of disorientation in architecture stood in opposition to his comfort when it came to disorientation surrounding hot-button issues of

Detail of fig. 149

conscience—whether masturbation was a sin, whether drinking chocolate violated a fast—on which the Catholic Church had not taken a definitive dogmatic stand. Caramuel proposed strict correctives to the vagaries of vision and environment in his writings on architecture, intended to prevent experiences of confusion and unclarity. And yet, in his theological and philosophical writings, he recognized and tolerated the ways that these same vagaries of human experience reflect and construct a relationship to the divine in which more than one moral choice could exist. Indeed, he held that two people could act in opposing ways on a single issue—for instance, one person could decide to drink chocolate during Lent, and another could decide not to do so—without either of them falling into sin. Caramuel's discomfort with architectural disorientation, and his comfort with theological disorientation, might have partly reflected his relative power to intervene in each domain: In the realm of architecture, he could claim the authority to advocate strict building correctives (even if few architects ultimately followed his architectural decrees). But when it came to making a moral choice on a topic where the Church had issued no definitive pronouncement, he could not, of course, claim the authority of the pope or take a dogmatic stand in his name.

A study of Caramuel's writings on architecture underscores that whether and how a theorist of the period recommended generating or avoiding visual and spatial deformations depended on how he understood visual discernment. This was as true for deformation's opponents as for its proponents. Caramuel believed his architectural reforms would help to prevent experiences of visual disharmony and disproportion, and he would therefore eschew deformation as we have thus far considered it, even recommending a kind of overcompensation for defects of human perception and irregularities in environment, such as the oblique slopes of hills and valleys, that he argued would most immediately deliver to the eye the classical balance he valued. Meanwhile, the French medical doctor, architect, and architectural theorist Claude Perrault argued—though without citing Caramuel explicitly—that architectural theories in line with Caramuel's recommendations would in fact render works "visibly deformed" (*visiblement difformes*).[2] Perrault thus agreed with Caramuel in that both were skeptical of visual and spatial deformation, but Perrault's architectural recommendations diverged from those of Caramuel because he ascribed to a different theory of human visual discernment, one that anticipated distortions caused by angle and distance. This chapter's discussion of these theorists, among others, shows that in this period the multiple ways of comprehending visual and spatial deformation produced directly contradictory guidance even where theorists were equally committed to preventing deformation. I will suggest, as a result, that some Catholic Reformation thinkers were effectively led to deformations by mistake, as they developed architectural theories intended to help avoid deformation.

In a portrait engraved by Nicolas Auroux (1630–1676), Caramuel, wearing a robe and biretta, gazes beyond the print's observer (Fig. 141). An annotation on a pedestal celebrates his encyclopedic learning and wide acclaim: "This is Caramuel, visible to the whole world, his great fame has raised his illustrious name to the stars. His face is engraved, his great mind occupies many books, and is equal to many; he holds that all things must be known."[3] The inscription proclaims his erudition, and a 1760 biographer piles on by describing his "omnipotence" (*onnipossanza*).[4] But Caramuel in fact devoted much of his career to debunking the idea that humans could accurately be characterized by terms like these. In his lavishly illustrated

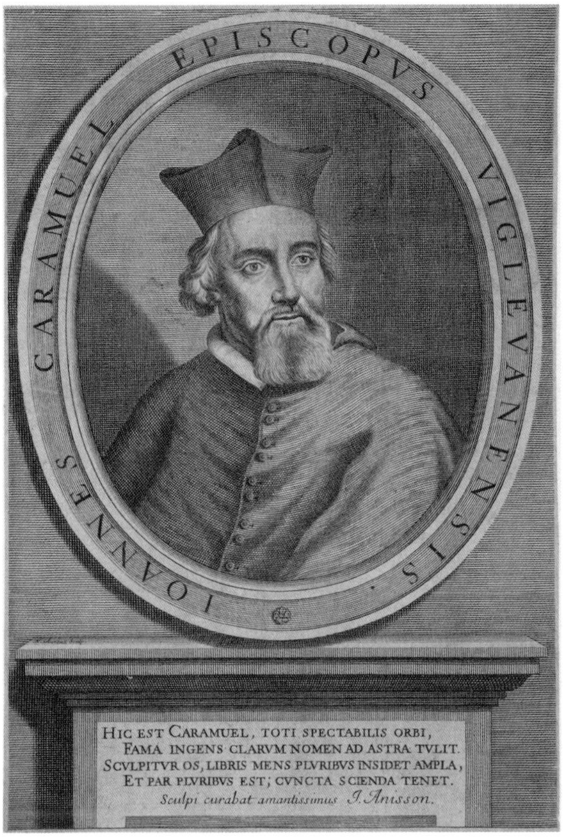

Fig. 141. Engraved by Nicolas Auroux, *Portrait of Caramuel*, after 1673. Engraving printed on paper, 11.7 × 7.7 in (29.6 × 19.5 cm). British Museum, London.

books, Caramuel emphasizes human miscomprehensions, our limited perspectives, and our deficient understanding across a wide range of domains, including architecture, which is my focus in the first section of this chapter, and theology, which I discuss in the second. He often betrays a particular concern with the limits of visual perception and a related need to correct for those limits—advocating, for instance, that elevated statues should be enlarged to compensate for our tendency to see as "smaller" objects that are farther away from us. At the same time, his persistent use of printed images throughout his numerous publications demonstrates his commitment to visual modes of disseminating information; the third volume of his pathbreaking treatise, *Architectura civil recta y obliqua* (*Straight and Oblique Civil Architecture*, 1678 [1679]; hereafter, *Architectura*), for example, is composed entirely of images.[5] In this work he underscores the mnemonic potential of visual communication (and would thus seem to share, at least to a point, the Jesuits' and Minims' belief in the didactic potential of the visual arts): "What is seen with the eyes is better imprinted in the memory than what is said with words."[6]

The elements of Caramuel's contribution to contemporary discussions on visually incomprehensible works of art and architecture and education can be divided into two interrelated domains. The first of these is theoretical. Caramuel underlines the ways in which entities, from buildings to moral ideals, are grasped at particular moments and from particular locations. He repeatedly stresses that vantage points and contingent relations inform how people observe the world and make sense of abstract notions. Throughout his *Architectura,* at least in a literal, physical sense, Caramuel appears to recognize the importance of viewpoint: he is careful to position observers at precise vantage points in relation to architecture. Caramuel's architectural theory thus far recalls the anamorphoses discussed in the previous chapters, which also privileged predetermined vantage points. But we should also note important distinctions: in the *Architectura,* Caramuel does not propose or anticipate that observers would depart from those predetermined vantage points, and he does not connect the literal experience of observing from a correct perspective to the Church's power to lead lost souls to revealed truths. In fact, as we will see, he devotes much of his career to revising ecclesiastical moral theology, aware that the Church will not have made a definitive pronouncement for each and every moral question or dilemma—a state of affairs that leaves room for some personal discernment. Caramuel's second contribution is optical. Like so many others before him, he dwells on the gap between how things appear and how they really are. Caramuel's impact on the theorization of visual incomprehensibility and discernment emerges from a key section of his architectural theory, which he terms "oblique architecture," as well as from other sections in his *Architectura* and his other writings. An oblique architecture, as Caramuel recommends it, adjusts the norms of classical architecture by means that (from precise vantage points) maintain the appearance of proportional harmony—breaking classical rules in practice to support the theoretical goals of classical architecture. In Caramuel's theory, this method would compensate for obliquities caused by environments as well as differences between appearance and reality, presenting a view of a seemingly undeformed and regular structure. Someone practicing Caramuel's architectural theory might, for instance, build a structure on a sloping site, but make design choices to ensure that the structure does not appear to be tilting; or, as we have already seen, they might adjust the proportions of elevated statues, so that they appear normally proportioned when seen from an angle and distance far below.

Caramuel describes oblique architecture in various forms: in inclined planes, in crossing, non-orthogonal planes, and in circular arrangements, or in some combination of these configurations.[7] In his dedication to Don John of Austria (1629–1679), the son of Philip IV (1605–1665), he claims that his architectural theory is entirely novel: "A new Art is born … of which no one else in the world has written."[8] Despite such bold proclamations, his was not, in fact, an entirely new art: Caramuel was not the first theorist or artist to anticipate—and try to correct for—deficits of human perception. Caramuel himself acknowledges that masons had adopted oblique building techniques to avoid producing structures that, in an imperfect world, would seem imbalanced, though he emphasizes that their practices have never been subject to theoretical study:

> I have seen many things well worked out in practice and in which the laws of oblique architecture have been well executed. But I have also observed many mistakes in existing buildings. As there is no other book that deals with oblique architecture and it is usually masons who wish to do away with architects when the occasion arises and they have to accommodate oblique places that are not amenable to straight delineations. They do so by filling the resulting empty spaces with irregular shaped elements.[9]

The contingent nature of architectural observation and optical correction is also perceptible in classical and Renaissance literature and practice, as Thomas Frangenberg has observed.[10] Vitruvius, for instance, presents optical corrections throughout his *De architectura libri decem* (*Ten Books on Architecture*).[11] References to optical correction also appear in Alberti's (1404–1472) *De re aedificatoria* (*On the Art of Building*), which dates from the mid-fifteenth century, and in sixteenth-century architectural treatises by Sebastiano Serlio (1475–c. 1554), Walther Hermann Ryff (c. 1500–1548), and Pietro Cataneo (c. 1510–c. 1574).[12] However, these works do not highlight systematically, as Caramuel does, the contingent nature of observation, human misapprehensions, and the need for optical corrections.

If Caramuel's theories built on concerns that had perplexed architects for centuries, the questions he raised were also consonant with those asked by other early modern practitioners. Visual artists and theorists including Bernini, Salomon de Caus (1576–1626), and Dürer, whose ideas I will discuss further below, also considered the impact of contingent factors on observation, and they recognized that observers were limited in their ability to distinguish between appearance and reality. But other seventeenth-century architectural theorists had more faith in the accuracy of individuals' visual discernment and thus did not believe that architects should adjust structures to counter alleged misapprehensions. (In this vein, Perrault, who praises observers' perspicacity, usefully counterbalances Caramuel's architectural theory.)

The details of Caramuel's biography help to explain why contemporaries labeled him illustrious or even omnipotent; he seems to have made remarkable achievements in fields including music, literature, architecture, and philosophy, while also distinguishing himself for bravery in Catholic-Protestant conflict. His broad learning supported his work in architecture, if we are to believe his claim in the *Architectura* that architects need to have some knowledge of a range of disciplines, such as medicine, astronomy, geometry, arithmetic, and perspective.[13] Caramuel was born in Madrid to a father who was a military engineer from Luxemburg and a mother from Bohemia.[14] According to his 1760 biography, he learned twenty-four languages, including Greek, Hebrew, Syriac, and a little Chinese.[15] He authored over sixty books on a broad range of subjects, such as the life of Saint Benedict, Gregorian chant, grammar, architecture, poetry, mathematics, logic, heraldry, cryptography, the cabala, and astronomy; in addition, a large number of his manuscripts survive in Italy in the Vigevano city archive.[16] He also maintained a rich correspondence with well-known scholars throughout Europe, including Pierre Gassendi, Marin Mersenne, and Athanasius Kircher.

Caramuel attended the universities of Alcalá de Henares, where he studied philosophy, and Salamanca, where he studied theology. When, in 1625, he joined the Cistercian order as a novice at the Royal Monastery of the Holy Spine (Monasterio de la Santa Espina), he was instructed by the blind music reformer and astronomer Pedro de Ureña; two years later, he took his religious vows.[17] He then spent some years teaching and living in monasteries in Spain and Portugal before moving around 1632 to Louvain in the Spanish Low Countries, where he was linked to the local Cistercian college. He remained in Louvain for a decade, earning a theology doctorate in 1638.

Thanks to the aid of King Philip IV of Spain and Fabio Chigi (1599–1667), he was made abbot of Disibodenberg in the Rhineland-Palatinate in 1643; however, when he arrived in the town of Bad Kreuznach, about eighteen kilometers northeast of Disibodenberg, in April 1644, he learned that the monastery had been burned down. In 1647, he was made abbot of the

Benedictine Emmaus Monastery in Prague. In 1648, he took up arms to protect Prague from an attack by Swedish forces; he was subsequently celebrated for his courage. During this period he authored a treatise, titled *Sacri Romani Imperii pax licita demonstrata* (*A Demonstration that the Peace of the Holy Roman Empire is Licit*, 1648), supporting peace with the Protestants, which helped to promote the signing in 1648 of the Peace of Westphalia, a collection of treaties that terminated the Thirty Years' War.[18] In 1655, after Chigi became Pope Alexander VII, Caramuel relocated to Rome, where he lived from 1655 to 1657 and then again for shorter periods in 1658, 1662, and from 1669 to 1671.[19] He also served as Bishop of Campagna and Satriano (1657–1673), an impoverished and remote diocese in the vicinity of Naples, and as Bishop of Vigevano (1673–1682), as he is identified in Auroux's portrait (see Fig. 141). In each of his dioceses he established a printing shop to produce his own writings; he published his *Architectura* in Vivegnano. Although Caramuel began his life in Spain and had experience in countries where reformation was contested, he would ultimately spend nearly three decades of his adult life in Italy, some of that in Rome, where, as a high-ranking Church leader and a correspondent of Kircher, he would have encountered the scholars who were then promoting deformation as, among other things, a pedagogical tool.

Caramuel's Straight and Oblique Architecture

Although the *Architectura* appeared toward the end of his life, Caramuel explains in the opening section of his treatise on oblique architecture that he had begun thinking about the subject as early as 1624, when a chapel was being erected in his monastery; since 1635, he had worked on the book's accompanying prints while in Brussels, Louvain, Antwerp, Vienna, Prague, Rome, and elsewhere.[20] Most of the prints are uncredited; a few are signed by the engraver Bernard de Bailliu (c. 1641–1694), who was born and trained in Antwerp and worked in Rome, and others by the Milanese printmakers Giovanni Francesco Bugatti (fl. 1660–1695), Cesare Laurentio (fl. 1657–1689), and Simone Durello (1641–1719). Caramuel asserts that he applied oblique architecture in designing the facade of the Cathedral of Vigevano—his sole surviving original built work—although he does not explain which details of the facade demonstrate his theory (Fig. 142). Luisa Giordano has speculated that Caramuel was responsible for giving the facade a different orientation from the body of the duomo, to make the facade functional in relation to the square (Fig. 143).[21] Today, the facade features two overlapping orders of pilasters surmounted by a curved pediment. No architectural drawings for the facade seem to have survived, and because it was likely completed posthumously and altered in restorations in 1910, 1965, and between 1998 and 1999, it is difficult to determine how closely the present structure conforms to both Caramuel's design and the precepts of oblique architecture.[22] I will thus make only the further observation that because the facade's concave surface covers not only the church but also a contiguous street, with one of its four portals leading onto this adjacent street, the facade causes the church to appear wider than it really is. The fictive entryway on the facade relies on observers' tendency to take appearance for reality.

In the *Architectura*, Caramuel notes the historical diversity of architectural approaches. In a section titled "How there has been, and will always be, a remarkable variety of opinions in architecture" (*Como en la Architectura ha havido y havrá siempre notable variedad de Opiniones*), he writes:

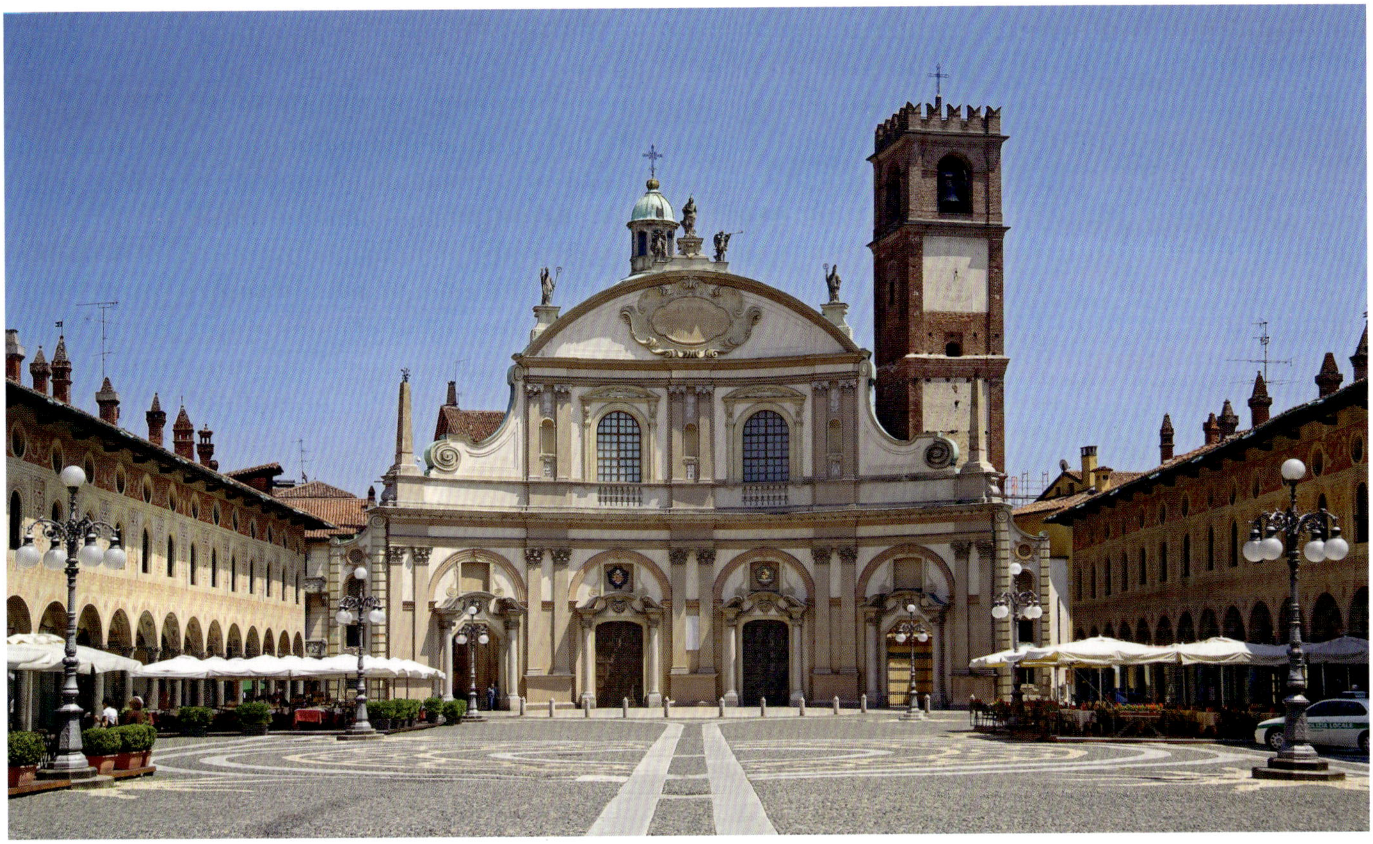

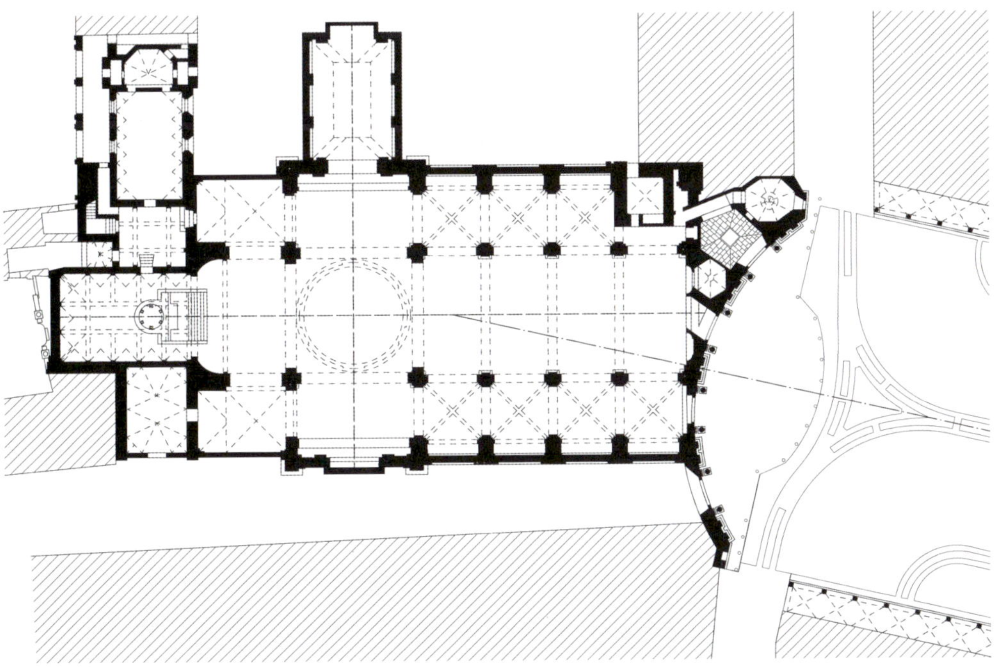

Fig. 142. Facade of the Duomo of Vigevano.

Fig. 143. Plan of the Duomo of Vigevano.

> In Latin these are called the doctrines of the philosophers, for where there is no convincing argument, free will comes in and *will stands in place of reasons*. All sciences are rife with this genre of opinions, because human wit demonstrates very little, and everything else is solved according to conjectures and suspicions. … Opinions of this kind are not scarce in architecture, defended not only by modern, but also ancient Masters, some of which have unfortunately entered the very books of Vitruvius.[23]

Note that Caramuel here does not shy away from attacking the most revered architectural theorist, the Roman author, architect, and civil and military engineer Vitruvius, whom he associates with classical teachings of the discipline.

In his own theorization, Caramuel is particularly attuned to obliquities found in nature or deriving from frailties in human perception; compensating for these is essential to avoiding unbalanced works. From his discussion of these categories, we can deduce that he defines the architect's role as reforming architecture to adjust to humans' visual frailty and the obliquity of divine creation, so that observers can see buildings that maintain the appearance of proportional harmony. God has designed the world with oblique lines and has created humans with faulty vision; the consequence for human vision is that without oblique architecture, buildings would appear unbalanced, slanted, out of proportion, or deformed; architects, by adopting the corrections of his architectural theory, can bring observers to perceive visual harmony. He writes:

> God is the first architect who made oblique lines in the heavens and Earth. Because moving in the heavens between the two Tropics and the Arctic Circle and Antarctic Circle, which are parallel to the Equinoctial Circle, He made the Sun's annual motion mark out the Ecliptic, which is a circle that cuts the Equinoctial Circle obliquely at the Zodiac, dividing it ultimately into two equal parts. … He erected mountains on Earth that grow obliquely, like the rivers and the valleys.[24]

Caramuel refers here to the celestial sphere, the sphere of fixed stars that includes all constellations and the zodiac, demarcated by great circles crossing at both poles and parallel circles at right angles to them. His verbal description resembles the celestial sphere as represented in a print, by an anonymous artist, integrated into the *Cosmographia* (*Cosmography*, 1539) of Petrus Apianus (1495–1552) (Fig. 144).[25] The primary parallels are the Celestial Equator (which Caramuel calls "the Equinoctial Circle"), the Tropic of Capricorn and the Tropic of Cancer (which he terms "the two Tropics"), and the Arctic Circle and the Antarctic Circle. The great circle, known as the Ecliptic, on which the Sun seems to travel against the backdrop of fixed stars, is angled at about 23.5 degrees to the Celestial Equator. Elsewhere in his treatise, Caramuel states his sympathies with the notion of fluid heavens presented by Tycho Brahe (1546–1601), who, like Caramuel, placed the Earth at the center of the cosmos.[26] In the third volume of his *Architectura*, Caramuel adds an anonymous print of the Tychonic system, and in his *Mathesis biceps* (*Two-Headed Knowledge*, 1670), he rejects the Copernican system explicitly and embraces the Tychonic system (Fig. 145).[27] When Caramuel proclaims to "have no need for that which the Church has condemned," we see him adhering strictly to Catholic ortho-

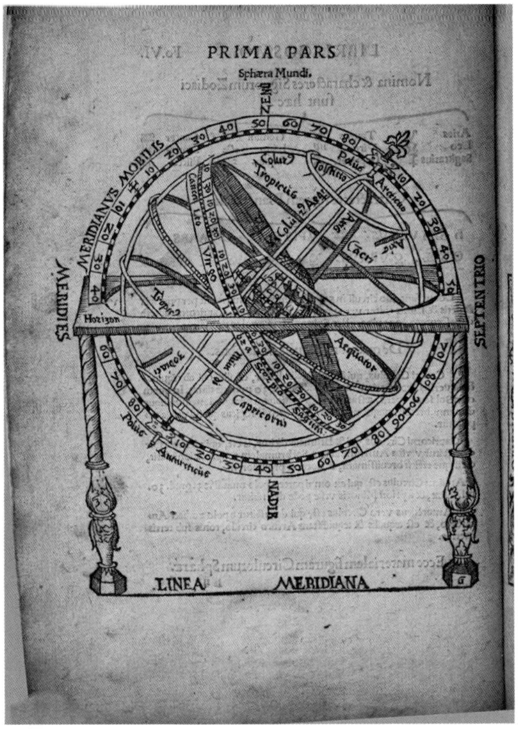

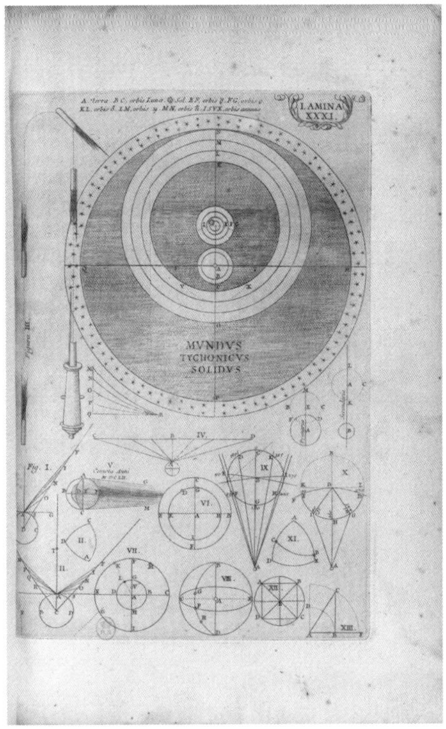

Fig. 144. Unidentified artist, celestial sphere from Petrus Apianus's *Cosmographia*, 1539, fol. VI^{v}. Woodcut printed on paper, 9.5 × 6 in (23 × 15 cm). British Library, London [General Reference Collection DRT Digital Store 10007.g.24].

Fig. 145. Unidentified artist, Tychonic system from Juan Caramuel y Lobkowitz's *Architectura*, 1678 [–1679], vol. 3, plate 31. Engraving printed on paper, measurements of the full page are 14 × 9.1 in (35.5 × 23 cm); measurements of image are 10.9 × 7.2 in (27.6 × 18.2 cm). BnF, Réserve des livres rares, Paris [V-2058].

doxy, even though, almost twenty years earlier, in his *Theologia moralis fundamentalis* (*Fundamental Moral Theology*, 1652), he had asserted that he could not present conclusive and certain reasons to reject Copernicanism.[28] In the *Architectura,* Caramuel encourages architects to recognize obliquity in the heavens and on Earth and to adapt to it by integrating some combination of inclined planes, crossing, non-orthogonal planes, and circular arrangements into their designs in such a way that structures would appear to be balanced from specific viewpoints, in other words, so that they would appear ideal according to classical standards.

Caramuel's architectural theory also addresses the frailty of humankind's embodied visual judgment. He aims throughout the *Architectura* and his other writings to make architectural theorists pay more attention to human perceptual errors; he urges architects to compensate in their designs for these perceptual failings as well as for obliquities occurring in nature. The theorizations of visual deceptions we considered in earlier chapters relied on the interpretations humans assign to their visual perceptions—for example, if humans interpret something that looks smaller as being more distant, then distorting the actual size of objects can falsely imply distance—to inform the intentional creation of deformations. We also saw artists and architects relying on misperception, at least temporary misperception, to communicate and reinforce doctrinal truth. Caramuel, like the Jesuits and Minims considered in earlier chapters, recognized humans' perceptual limits. But whereas the artists and architects we examined previously applied their understanding of perceptual limits to craft deformations that would resolve into clarity, Caramuel wanted to compensate for these perceptual limits and hoped to avoid altogether the appearance of deformation and imbalance.

Fig. 146. Unidentified artist, columns in an ellipse from Juan Caramuel y Lobkowitz's *Architectura*, 1678 [1679], vol. 3, plate 23. Engraving printed on paper, measurements of the full page are 14 × 9.1 in (35.5 × 23 cm); measurements of image are 10.4 × 7.2 in (26.3 × 18.2 cm). BnF, Réserve des livres rares, Paris [V-2058].

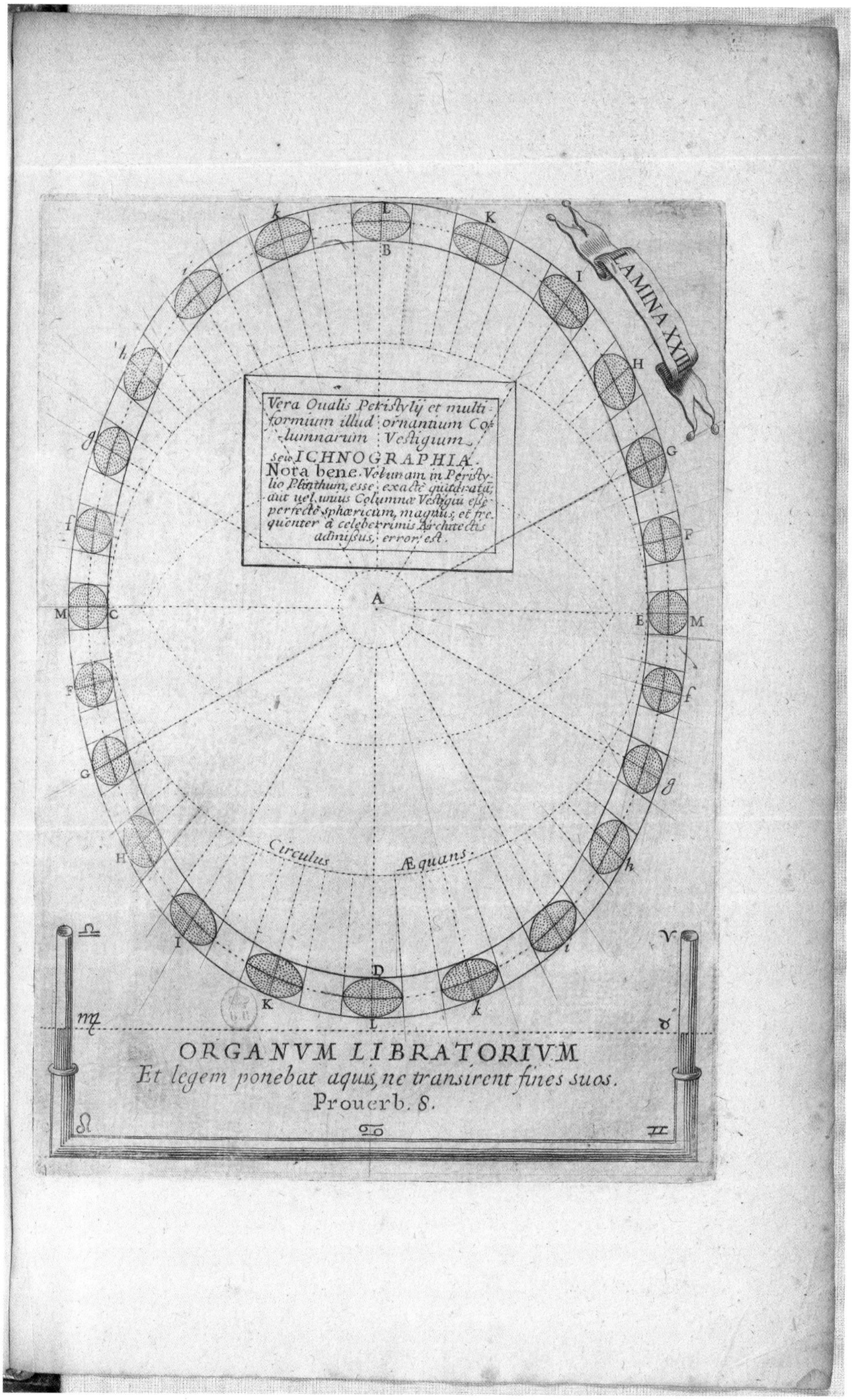

Skepticism, as it was simultaneously developing among thinkers in Europe—and especially following the publication by Henri Estienne (1528–1598) in 1562 of Sextus Empiricus's account of Pyrrhonist skepticism in Latin translation—helps to contextualize further Caramuel's interest in perceptual errors.[29] The second- or third-century CE philosopher Sextus Empiricus, for instance, famously describes water's deceitful effects on perception, noting that "the same oar appears bent in water but straight when out of the water."[30] When we see Caramuel's discussions of human perceptual limitations beside the creative works of other cosmopolitan Europeans of the period—who were producing novels, plays, paintings, philosophical tracts, and other cultural objects that questioned basic assumptions and beliefs and urged readers and audiences to reconsider traditional perspectives with reference to vantage points and contingent relationships—we can see that Caramuel's architectural, theological, and philosophical ideas contributed to these discussions.[31] To accentuate the gap between appearance and reality Caramuel draws a distinction between "things as they appear to us" (*quoad nos*) and "things as they are in themselves" (*quoad se*); with this vocabulary, he reframes in linguistic terms the theorization of visual incomprehensibility and discernment.

The distinction between "things as they appear" and "things as they really are" is manifest as Caramuel describes plate 23 of his book (Fig. 146):

> The vista should be placed on A, which signals the center. If placed in the equant circle, all columns should be equal, and have the same borders on their limbs; but having them placed in the oval or elliptical periphery, where some are distanced from sight and others are less so, in order for them to be equal *quoad nos* (before our eyes), they should be unequal *quoad se* (by their own accord) and their bases and limbs should have various borders, as is clearly illustrated in the engraving.[32]

The accompanying plate, engraved by an anonymous hand, offers a bird's-eye view of columns in an ellipse (see Fig. 146). Caramuel argues that in such an arrangement, if the observer is at point A, the architect should integrate inequalities that violate the expectation that columns in a peristyle will be identical. Caramuel holds that distortions in some of the columns' bases and shafts (depending on where they are positioned along the circumference) would not be visible from point A as they would compensate for the distortions generated by the columns being unequally spaced from that point. He repeats this lesson in the inscription on plate twenty-three (see Fig. 146): "True drawing and plan of an oval peristyle and of the multiform columns that adorn it. Note well: it is a great error, frequently made by the most famous architects, who accept either the plinth of the peristyle as perfectly square, or the base of a column as perfectly circular."[33] This aspersion against "the most famous architects" may refer to Bernini, among others. Elsewhere in the *Architectura*, Caramuel criticizes Bernini's colonnade in the Piazza di S. Pietro for what he deems its "errors" (Fig. 147):

> It forms an Oval or Elliptical plaza in front of the church of Saint Peter. It truly is a very sumptuous and beautiful work before the eyes of the common people, who ignore the foundational principles of Architecture, and hence think with the outer

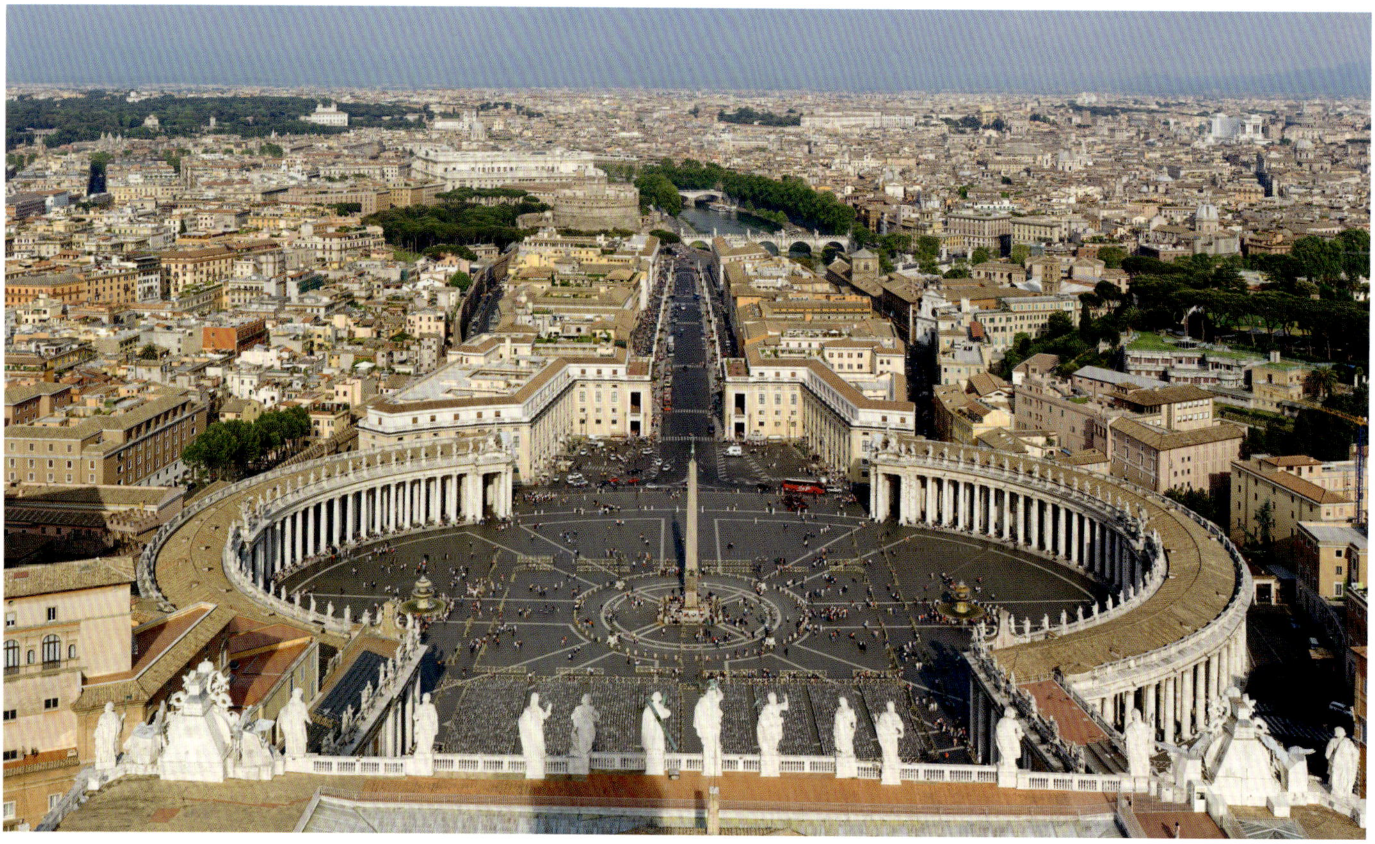

Fig. 147. Gian Lorenzo Bernini's colonnade in the Piazza di S. Pietro.

> senses; but those who know this Art look at it with discontent, for there are as many errors as stones in it.[34]

Caramuel differentiates between the "common people" (*vulgo*), misapprehensive viewers who apparently lack visual education and discrimination, and those perspicacious viewers with expertise, who can see the problems with the square that Caramuel recognizes. Bernini, in fact, integrated some modifications into the forms of column bases in the Piazza, but he did not assign diverse shapes to columns according to their location on the configuration's circumference, as the precepts of Caramuel's oblique architecture would have mandated.[35] In asserting that the *vulgo* "think with the outer senses," Caramuel implies that they do not think at all.

By drawing attention to the limited and flawed embodied perspectives of humans, Caramuel contributed to refining the theorization of visual incomprehensibility and discernment for readers of his book—even though in discussing ancient constructions, he also acknowledged that a select group of elite Europeans had already developed unprecedented visual perspicacity:

> *Those that were seen as miraculous in ignorant and rude times, today they would not be considered as such, because our eyes are taught to see greater and better things.* And

here we return to the *quoad se* and *quoad nos* distinction mentioned in the beginning. Because those that in themselves are only but average, are deemed marvels by those who have not seen anything great. *The Ancient Marvels were considered so because they precede us in time, and whatever novelty emerges from a rude era is rightly deemed excellent in the eyes of men.* These words are Cassiodorus's, and they are so true to my purpose and opinion, that I have included them twice.[36]

That Caramuel cites the Roman statesman and renowned scholar of antiquity Cassiodorus (c. 487–c. 585) on the rise in visual learning in his era clearly demonstrates that Caramuel was aware that thinking about visual expertise went as far back as antiquity. On the one hand, in calling attention to flaws in visual perception, Caramuel complements the work done by other theorists discussed in earlier chapters, like Niceron and Maignan—recall the latter's remark "All of this work is composed in order to deceive vision." On the other hand, unlike those theorists, Caramuel does not embrace deformation for its didactic and rhetorical potential. He wants to keep his observers at a single point from which structures appear ideal and does not appreciate a pedagogical potential in moving from deformations to reformations.

Caramuel's criticisms notwithstanding, Bernini also spoke of his concern for the difference between how things appear to observers and things as they are in themselves. In particular, he discussed *i contrapposti*—shapes sited near freestanding works that influence how the entities' proportions would appear.[37] The French collector and patron Paul Fréart de Chantelou (1609–1694) recorded a comment Bernini made in 1665 on how objects might be observed in certain visual juxtapositions: "The Cavaliere added that one of the most important things was to have a good eye in assessing the *contrapposti*, so that things should not only appear simply to be what they were, but should be drawn in relation to objects in their vicinity that change their appearance."[38] Chantelou reports an example Bernini offered to illustrate such an experience of visual incomprehensibility and discernment:

> He had once made a statue the head of which when completed seemed too small, although it was the right size, one-ninth of the whole, which was customary in a Christ. … It looked too small not only to him but to one of his friends. … Finally, however, by dint of examining it and thinking about it he discovered that the effect was caused by a piece of drapery on the shoulder; he had made it very much smaller, and the appearance of the head was quite changed. When his friend returned and saw the figure with the head apparently in its true proportion, he was astonished and could not understand how he had corrected it.[39]

The incident underscores Bernini's understanding that an entity's proportions appear correct based not on measurement, but rather on its relationship to adjacent entities. Here we might recall the cat that seems enormous when it is at the far end of Spada's perspectival colonnade. Bernini's emphasis on visual discrimination of the relationships existing among parts and wholes and among adjacent entities recalls Michelangelo's notion of the *giudizio dell'occhio* ("judgment of the eye"), as Carolina Mangone has elaborated.[40] In an often-quoted passage, the Italian painter, architect, and art theorist Giorgio Vasari (1511–1574) describes Michelangelo

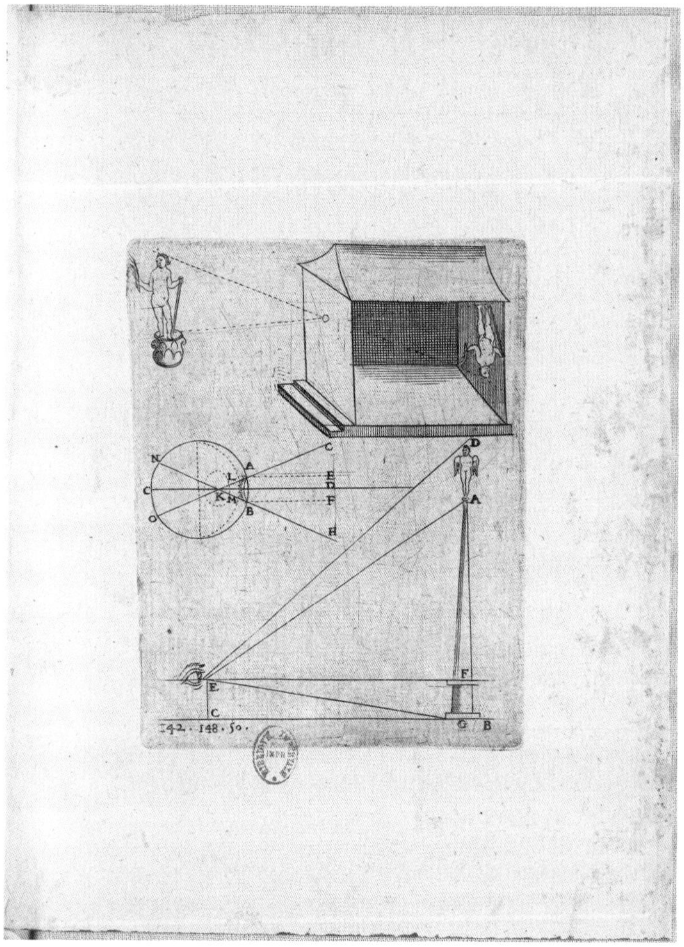

Fig. 148. Unidentified artist, distortions to raised statues from Juan Caramuel y Lobkowitz's *Mathesis biceps*, 1670, facing p. 148. Copper engraving printed on paper, 4.25 × 3.15 in (10.8 × 8 cm)"). BnF, Réserve des livres rares, Paris [RES-V-140 and RES-V-141].

expounding on this concept: "He said that one should have compasses in one's eyes, not in one's hands, because the hands execute but it is the eye which judges."[41] Michelangelo contends that artists should follow the judgment of their eyes, trained with practice. The painter and writer Gian Paolo Lomazzo (1538–1592) underscores Michelangelo's commitment to training the visual judgment of artists and architects: "Michelangelo, that very great sculptor, painter, and architect used to say that all the reasons of geometry, or arithmetic, or proofs of perspective were of no use to men without training the eye in knowing how to see and in making the hand do."[42] Bernini referred to the facade of St. Peter's when he defended his own theory to Chantelou: "He mentioned the facade of St. Peter's and the much lower colonnades he had placed on either side of it in order that the facade should look taller in contrast. This design had succeeded just as he had imagined."[43] Chantelou's diary makes clear that Bernini was concerned with optical correction, even if his practice differed from that which Caramuel articulated in his theories of architecture.

As I have noted above, Caramuel's theory of architecture dictates particular handling for raised statues, designed to make these more distant objects look ideal. His strategy reflects his underlying belief that visual discernment does not correct for distortions of scale, thus the architect who fails to intervene will actually be distorting in practice. A passage from his text *Mathesis audax* (*Audacious Knowledge*, 1644) that is helpful to read in conjunction with its accompanying print explains why and how to correct apparent scalar distortions (Fig. 148):

> *In this way, the Architect ought to lay out the symmetry of the building, so that statues seem to have a natural height.* It is demonstrated. Because statues are not the signs but the figures of men; therefore, they ought to be of such size as is required so that they are seen from the angle which corresponds to tall men. To consider distance and height will follow. For in the present plan, if the height of the column is AB and the distance of the eye CB, I say that the statue will be much larger than human, and I show the excess. … Therefore, I say that the statue AD, seen by C, appears equal to the man FG, because it is seen from an equal angle; and at greater distances, the apparent size of things depends upon the angle of vision.[44]

Architects who plan to elevate statues should thus produce figures that are larger than life size, to correct the apparent diminution that will result when the statue is viewed from a distance and at an angle. The engraving by an anonymous artist that accompanies Caramuel's explanation presents the reasoning in an easily understood visual form. The print embodies Caramuel's investment in visual modes of communication, even when he is detailing visual deceptions; to the viewer who considers the print, the figure atop the obelisk appears larger than life size, even if Caramuel holds that it would seem smaller—life size—to the observer at C.

As we have seen, practitioners like Bernini and Michelangelo were aware of the distorting effects that concerned Caramuel. Nor was he the first theorist to remark on distortions when observing elevated entities.[45] Salomon de Caus, writing in 1612, provides textual and visual instructions for how to place writing on a wall in such a way that all the words appear to be the same size to an observer, even though in reality the text increases in scale as it is positioned higher on the surface (Fig. 149).[46] In the unsigned engraving, the dictum (*Congnois toy mesme*, or "Know thyself"), inscribed three times onto the wall, reminds observers to follow the precept and to examine how they themselves contribute to how things appear, whether by misjudging scale or some other intervention.[47] De Caus, in turn, notes that Dürer had already discussed and visualized this kind of typographic setup in his *Underweysung der Messung* (*Treatise on Measurement*, 1525), where he offered young artists an understanding of mathematics to help them avoid making mistakes in their work (Fig. 150).[48] In Dürer's image, the wall inscription reads, "Das Wort Gotes bleibt ewiglich dieses Wort ist Christus aller Christglaubigen Heil" ("The word of God lasts forever, this word is Christ, the salvation of all Christians"). Centuries earlier, Plato dealt with the question of how visual artists should react to apparent distortions when elevated entities are observed from a distance. In his dialogue *The Sophist*, a philosopher visiting from the town of Elea tells the student Theaetetus:

> If [the ones who sculpt or draw very large works] reproduced the true proportions of their beautiful subjects, you see, the upper parts would appear smaller than they

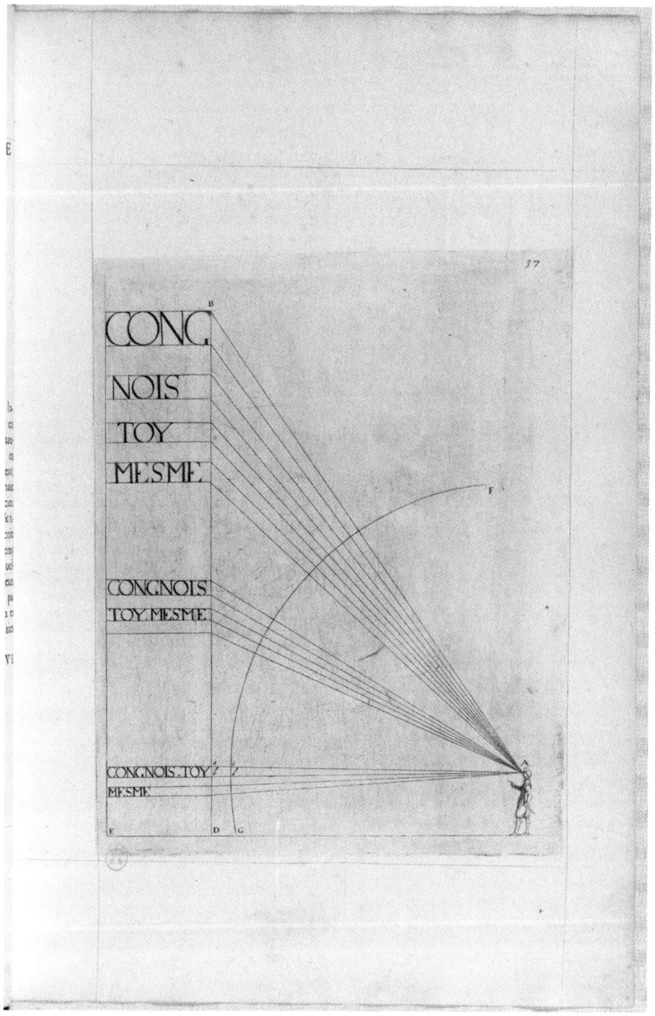

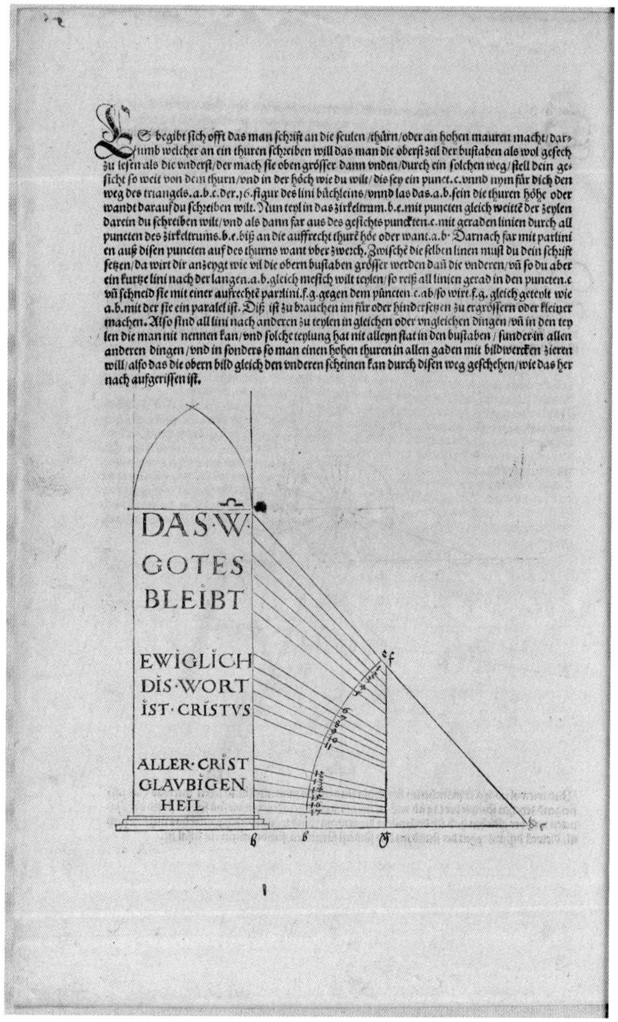

Es begibt sich offt das man schrifft an die seulen/thürn/oder an hohen mauren macht/darumb welcher an ein thuren schreiben will das man die oberst zeil der bustaben als wol gesech zu lesen als die vnderst/der mach sie oben grösser dann vnden/durch ein solchen weg/stell dein gesicht so weit von dem thurn/vnd in der höch wie du wilt/dis sey ein punct.c.vnnd nym für dich den weg des triangels.a.b.c.der.16.figur des lini büchleins/vnnd las das.a.b.sein die thuren höhe oder wandt darauf du schreiben wilt. Nun teyl in das zirkeltrum.b.e.mit puncten gleich weitte der zeylen darein du schreiben wilt/vnd als dann far aus des gesichts punckten.c.mit geraden linien durch all puncten des zirkeltrums.b.e.biß an die auffrecht thurē höe oder want.a.b. Darnach far mit parlinien auß disen puncten auf des thurns want vber zwerch. Zwischē die selben linen must du dein schrifft setzen/da wirt dir anzeygt wie vil die obern bustaben grösser werden dañ die vnderen/vñ so du aber ein kurtze lini nach der langen.a.b.gleich mesich wilt teylen/so reiß all linien gerad in den puncten.c vñ schneid sie mit einer aufrechtē parlini.f.g.gegen dem püncten.c.ab/so wirt.f.g.gleich geteylt wie a.b.mit der sie ein paralel ist. Diß ist zu brauchen im für oder hindersetzen zu ergrössern oder kleiner machen. Also sind all lini nach anderen zu teylen in gleichen oder vngleichen dingen/vñ in den teylen die man nit nennen kan/vnd solche teylung hat nit alleyn stat in den bustaben/sunder in allen anderen dingen/vnd in sonders so man einen hohen thuren in allen gaden mit bildwercken zieren will/also das die obern bild gleich den vnderen scheinen kan durch disen weg geschehen/wie das hernach aufgerissen ist.

should, and the lower parts would appear larger, because we see the upper parts from farther away and the lower parts from closer. ... So don't those craftsmen say goodbye to truth, and produce in their images the proportions that seem to be beautiful instead of the real ones?[49]

Caramuel also singles out the baluster and explains how it should be reformed to appear ideal. He explains how to adjust this structure according to the precepts of oblique architecture, representing the principles in plate 6 and discussing its application in the accompanying inscription (Fig. 151):

To the reader, free from prejudices. We represent the two orders of the small columns (a Spaniard calls them *Balaustres*). The lower ones are unsuitable and nevertheless fashioned on stairs by architects, all over the world. But they ought to be rejected

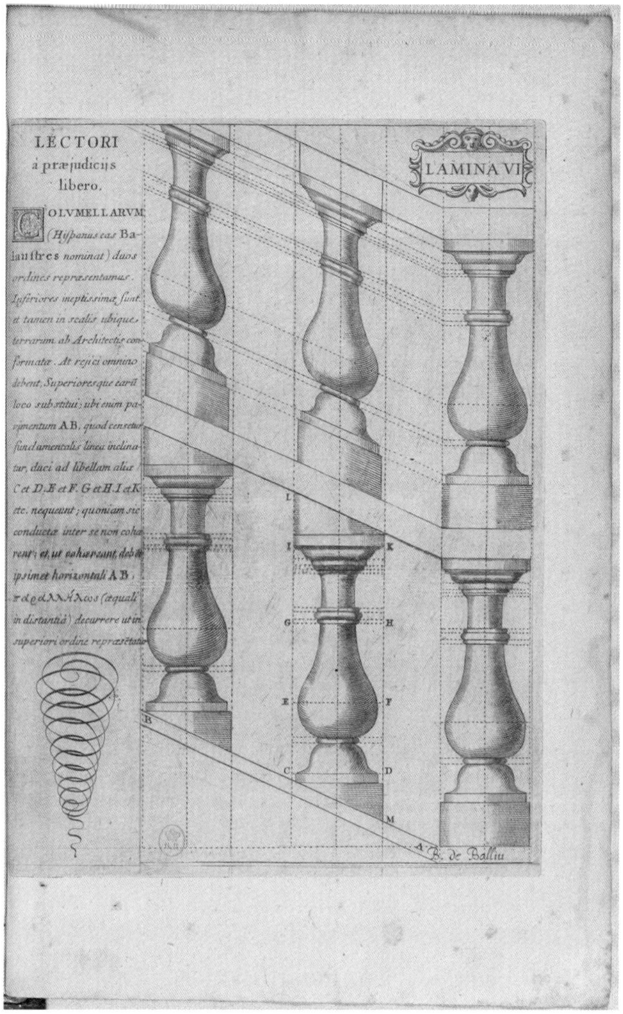

OPPOSITE

Fig. 149. Unidentified artist, how to place writing on a wall from Salomon de Caus's *La perspective, avec la raison des ombres et miroirs*, 1612, plate 37. Engraving printed on paper, the measurements for the entire page are 16.54 × 10.87 in (42 × 27.6 cm). BnF, Réserve des livres rares, Paris [RES-V-442].

Fig. 150. Albrecht Dürer (designer), how to place writing on a wall from Dürer's *Underweysung der Messung*, 1538, $K2^{v}$. Woodcut printed on paper, measurements of the full page are 12.6 × 7.87 in (32 × 20 cm); measurements of print are 5.91 × 5.51 in (15 × 14 cm). BnF, Réserve des livres rares, Paris [RES-V-455].

LEFT

Fig. 151. Bernard de Bailliu (engraver), balusters from Juan Caramuel y Lobkowitz's *Architectura*, 1678 [1679], vol. 3, plate 6. Engraving printed on paper, measurements of full page are 14 × 9.1 in (35.5 × 23 cm); measurements of image are 10.9 × 7.7 in (27.7 × 19.6 cm). BnF, Réserve des livres rares, Paris [V-2058].

> altogether, and the upper ones ought to be substituted in their place. Where the pavement AB [is], what is judged the foundational line is inclined; the others, C and D, E and F, G and H, I and K, and so on, do not need to be brought to the level; because, assembled in this way, they are not level amongst themselves; and in order for them to be level, they ought to go down in parallel (in equal distance), to that very horizontal AB, as the representation in the upper order.[50]

In the engraving, signed by de Bailliu, the lower baluster corresponds to what Caramuel would associate with traditional theory, which he calls *Architectura recta* ("straight architecture"), whereas the upper baluster illustrates oblique architecture, conforming to Caramuel's recommendation that the baluster be distorted so that its parts appear parallel not to the floor but to the slanted plane of the stairs. Although few oblique balusters exist, one by the Italian architect Filippo Juvarra (1678–1736) survives in Palazzo Madama in Turin, as Jasmin Mersmann has

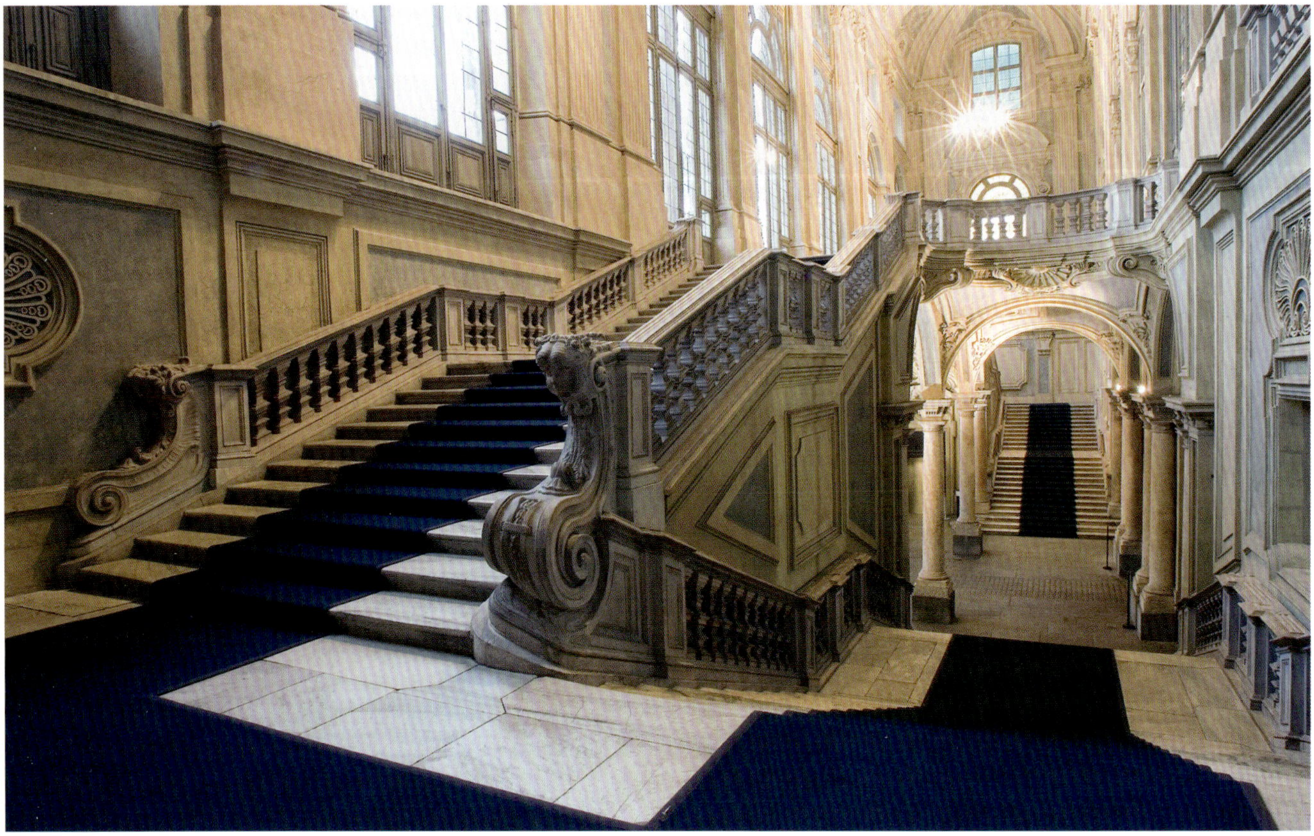

Fig. 152. Filippo Juvarra's oblique balusters in Palazzo Madama in Turin.

noted (Fig. 152).[51] Examples of straight and oblique balusters made in 1563 and 1573, that is, well before the *Architectura* appeared, can also be found in the pulpits of the Arezzo Duomo, as Jorge Fernández-Santos Ortiz-Iribas has observed.[52]

Caramuel's *Mathesis audax* of 1644 offers further recommendations for how to adjust another structure, a tower, according to the precepts of his architectural theory, so that it too will appear ideal:

> *Towers and universal edifices which are erected at a perpendicular, are wider above than below.* It is shown. Walls of towers are not parallel, nor do they make an approach against the sky, but against a center: therefore: the assumption is proven. Because the perpendiculars sent from P and from Q coincide at center I, therefore, there is a greater distance between these lines than the length from the center.[53]

In the unsigned print that is paired with this passage, the reader finds such a tower, wider at the top than the bottom (Fig. 153); the effect, which appears exaggerated for didactic purposes, is presumably meant to compensate for the human tendency to see parallel lines as converging.[54] Caramuel effectively integrates a distortion into the structure of the tower so that it will appear ideal to individuals viewing it from the ground. Here and elsewhere, Caramuel calls attention

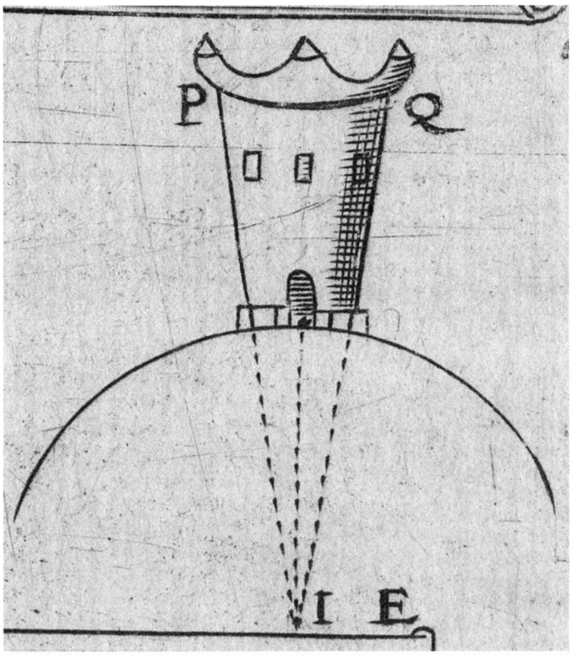

Fig. 153. Unidentified artist, tower whose upper region is wider than its lower area from Juan Caramuel y Lobkowitz's *Mathesis biceps*, 1670, detail from plate facing p. 25. Engraving printed on paper, 4.21 × 3.31 in (10.7 × 8.4 cm). BnF, Littérature et art, Paris [V-2058].

to the ways that appearances differ from reality and encourages an awareness of the deceptions inherent in visual experience. At the same time, however, because he uses an image to highlight deception, he also betrays a commitment to using visual data to transmit knowledge. While he presumes that unsophisticated observers of the architecture will not be able to understand constraints on their visual perception, and thus need corrective support to prevent them from concluding that the architecture is distorted, he apparently presumes educated, perspicacious observers as readers of his book: these observers will be sufficiently sophisticated to use their discernment as they consider what is actually on the page, and they will not be confused. Caramuel may have also thought that the illustrations for his book already corrected for the perceptive limitations he anticipated, because the conventions for representing three-dimensional objects on the flat page already compensate for the limits of human visual perception. In showing how observations could be deceptive, readers of Caramuel's books hone their capacity to discern—and thereafter to have greater faith that what they perceived, corrected via the knowledge Caramuel had provided, corresponded with the way things really were. Yet this visual education creates a potential problem: would not readers, armed with an understanding of how these deceptions arise, be confused when they encountered architecture that preemptively countered misapprehensions that these viewers were already anticipating and correcting for? It is precisely this issue of supererogatory correction that Perrault would take up a few years after the *Architectura* appeared.

Perrault, who had made an important French translation of Vitruvius's *De architectura*, deals with optical corrections in chapter seven of the second part of his architectural treatise, titled *Ordonnance des cinq espèces de colonnes selon la méthode des anciens* (*Ordonnance for the Five Kinds of Columns after the Method of the Ancients*, 1683).[55] Beyond his contributions to architectural theory, Perrault was a founding member of the Académie Royale des Sciences (1666) and focused much of his energy on biological research; he died of an infection he caught from a camel he was dissecting.[56] Without ever mentioning the *Architectura* directly, Perrault unequivocally opposes the employment of adjustments in architecture to make the proportions of a building, a relief, or a statue look ideal from particular vantage points.[57] He contends that instances of optical correction in ancient and early modern architecture are not only unusual but also unsuccessful.[58] He then praises the human capacity to judge visual objects: "This judgment never fails ... to prevent our being deceived by the distortions and the adverse effects that we imagine can be caused by distance and varying relative position."[59] He goes on to stress that people are not aware of the effects of their sensory faculties—in other words, that they will correct automatically for the potentially distortionary effects of angle or distance: "The judgment with which all the senses are endowed is something that we possess without knowing it and that we exercise without being aware that we do so, as if it were habit or second nature."[60] Perrault's remarks acknowledge our lack of self-awareness when it comes to sensory experiences, something we already noted in the introduction in relation to period discussions of the *je ne sais quoi*. Perrault argues that the senses of sight and hearing in particular are sufficiently acute not to succumb to deceptions:

> Habit also makes the judgment of sight and hearing, as the senses we use most often, much more precise than that of the other senses, and habit makes it rare for them to be deceived by the discernment of situations that might entail deception. As a result, we judge the distance, height, and strength of objects with great certitude through sight and hearing and discern such matters less easily through the other senses.[61]

If Caramuel worried that humans would not be able to distinguish with certainty between "things as they appear to us" and "things as they are in themselves," Perrault, to the contrary, celebrates the sharpness of our visual and auditory discernment. Perrault's faith that people could depend on their senses for judgment and his belief that they were able to reach legitimate inferences based on attentive observation resonates with the views of empiricists like John Locke, who claim that the extent of knowledge is limited to the sphere of simple ideas people have experienced.

Perrault draws on his biological research to posit that as animals grow older their "judgment of sight" improves:

> It is very likely that animals see badly when they are born and that they judge distant objects to be as small as the image of them represented inside their eye makes them appear. By letting them know they were mistaken, experience corrects the error of this first judgment. Subsequently, judgment becomes so apt in the use of all possible means to ward off such deception that finally it achieves the perfection it has when we begin to see properly. This perfection is such that no one believes that a distant tower is smaller than a finger because it can be covered by a finger placed close to the eye, or that a circle viewed obliquely is an oval, or that an oval is a circle, although these are the actual images of things in the eye.[62]

Whereas Perrault saw acute visual judgment as an inevitable, and largely imperceptible, result of lived experience, many Jesuits and Minims saw a sophisticated visual discernment as emerging, at least among educated observers, via careful instruction and exposure to specific information and devices. We can see, then, in Perrault a trust that humans will naturally acquire the ability to interpret the information collected through visual perception—a trust that seems opposed to the views of figures like Niceron or Maignan, though there is little evidence that Perrault was directly in dialogue with these theorists. Perrault would likely have also been skeptical of claims like Virgilio Spada's, that men in Bologna had "decided to climb up with ladders and touch" illusionistic frescoes to confirm that they were not real (as discussed in chapter 1). Indeed, Perrault trusts humans' interpretive ability so far that he ultimately dismisses illusionistic art's power to deceive: "What proves that the judgment of sight is both accurate and infallible and what imparts certain knowledge that this sense is not as susceptible to surprise and deceit as people claim, is the difficulty even the most perfect and ingenious art has in succeeding in the attempt to delude us."[63] Perrault finds auditory judgment equally precise, noting that ventriloquists also do not deceive.[64]

Because Perrault views people's visual judgment as so capable, he argues that attempts at optical correction fail and result in deformations instead of improvements. Perrault thinks such attempts fail for any observer; he does not make the distinction here Caramuel would seem to make between less educated misapprehensive observers who cannot figure out what they are observing and the more sophisticated perspicacious individuals he imagines as consuming the explanatory prints in his book. Moreover, although Caramuel intended that by following his recommendations a practitioner would produce works of art and architecture that appear ideal, Perrault would posit, because he understood human visual discernment so differently, that Caramuel's recommendations would cause works to appear deformed. And Perrault disapproved of deformations. He would argue that a Church that has leaned toward the adoption of deformations should end this practice—not because he feared viewers would be deceived, but because he knew they would not. For instance, he dismisses the notion that elevated statues should be enlarged:

> If we make crude and massive the parts of sculptures that are placed high up in order to avoid their appearing indistinct and confused because of their great distance from the viewer, the eye will perceive them as being crude and massive. This is because when the eye compares the distance it knows with the confusion it knows ought to be present in things that are far away, it will be offended if it detects a clarity that it judges should not be there.[65]

Perrault notes that observers would likewise be disturbed by a painting "in which the painter made distant things as forceful and distinct as those that are near."[66]

He explains how using distortions to make up for the weaknesses of vision, as Caramuel recommends, will only create "visibly deformed" works:

> I believe that once people have thought the matter over carefully, they will find no reason at all for distorting or spoiling proportions in order to prevent them from

> appearing distorted or for making something defective in order to correct it. Since the appearances created by distance and position that people take to be shortcomings and adverse effects are actually the true state and natural form of things, they cannot be changed without making them visibly deformed. For all we have said, or can say, on this subject is that distance is less likely to distort proportions than is the alteration of those proportions and that there is greater danger in making a proportion appear distorted when, in fact, it is than in having it appear distorted when, in fact, it is not.[67]

Moreover, he claims that such optical adjustments are ill-advised, because they would work only from a single observational perspective; as noted earlier, Caramuel often specifies the precise location from which oblique architecture should be regarded.[68]

Perrault later uses language that evokes Caramuel's theory of oblique architecture—while also criticizing anamorphoses of the sort we have studied in the previous chapters—to argue against deformations in architecture:

> There are optical figures whose proportions are modified in such a way that their effect is favorable only if they are viewed from a specific location but become immediately deformed as soon as the eye changes place. If we alter the proportions in a building in order to make them create a favorable effect in an eye situated in a specific location, these proportions, like those of the optical figures, will also appear totally defective as soon as the viewer changes place, because an aspect that is oblique when a person is near becomes progressively less so as he moves away. Thus, enlarging or inclining the fascia of a corona, in order to avoid its appearing too small due to the obliqueness of the aspect, would make it appear too large as soon as a change in aspect obviated this obliqueness.[69]

Instead of shifting established proportions and rules, Perrault holds that architects should follow mathematical reason. If those Minim and Jesuit scholars who advocated for the calculated use of deformations trusted they could use them effectively to direct viewers' attention in particular ways, to guide wanderers, then Perrault, in effect, seems to be asking what happens if an observer decides to walk in an unintended direction or insists on observing Caramuel's buildings from somewhere other than the single point he recommended.

Perrault's attitudes to optical corrections and visual discernment are diametrically opposed to Caramuel's. Whereas Perrault is confident in the human capacity to see things with certainty, Caramuel calls attention to the uncertainty of visual observations, even as he educates his readers to become aware of the ways in which they misapprehend the world in all its obliquities. The tension between Perrault's and Caramuel's positions underscores three main points. First, their respective attitudes clarify how significantly architectural theorists, in determining the architectural forms that they ultimately advocated in their treatises, were influenced by their underlying conceptions of observers' visual expertise. For both Perrault and Caramuel, how one should build is directly related to how one sees, and how one thinks others see. Second, Perrault's faith in human visual discernment indicates that Caramuel and other artists' and architects' distrust in the visual discernment of observers was not, in this period, a given.

And third and relatedly, Perault's skepticism that observers should and could always follow the pathways intended by artists and architects indicates that some questioned these assumptions around necessary progressions and "correct" vantage points. Even in Caramuel's own attitude to visual data, I have noted some ambivalence: although he accentuated the gap between how things appear and how they really are, he also relied heavily on visual modes of transmitting data. To understand Caramuel's relative pessimism regarding human visual discernment, it is helpful to consider his writing on theology.

Theology

Caramuel's writings on theology help to explain his focus, in his architectural theory, on exposing human finitude and misapprehensions, and also the significance he attached to vantage points and to how contingent factors inform what is observed. In his theological writings, Caramuel uses observational errors and impediments to seeing to model uncertainty regarding human doctrines.

As he develops his theology, Caramuel often accentuates the individual's contingent perspective and deficient knowledge, contrasting those with divine omniscience. He presents strategies for adapting to these limitations of perspective and knowledge, much as his oblique architecture presents strategies for adapting to the limitations of embodied perspective within irregular environments. Of particular interest are his discussions of grace and free will, which he considered essential to the human experience, and his participation in debates over choosing correctly when the sinfulness of a possible action is unknown.[70] In his *Mathesis audax*, Caramuel introduces a print of a cosmological diagram to contrast different perspectives on grace and free will (Fig. 154).[71] In the accompanying text, he explains that the Sun denotes grace and the Moon denotes nature: "The Sun, the symbol of grace, rises as God permits, above the good and the bad. … The Moon is the symbol of nature; inconstant to the inconstant, various to the various. What is the Moon, unless sprinkled with the Sun? What is nature, unless anointed with grace?"[72] He then goes on to explain the letters with which the diagram is marked:

> With this doctrine set out in advance, let us contemplate the spheres and motions of the mystical Sun and Moon. Let A be the Earth, GFC the orbit of the Moon, HEI the orbit of the Sun. KDL the sphere of Mars. Let L be the east, M the west, K the south. Therefore, when the eye is positioned at B, only the Moon F will be seen; for snatching away the whole Sun E with its mass, it will make it so that it cannot be seen by those who have been stationed at B. When the sphere of Mars has been positioned at D, doubtless in a place opposite this B, only the Sun is visible, the Moon invisible; for E lighting up more with its mass, and lighting up less with its rays, hides F. But when the eye is positioned at C, the orbit of the Moon, the Sun will be seen and also the Moon, the one at P, the other at Q. The one leads, the other follows. Different appearances, surely, because different parallaxes. Nothing is true, because all are eccentric; for every eccentric appearance is a deviation.[73]

Here and elsewhere Caramuel embraces a geocentric model of the universe with the Earth at A, the center of the diagram. He urges readers to consider how the various perspectives of

Fig. 154. Unidentified artist, cosmological diagram from Juan Caramuel y Lobkowitz's *Mathesis biceps*, 1670, detail from plate facing p. 195. Engraving printed on paper, 4.33 × 3.27 in (11 × 8.3 cm). BnF, Littérature et art, Paris [V-2058].

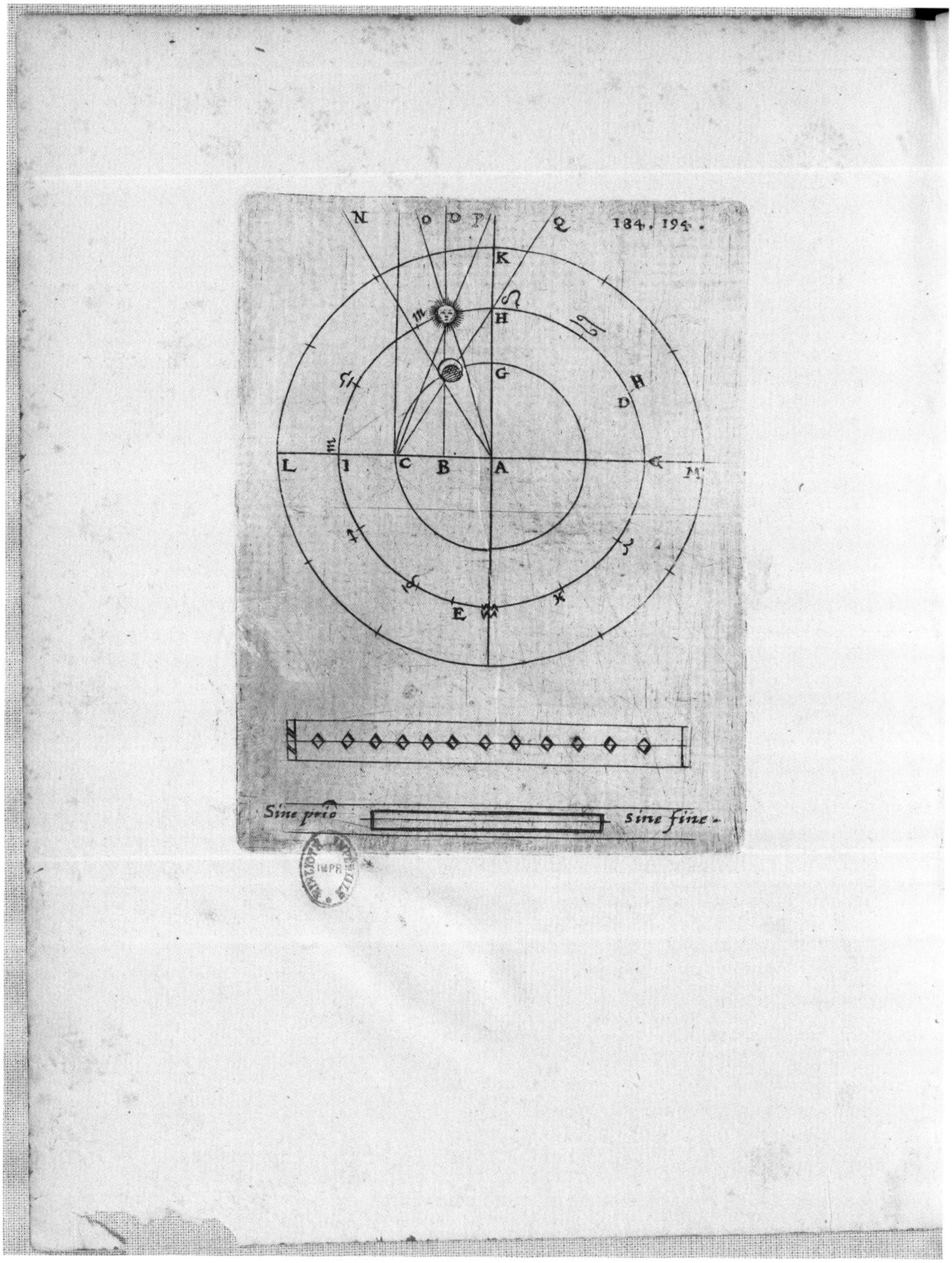

observation dramatically alter what is seen, such that entire celestial bodies become invisible or visible depending on the beholder's point of observation. Without this print, it would be more difficult for the readers of the *Mathesis audax* to understand the relations that Caramuel describes. The image enables the reader to grasp the model of the universe Caramuel advances in his text as well as how vantage points shift the ways in which individuals perceive. Caramuel relies on the reader's experience of an image to demonstrate how visual discrimination is dependent on vantage point.

He then ties the diagram to debates about grace and free will, placing adherents to different doctrines at distinct locations on the diagram where their viewpoints are obstructed to varying degrees and effects:

> The Pelagians, established at B, take pride in the atmosphere, in the region of vanity: they observe and speculate about the nature of the Moon, or the Moon of nature; they see neither the Sun of grace, nor the grace of the Sun: the severe cruelty of the Manicheans, stirring up the Church with arms and massacres, sitting at D, the sphere of Mars, sees the Sun of grace, the Moon of nature: (that is, increases and decreases of free judgement) it does not see. We at C, the orbit itself of the Moon, will put into practice and repeat that the Moon is observed going before, the Sun following, nature going before, grace following.[74]

Caramuel positions followers of Pelagianism—the heresy that humans may of their own accord take the first step toward salvation without divine grace—at B, since they cannot see the Sun, that is, the symbol of grace, from this vantage point. He places adherents of Manichaeanism—the religious doctrine that saw the world as controlled by a conflict between forces of good and evil, rather than governed by a single ideal being, a doctrine that Caramuel associates with Luther and Calvin (1509–1564)—at D, where they observe only grace and not nature and human freedom. At point C, both the Moon and the Sun can be observed, although the latter appears to follow the former. Finally, Caramuel describes point A as the vantage point with the most advantageous perspective: "Let us go just once to A, the true and legitimate center of all orbits, where those instructed by the best theoscopes, the early Church Fathers, doctors, and scholastics, observe heaven most accurately and put down their celestial observations into tables and numbers."[75] In sum, with this cosmological diagram and its symbolic resonances, Caramuel shows how theological positions are limited by their perspectives. Yet he also postulates that one perspective, that from A, is correct; he is not a relativist. Man is free in that he can position himself at various points, but he is not free to decide which point is the correct one. Indeed, his diagram brings to mind the actualized allegories we have seen in earlier chapters, which also used positions in spatial systems to locate theological concepts and insights. The observer of the diagram, not unlike someone walking down an anamorphic corridor in a Minim convent, can imagine moving from the more limited state of understanding he possesses at one location on the diagram to a new standpoint that offers a clearer perspective. This cosmology would then seem to be in accord with the idea—which Caramuel seems to share with some of the religious practitioners we have discussed—that there is a single vantage point and that leaders of the Church can show us where it is.

Caramuel is orthodox in holding that man is not free to decide which observation point in Figure 154 is the correct one. But at the same time, like other probabilistic moral theologians, he stresses that there are distinct opinions among Church authorities on certain topics in moral theology and recognizes that the Catholic Church has not always, on every single topic, presented absolute moral clarity. Moreover, as he debates with others how to avoid falling into sin, Caramuel calls further attention to the contingent and limited nature of human perspectives. Following the traditional doctrine inspired by Augustine, European scholastic theologians of the Middle Ages generally held that when an individual, uncertain whether an action would be sinful, pursues the action in question, that person sins and is morally culpable: they were ready to submit to the danger of sinning, and in this readiness they sin in actuality. Against this Augustinian stance, Caramuel promotes probabilism, a doctrine holding that the choice of a probable opinion (a stance supported by a solid argument), even in the face of a more probable contrary opinion (a stance supported by an even more solid argument), does not result in moral culpability.[76] The Spanish Dominican theologian Bartolomé de Medina (1527–1580), who taught initially in Alcalá and then at the University of Salamanca (from 1576 to 1580), articulated this position in his commentary of 1577 on the *Summa theologiae* of Saint Thomas Aquinas (1225–1274): "It seems to me that, if an opinion is probable, it is licit to follow it, even though the opposite opinion is more probable."[77] Caramuel himself never employs the term "probabilism" but asks instead whether one's conscience can be in good standing if one depends on probable opinion.[78] Not long before, the Spanish Jesuit Tomás Sanchez (1550–1610) had shocked theologians by contending that it was probably legitimate for a husband and wife to engage in fondling and other gratifying sexual activities without intercourse, in order to nurture their relationship.[79] Caramuel drew still more condemnation by proposing that masturbation was probably not a sin by nature.[80] He also participated in less salacious debates, including over chocolate's status as a beverage, as Stefania Tutino has explained.[81] If chocolate was deemed likely to be a drink, then anyone, according to Caramuel, would have been permitted to respond to this probability by consuming it during fasts without sinning. He ultimately concluded that chocolate could be drunk during fasts without falling into sin.

While within his theology, as in his architectural theory, Caramuel argued that human perception was limited, his moral theology was more skeptical of the possibility that a single ideal vantage point might be identified on matters where the Church had made no definitive dogmatic pronouncement, to the point that he countenanced departing from what others considered a moral "safe path." In his *Apologema pro antiquissima et universalissima doctrina de probabilitate* (*Defense of the Most Ancient and Universal Doctrine of Probability*, 1663; hereafter *Apologema*), Caramuel takes issue with a thesis pertaining to probable opinions presented by the influential Louvain professor Libert Froidmont (1587–1653). The thesis as quoted by Caramuel reads: "The use of probabilities is new. He who leaves behind the safe path and relies on probable opinions must be condemned before God. Opinions that are said to be or are probable for us will not be probable for God."[82] Froidmont accentuates the stakes involved in the employment of probable opinions: only the "safe path" is righteous to God. Caramuel, in his *Apologema*, asserts his disagreement with this viewpoint: "I opposed him in the *Theologia regularis*, which I published in Brussels in 1639 … and more copiously and more strongly in the second and third editions of the same work."[83] His resistance to Froidmont's single "safe path"—in other words, his commitment to probabilism—is unequivocal.

Caramuel's *Apologema* was placed on the Index in January 1664, and thereafter Roman Catholics were forbidden to read or possess it. And yet, reactions to probabilism within the Catholic Church were complex and varied over time.[84] From the second half of the sixteenth century, some Jesuits had embraced the doctrine. Serious opposition to probabilism then emerged in the 1640s, with a carefully argued attack on probabilism by the Jesuit Andrea Bianchi (1587–1657), in his *De opinionum praxi* (*On the Practice of Opinions*, 1645), written under the pseudonym Candidus Philalethus.[85] In 1656, the Dominican Order forbade its friars to employ probabilism, and Blaise Pascal mocked it in his *Lettres provinciales* (*Provincial Letters*).[86] Subsequently, a slew of anti-probabilist texts appeared. The Spanish theologian Tirso González de Santalla (1624–1705), elected the thirteenth Superior General of the Society of Jesus in 1687, opposed probabilism and developed an alternative doctrine called Probabiliorism: Probabiliorism did not make certainty a precondition for moral action, but it stipulated that in circumstances of uncertainty one should choose to act only according to the opinion that was more probable.[87]

I am especially interested in the ways that Caramuel's probabilistic theory emphasized the limits on human understanding of correct or moral behavior—and in the ways that this theory intersected with his theories of visual incomprehensibility. Without eliding the difference between discernment in regard to moral questions and discernment in relation to the visual and spatial perception of architecture (whether or not deformed), it is possible to see commonalities in the state of uncertainty and the potential for deception that exist in both contexts; the parallel is hard to avoid when reading Caramuel's writings, although he never directly analogizes one to the other.[88] His discussion of probabilism rests on the premise that sometimes humans will not know if an action is sinful. He thus begins at much the same place that proponents of anamorphosis occupy in the earlier chapters—within an experience of uncertainty, in the absence of clarity, facing circumstances that are necessarily deformed, as human existence is by default. But where the anamorphosists would trust the Catholic Church always to guide men through the space of uncertainty toward clarity, Caramuel sees continuing confusion on matters of conscience where the Church has not taken a definitive dogmatic stand: that laypeople will inevitably find themselves in some circumstances of obscurity where there is not a single preexisting path forward. Although the Church has the theoretical ability to resolve each issue, logistically it could not have already issued a definitive pronouncement for each and every question. An example that I have already mentioned would be the dilemma Catholics faced during fasts in relation to drinking chocolate, which was introduced to Europeans in the sixteenth century and became popular among the Spanish by the early seventeenth century: could they consider chocolate a beverage, for the purposes of conforming to rules around fasting, or would it be considered a food? It was not that the Church would be unable to resolve that issue; but because the institution had not previously encountered chocolate, the issue remained one on which there was no definitive pronouncement.[89] One wonders if these logistical obstacles to the path of perfect moral clarity in matters of regulating moral dilemmas did not explain, at least in part, why Caramuel evaded his peers' embrace of the movement from deformation to reformation as a didactic strategy.

Caramuel had previously written about the implications of limited knowledge in his *In regulam divi Benedicti commentarius* (*Commentary on the Rule of Saint Benedict*, 1640; hereafter *Commentarius*): "Since we are not angels, but human beings, we have evident knowledge of scarcely four things, so we are required to act according to probable opinions."[90] He defines the domain of probability in his *Apologema*:

> Between night and day, between manifest falsehood and manifest truth, [there is] a middle [way] that we call *probability*. Saint John seems to show this light of truth in his Apocalypse, under the form of the book closed with seven seals. As long as no seal is opened, we walk in darkness, and we are entirely ignorant of the truth. When natural demonstration or divine faith in this life, or beatific vision in the next, opens the seventh, we follow truth infallibly, and we depart from every fear of error. But when some are opened, and some are not, shadows and lights conflict in our intellect, and we are led to truth by probable footpaths.[91]

The metaphorical relationship between changing one's state of knowledge and moving physically along a path was widely accepted in this period. We saw this image at work in the Trinità dei Monti convent, as walking in the corridor resolved a confused landscape with strange intruding shapes (like the cross and prayer beads) into an image of Saint Francis of Paola, the observers' steps walking them toward clarity—and, indeed, we have seen this image operative even in an anamorphic fresco of Saint John receiving the Revelation. The passages above evince Caramuel's recurrent concern with acknowledging the limitations of perspectives in this life as opposed to the hereafter.

In another passage of the *Apologema*, Caramuel underlines how hard it is for humans to differentiate among degrees of probability, asserting that "the distinction is not always perceptible and recognizable."[92] He offers a formula for weighing probable opinions in the *Commentarius*: "If you have one authority on your side, against eight of the same quality, you may say your side is still probable; but against nine, you may not."[93] Yet he clarifies that opinions necessarily vary in quality: the opinion of a single person "of the greatest reputation" was equal to five "of great name," as Caramuel tries to quantify the value of individuals' varying powers of discrimination.[94] Even if we leave aside the slippery effort to value authorities by reputation, we might observe that Caramuel's formulas for measuring probable opinions were not set in stone: in the *Theologia regularis*, which first appeared two years before the *Commentarius*, he writes that one authority on your side is equivalent to only seven (not eight) of the same quality on the opposite side of the issue.[95]

We might still wonder why Caramuel was drawn to consider the significance of restricted perspective in architecture and theology. In a summary printed at the start of his *Apologema*, he implies that religious pressures related to the global conversion project motivated his support of probabilism: he has identified what he sees as a theological error that he believes will undermine the global progress of Christianity: "At length, it is demonstrated, that in this way, theologians have constrained consciences for the last 100 years, so that if they are permitted to continue with the same force for another 100 years, the conversion of the infidels will be rendered most difficult, and among the orthodox themselves, the greatest difficulties may most certainly be feared."[96] He contends that the traditional opposition to probabilism alienates potential converts, not to mention members of the Catholic Church, insofar as it puts an unnecessary burden on the conscience. The Church's position that it is sinful to follow a probable opinion whenever a more probable one exists may well have been unpopular: to constrain choice so narrowly in the presence of uncertainty would predictably have increased "moral scruples" and feelings of remorse and guilt. Here, Caramuel accentuates the debate's practical, political stakes, in the aftermath of the Protestant Reformation, which had irrevocably divided Christian Europe. As a Catholic Reformation religious leader, Caramuel had frequently confronted the challenge to the Catholic Church; at one point he claimed to have converted as many as thirty thousand souls.[97] His approach to probabilism lifts a burden on the conscience that he thought could alienate some individuals.[98] Yet if probabilism, like the experience of deformation, could be imagined as a journey beginning in uncertainty and confusion, Caramuel's recognition that there could exist multiple moral choices denied the single path to clarity; his theology on this point was thus contrary to the single-vantage-point certainty offered by the anamorphoses and illusionistic frescoes analyzed in the previous chapters (visual constructions that implied the Catholic Church had a cohesive moral vision), and even, perhaps, contrary to his own architectural theory. And Caramuel would have agreed as a matter of course that the Church had a clearly established position on matters such as the questions of grace and free will; he proposed his probabilism as a solution only where the Church had not yet delineated a clear path.

Beyond connecting his interest in probabilism to the religious schism, Caramuel also situates it in the context of European voyages of conquest and colonial expansion. In remarks in the first edition of his *Theologia moralis fundamentalis*—a text that he revised in later editions to evade prohibition—he acknowledges that something not probable or provable in the past can become probable or provable in the future. This perspective modifies the imagery around intellectual journeys that is embodied in encounters with the corridor anamorphoses. As Minims and Jesuits progress in their metaphorical journey toward a preconstructed anamorphosis, they arrive at a predetermined certainty; by contrast, Caramuel describes other metaphorical and physical journeys here that will take men toward situations where the results are uncertain and may require revision of things they believed "known." Though he does not mention anamorphoses directly, in reading Caramuel's remarks his readers may nevertheless conclude that the quest for knowledge is not always accurately susceptible to being modeled by an anamorphosis that resolves into a clear truth. To support his point that humans' knowledge of the world is also bounded by our limited perspective, he cites debates over the existence of antipodal humans (located on the opposite end of the globe) and over Christopher Columbus's (1451–1506) theories of world geography:

> *The proposition which today is neither probable nor improbable could be probable tomorrow, and either evidently true or evidently false the day after tomorrow.* I provide an example. This proposition: *there are antipodes,* it was not positively probable in the age of Augustine and the other ancient Doctors [of the Church], for evident reasons were not given which proved that they exist, or do not exist. ... Columbus learned that there were some lands beyond the sea from the fact that, having been

> driven there by a storm, he surveyed them; he believed it with probability, since it was grounded in serious evidence. He went to princes, he went to kings, and because they were prejudiced by ignorance or an opinion to the contrary, he was ridiculed; at length, he persuaded Queen Isabella of Spain of his opinion: he went, he saw, he returned, he brought back new men, birds, and animals, and he transformed matters which had been denied by everyone into certainty, which today no one is ignorant of. It happens similarly in moral theology, there are many probable opinions today, which were not seen as such yesterday, and many which were judged probable yesterday, which today, since the matter has been examined better today, have crossed into certainty—that is to say, either certainly false or certainly true.[99]

Caramuel reminds his readers that Columbus's findings dislodged positions previously taken for granted in the natural sciences and geography, and his claim that "today no one is ignorant of" a new world beyond the ocean would, by the middle of the seventeenth century, have been uncontroversial.

In another passage from his *Theologia moralis fundamentalis,* he looks at humans' error-prone habits of observation and suggests that we decide the "certain things" of human doctrine under imperfect conditions:

> The foundations of humane doctrine! Under the doubtful light of the evening star, the most certain things are decided: while the clearest mind operates in the dark night: and if, in the morning, when the morning star of truth shines, I ask the same things, they do not stand firm. Perhaps these things stand, although they certainly seem to waver. Perhaps we will affix the errors of our eyes to things; for if we use a yellow telescope, we will say that snow is golden. Perhaps our eyes, overcome with too much clarity, hallucinate; but I have no knowledge of this; and because I recognize clearly what is neutral, I pronounce it neutral. And because I wish to find a point where I may indeed securely fix my foot, and I do not find this on land, *I flee to you, and as a suppliant I beg your divine favor.*[100]

Caramuel draws on his theories concerning visual perception and flawed discernment to challenge the relationship between observed reality and truth. He associates uncertainty regarding the basis of human doctrines with observational errors and impediments to seeing accurately, whether that is the error of confusing stasis and motion or that of describing snow as golden because one has inspected it through a telescope with a yellow-tinted lens. The three-fold rhetorical anaphora of *forte* ("perhaps") adds to the overall impression of uncertainty.

The suggestion that seeing "with too much clarity" could indicate hallucination—the apparent perception of an external object that is not actually present—connects us again to Caramuel's research on visual discernment. Beginning from his theory that human visual perception is subject to error, Caramuel challenges the idea that human observation can be a valid basis for truth claims. I turn here to a passage in his *Animadversiones in Meditationes Cartesianas, quibus demonstratur clarissime nihil demonstrari a Cartesio* (*Notes on the Cartesian Meditations in Which It Is Very Clearly Proven That Nothing Was Proven by Descartes,* 1644), a manuscript that (as its title implies) responds to Descartes's *Meditationes de prima philosophia* (*Meditations on First Philosophy*, 1641).[101] Caramuel paraphrases Descartes's proposition from the third meditation concerning the twin predicates for certainty: "All that is very clearly and distinctly apprehended is true" (*Illud omne est verum, quod valde clare et distincte percipitur*).[102] He then undermines the Cartesian principle by asserting that human understanding of the truth values of things is subject to revision on further observation:

> You admitted many things before as entirely clear and manifest, which nevertheless you discovered later were uncertain, dubious, or false; therefore, it happens, if I may use the old terms, that you thought that your ideas recognized the matter, as it was; if I use your terms, that you think that you have a very clear and distinct understanding, although you have only an [understanding] which is obscure, deformed, and different from the object.[103]

Caramuel's argument that our perceptions are limited—that we may think we can deduce truth from the information before our eyes, but might then, on further information, realize we are mistaken in our conclusions—is connected to the operations of visual deformations, which play with those limitations and misperceptions. His overarching premise is that humans are subject to misapprehension and misinterpretation of visual information. The Minims and Jesuits we have seen might agree with that premise but then conclude that they could capitalize on this state of affairs to direct observers toward the truth. Caramuel concludes instead that our perceived truth might itself be subject to revision, because our capacity for moral discernment is as limited as our capacity for visual discernment.

Caramuel posited his philosophy of moral uncertainty in an era of religious and political change in Europe, and as the "new" sciences and novel technologies—from the microscope to the telescope—were shifting methods of observation and increasing epistemological uncertainties.[104] I do not, following Stefania Tutino's history of probabilism, see Caramuel's (and others') interest in probabilism as a direct or necessary consequence of these broader cultural and social changes.[105] Reformations and the rise of the nation-state did not "cause" probabilism. We might instead more neutrally observe that Caramuel's theories of moral choice emerge from the same confrontation of old and new structures that, in this period, produced other reexaminations of previously established truths.

Caramuel sets humans' misapprehension of things that they think they understand in contrast to God's perfect knowledge. Caramuel dwells on God's epistemological prowess not only in the passage cited above from his *Theologia moralis fundamentalis* but also in an earlier text, titled *Rationalis et realis philosophia* (*Rational and Real Philosophy*, 1642). There, he writes about divine power, discernment, and the being of souls in the past or future: "In this 'eternal now' God can decide that they [i.e., the souls] will have existence in the year 2000; and if He should do this, they would be to come in this 'temporal now' [i.e., the present moment as perceived by human beings in time]."[106] He inserts himself here into a long-standing debate concerning whether time should be conceptualized as materially and formally real, as Jacob Schmutz has observed.[107] And he uses an architectural metaphor to explain the ontological standing of temporal things:

> For those things which are past, present, and future, are objectively and intentionally present for God: and in this sense, the present has eternally existed for God and will be for perpetuity. … Just as, for example, stone coexists with stone in a building, so also past and future things already coexist physically and really and are present for God in the *now* of eternity.[108]

This material analogy illustrates that Caramuel's theological insights were never far removed from his musings on architecture. Indeed, in his *Architectura*, he reflected on the relation between his work in both fields: "As an architect and a theologian, theology provides me with matter to exercise my architectural contemplations."[109]

Caramuel was far from alone in finding an interrelation between theology and architecture; theological-architectural connections were something of an obsession among Europeans of the sixteenth and seventeenth centuries, in part because of the powerful analogy between moving through a space and moving toward understanding that we have traced in earlier chapters. We can see the intensity of this fevered relationship in the influential three-volume work *In Ezechielem explanationes et apparatus urbis ac templi hierosolymitani* (*Explanations on Ezekiel and the Layout of the City and of the Temple of Jerusalem*, 1596–1605), produced by the Spanish Jesuit architect Juan Bautista Villalpando (1552–1608) and the Jesuit priest Jerónimo de Prado (1547–1595) in collaboration with Flemish and Italian engravers.[110] Villalpando holds that classical architecture was based on Solomon's Temple, a building that he reconstructs and takes to be identical to the eschatological temple of the Last Days in Ezekiel's vision.[111] Caramuel, who considers the Temple to be the most perfect work of architecture, likewise situates it at the beginnings of classical architecture.[112]

✚

Caramuel authored a treatise of architecture designed to avoid deformations. His recommendations for architectural reforms compensate first for the obliquities caused by environments and second for differences between things as they appear to observers and things as they are in themselves. Using his building techniques could counteract the apparent tilt of a building erected on a hill, while adjusting the proportions of a raised statue could correct for scalar distortion. Caramuel's reformations of architecture and theology have their roots in his conviction that viewpoints and contingent relationships shape how humans come to observe the world. Although his *Architectura* contained passages promoting visual education and although his books are filled with instructive images—apparently communicating a faith that some observers could interpret visual information correctly—he ultimately had less faith than some architectural theorists, not least of all Perrault, in human visual discernment and in the human capacity to form certain visual observations, much as he recognized uncertainty to be a fundamental component of human knowledge in matters of theology. Because Perrault subscribed to a different theory of vision, he would have perceived the sort of architectural reforms that Caramuel recommended—intentionally departing from classical norms in pursuit of a more immediately ideal viewing experience—as a process of deformation, not reformation. If we can imagine our way back to the *piano nobile* of the Palazzo Spada, we might suggest that Caramuel would have expected visitors to perceive the *prospettiva* as a Farnese-type atrium with a regular recession. Perrault, by contrast, would have likely expected visitors to recognize immediately the irregularities of the structure—for instance, that its columns are of different sizes and its floor is sloped. In other words, if Caramuel's imagined visitor would have seen the *prospettiva* as undeformed, Perrault's imagined visitor would have seen it as deformed. Both theorists are, of course, correct, as both types of visitors likely existed (and the admiration of either type would have met Bernardino Spada's goals). But these figures' divergent perspectives show us how period conceptions of what constituted a deformation were dependent on how humans were understood to perceive. If this chapter has shown how some Catholic Reformation thinkers were aware of human limits—and even of logistical and chronological limits on the Catholic Church's ability to anticipate and answer every possible moral question—the next chapter will consider the corresponding appeal, to Catholic sovereigns and aristocrats, of certain artworks that exceeded ordinary notions of human limitations.

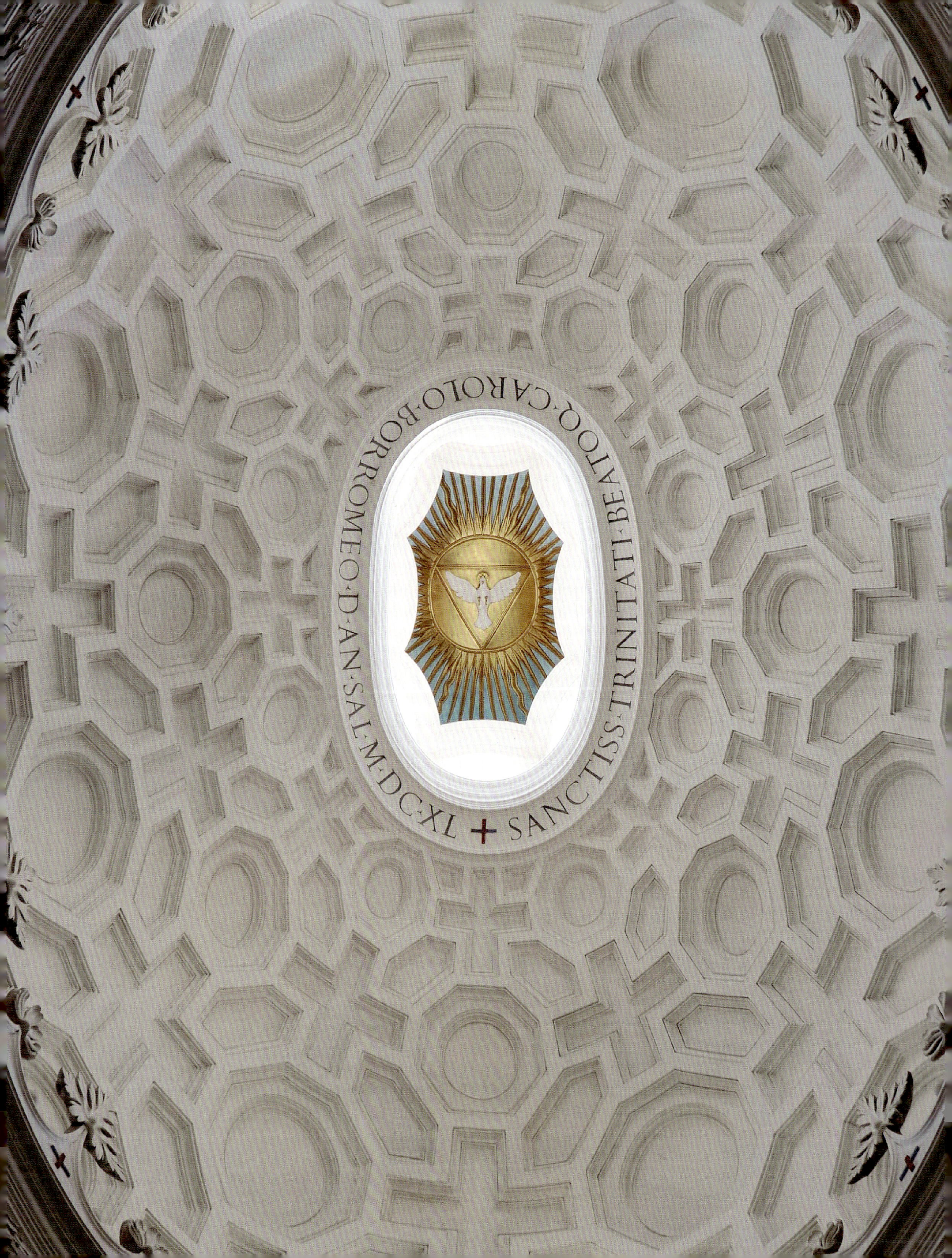
SANCTISS·TRINITATI·BEATOQ·CAROLO·BORROMEO·D·AN·SAL·M·DC·XL

6 PLUMIER'S IVORIES AND THE INCOMPREHENSIBLE

When I imagine a triangle, for example, I do not merely understand that it is a figure bounded by three lines, but at the same time I also see the three lines with my mind's eye as if they were present before me; and this is what I call imagining. But if I want to think of a chiliagon, although I understand that it is a figure consisting of a thousand sides just as well as I understand the triangle to be a three-sided figure, I do not in the same way imagine the thousand sides or see them as if they were present before me.[1]

—Descartes, *Meditations*

AMONG THE MOST finely detailed sculptures from the Medici collections is an early seventeenth-century turned ivory tower measuring just over four inches high. The artwork is attributed to an artisan from Nuremberg, a city famous for its turned objects, particularly those produced by members of the Zick family (Fig. 155).[2] It disposes a flattened sphere and then a vase, both pierced with flourishes, above its circular base; higher still rests a polyhedron within whose circular openings six further polyhedra are nested by a method that seems to resist explication. These are surmounted by a tiny vase, which, in turn, supports a spiral form so thin that it seems to test the limits of ivory's structural strength. The helicoidal form, made of discs with 0.16-inch diameters, leans toward one side and terminates in another minute vase.

In a portrait of about 1620 by Daniele Crespi (c. 1597–1630), the Italian collector Manfredo Settala (1600–1680) elegantly holds in his fingertips a turned ivory featuring another absurdly thin spiral form as well as sequences of spheres nested one within the other, known among turners as *contrefait* spheres (Fig. 156).[3] The painting's strongly contrasted Caravaggesque chiaroscuro brings into relief Settala's face, neck, and left hand, as well as the piece of turnery he

Detail of fig. 187

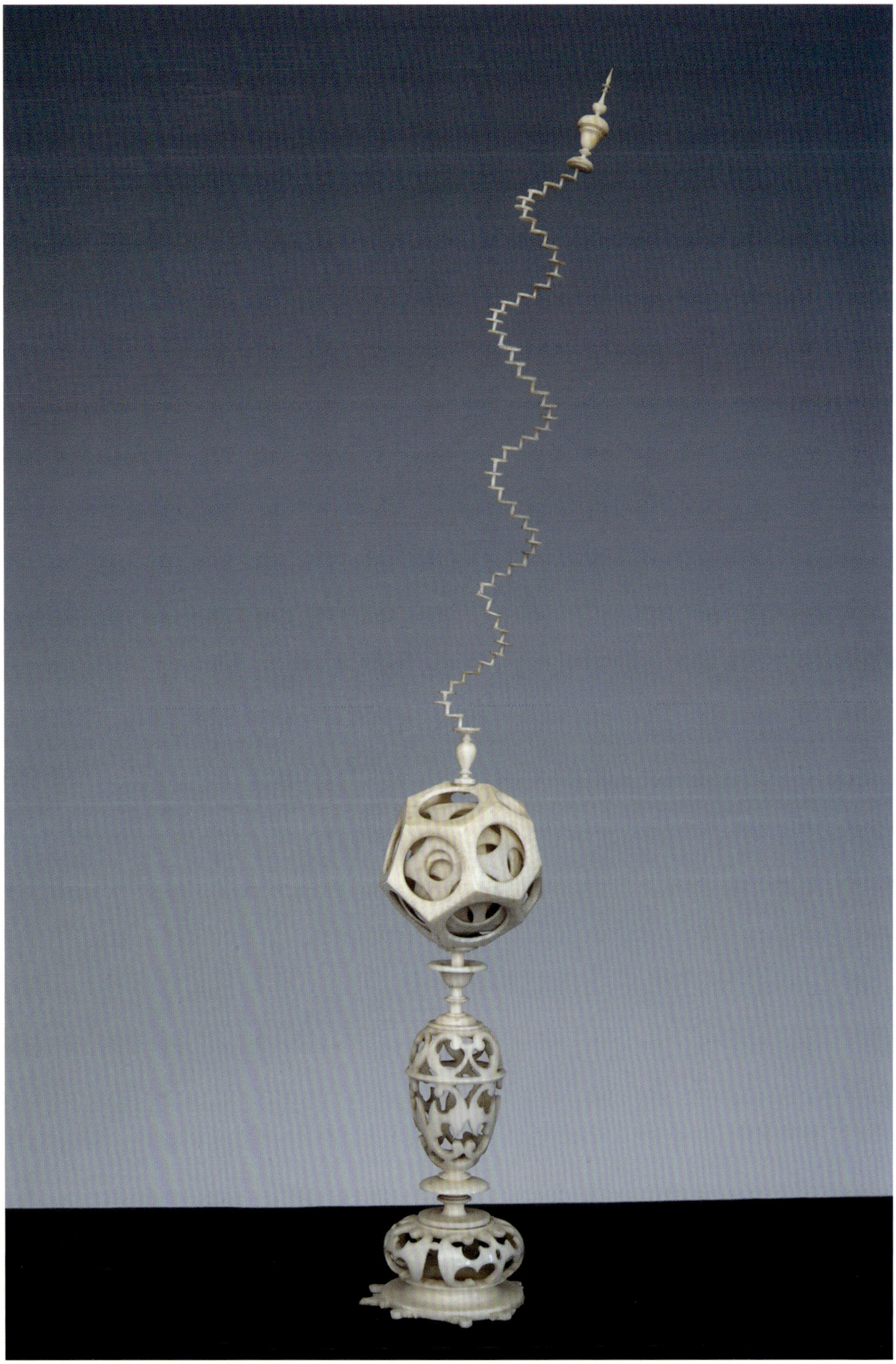

Fig. 155. Tower attributed to an artisan from Nuremberg, early seventeenth century. Turned ivory, height 4.3 in (10.9 cm). Tesoro dei Granduchi di Palazzo Pitti, Florence [Bargello inv. 1879 no. 207, inventory number 654067].

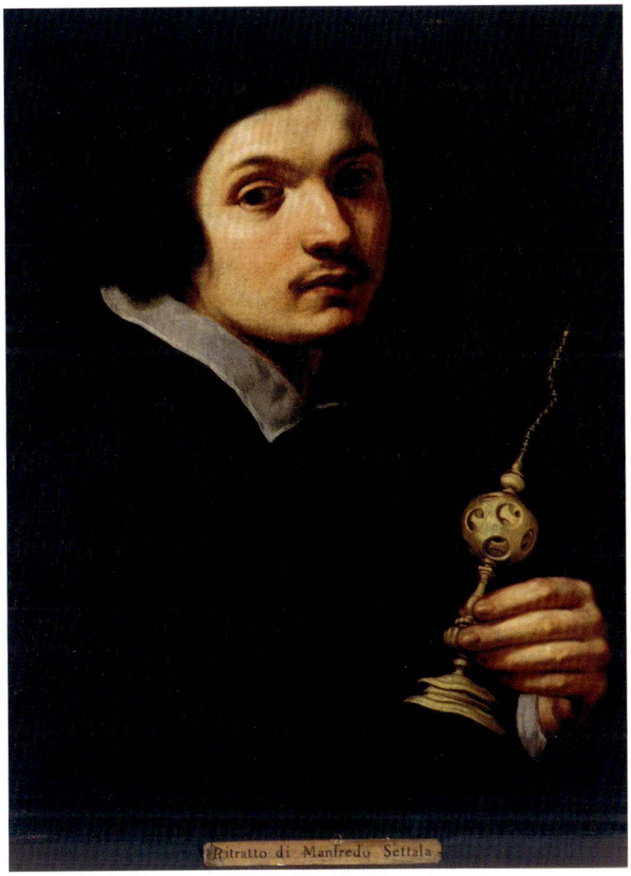

Fig. 156. Daniele Crespi (attributed to), *Portrait of Manfredo Settala*, c. 1620. Oil on canvas, 17.3 × 13 in (44 × 33 cm). Biblioteca Ambrosiana, Milan [Inv. 492].

holds, while the rest of his figure recedes into the black background. Settala was both a turner and a collector of refined pieces of turnery.[4] In Crespi's portrait, he presents his miniature tower as an emblem of his refined taste and of his qualities as a craftsman: his capacity to maintain sustained concentration and his reverence and exacting aptitude for applied geometry. The last chapter was ultimately pessimistic about patrons' capacity to control the reception of architectural wonders—given that such opposite opinions could exist regarding the limits of human visual perception and the capacity for discernment, and thus around the possibility of anticipating and correcting for those limitations. This chapter will be more optimistic, suggesting that turned ivories, in their resistance to explication, offered patrons an apparent solution: unlike most (though not all) deformations, a turned ivory does not ultimately yield itself up to interpretation; the progression from stupefaction to comprehension may be sustained indefinitely, holding the observer in the moment when he is most taken by the inexpressible wonder of the object and, by extension, the social power of its patron or creator.

When a prominent visitor to his celebrated collection expressed skepticism that a craftsman outside of Germany could have fabricated such excellent pieces of turnery, Settala began working on his lathe in front of his incredulous guest to show that "even in Lombardy men had the patient genius of Germans."[5] This anecdote not only reflects the respect afforded to

German turners, many of whom were Protestant, but as a moment in the oral culture of early modern turnery it offers insight into the values of turnery's patrons, who judged the worth of these sculptures based on the perceived challenges inherent in their production. Joseph Moxon (1627–1691), an English creator of globes and mathematical instruments, succinctly captures this system of valuation when he praises turners' dexterity and, significantly, their ability to perplex observers, captivating them in a state of confusion resistant to resolution and in this way, unlike those deformations that lead observers from unclarity to clarity: "Some *Turners* to shew their Dexterity in *Turning*, and make others that know not the way how it is done admire at their Skill, *turn* long and slender springs of *Ivory* as small as a Haystalk, and perhaps a Foot or more long."[6] An experience of turnery is less appropriately compared to an experience of an anamorphosis than to a visit to San Carlo alle Quattro Fontane, generating stupefaction (or perhaps, for a more sophisticated observer, an experience of the *je ne sais quoi*) and—crucially—refusing to yield to explanation.[7]

Joseph Connors has suggested a comparison between this signal structure of architectural deformation and the delicate art of turnery in Bernini's attempt to characterize Borromini's work (at the request of Virgilio Spada): Bernini seems to liken the architect to a turner forming a *contrefait* sphere, as one who "was never satisfied" but "wants to hollow out one thing inside another, and that inside another [*una cosa cavare un'altra, e nell'altra l'altra*], without ever coming to an end."[8] Connors describes how San Carlo, the church Bellori branded "ugly and deformed," evokes a piece of turnery: "Inspired by the double-shell structures of Milan, especially S. Lorenzo Maggiore, Borromini would design S. Carlino as a shell within a shell, a screen of columns set inside a quatrefoil plan that is itself plastically moulded to fit inside the box of the outer walls, *una cosa cavata dentro l'altra, e nell'altra l'altra*."[9] Entering a white interior of Borromini's design, we might imagine that we are stepping inside a miniature turned ivory expanded to a gigantic scale, but in what sense is San Carlo really similar to or different from a work of turned ivory? Should we primarily locate their similarities in that they are both white and characterized by embeddedness? Should we stress the role of geometry in the design of both works? Do items of turnery deform in the same way that Borromini was seen by some critics to depart from classical norms? What is the true point of commonality between a turned sculpture and this architectural deformation?

Turned ivories, while not labeled as "deformations" by seventeenth-century audiences, initiated many of the same mental operations we have seen triggered by the deformations discussed in this book. Especially when first considered, they demanded close attention and challenged observers' visual perception and capacity for discernment, producing an initial effect of wonder, perhaps of disorientation, that resonates with the effect produced by a deformation. The impact—even to some extent the existence—of turned ivories depended on the history of social power and aesthetic theory that was developed via the sustained investment in deformations. We have seen in chapter 1 a rather straightforward exercise of a patron's power—displaying marvels and a visual and spatial deformation to control what guests experienced as they moved through his residence, and reinforcing his power as they then deciphered the deformation and wonders he incorporated into the architecture and decorative program. Bernardino's palace and his deformation determined what visitors experienced as they walked along specific paths, how they constrained their bodies, and what emotions—such as wonder and awe—they might have felt while doing so. The world of turned ivories similarly played on wonder and awe—and the makers of these objects were ascribed powers approaching the divine precisely because turned works were impossible to comprehend; the connoisseurs who displayed these incomprehensible objects to their guests enhanced their own social status accordingly. Turnery, like anamorphosis, depended on geometrical understanding, but the mathematical explanations for turnery were harder to conceptualize: if creating an anamorphosis was like imagining a triangle, turning an ivory on a lathe was akin to imagining a chiliagon.

Both the Medici tower and the turned work in Crespi's portrait would have made viewers marvel at their creators' technical prowess. How could such exquisite sculptures have been crafted by cutting a piece of ivory while rotating it on a lathe?[10] How could such "slender springs" not breach ivory's structural strength? Their creators' imaginative powers also drew admiration. Descartes's mind resists his effort to imagine a chiliagon's thousand sides—even though he can understand a thousand-sided object conceptually—and the complex geometric forms of the Medici tower or the turned sculpture in Crespi's portrait are similarly resistant to the imagination. How could turners have visualized such intricate nested polyhedra with sufficient clarity to make them? We can ask similar questions on viewing a late sixteenth- or early seventeenth-century German piece in the Museo degli Argenti in Florence: a sphere with circular openings within which another eight spheres and polyhedra are nested, and which, in turn, enclose a polyhedron with outward-protruding spikes (Fig. 157).[11] How could a turner have imagined such a labyrinthine combination of spheres and polyhedra, let alone constructed it from a single piece of ivory?

This chapter focuses on the French Minim Charles Plumier's groundbreaking folio volume *L'art de tourner en perfection* (*The Art of Turning in Perfection*, 1701), the first thorough study of the practice, theory, and history of turning.[12] In his preface, the friar particularly notes this pastime's suitability for churchmen: "To the people of the Church in their benefices, both to exercise, to use part of their time, and to banish idleness."[13] Employing Plumier as a guide, I attempt to parse the similarities between the experience of interpreting or trying to discern turned ivories and experiencing—and trying to understand—deformations like San Carlo. That encounter with a turned ivory expands in time (perhaps indefinitely) the transition from confused initial perception to understanding that is characteristic of so many of the deformations discussed in the previous pages. I begin from Plumier's demand that artists and connoisseurs cultivate observant care and scholarly perspicacity, which returns us to the *je ne sais quoi* experienced before Borromini's undulating facades or to the stupefaction elicited when first encountering the large-scale anamorphoses of the Trinità dei Monti convent.[14]

Spoils of a Religious War

Whereas the visual deformations of the previous chapters were primarily produced in Catholic geographies, the turned forms examined here include not only those made in Catholic regions by Catholic practitioners but also objects made in Protestant areas that were subsequently integrated into Catholic collections. A cup made in 1626 by the German turner Johann Eisenberg (fl. 1618–1660) measures twelve inches in height and incorporates an astonishing chain with double rings that have been cut from a single piece of ivory, as opposed to being put together

Fig. 157. Sphere with polyhedra, late sixteenth- or early seventeenth-century German. Turned ivory, painted in part in gold, height 11.3 in (28.6 cm). Tesoro dei Granduchi di Palazzo Pitti, Florence [Bargello inv. 1879 no. 54, inventory number 653606].

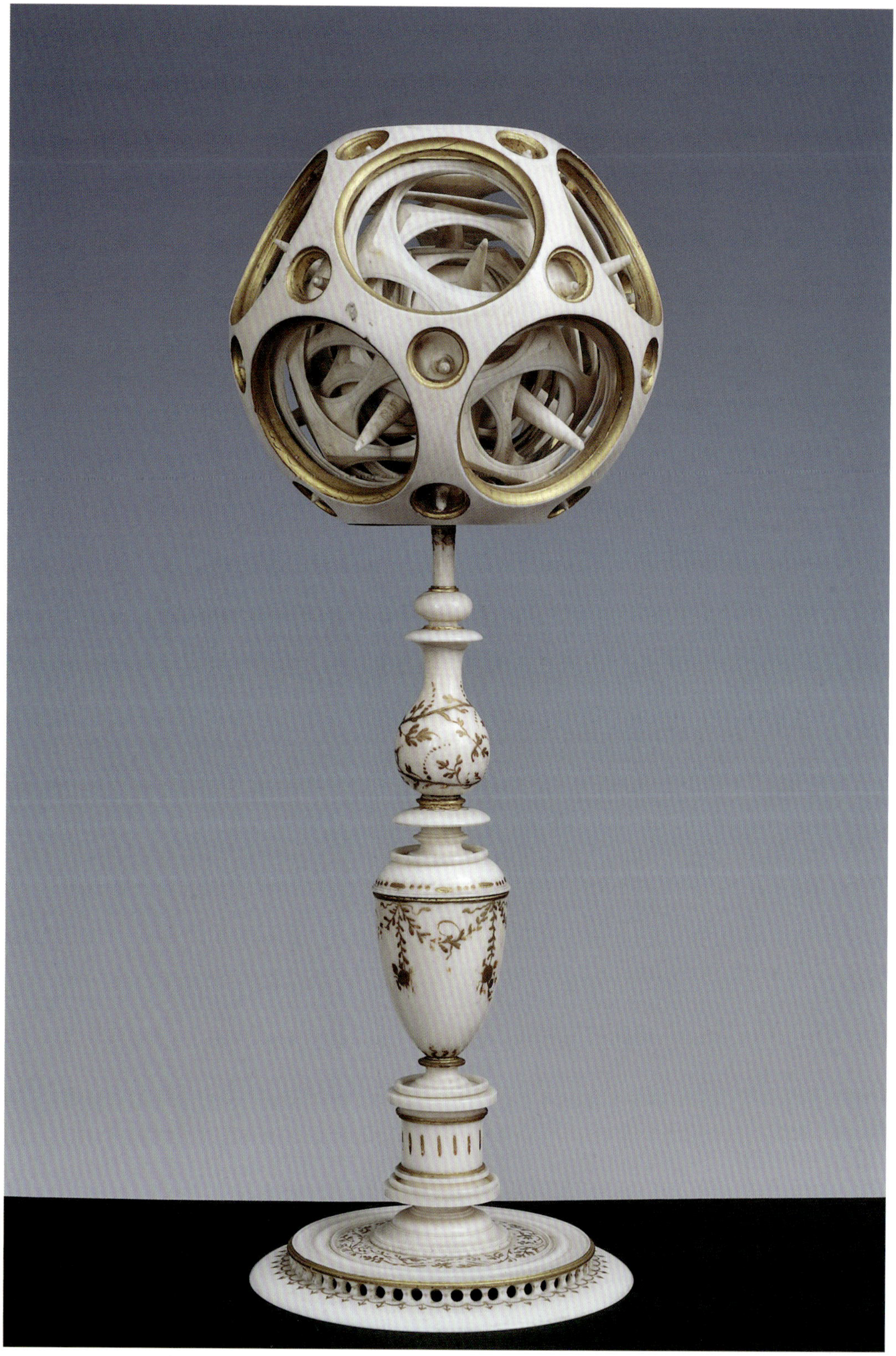

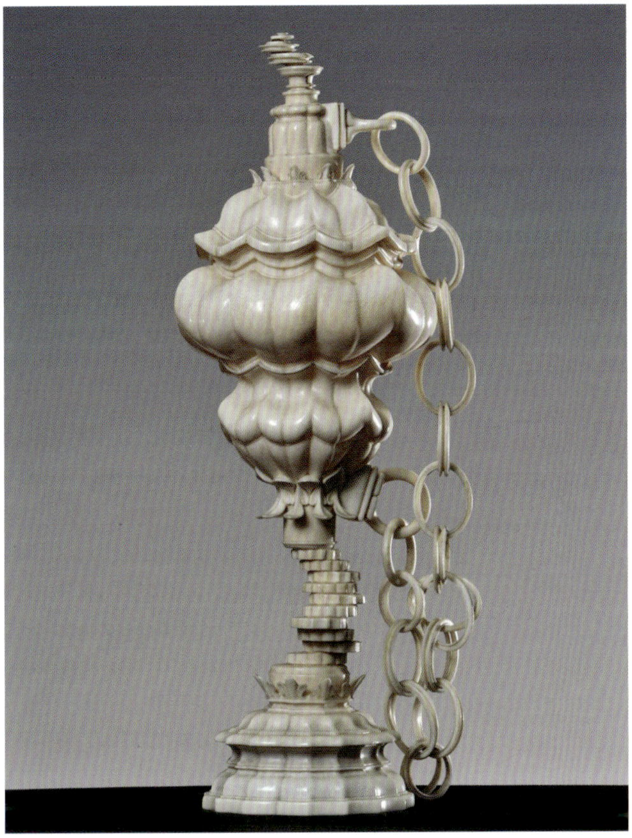

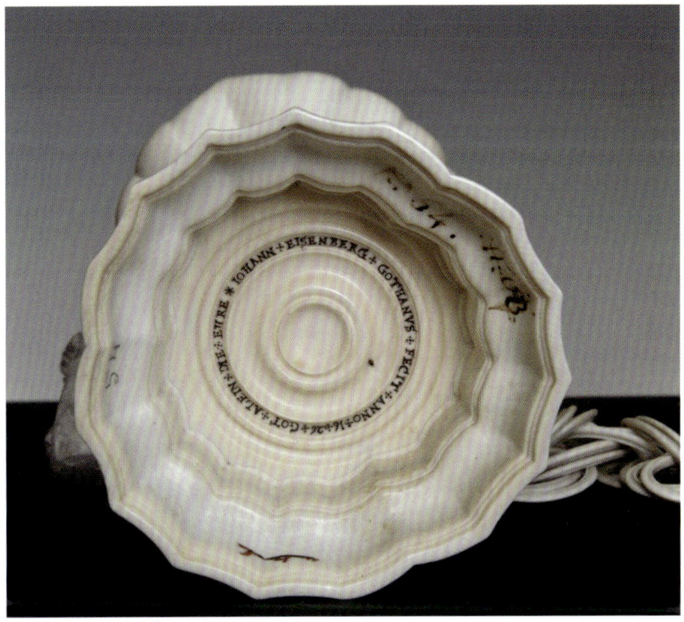

from multiple, smaller pieces (Fig. 158). The chain connects a lid to the chalice. The cup itself, constructed by cutting the material into a form while rotating it on a lathe, seems to be a product of legerdemain. An annotation inscribed under its base gestures toward an explanation. Next to Eisenberg's signature and the date lies the Lutheran creed "GOT ALEIN DIE EHRE," or "All glory to God" (Fig. 159). The inscription implies that God made it possible to make such an exceptional work and that only God, not the artist, should be exalted for the achievement. Other turned works offer images and similar inscriptions that likewise stress the spiritual dimensions of the art.[15] For example, an ivory now in the Royal Danish Collection features a *contrefait* sphere that houses gouache images of Christ and the Virgin (Figs. 160, 161, and 162).[16] Much as Minim and Jesuit anamorphic wall frescoes directed the viewer's attention to particularly important religious images for contemplation and veneration, the images in the spheres would have elicited meditation and presented models for imitation. A German verse published in 1705 even went so far as to characterize God as a turner; God is described as "the master of the world's first sphere. ... who Himself turned the vaults of the heavens so delicately."[17] On the one hand, Eisenberg's inscription could lead one to believe that the cup defies ordinary human limits because God was involved in its production. On the other hand, Eisenberg felt sufficiently conflicted about ascribing all glory to God to record his own creative role, inscribing his signature on the cup's base. Eisenberg thus associates his own dexterity with the divine

Fig. 158. Johann Eisenberg, cup with lid, 1626. Turned ivory, height 12 in (30.5 cm). Tesoro dei Granduchi di Palazzo Pitti, Florence [Bargello inv. 1879 no. 133, inventory number 636579].

Fig. 159. Johann Eisenberg, base of cup with lid, 1626. Tesoro dei Granduchi di Palazzo Pitti, Florence [Bargello inv. 1879 no. 133, inventory number 636579].

Fig. 160. Lidded Nuremberg goblet with a counterfeit sphere, c. 1630–1665. Turned white and yellowed ivory and traces of gilding and green staining, height 28 in (71.0 cm). The Royal Danish Collection, Rosenborg Castle.

Fig. 161. Gouache image of Christ within counterfeit sphere in a lidded goblet. The Royal Danish Collection, Rosenborg Castle.

Fig. 162. Gouache image of the Virgin within counterfeit sphere in a lidded goblet. The Royal Danish Collection, Rosenborg Castle.

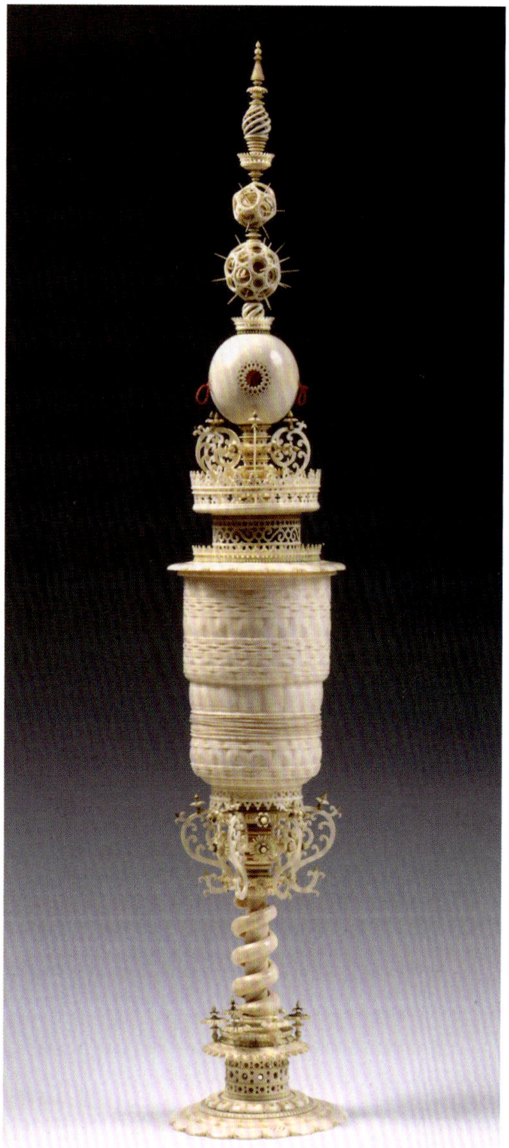

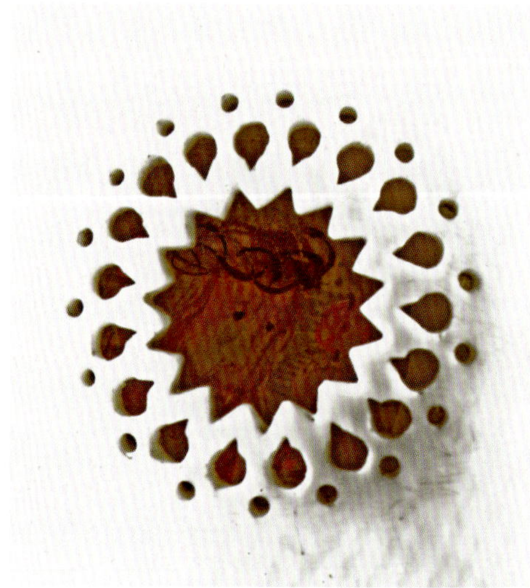

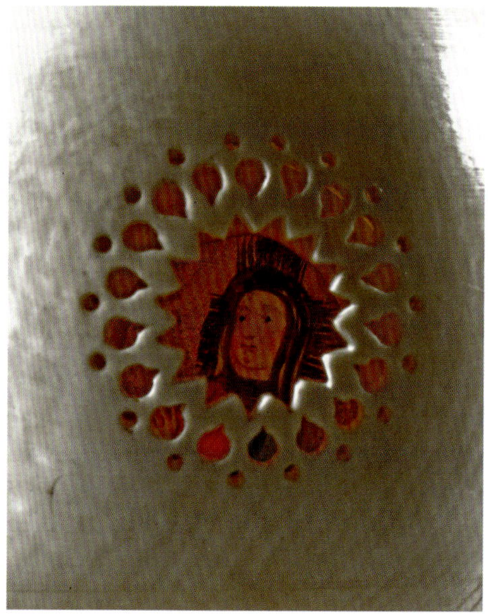

and points to a universal quality of turned ivories, namely that their extraordinary form brings them, their makers, and their owners close to God.

Eisenberg's cup is one of twenty-seven works pillaged from the palace of Ehrenburg in the sack of the Protestant territory of Coburg on September 28, 1632, during the Thirty Years' War.[18] The Catholic troops were led by Mattias de' Medici (1613–1667), who shipped the ivories to his brother in Florence, Ferdinando II de' Medici. Plumier writes that he traveled to Florence to examine the turned objects in the Grand Duke of Tuscany's cabinet, now in the Museo degli Argenti.[19] Another work among those plundered from Coburg and held today in Florence is topped by a crescent moon with a human face and proffers a cup that is carved so finely that

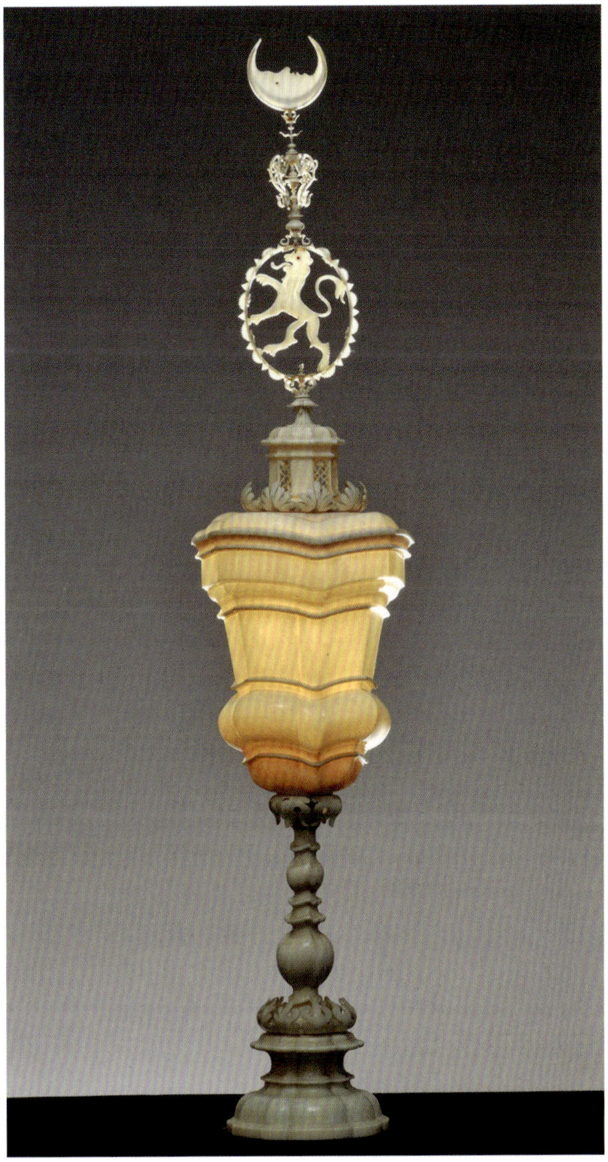

Fig. 163. Chalice with lid from Coburg, 1618–1631. Turned ivory, height 20.3 in (51.5 cm). Tesoro dei Granduchi di Palazzo Pitti, Florence [Bargello inv. 1879 no. 74, inventory number 636706].

when illuminated it accentuates ivory's translucency (Fig. 163). An object of such remarkable delicacy, seeming to exceed the limits of human creative capacity, may well have led its observer and owner to feel closer to God.

Eisenberg's teacher Markus Heiden (fl. 1618–1664) authored a book in 1640 that discusses the siege of Coburg and also emphasizes the spiritual message of his art.[20] He explains how he sought above all else to honor God with his turnery; in this vein, he cites Saint Paul's First Epistle to the Corinthians 10:31: "Therefore, whether you eat or drink, or whatsoever else you do, do all to the glory of God." Given this spiritual application of turnery, it is perhaps ironic that art that one side described as spiritually motivated transferred to the other side in a religious war. If these objects were initially created to display the power of God, once appropriated in the course of the religious wars they were displayed to enhance the power of men instead.

Plumier's Art of Turning

Catholics, including members of the Minim order, also fostered a strong tradition of turning. Among these craftsmen Plumier was authoritative, primarily as the author of *L'art de tourner,* in which he carefully described and categorized an art that was otherwise understood to defy language, to stupefy, to elicit the experience of the *je ne sais quoi*. Plumier asserts that he judged the Minim friar Emanuel Maignan, who created the monumental anamorphic wall frescoes in Trinità dei Monti in Rome and in the Place Royale convent in Paris, to be among the most skillful in the craft, someone "known throughout Europe for his rare knowledge, & his happy & excellent genius."[21] As a result, after initially learning how to turn from his father, Plumier sought out Maignan's guidance. Although he was at first primarily interested in mathematics—a discipline essential to turnery's geometric reveries—his focus shifted to plants (an interest that we saw in chapters 2 and 4) while he was studying in the Minim convent of Trinità dei Monti in Rome, where he would have seen the convent's corridor anamorphoses, painted by Maignan and Jean-François Niceron.[22] Plumier went on to travel to the French Antilles in 1687, 1689, and 1694 at the behest of King Louis XIV to conduct research into natural resources and their use; this knowledge would have been vital to French economic interests (and was acquired despite indigenous rights and claims). Plumier's passion for turning, which predated his fascination with natural history, continued in tandem with these three transatlantic natural history experiences. After returning from his final Caribbean voyage, Plumier published his book on turning in 1701 in Lyon.

The book, reissued in 1706 and 1749, was evidently met with interest.[23] A French-German bilingual edition was printed in 1776 in Leipzig.[24] In addition, according to P.J.S. Whitmore, the text was translated into Russian under the guidance of Czar Peter I of Russia (1672–1725), who practiced turning, and a Russian-Dutch translation reportedly survives in the Russian Academy of Sciences in St. Petersburg.[25] In a letter of 1702, Plumier wrote of his plans to bring out a second volume on the subject and to travel to Nuremberg and Milan to study more turned objects.[26]

L'art de tourner contains seventy illustrations, a large number that would have come with high production costs. The engravings visualize machines and tools for turning; some depict exemplary turned objects; one presents designs for moldings.[27] Plumier encourages the book's audience to learn about turning and to refine their understanding of turned objects through reading as well as acts of observation, in comments that argue for the explanatory force of the images he includes: "I represented the geometric plans, the profiles, the perspectives, and I marked all the parts of the pieces with letters and numbers, so that, relating to the explanation, one could, in reading and seeing the figures, conceive without difficulty the entire construction of the machines, there being even a quantity that one could understand by looking only at the plates."[28]

Plumier's images of turning machines and their parts typically show these entities within framed plain backgrounds (Fig. 164). He and the anonymous engraver provide naturalistic accounts of these objects; differences in light and shadow magnify the vividness of the objects

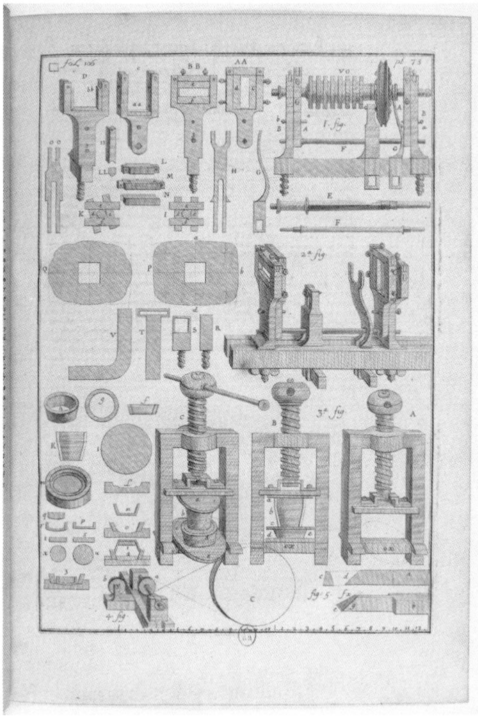

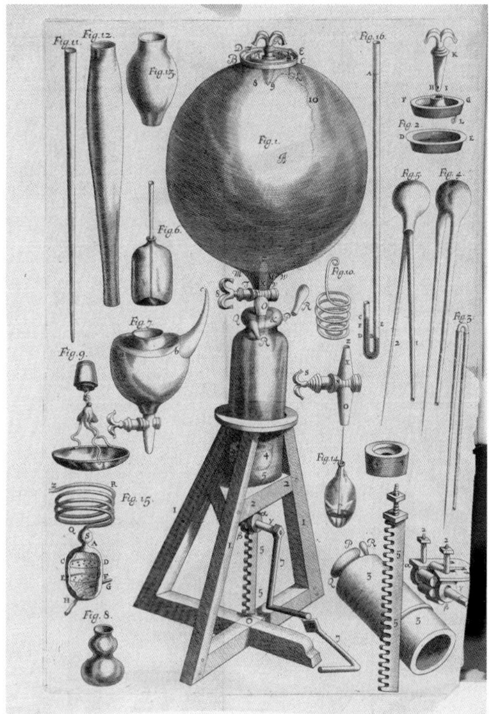

Fig. 164. Charles Plumier (designer), turning machines, their parts, and tools from Plumier's *L'art de tourner*, 1701, following p. 106 [mis-numbered as plate 73]. Printed on paper, 13.94 × 9.37 in (35.4 × 23.8 cm). BnF, département Estampes et photographie, Paris [RESERVE LD-12-PET FOL].

Fig. 165. Unidentified artist, the air-pump from Robert Boyle's *New Experiments*, 1682, following p. 4. Printed on paper, 12.36 × 8.66 in (31.4 × 22 cm). The Huntington Library, San Marino, California [432407].

such that readers might momentarily imagine that they were actually before them. The format is familiar from prints in natural philosophical tracts and other *Kunstbücher* (books providing how-to knowledge) produced throughout the sixteenth, seventeenth, and eighteenth centuries in Europe.[29] For example, the engraving of the air pump that was employed in experiments described in Robert Boyle's (1627–1691) *New Experiments Physico-Mechanical* of 1660 also sets forth various naturalistically shaded devices against a plain background (Fig. 165).[30] And Plumier's presentation of tools against a framed blank background in such images as plate 60 recalls the instruments of deformations and reformations featured in Niceron's *La perspective curieuse* (see Fig. 127), the study of eye treatments by the Dresden artist and oculist Georg Bartisch (1535–1607), or the study of surgery by Hans von Gersdorff (d. 1529) (Figs. 166, 167, and 168).[31] Much as Niceron's *La perspective curieuse* or Mario Bettini's *Apiaria* explain how anamorphoses are created, Plumier's treatise elucidates how particular turned objects are produced.

The first edition of *L'art de tourner* appeared in a parallel-column French and Latin translation. The Latin text would have helped to authenticate the project by giving it a scientific air; the friar also indicates that the Latin text is "for the benefit of foreign readers" (*en faveur des Etrangers*).[32] He requests that readers not focus on the purity of one translation or the other, writing that he is paying less attention to correct language than to the function of turners' machines.[33] Indeed, he complains that it was challenging for him to develop an intelligible terminology for an art that had not previously been discussed in such detail: that practitioners had not yet devised a robust vocabulary around the craft contributed to turnery's incomprehensibility to seventeenth-century audiences.[34] In 1644, the English diarist John

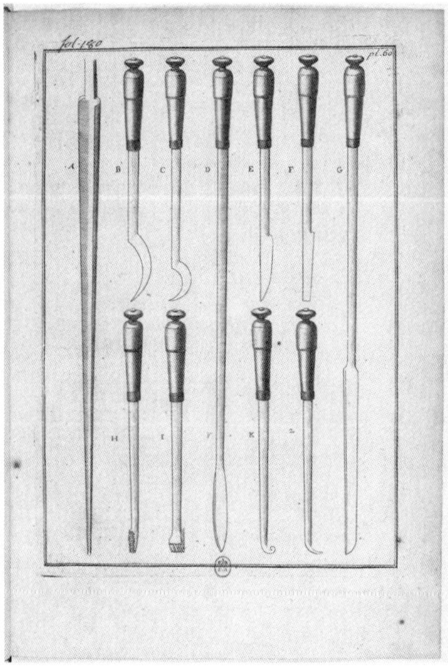

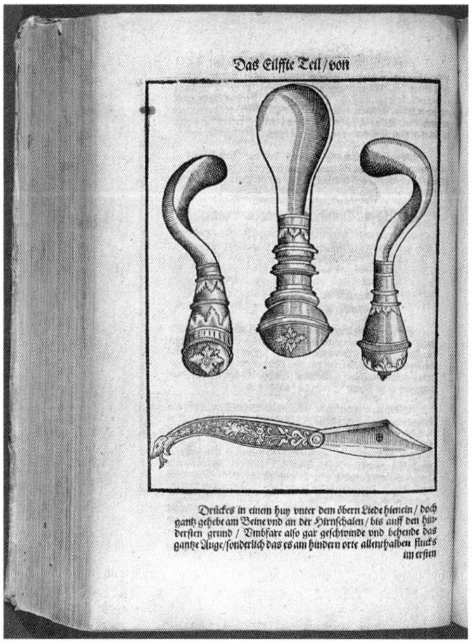

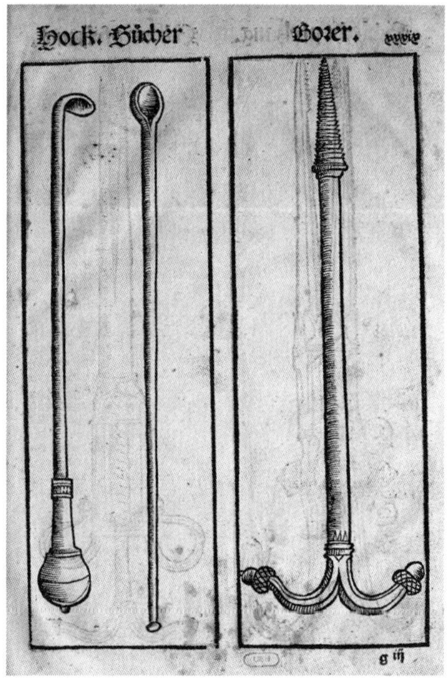

Fig. 166. Charles Plumier (designer), tools from Plumier's *L'art de tourner*, 1701, following p. 180 plate 60. Printed on paper, 13.94 × 9.37 in (35.4 × 23.8 cm). BnF, département Estampes et photographie, Paris [RESERVE LD-12-PET FOL].

Fig. 167. Georg Bartisch (designer), instruments from Bartisch's *Ophthalmodouleia*, 1583. Woodcut printed on paper, 11.88 × 7 in (30.2 × 17.8 cm). San Marino, Huntington Library.

Fig. 168. Unidentified artist, instruments from fol. g3ʳ in Hans von Gersdorff's *Feldtbuch*, 1517. Woodcut printed on paper, height 10.2 × 6.7 in (26 × 17 cm). Heidelberg University Library, P 7584 Folio RES, fol. XXXIXʳ.

Evelyn (1620–1706), after inspecting the collections of the Grand Duke in Florence, wrote of "such rare tourneries in Ivory, as are not to be describ'd for their curiosity."[35] Turned ivories left observers stupefied and at a loss for words.

A Princely Art

Even as Plumier emphasizes throughout *L'art de tourner* that these sculptures are "incompréhensible," the treatise is like other *Kunstbücher* in training observers to comprehend the craft it introduces.[36] Precisely because turned ivories were deemed impossible to comprehend, those who could claim to make sense of them could, to paraphrase Bourdieu, distinguish themselves through the distinctions they made.[37] Turned ivories epitomize a dynamic—pervasive in the seventeenth century and visible, as we have seen, in the Palazzo Spada and in other formal spaces—where attending to and discerning that which was not readily apparent was not only associated with establishing truths and with ongoing efforts to interpret the natural sciences, but also worked to strengthen social hierarchies. If once a tool primarily used by craftsmen, the lathe quickly became the beloved instrument of nobles, sovereigns, clerics, and scholars, who could thus display their capacity to comprehend the incomprehensible. Technological enhancements by the late sixteenth century, such as the institution of continuous drive of a lathe with the substitution of a turner's attendant with waterpower, facilitated the adoption of turning by elites.[38] Modernization allowed a nobleman to work at his turnery at any time, without having to retain a servant solely to turn the lathe, making his work more of a contemplation and less of a performance. Continuous drive augmented the rotation velocity and refined the precision of turning; by adding a sizeable driving wheel in conjunction with pulleys, a turner could also vary velocity, which increased the adaptability of the machine.[39]

The many sovereigns drawn to turnery included Duke Alfonso d'Este of Ferrara (1476–

1534), Grand Duke Cosimo III de' Medici (1642–1723), Ferdinando de' Medici, Grand Prince of Tuscany (1663–1713), Carlo Emanuele II of Savoy (1634–1675), Emperor Maximilian I (1459–1519), August, Elector of Saxony (1526–1586), Emperor Rudolf II (1552–1612), Peter the Great of Russia, King Frederick III of Denmark (1609–1670), King Christian VI of Denmark and Norway (1699–1746), King Frederick V of Denmark and Norway (1723–1766), Louise of England, Queen of Denmark and Norway (1724–1751), and Juliana-Maria, Queen of Denmark and Norway (1729–1796). High-ranking members of society and families hoping to ascend the social ladder embraced the pastime as well. Palazzo Spada's garden for some time made available a space in which Bernardino Spada's grandchildren could practice the princely art.

The great expense of ivory undoubtedly contributed to turning's princely character. Although turners did not create only in ivory—Plumier also mentions wood, gold, silver, iron, and copper—elephant tusks were especially well suited to the demands of working fine details on a lathe.[40] Ivory is composed of dentine, which is not as dense as bone.[41] Specialized cells called odontoblasts, which make up the dentine, are ordered in a columnar arrangement and also in a further arrangement emanating from the interior of the tusk outward, resulting in a series of nested cones. The tension between these two arrangements generates the tusk's resilience and elasticity. The ivory's stretchiness keeps the tusk from fracturing when its exterior layers broaden to adapt to the tusk's ongoing interior growth—and thus from fracturing due to the centrifugal force of the lathe.[42] Ivory is also oily, and thus can be readily polished.[43]

Pliny the Elder had long ago described elephant tusk as "the most costly product found on land."[44] And European commercial interest in ivory was stimulated in the fifteenth and sixteenth centuries by Portuguese expeditions to Africa and India that brought back to Europe raw ivory as well as carved pieces, such as salt cellars and small boxes, made by artists outside the European continent.[45] Ivory continued to be imported into Europe in large quantities in the seventeenth century. The turner Markus Heiden writes of selecting one tusk in Amsterdam for a piece of turnery—out of 300,000 tusks on offer.[46] Eighteenth-century travel accounts document the brutal ways hunters killed African elephants and then used saws or axes to dislodge the tusks from the animals' corpses before leaving them to rot.[47] Yet, as Noam Andrews has observed, the glimmering and pristine white surfaces of turned ivories betray no trace of their cruel provenance, as if their relationship to slaughtered elephants has been erased.[48]

Turned ivory was restricted to the elite not only because the material was so expensive, but also because it was understood as an art that should be practiced and not just appreciated, and that was very time consuming. In his preface, Plumier notes on three occasions that turning serves as an antidote to idleness—a condition more likely to afflict privileged sectors of society than others. Insofar as turning offsets the state or condition of being unoccupied, it focuses the mind and thus gives rise to an attentive state of being. He initially contrasts the state of idleness and the many qualities of turning as its remedy: "It is clear that this art is currently in Europe the most serious occupation for people of spirit and merit, and between the entertainment and the reasonable pleasures, that which is the most highly considered by those who seek in some honest exercise the means to avoid the faults into which too much idleness in life throws one."[49] He underscores this same function a second time when he explains that together with optics, turning is the primary pastime "to banish idleness from life" for the Abbé de Perichon, Provost of S. Salvadour.[50] One wonders if the Abbé's studies in optics included examinations of anamorphoses. Still later in his preface, Plumier brings up idleness yet again in the passage cited earlier in which he asserts that his book is suitable for churchmen.[51]

Delicate and Inimitable

L'art de tourner refines its readers' capacities to attend to, perhaps even to contemplate, lathe-turned objects by showing them how these objects are created, which, as Plumier says toward the end of his preface, leads to understanding: "Finally, curious people will see there the machines and inventions that are used to make these beautiful lathe works, which appear incomprehensible to those who do not know them."[52] He explains that part of what makes these objects so hard to understand is their "surprising delicacy" (*delicatesse surprenante*).[53] Slender turned details stretch the boundaries of belief and imagination, as seen in the helicoidal structures mentioned earlier or the very fine ornamentation on the turned lidded goblet in the Royal Danish Collection (see Figs. 155, 156, and 160).[54] How could a turner have brought forth such delicate works? Plumier brings up this point again later in his preface: "Sir du Rosay Bourgeois of Lyon keeps in his home two candlesticks and one crucifix made of ivory variously figured, and so artistically worked that it seems almost impossible that the hand of a man has been able to produce a work so delicate on a lathe."[55] The trope reappears in a review of Plumier's book that speaks of turned objects as having "a delicacy that surprises, and that seems inconceivable."[56]

In addition to "delicate" and its cognates, another term used to describe the perception of lathe-turned objects is "inimitable."[57] With respect to the seventeenth-century European ideal of knowledge-by-doing or by making, the term "inimitable"—"impossible to imitate"—conveyed that the objects could not be known.[58] For Plumier, as for many other craftsmen and scholars, comprehending a form involved making it, whether in two or three dimensions. Plumier, in fact, affirms that he imitated or recreated every single turned object he had seen.[59] This was an impressive feat if true, since he traveled extensively throughout Europe to learn from the greatest exemplars of the craft, and some objects he imitated must have indeed been considered inimitable.[60] Plumier not only speaks of his own three-dimensional imitations of turned objects but (as noted earlier) presents readers with two-dimensional images of those objects and the machines used to carve them. In one plate, for instance, he visualizes turned objects in the Lyon cabinet of the French ornamental turner Nicolas Grollier de Servière (1596–1689), which he drew "after the originals" (*d'après les Originaux*) (Fig. 169).[61] He underlines how difficult the globes appear to those who do not practice the craft: "A pointy star is seen in one … but besides this several globes, one inside another, which seems perplexing to those who do not know turning, but of little difficulty to those who practice the craft."[62] And yet, even as *L'art de tourner* aims to render the incomprehensible comprehensible and the inimitable imitable, Plumier admits elsewhere that there are limits to the instruction his book can provide: "Practice and use will teach it to you better than any theory."[63] As Pamela H. Smith has shown, it was typical for artisans to stress in their books of art that writing could not fully convey practices.[64] And yet beyond this convention, ivory turning tested the limits of description in artisanal books because the art itself exceeded descriptive capacity. Plumier's conclusion that only by making a particular turned ivory could one achieve knowledge shifts the power dynamic between the collector—who was often (although not always) the producer

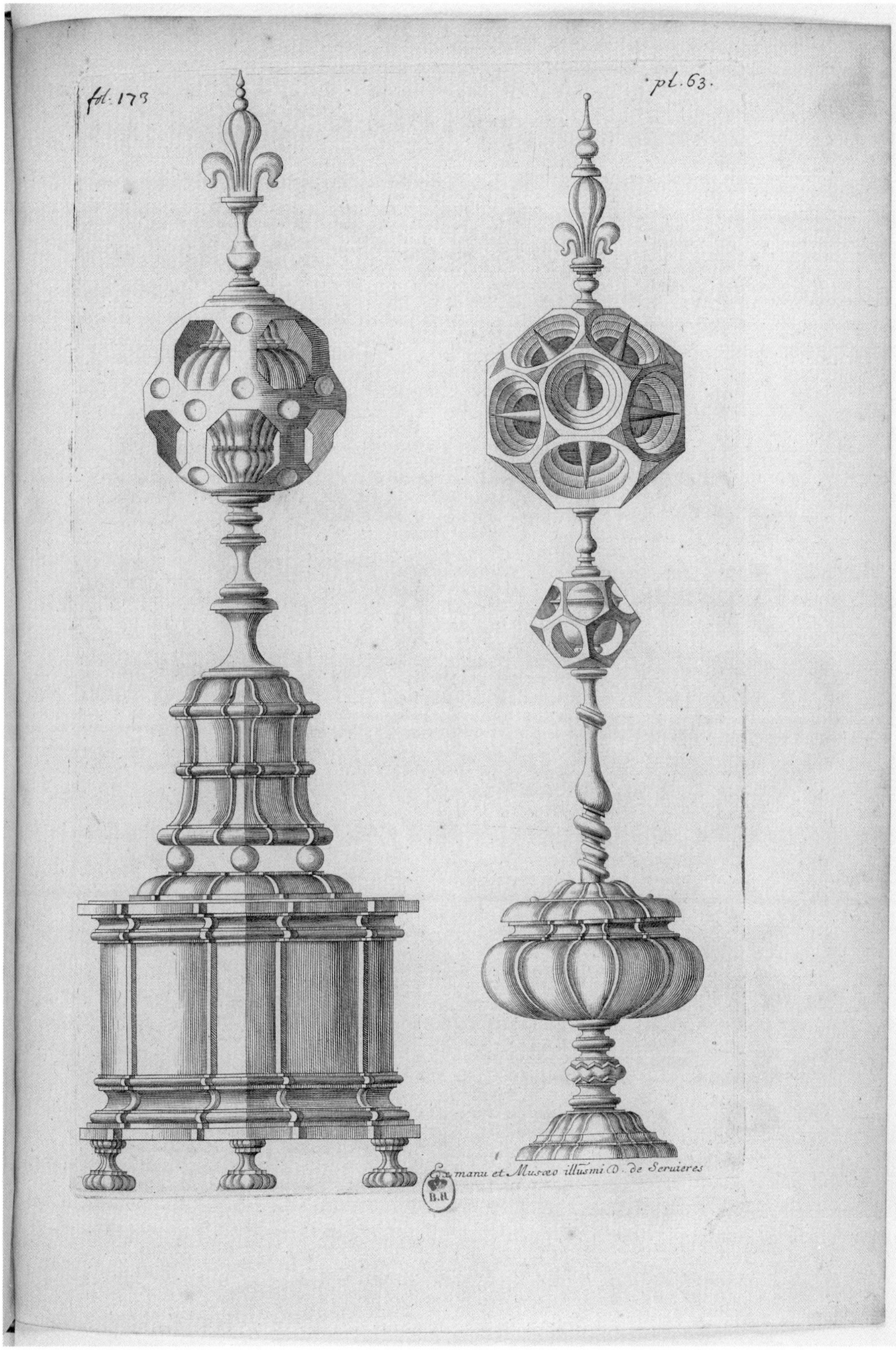

Fig. 169. Charles Plumier (designer), turned objects from Plumier's *L'art de tourner,* 1701, following p. 172 (mis-numbered as plate 63). Printed on paper, 20.39 × 14.25 in (51.8 × 36.2 cm). BnF, département Estampes et photographie, Paris [RESERVE LD-12-PET FOL].

of the turned ivory and therefore knew the secrets of its creation—and the guest or observer of the work. Plumier is playing both sides in his treatise, claiming to make these difficult works comprehensible but then simultaneously insisting that they are inimitable and unknowable: it is a bit like he has offered to hold up a mirror to an anamorphosis but then, when we look through it, we see further deformation. That the moment of clarification is, for the viewer, deferred (perhaps indefinitely) leaves the creator, as the possessor of knowledge, in a position of sustained power.

The apparent contradiction in Plumier's account of turned objects as incomprehensible forms that his treatise is rendering comprehensible is inextricable from questions of access. The large anamorphoses in Minim convent hallways were relatively accessible to scholars, including men who were not of particularly elevated origin. Turned ivories were not. The high cost of materials and the significant time that would go into understanding the art or producing a work on the lathe—endeavors Plumier defined as inextricably linked—expanded almost indefinitely the timeframe between first perceiving a turned ivory and comprehending its design, for any but the most elite of observers. Turned works do not yield themselves to direct visual inspection and do not clarify into a resolved self-evident form; despite a book like Plumier's *L'art de tourner*, all but an elite few would have continued to find these objects very difficult to parse.

Grollier's cabinet, which contained not only turned objects but also anamorphoses, extraordinary clocks, burning mirrors, and a range of further instruments, was memorialized in a book by his grandson Gaspard II Grollier de Servière (1676–1745). In this work, Gaspard II repeatedly asserts that his grandfather's collection features "inimitable masterpieces" (*des chefs d'oeuvre inimitables*).[65] He notes, moreover, that his grandfather would invite his cabinet's visitors, who included Louis XIV, to conjecture how he had created the works before them.[66] This guessing game suggests that objects whose facture appeared particularly enigmatic would have served as powerful conversation pieces, with a turned ivory perhaps even better suited to this purpose than the colonnade in Palazzo Spada, whose basic principle of design (if not the precise math necessary to duplicate it) can be understood by anyone who walks into the colonnade. One wonders about how the social dynamic would have played out during a visit from the sovereign. Would Grollier have invoked God, as the Lutheran turners did, to ensure that Louis XIV did not feel threatened by an artisan's skill? Gaspard II calls attention to the fragility of objects in the collection; one sculpture "is reduced to the thickness of a strand of hair" (*se réduit à la grosseur d'un cheveu de tête*).[67] In another turned work, "in which there are small, almost imperceptible figures" (*dans laquelle il y a de petites Figures presque imperceptibles*), he reflects on the now-familiar challenges of visual perception, in this case, on how hard it is to discern the details.[68] Incomprehensibility comes up in the evaluation of another work: "The beauty of this piece consists in the difficulty one can imagine having had in making it from one single ball of ivory."[69] Whereas the incomprehensibility of anamorphic deformations was temporary, and centered around not knowing how the images would resolve, here questions revolved around the making of the sculpture—not "what does it represent?" but "how did someone create this?"—a question that might remain unanswered. The unresolvable incomprehensibility of a turned sculpture recalls the disorienting experience of visiting San Carlo: early visitors wanted to understand how the space was put together. Knowledgeable visitors might now understand from plans that Borromini designed the space according to geometrical princi-

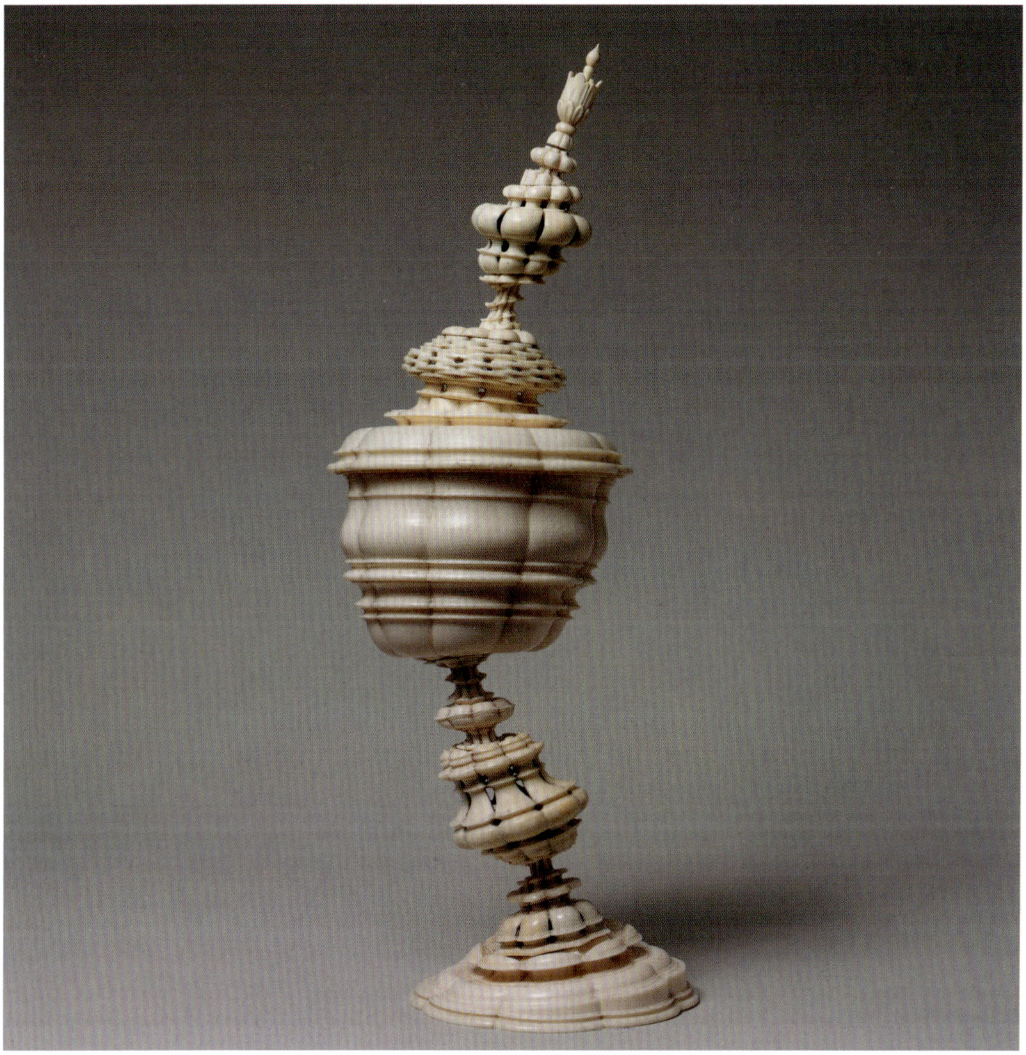

Fig. 170. Possibly workshop of Lorenz Zick, Standing cup with cover, mid-seventeenth century. Turned ivory, height 14 in (35.6 cm). The Metropolitan Museum of Art, New York.

ples, but without the plans they cannot grasp this building's structure. If observers of complex turnery struggled to make sense of the forms before them, so too visitors to San Carlo struggled to resolve for themselves the underlying principles or the specific calculations that went into producing the church. Gaspard II singles out a category of turned objects, the so-called eccentric pieces (*pieces excentriques*), which are "turned structures worked on different centers" (*des ouvrages de Tour travaillez sur differens centres*).[70] We have already encountered this category of turnery in Figure 155, as well as the turned sculpture represented in the painting in Figure 156. Another eccentric cup, which has been attributed to the workshop of Lorenz Zick (1594–1666), a turner who instructed Ferdinand II, the Holy Roman Emperor (1578–1637), or to the workshop of his son, Stephan Zick (1639–1715), appears to be made of a material softer and more elastic than ivory (Fig. 170). It seems to sway in strong wind, although it is solidly grounded. At the same time, its curve echoes that of an elephant's tusk. Of yet another eccentric

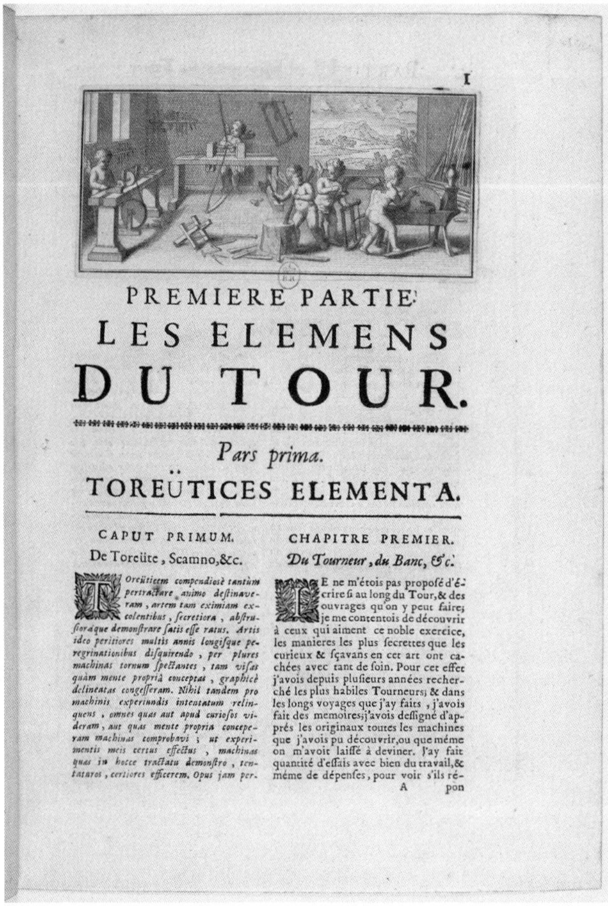
I

PREMIERE PARTIE.

LES ELEMENS

DU TOUR.

Pars prima.

TOREÜTICES ELEMENTA.

CAPUT PRIMUM.

De Toreüte, Scamno, &c.

Toreüticem compendiosè tantùm pertractare animo destinaveram, artem tam eximiam excolentibus, secretiora, abstrusioráque demonstrare satis esse ratus. Artis ideo peritiores multis annis longisque peregrinationibus disquirendo, per plures machinas tornum spectantes, tam visas quàm mente propriâ conceptas, graphicè delineatas congesseram. Nihil tandem pro machinis experiundis intentatum relinquens, omnes quas aut apud curiosos videram, aut quas mente propria conceperam machinas comprobavi; ut experimentis meis certus effectus, machinas quas in hocce tractatu demonstro, tentaturos, certiores efficerem. Opus jam per-

CHAPITRE PREMIER.

Du Tourneur, du Banc, &c.

JE ne m'étois pas proposé d'écrire si au long du Tour, & des ouvrages qu'on y peut faire; je me contentois de découvrir à ceux qui aiment ce noble exercice, les manieres les plus secrettes que les curieux & sçavans en cet art ont cachées avec tant de soin. Pour cet effet j'avois depuis plusieurs années recherché les plus habiles Tourneurs; & dans les longs voyages que j'ay faits, j'avois fait des memoires; j'avois dessigné d'apprés les originaux toutes les machines que j'avois pu découvrir, ou que méme on m'avoit laissé à deviner. J'ay fait quantité d'essais avec bien du travail, & méme de dépenses, pour voir s'ils ré-

A pon

Fig. 171. Charles Plumier (designer), putti operating machinery and preparing materials in a turner's workshop from Plumier's *L'art de tourner*, 1701, p. 1. Printed on paper, 20.39 × 14.25 in (51.8 × 36.2 cm). BnF, département Estampes et photographie, Paris [RESERVE LD-12-PET FOL].

piece, Gaspard II declares that this species of distortion "makes it impossible to imagine how the Author was able to execute this work" (*ce qui fait qu'on ne peut s'imaginer comment l'Auteur a pû executer cet ouvrage*).[71] Needless to say, Gaspard II persistently highlights the dumbfounding nature of his grandfather's ingenuity.[72]

And because turned objects were deemed incomprehensible, they were thought to testify to their makers' mental powers; the artwork enhanced the authority of its creator. Plumier writes that the pieces in the cabinet of de Servière, such as those reproduced in Figure 169, "are even after his death the proofs of the beauty of his mind, of the excellence of his genius, and of his skill in this art."[73] Elsewhere, he celebrates the Abbé de Perichon's "subtlety of mind in conceiving and executing what is most rare and most mysterious in this craft."[74] The implication for both men is that their works exhibit mental and physical acuity. It is to illustrate the essential role of genius and intelligence in the practice of turning that Plumier, in one of his illustrations, features putti—that is, representatives of creative minds—operating machinery and preparing materials in a turner's workshop, as Paola Bertucci has noted (Fig. 171).[75] For the same reason, in the frontispiece to Niceron's *La perspective curieuse*, putti experiment with different genres of deformations and reformations (see Fig. 129).

Turned Geometry

What kind of genius did turnery manifest? The polyhedra integrated into miniature ivory towers indicate their creators' talents for practical geometry. Only a finely trained visual imagination would have been capable of both envisioning and executing such complex geometric variants. Geometric turnery must be situated in the context of the enthusiasm, which lasted for over a hundred years, for experimenting with permutations of the five Platonic solids—the pyramid or tetrahedron (four faces), the cube or hexahedron (six faces), octahedron (eight faces), dodecahedron (twelve faces), and the icosahedron (twenty faces).[76] Attending to and mentally grasping polyhedra (or other forms) involved making them, in either two dimensions in perspective or in three dimensions.

Fifteenth- and sixteenth-century European artists and artisans visualized geometric bodies in paintings, prints, drawings, cabinetry, illustrated books, and other media, as Andrews has demonstrated.[77] The interest in three-dimensional geometry as a topic of two-dimensional picturing is apparent in a broad range of works, from intarsia wood paneling by Giuliano da Maiano (c. 1432–1490) in the Palazzo Ducale *studiolo* in Urbino and drawings by Leonardo, Parmigianino (1503–1540), Dürer, and Carpaccio (c. 1460–1525), to printed works such as the polyhedral visualization of the universe in the *Mysterium cosmographicum* (*The Secret of the Universe*, 1596) by Johannes Kepler (1571–1630).[78] This mathematical fascination is also manifest in the fifteenth-century treatise *Libellus de quinque corporibus regularibus* (*The Little Book on the Five Regular Solids*) by Piero della Francesca (c. 1415–1492), as well as in *De divina proportione* (*On Divine Proportion*), a manuscript by Luca Pacioli (c. 1445–1517) dated to around 1498 and printed in 1509. Colored drawings for Pacioli's treatise, attributed to Leonardo, skillfully render perspectival views of corpora, often in skeletal configurations, furnished with indications of shade that suggest that they represent actual objects as opposed to being mathematical diagrams (Figs. 172 and 173).[79] Much like the geometric bodies in turned ivories, these drawings display their makers' talents for spatial visualization. Although the clumsiness of the woodblock prints included in the 1509 printed edition of Pacioli's treatise do not do justice to the original drawings, they nevertheless stimulated further excitement for geometric visualization.[80]

Dürer contributed to the craze for fabricating three-dimensional geometric bodies in the final part of his *Underweysung der Messung*, and in the *Lehrbücher* (textbooks) he published in its wake. Instead of offering two-dimensional images of the Platonic solids in perspective, Dürer outlined surfaces that could be cut out and folded to form paper models of polyhedra.[81]

The concern with imagining and representing three-dimensional polyhedra in perspective on a printed page reached a high point in the *Perspectiva corporum regularium* (*Perspective of the Regular Bodies*, 1568), which contained etchings by Jost Amman (c. 1539–1591) after designs of the renowned goldsmith Wenzel Jamnitzer (1508–1585). Five parts of this work on perspectival geometry present twenty-four variants of the five Platonic solids. One page, for example, displays six reflections on the dodecahedron, each integrated within six circular forms (Fig. 174). Jamnitzer places the form he regards as the simplest in the upper left corner and the variant he deems most changed in the bottom right.[82] In this manner, more than visualizing difficult polyhedra, he shows how one form could shift into another both within a single sheet and from one page to the next. After his twenty-four polyhedral variants come forty further

Fig. 172. Empty [shape] with seventy-two sides (*Septuaginta duarum basium vacuum*), from Luca Pacioli's *De divina proportione*, manuscript completed in Milan, 1496–1498, fol. 111r, fig. 40. Watercolor on parchment, 11.2 × 7.9 in (28.5 × 20 cm). Venerando Biblioteca Ambrosiana, Milan.

Fig. 173. Augmented rhombicuboctahedron (*Vigintisex basium elevatus vacuus*), from Luca Pacioli's *De divina proportione*, manuscript completed in Milan, 1496–1498, fol. 110r, fig. 38. Watercolor parchment, 11.2 × 7.9 in (28.5 × 20 cm). Venerando Biblioteca Ambrosiana, Milan.

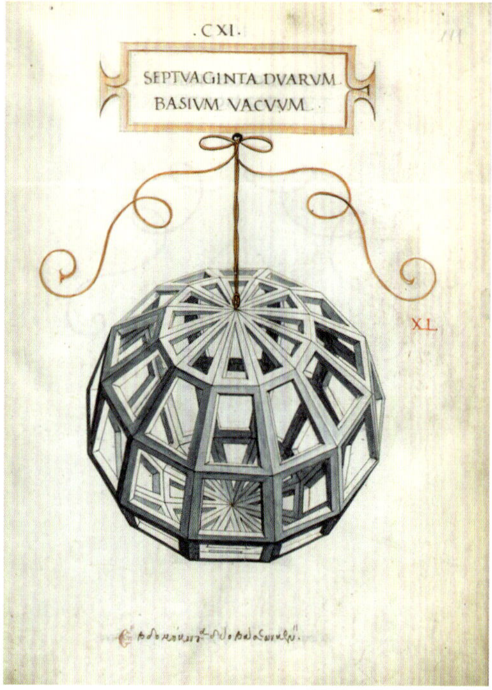

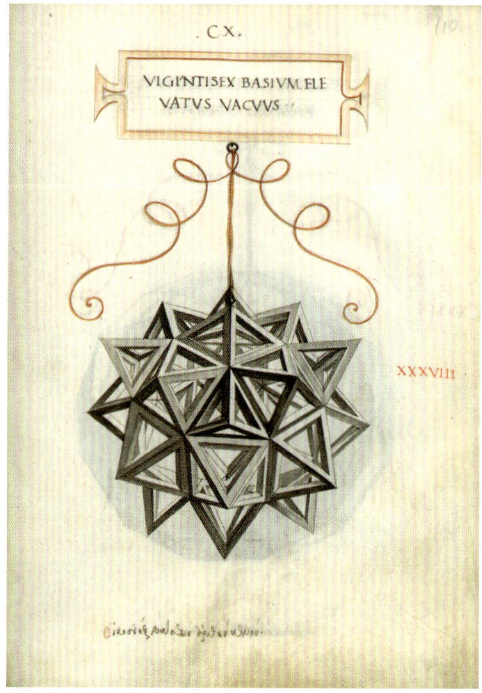

models of geometric forms, displayed on bases, which again exhibit stunning complexity (Fig. 175). The geometric corpora in ivories turned in the late sixteenth and seventeenth centuries should be seen as three-dimensional expressions of the mathematical experiments originated on paper by Jamnitzer and his peers. *Lehrbücher* like the *Perspectiva* raised the ability to visualize and render three-dimensional polyhedra in two dimensions into a pedagogical requirement for artisans and those keen to demonstrate a command of perspective.

The geometric forms included in Niceron's writings on perspective indicate that this didactic norm thrived well into the seventeenth century, an era that, as we have seen throughout this study, witnessed a surge of interest in perspectival treatises.[83] In regard to a star-like geometric body depicted in plate 13 of his book (Fig. 176), Niceron points out, "This perspective seems extremely difficult because of the great diversity of the planes and their inclinations and terminations."[84] He explains, however, that as soon as one understands how the shape is composed, it becomes possible to devise it.[85] In a portrait engraved by Michel Lasne (c. 1590–1667) and integrated into the 1646 edition of Niceron's *Thaumaturgus opticus* as well as into the 1652 edition of *La perspective curieuse*, both of which appeared after the friar's death, Niceron grasps an image of this same geometric body (Fig. 177). This print appears to be based on an anonymous oil painting of the friar reproduced in chapter 2 (see Fig. 64). Much as the complex geometric drawing that Niceron displays in his portrait introducing his studies of deformations proves his competency in perspective and practical geometry, geometric fancies turned on the lathe attest to their creators' competency in the same arenas. If, then, we can see the passion for turned ivories and the interest in deformations as deriving from similar social

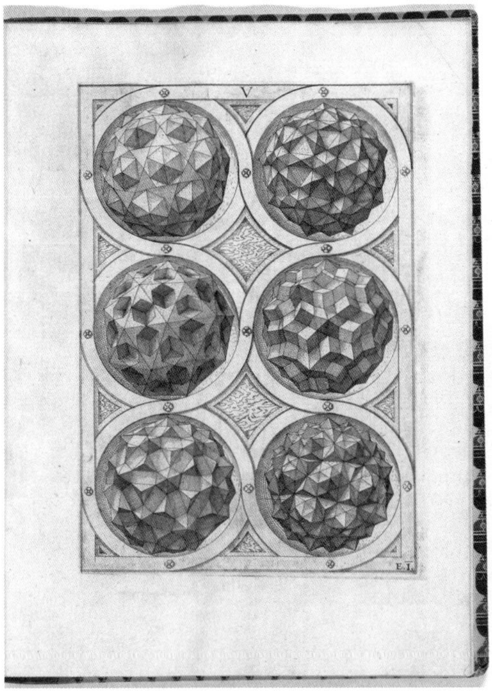

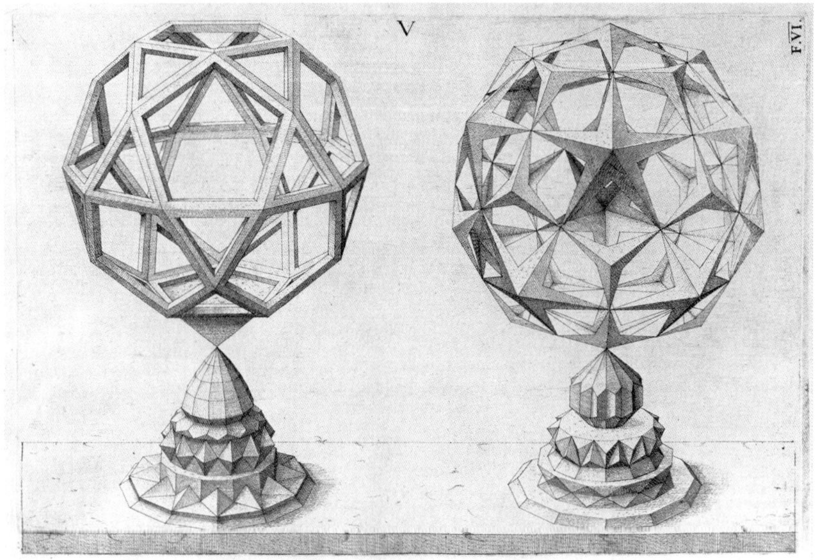

imperatives—the desire to display one's "genius" and educational accomplishment—and likewise resting on the era's enthusiasm for mathematics, then a structure like San Carlo, with its evident stylistic deformation coupled with the incomprehensible geometries of its plan, combined elements of both.

Fig. 174. Jost Amman and Wenzel Jamnitzer, polyhedral bodies from *Perspectiva corporum regularium*, 1568. Etching printed on paper, 13.75 × 10.1 in (35 × 25.5 cm). The Metropolitan Museum of Art, New York.

Fig. 175. Jost Amman and Wenzel Jamnitzer, polyhedral bodies from *Perspectiva corporum regularium*, 1568. Etching printed on paper, 13.75 × 10.1 in (35 × 25.5 cm). The Metropolitan Museum of Art, New York.

From Astronomy to Anatomy

Turned ivories were incorporated into the study of the natural sciences much as Bernardino Spada included illusionistic frescoes in his sundial gallery—in a gesture that associated the genius of the patron (here, embodied in the turned ivory) with the natural sciences, and thus with a kind of higher-order understanding. The turned ivory telescopes produced in the sixteenth and seventeenth centuries offer a compelling contrast with the close looking that miniature ivory works tend to invite. One such telescope, created by an unknown turner in Germany during the first half of the seventeenth century and held today by the Kunsthistorisches Museum in Vienna, provided the advantages of both (Fig. 178).[86] It enabled its owner to demonstrate the capacity to discern distant entities, while its inclusion of a miniature writing set featured such accessories as an inkwell and a knife, as well as a sculpted elephant, reminding observers of the ivory's elephantine origins.

Other particularly captivating categories of turnery would have evinced not only a powerful visual imagination and mathematical knowledge but also, and especially, conversance with ocular anatomy. Eyes turned on the lathe can be subsumed within the late seventeenth-century German genre of anatomical models and miniature statues of human figures that functioned both as medical teaching aids and as works to be exhibited in cabinets of wonder.[87] Typically,

Fig. 176. Jean-François Niceron (designer), plate 13 from *La perspective curieuse*, 1652. Engraving printed on paper, in folio. BnF, Paris [département Réserve des livres rares, V-1663].

Fig. 177. Michel Lasne, *Portrait of Jean-François Niceron*, 1626. Engraving printed on paper, 12.9 × 8.9 in (32.8 × 22.7 cm). British Museum, London.

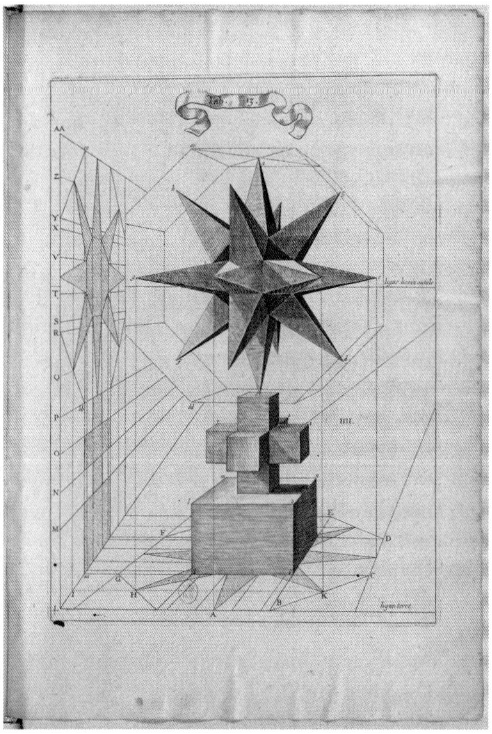

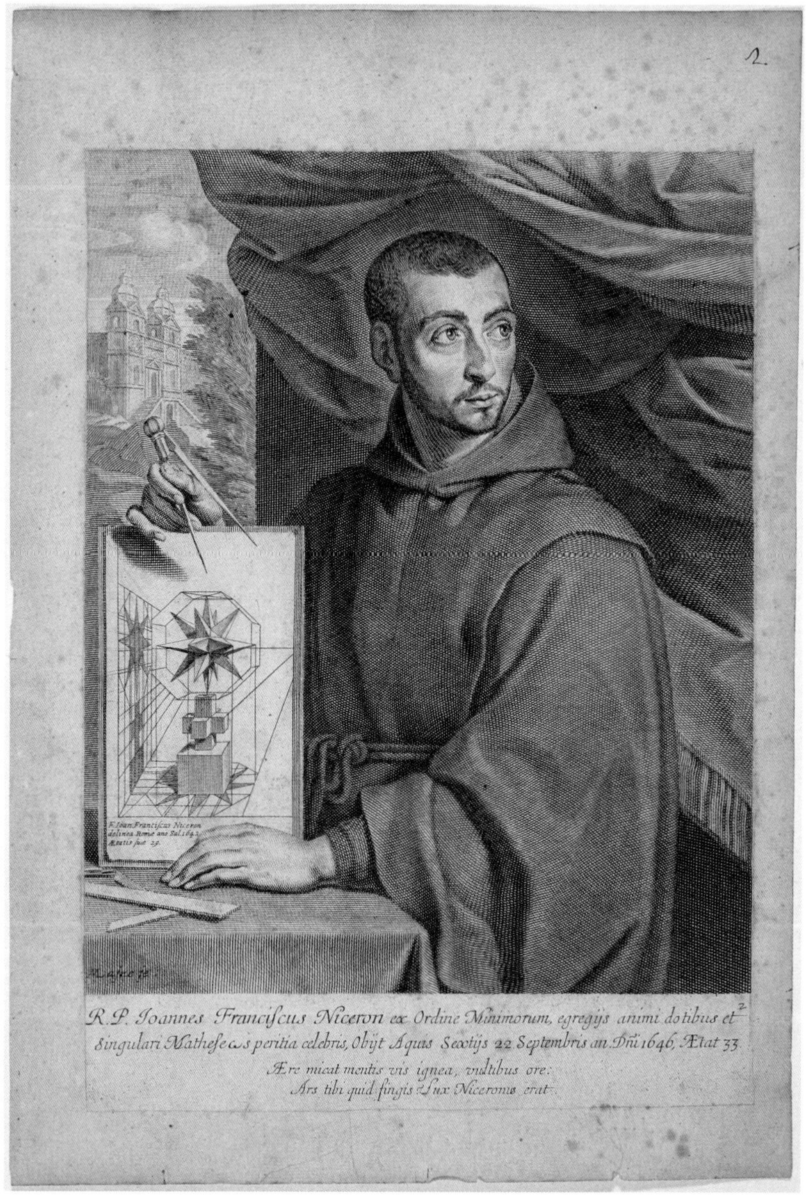

only eye models were created on the lathe, although both eye and ear models were presented on turned pedestals.

Stephan Zick is among the best-known creators of works in this genre. A page from a historical account of notable Nuremberg mathematicians and artists features turned ivory eyes by Stephan alongside a turned ivory sculpture with a *contrefait* sphere and another turned ivory tower with nested geometric forms made by Stephan's father, Lorenz Zick (Fig. 179).[88] The integration of these works on a single page suggests a classificatory kinship between ocular and geometric turnery.

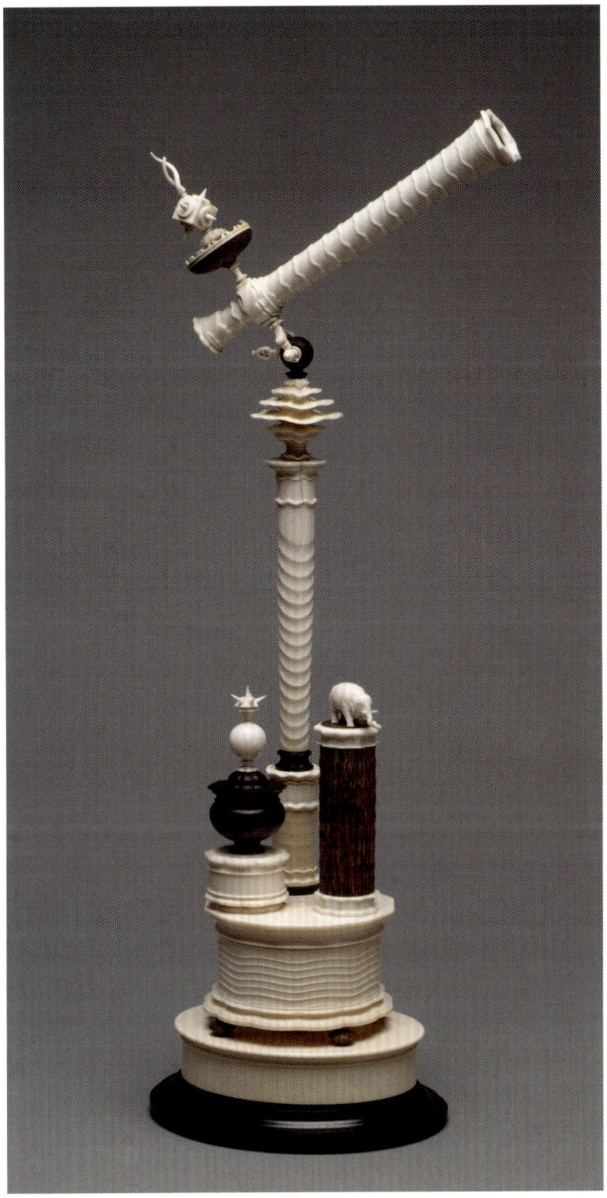

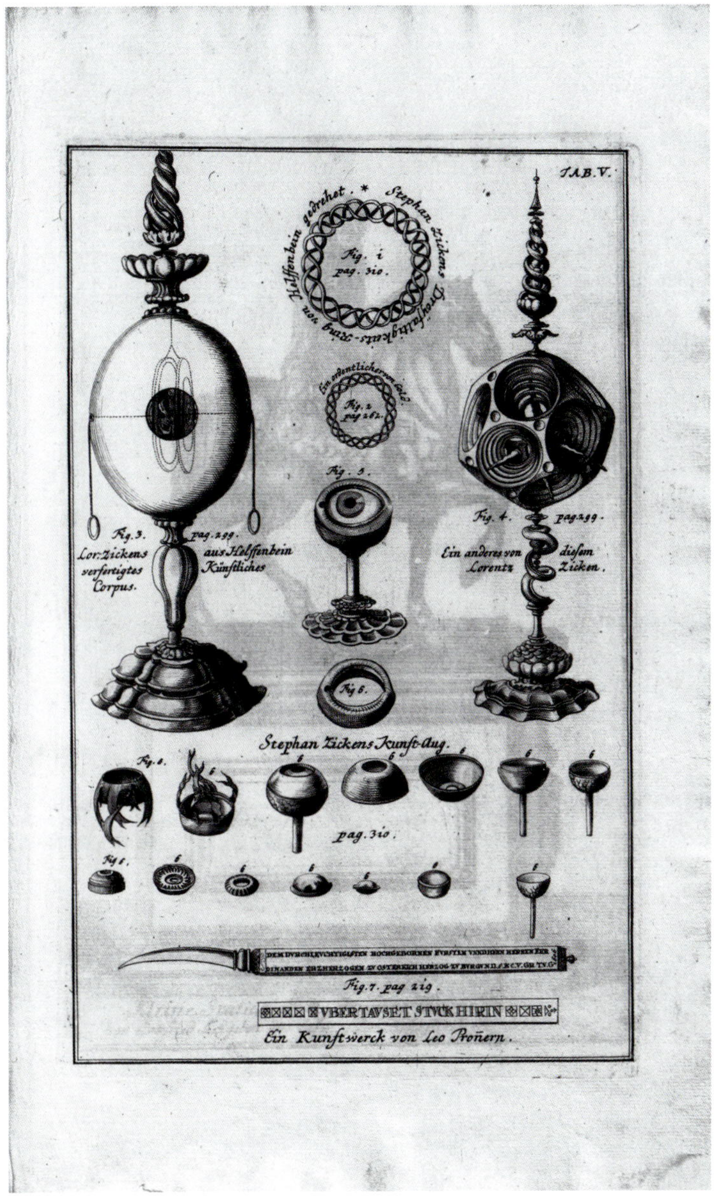

Fig. 178. Telescope created by an unknown turner in Germany during the first half of the seventeenth century. Turned ivory, height 22.6 in (57.5 cm). Vienna, Kunsthistorisches Museum.

Fig. 179. Johann Gabriel Doppelmayr, works of turnery by Stephan and Lorenz Zick in *Historische Nachricht von den Nürnbergischen Mathematicis und Künstlern*, 1730. Engraving printed on paper, 11.6 × 7.1 in (29.5 × 18 cm). Regensburg, Bayerische Staatsbibliothek [999/Bav.957].

RIGHT

Fig. 180. Attributed to Stephan Zick, anatomical eye model, c. 1700. Turned ivory, height 4.9 in (12.5 cm). Germanisches National Museum, Nuremberg [LGA600].

FAR RIGHT

Fig. 181. Attributed to Stephan Zick, anatomical eye model, c. 1700. Turned ivory, glass, possibly paper or parchment, height 3 in (7.5 cm). Germanisches National Museum, Nuremberg [WI2282].

BELOW RIGHT

Fig. 182. Attributed to Stephan Zick, anatomical eye model, c. 1700. Turned ivory, glass, possibly paper or parchment, height 1.9 in (4.8 cm). Germanisches National Museum, Nuremberg [WI2280].

BELOW FAR RIGHT

Fig. 183. Attributed to Stephan Zick, anatomical eye model, c. 1700. Turned ivory, glass, possibly paper or parchment, hair, height 1.5 in (3.8 cm). Germanisches National Museum, Nuremberg [WI2284].

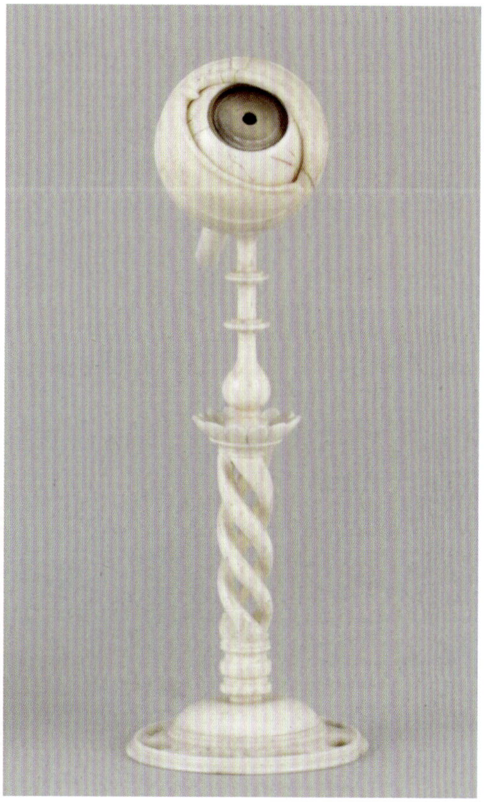

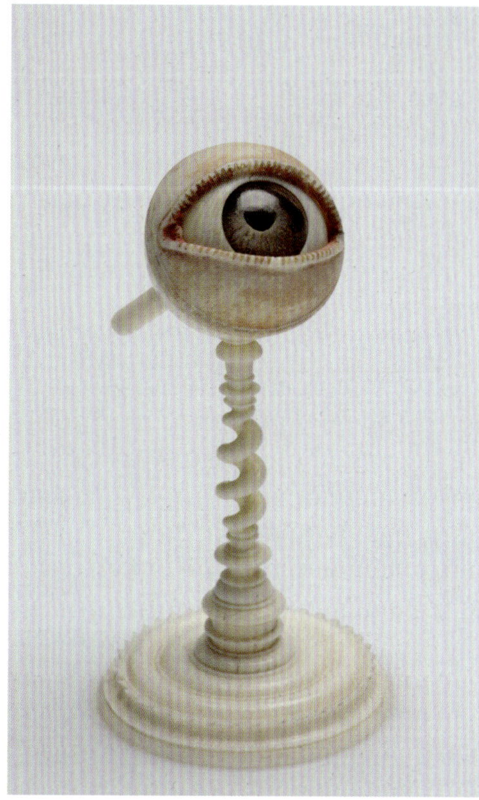

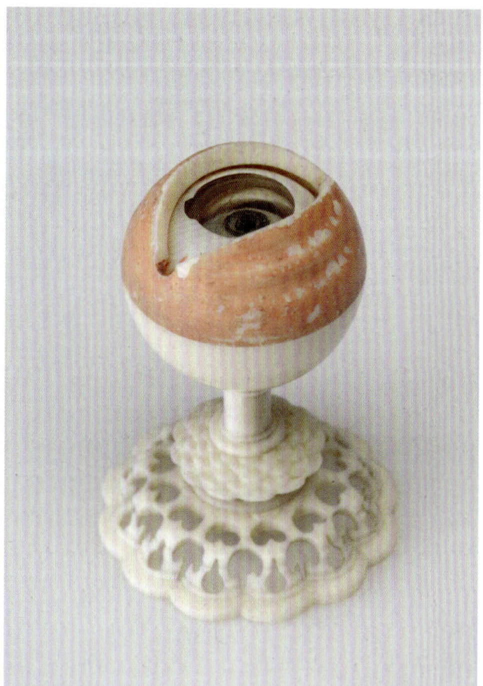

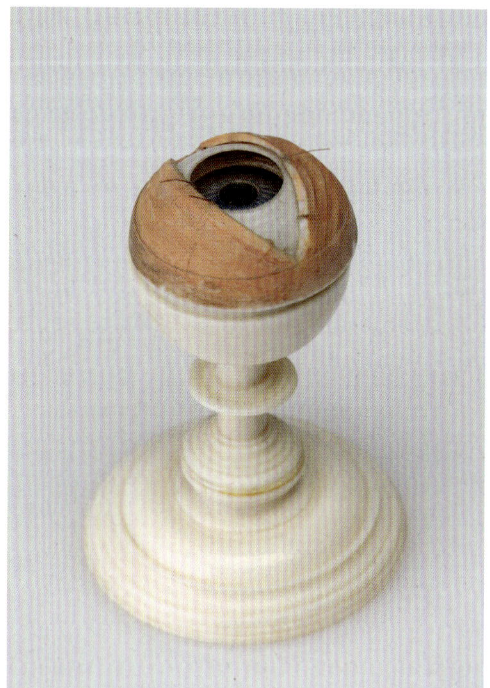

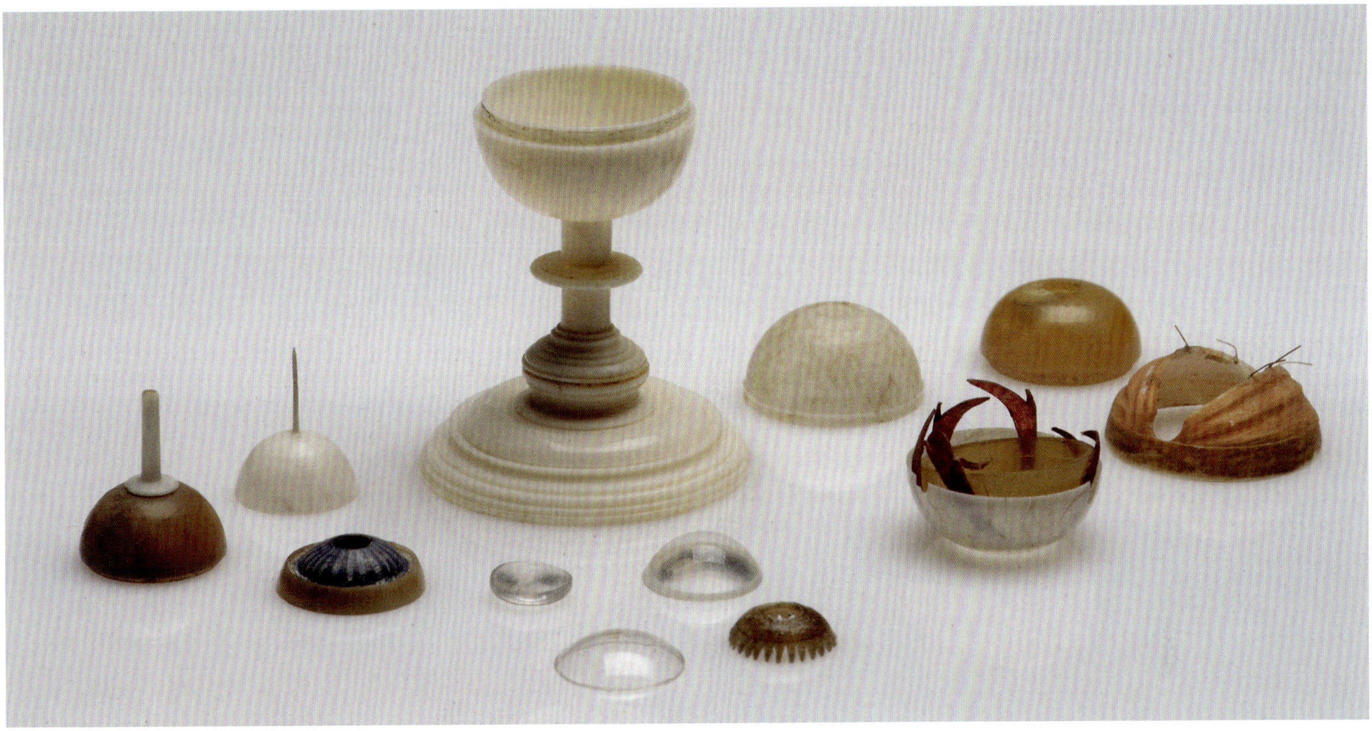

Fig. 184. Attributed to Stephan Zick, unassembled anatomical eye model, c. 1700. Turned ivory, glass, possibly paper or parchment, hair, height 1.5 in (3.8 cm). Germanisches National Museum, Nuremberg [WI2284].

Four anatomical ivory eye models, dated to around 1700 and attributed to Stephan, survive in the Germanisches Nationalmuseum in Nuremberg. The tallest, which measures 4.9 inches, has lost its iris and rests on a base with a coiled stem (Fig. 180). Another model, about 3 inches in height, presents a grey-toned iris and also rests on a coiled stem (Fig. 181).[89] A third model, measuring 1.9 inches in height, likewise features an ornately decorated base (Fig. 182). The fourth model measures about 1.5 inches in height and showcases a blue eye on which four eyelashes survive (Fig. 183). In another south German anatomical model of a blue eye, dated to the late seventeenth century (and measuring 3.35 inches in height), the artist indicated skin folds beneath the lower eyelid to increase the model's verisimilitude.[90] It is tempting to imagine how realistic these eyes would have originally appeared with their lashes and colors intact. The eye models can be disassembled to instruct students of anatomy, as is apparent in the twelve parts that constitute the larger of the two blue eyes mentioned above or the twelve parts that constitute the eye shown in Figure 184. In addition to ivory, ocular models incorporated such materials as glass, paper / parchment, hair, and light and dark horn. A pamphlet dated to 1680 and written by Daniel Bscherer (1656–1718) in collaboration with the anatomist Johann Georg Volkamer (1616–1693) and under the direction of Stephan Zick offers a description of a turned eye to instruct students on ocular anatomy.[91] The work concludes with a fold-out broadside containing engraved representations of an assembled turned eye, followed by its parts, annotated with letters to refer readers to descriptions in the main text (Fig. 185).

Even though eye models were teaching aids, designed to assist comprehension, they were also labeled "incomprehensible." An inscription on a broadside celebrating turnery that was

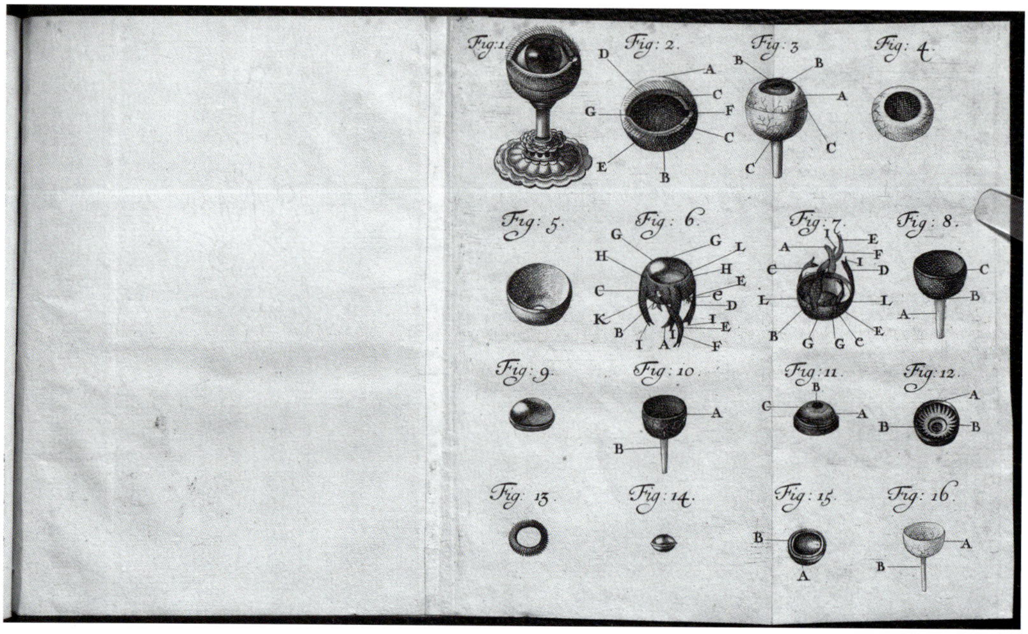

Fig. 185. Daniel Bscherer, Johann Georg Volkamer, and Stephan Zick, *Kurtze und mechanische Beschreibung dieses Kunst-Auges,* 1680, 21. Engraving printed on paper, 7.6 × 13 in (19.3 × 33 cm). Universitätsbibliothek Johann Christian Senckenberg Frankfurt am Main [Ms. germ. fol. 15 Nr. 10].

printed in 1760, 1781, and 1801 by members of the Zick family asserts: "The human ear and eye of the finest ivory seem quite incomprehensible, turned, of course, arranged without lacking for use and teaching in anatomy; These are pieces of rarity!"[92] Above the inscription is a print in which Athena sits enthroned and rests her left forearm and hand on her shield, which bears the head of Medusa; to the right, a turned eye seems almost to peer out at viewers above a pair of ivory ears that seem to listen (Fig. 186).[93]

Taste

The experience of a turned ivory as "incomprehensible" was not a simple unthinking stupefaction but, perhaps, more akin to an experience of the *je ne sais quoi*, an experience that transcended the expressive or explanatory capacity of a sophisticated observer. That experience may have resembled a visit to San Carlo in that the church similarly resisted knowledgeable observers' explanatory efforts in its sustained refusal to "resolve" into something knowable and explicable. Both turnery and San Carlo belonged to a rule-transcending space of aesthetic achievement that was defined ultimately by taste—knowing the rules of an artform but also recognizing that which might transcend them and applying the rules in a way that was so clever that it could not readily be apprehended by observers. Both existed in a kind of transcendent and unknowable condition that could be characterized as a triumph of taste over rules. Plumier argues that one who would be an able turner needs more than in-depth knowledge of machines and skill with tools. The turner must also understand the simple contour of objects in order "to give good taste to works" (*pour donner le bon gout aux ouvrages*).[94] He follows this with a definition: "I call ... good taste this pleasure to the sight which first satisfies the mind by only the impression and aspect of the work."[95] Then he qualifies his definition:

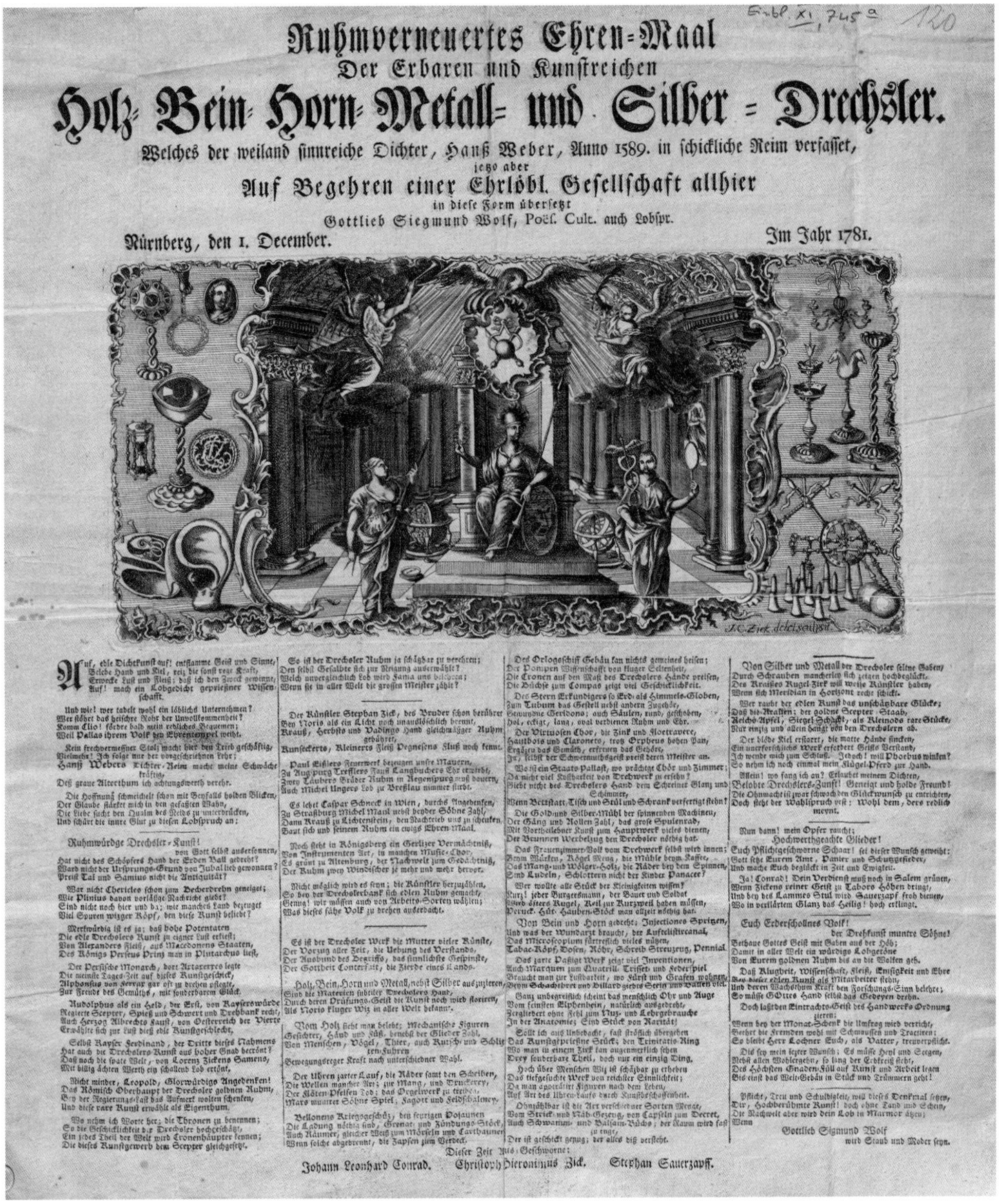

Fig. 186. Hans Weber, Gottlieb S. Wolff, and Johannes Christoph Zick, *Ruhmverneuertes Ehren-Maal Der Erbaren und Kunstreichen Holz-Bein-Horn-Metall-und-Silber-Drechsler*, 1781. Engraving printed on paper, 12.4 × 6 in (31.5 × 15.3 cm). Bayerische Staatsbibliothek, München [Einbl. XI,745 a].

In truth, it is very difficult to be able to explain this good taste, and to establish precise rules for it, since it depends more on the idea and the genius of people than on any certain method. The eye alone must prescribe the rules and the laws, as in the art of painting and in architecture, where all the rules that one could have given were never able to form works as agreeable as those that very often the idea alone or the whim alone imagines.[96]

The eye alone must prescribe taste's rules and laws: taste cannot be reduced to preexisting formulas, because it must be open to the capricious fancies of artists. Plumier's remarks on visual discernment recall the disagreement between Charles Le Brun and Abraham Bosse cited in the introduction to this book. While Bosse held that good painting could be reduced to his rules of perspective, Le Brun considered perspective to be important to artistic practice but not the sole component that mattered; there were also important empirical components to art making. One wonders how engaging on a daily basis with the Trinità dei Monti anamorphic wall frescoes shaped Plumier's theory of art. Walking down the hallways of his convent, Plumier would have been led to think about and physically experience the transformative force of deformations that break free from traditional perspectival rules. (Of course, the anamorphic wall paintings introduced new sets of rules even as they broke old ones, much as Borromini followed rules as he sometimes appeared to defy conventional practice.)[97] It is tempting to speculate that the optical tricks of these deformations would have encouraged future generations of Minims to continue to value deformation and the breaking of norms.

Plumier explains that the French painter, sculptor, architect, and engineer Pierre Puget (1620–1694) helped him to recognize the importance of breaking free from rules:

> Our illustrious Pierre Puget of Marseille, great painter, great sculptor, and at the same time great architect, seeing a book of architecture that I had drawn after Palladio and Vignola, showed great pleasure in seeing it, but at the same time, he told me that all the rules of these authors were hardly necessary, and that the architect himself had to formulate designs suited to the places and situations where one was to build.[98]

Puget's emphasis on accounting for precise locations recalls Juan Caramuel y Lobkowitz's argument that his oblique architecture supported a scenario where the architect's discriminating taste might determine the correct balance between the fixed rules of classical architecture and the conditions of a particular site. Plumier further cites Michelangelo as agreeing that good taste could require modifying rules, claiming to have learned from the French painter and engraver Dominique Barrière (c. 1622–1678) that Borromini kept close at hand an architectural manuscript by Michelangelo.[99] Michelangelo offered several lessons of architecture but concluded nevertheless that "the best [lessons] were those of the genius of the architect himself" (*les meilleures étoient celles du genie de l'Architecte même*).[100] Plumier himself thus encourages us to locate another commonality between turnery and Borromini's work, in this valuation of his openness to moving beyond preexisting rules—the very tendency that led to his San Carlo being branded as "ugly and deformed." Also relevant in this context is the *giudizio dell'occhio* ("judgment of the eye") mentioned in the previous chapter, which was closely linked to Michelangelo and came to denote a discerning eye. It is somehow apt that in his book on turnery, Plumier references *ce bizarre Architecte Il Borromino* ("this bizarre architect Borromini")—an artist who created plans from geometric forms.[101]

Plumier stresses the transcendent, even the infinite, as he argues that taste in turnery cannot be reduced to rules (any more than would be possible for painting or architecture): "One can judge the difficulty that there is in establishing precise and particular rules for works on the lathe—especially being almost infinite and all of a different nature."[102] He thus alludes to variety, an ideal of art and nature advanced by Pliny and, in the Renaissance, among the most highly valued aesthetic attributes of paintings, music, literature, and natural specimens.[103] There are so many different turned objects that one cannot reduce their virtues to simple laws. Rather, judging the taste of turned objects requires a good eye, one with a keen sense of attention. Insofar as taste cannot be reduced to a set of prescriptions, it takes on an ineffable quality. Like turned objects themselves, good taste becomes something that cannot be fully understood or reduced to language; those who possess this incommunicable quality, whether they display it by turning works of ivory or by expressing their admiration for the incomprehensible form of San Carlo, become worthy of admiration.

+

Turned ivories derived their power from relationships among creators, patrons, and observers that had been crafted through a century of observations and conversations around optical devices and effects; while turned ivories are not deformations and were not so labeled by contemporaries, the aesthetic appreciation of turnery relied on similar theoretical and educational frameworks and engaged many of the same mental operations as those we have observed for deformations. Turned ivories exceeded ordinary human limits of visual perception and discernment—more than did most deformations—since even to the most sophisticated observers the delicate ivories could remain inimitable and thus unknowable. The delicacy and incomprehensibility ascribed to these works imbued them with a power that some found overwhelming, particularly those who had not studied the art of turning and could not fathom how they had been produced. Turners and collectors of turnery aligned themselves with these objects' power to overwhelm, hoping to convert it into social power.

Among the many reasons Catholic Church leaders were drawn to deformations was that the attention they focused on facture could support pedagogical aims: they taught observers how to look and, in the process, supported contemplation; where process was associated with doctrinally significant subject matter, deformations could reinforce institutional power and theology. Visual and spatial deformations were used toward multiple ends, social and religious, throughout the period, much as the tools of optics were used to support religious teaching and meditative practice and also to support the natural sciences, which contemporaries would have understood as furthering knowledge of God's revealed creation. The experience of many religious deformations, by some descriptions, had an explanatory or revelatory function that made that which was unclear (visually or theologically) clearer, and did so in a way that seemed miraculous or capable of being associated with divine revelation, insofar as it transcended humans' limited capacity for visual discernment. If the Minims had privileged deformations within a religious culture that contrasted humans' limited visual capacity and discernment against God's power to reveal the unseen to humans, then, when they proclaimed God as a fellow-turner, enthusiasts of turnery by extension seemed to claim for themselves a power approaching the divine, that is, the power to create the incomprehensible.

When we remember the Marquis de Seignelay's experience visiting San Carlo—a truly complex production of deformation—we will recall how this building, unlike Minim and Jesuit anamorphoses, captured the vertiginous sensation of not-knowing without ultimately resolving into a simple doctrinal message. Entering the building, the marquis would have noticed the undulating facade—completed by Borromini's nephew only four years earlier—which violated the structural intelligibility and apparent stability of classical rectangular grid schema and parallel planes (see Fig. 2). The day before visiting San Carlo, on April 9, 1671, the marquis had gone to the Church of Trinità dei Monti, where, he records in his diary, he saw Volterra's *Descent from the Cross* (see Fig. 79).[104] As he walked past San Carlo's undulating facade, whose appearance seems to shift as one moves before it—more dynamically than would a regular facade of parallel planes—would he have recalled the optical effect Niceron described in Volterra's painting, where the figure of Christ seems visually to rise as the observer moves from the left to the right side of the canvas? Once within San Carlo, the marquis—like countless other visitors—would have been baffled in his efforts to make sense of its plan, whose complexity signals a departure from the legibility of classical designs like the Pantheon.

It is tempting to imagine the marquis wandering into San Carlo during a polyphonic performance, especially given frequent seventeenth-century comparisons between the making of visual art and musical composition. Particularly relevant here, possibly, is an observation by the Minim friar Marin Mersenne, that "music could perhaps be called the perspective of hearing, just as perspective could be called the music of the eye, and these two senses are intended for the sciences, and for the service of the spirit, just as the three others are intended for the service of the body."[105] Even more relevant are musical trends that paralleled the visual patterns of deformation and reformation.[106] The period's embrace of sonic deformation and confusion can be heard, for example, in a motet for six voices titled *Cum essem parvulus* (*When I Was a Child*) by the prominent Netherlandish Catholic composer Orlando di Lasso (c. 1532–1594), who was active in Italy and Germany.[107] This piece, which was composed in 1579 and published in 1582, was likely meant for celebrations in the court chapel of Duke Albrecht V of Bavaria (1528–1579) in Munich. While I have found no evidence that it was performed in San Carlo in the seventeenth century, it is tempting to imagine the marquis in this perplexing physical space, listening to this piece (attractive to the art historian for its explicit engagement with visual revelation) or to some analogously perplexing polyphonic work.[108]

Cum essem parvulus intertwines sonic and visual confusion, through its compositional techniques, on the one hand, and, on the other, its text, based on the famous image of observational difficulty from Saint Paul's First Epistle to the Corinthians: "We see now through a glass in a dark manner; but then face to face. Now I know in part; but then I shall know even as I am known."[109] Moreover, much like an anamorphic image that confuses observers before they move to an ideal viewing point, the composition moves from sonic confusion to sonic clarity. The sense of observing with limited understanding communicated by the line *Videmus nunc per speculum aenigmate* ("We see now through a glass in a dark manner") is aptly captured through the counterpoint full of dissonant suspensions and culminating in the E-flat/F-sharp cross-relation when the line ends with a half cadence (in measure 34). This unclarity suddenly changes when we come to the second half of the line, *Tunc autem facie ad faciem* ("but then face to face"): at *facie ad faciem*, the singers abandon the complex polyphony of rhythmically independent lines and declaim the text simultaneously, making it much easier to comprehend; what had previously seemed confused is now clear. To intensify the effect, Lasso repeats the phrase in triple meter (at the time, duple meter was considered imperfect and triple meter perfect), as if to underscore the passage from the imperfect to the perfect state.

The transition from confusion to greater clarity in the music of Lasso is akin to the move from confusion to clarity when one moves from faith and revelation on earth to direct viewing of the divine in the hereafter. As we have seen with anamorphoses, Lasso's composition uses the deliberate obscurity of a deformation—and then its reformation—to strengthen Catholic belief. The works also share a parallel temporal element: the initial frustration, the wait, and then the resolution. In view of the close interrelations among music and rhetorical principles in sixteenth- and seventeenth-century Catholic geographies, it would be interesting to explore how in music (as in visual art and architecture) a rhetorical imperative for clarity coexisted with other rhetorical imperatives, and how the contemporary value on clarity did not necessarily require that it be delivered immediately—or indeed at all.[110]

Juxtaposed against this polyphonic experience, we can picture the marquis raising his eyes, first noticing the columns with inverted volutes and then gazing at the coffered vault of San Carlo's cupola, its pattern drawn from a woodcut devised by Sebastiano Serlio (1475–1554) (see Fig. 8; Fig. 187).[111] The perspectival vault coffering makes the lantern with the dove of the Holy Ghost appear higher than it really is. If the marquis was as perspicacious as Perrault envisioned most observers to be, he would have recognized the coffering's illusionism and may have been unimpressed by the cupola's upward extension; if his skills in visual discernment were lacking, if perhaps he resembled Caramuel's misapprehensive viewer, then he might have succumbed fully to the illusionism of the vault and relished the sense it conveys of relating San Carlo's visitors to a transcendent realm above. Whether or not his perception of the perspectival illusion resolved neatly into comprehension, the vault's pattern is so intricate that it resists observers' attempts to take it in and thus accentuates the vertiginous sensations elicited by other elements in the church architecture.

The experience of listening to *Cum essem parvulus* recalls the experience of observing an anamorphic deformation: the initial incomprehension is sustained only briefly—as long as it takes to hold up the mirror or walk down the corridor. But for a visitor to San Carlo or a guest offered a chance to consider a turned ivory, that incomprehension might be resolved only painstakingly (for an experienced observer, such as one with Plumier's skills) or it might remain unresolved (for many, even most, beholders). Perhaps, in the end, we would not want the marquis to listen to *Cum essem parvulus* in San Carlo, as its resolution of polyphonic complexity contrasts with the experience of a church that refuses to resolve into a clear vision. San Carlo suspended its visitors in a state of sustained incomprehensibility, proposing itself as aesthetically and spiritually moving in its apparent transcendence of rules, in its ultimate unknowability. It is that quality of transcending rules and comprehension (or following rules in ways that were not readily comprehensible), that would appeal to Plumier and make him admire Borromini. Readers might recall here that we began this book with the observation that deformation was defined in the period "against an ideal that an entity fails to meet" and that Borromini was criticized for showing "too much desire to go beyond the rule." Whereas the anamorphic deformations of Minims and Jesuits used the transition from obscurity to clarity to shed light on theological mysteries, San Carlo, much like turned ivories, holds observers in states of seemingly unresolvable incomprehensibility to suggest that transcendence—whether of a spiritual or aesthetic variety—can be achieved even when or especially when the disorientation is not ultimately solvable.

Fig. 187. Francesco Borromini, San Carlo alle Quattro Fontane, 1638–1641, interior view of perspectival vault coffering on dome.

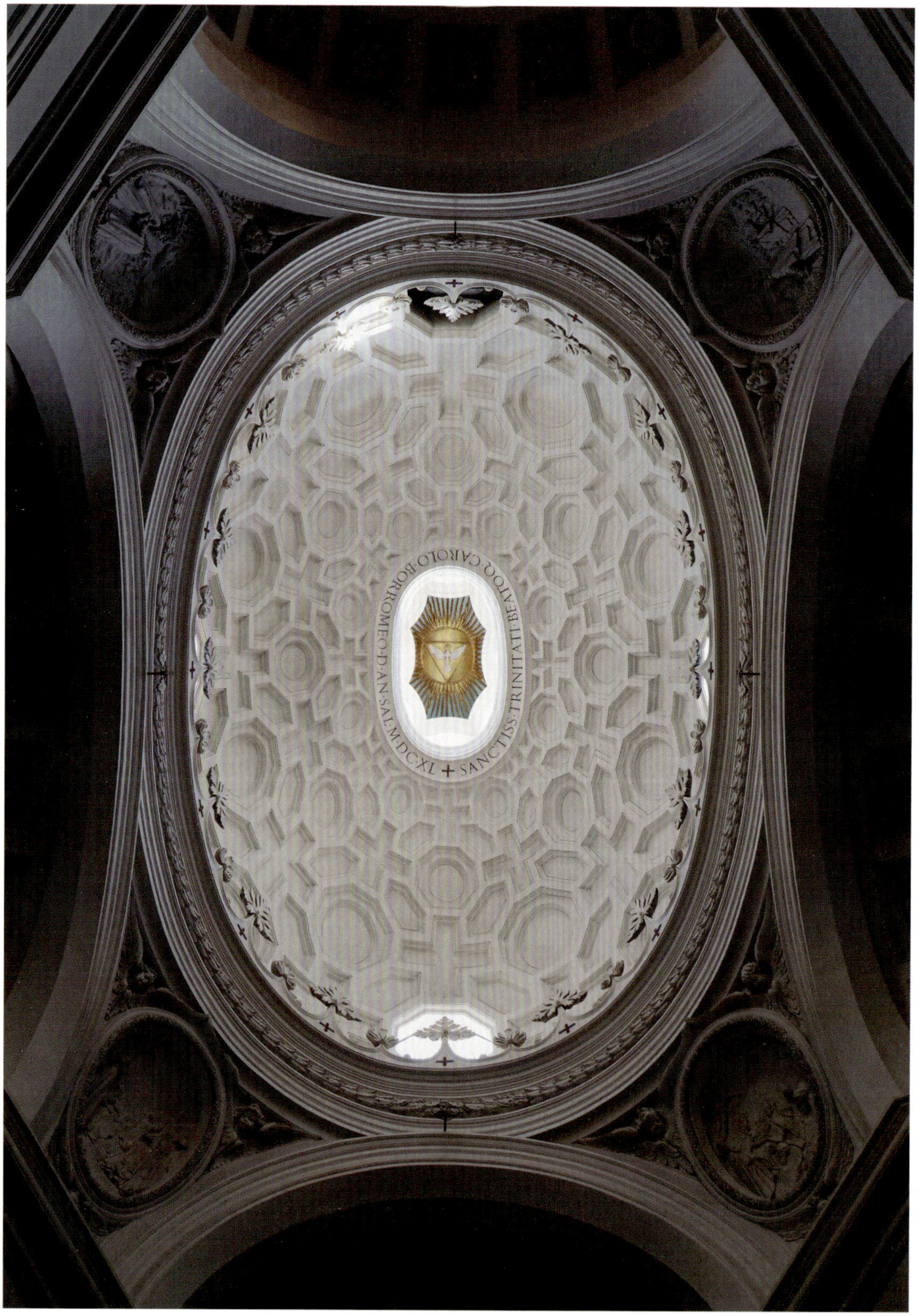

ACKNOWLEDGMENTS

IT IS A GREAT PLEASURE to thank my editor Michelle Komie for her enthusiasm for this project, as well as her colleagues, Annie Miller, Mark Bellis, David Luljak, Jeff Wincapaw, Jessica Massabrook, Steven Sears, and Lachlan Brooks for their work in bringing this book into print. I owe great debts to Peter Parshall and an anonymous reviewer for Princeton University Press who offered numerous thoughtful and in-depth comments that led me to revise the book extensively. I am also grateful to the Barr Ferree Foundation Publication Fund, Department of Art and Archaeology, Princeton University for its financial support of this book's production.

Several fellowships and institutions made it possible for this project to develop in unexpected directions. My thanks to St. John's College, University of Cambridge, and to my host Ruth Abbott for welcoming me for a term to an enchanting, deeply inspiring environment. Villa I Tatti: The Harvard University Center for Italian Renaissance Studies, the American Council of Learned Societies, the John Simon Guggenheim Memorial Foundation, the Renaissance Society of America, the Office of the Provost and Advancing Scholarship in the Humanities and Social Sciences grant program at the University of Southern California, and the USC-Huntington Early Modern Studies Institute provided invaluable monetary and institutional assistance. My thanks to the Alexander von Humboldt Stiftung and to my host at the Freie Universität Berlin, Klaus Krüger, for welcoming me to Germany; I am also grateful to Sunita Scheffel for her aid in navigating administrative matters in Berlin. My colleagues in the Department of Art History at USC have likewise provided precious guidance: I am indebted to Daniela Bleichmar and Vanessa Schwartz for their wise mentorship when I first embarked on this project; I thank Suzanne Hudson for her astute and careful advice on the introduction and framing of this book; my thanks to Lisa Pon for productive writing exchanges and to Kate Flint, Amy Ogata, Rebecca Lemon, and Amy Knight Powell for their support as chairs.

I am delighted to acknowledge Marina Cotugno, who photographed artworks and manuscripts in Italy when COVID and a newborn baby made it impossible for me to travel. My thanks as well to Adriana Capriotti and Walter di Domenico for their introduction to the rooms, staircases, and marvels of Palazzo Spada and for facilitating the publication of photographs of the palace, for which I am also grateful to Fabiana Bartorilla. I am likewise indebted to Léa Ruiz-Pacouret and Mathis Goddet for making it possible for me to visit Trinità dei Monti on multiple occasions. I owe a great debt to Agostino De Rosa, who not only provided inspiration in his and his students' scholarship on anamorphosis and Minim visual culture, but also generously gave me multiple high-resolution photographs that he and his colleagues had taken at Trinità dei Monti. I thank Giuliana Forti and Cosma Alessandro at Palazzo Barberini. I am grateful as well to Nick M. L. Mols, who created two architectural plans for this book.

My thanks to librarians and curators at the following institutions for their help: the Accademia Nazionale dei Lincei, the Pii Stabilimenti della Francia a Roma e Loreto, the Gabinetto dei Disegni in the Castello Sforzesco in Milan, the Archivio di Stato di Roma, the Hanna Holborn Gray Special Collections Research Center in the University of Chicago Library, the Cambridge University Library Rare Book collections, the Warburg Institute Library and Photographic Collection, the Houghton Library of Harvard University, the Huntington Library in San Marino, California, the British Museum, the British Library, the Bibliothèque nationale de France, the Bibliothèque Sainte-Geneviève, the Firestone Library of Princeton University, the Staatsbibliothek zu Berlin, the Bayerische Staatsbibliothek in Munich, the Graphische Sammlung der Albertina in Vienna, the Museo di Roma, the Pinacoteca Civica di Palazzo Chiericati in Vicenza, the Heidelberg University Library, the Royal Danish Collection in Rosenborg Castle, the Museo Diocesano Bitonto, the Palazzo Barberini, the Muséum national d'Histoire naturelle, the Staats- und Stadtbibliothek Augsburg, the John Carter Brown Library, the National Gallery in London, the Biblioteca Ambrosiana, the Museo Galileo, the Bibliothèque de la Sorbonne, the Palazzo Pitti, the Staatliche Bibliothek Regensburg, the Zentralinstitut für Kunstgeschichte Munich, the Kunsthistorisches Museum Wien, the Universitätsbibliothek Johann Christian Senckenberg Frankfurt am Main, the Metropolitan Museum of Art in New York, the University of Southern California Libraries, the Noord-Hollands Archief, and the Germanisches National Museum in Nuremberg.

In writing this book, I have relied on the guidance of numerous scholars. I thank Jacob Schmutz, Daniel Garber, Steven Nadler, Frederic Clark, Rachael Z. DeLue, Michael D. Gordin, Larissa Brewer-Garcia, Richard J. Oosterhoff, Adam Jasienski, and José Ramón Marcaida for offering insightful feedback on the introduction or chapters in the manuscript. Jill Kraye and Deborah Howard deserve special thanks for reading an early draft of the entire manuscript and for their constructive advice. My writing partner, Elizabeth J. Petcu, is a treasured interlocutor who offered exhaustive notes on several drafts of chapters in the manuscript as well as the introduction. Anthony Ossa-Richardson also gave brilliant feedback on multiple chapters; I thank him warmly for his erudition and for his kindness. I thank Sara J. Schechner for her generosity in sharing her learned knowledge of sundials and other matters with me. I am most grateful to the discerning Lisa Regan for productive conversations that helped to nuance my thinking and argumentation. Lory Frankel offered astute editorial advice. For further suggestions and assistance in the process of putting this book together, I thank Susan Stewart, Joseph Connors, Larry Silver, Christopher S. Wood, Victor I. Stoichita, Eugene J. Johnson, Maria Elisa Navarro

Morales, Anne Goldgar, Stella Nair, Kévin Daligault, Daniel J. Sheffield, Sophie Roux, Domenico Collacciani, Hélène Lo Iacono, Bernadette Russo-Amoros, Jacques Hancy, Jacob Soll, Roberto Bazzuoli, Ralph Dekoninck, Ioannis Mylonopoulos, Robert Kirk, Guillaume Champion, Glenn W. Most, Stephanie Koerner, Michael Koortbojian, Katharine Park, Alessandro Giammei, Mira Siegelberg, David Albertson, Christophe Litwin, Jack Hartnell, Eileen A. Reeves, Anthony Grafton, Jennifer M. Rampling, Mattia Mantovani, Hanneke Grootenboer, Emanuele Lugli, Harleen Kaur Bagga, David M. Stone, Ulrich Pfisterer, Annalena Brandt, David Freedberg, Michael Cole, Diane Bodart, Leonard Barkan, Giulio Pertile, Carolina Mangone, Mordechai Feingold, Jed Z. Buchwald, Hanna Park, Doogab Yi, Peter Mancall, Catherine Hofmann, Charlotte Denoël, Nathalie Charrier-Arrighi, Quentin Skinner, Marco Faini, Ruth Ezra, Alina Payne, James Grantham Turner, Paula Findlen, Ann Blair, Fanny Camara, Nathan Perl-Rosenthal, Alexander Bevilacqua, Helen Pfeifer, Alexander Marr, and audiences at the Zentralinstitut für Kunstgeschichte in Munich, KU Leuven, Princeton University, USC, Columbia University, Caltech, Seoul National University, UCL, Rutgers University, and Stanford University. I thank Timon Screech, Rebecca Salter, J. P. McDermott, Amy Raphael, Lihong Liu, Andrew M. Watsky, Florence C. Hsia, Kristina Kleutghen, Marco Musillo, Jerome Silbergeld, Noel Golvers, Jennifer Purtle, and Stephen Whiteman for their thoughtful advice on Chinese and Japanese anamorphoses. My thanks to David Crook, Daniele V. Filippi, Bernhold Schmid, Noel O'Regan, Craig Monson, and Arnaldo Morelli for discussing musicological matters with me. I am deeply grateful to the historians of early modern Catholicism Camilla Russell, who read sections of the manuscript, and Stefania Tutino who read the entire manuscript; both Camilla and Stefania gave me constructive feedback that I have integrated into this book. My deepest thanks to Julia Boss for her wise and detailed recommendations on the manuscript. My thanks to Andrea Lazzarini, Marco Faini, and Luca Zipoli for their productive suggestions on Italian translations. Latin translations were made with the help of Mali Skotheim. I thank Daniela Gutierrez Flores for her assistance with Spanish translations.

I thank my family and friends for sustaining me. Particular thanks to my siblings and my parents as well as to Isabel Buchanan, Alex Cole, Mira Siegelberg, Clara Botstein, Laetitia Drusch, Johanna Da Silva Rosa, Benjamin Seidler, Jacob Burda, and the late Anthony Newcomb, who generously left me portions of his library on early modern art and music which have been a continual source of inspiration. Finally, I dedicate this book with love to Torbjørn, Theodore, and Thomas.

NOTES

INTRODUCTION

1 Bellori wrote these words into his copy of Giovanni Baglione's (1566–1643) *Vite* (*Lives*, 1642). Baglione 1935, 180. On the term "goth" and the branding of Borromini's architecture, see Gombrich 1966, 83–84; and Connors 2000, 191–204. Unless otherwise indicated, all translations are my own.

2 On Borromini's reputation, see Blunt 1979, 211–22. Milizia 1768, 347.

3 Milizia 1768, 347: "Tanti retti, concavi, e convessi, con tante colonne sopra colonne di diversa sagoma, e finestre e nicchie e sculture in si poca facciatina, son cose che fan pietà."

4 Baldinucci 1948, 149.

5 On the rectangular grid schema, see Tzonis and Lefaivre 1986, 9.

6 Borromini discovered ancient Roman archaeology and facades with a sole concave curve, or with both concave and convex curves, via engravings and drawings by Giovanni Battista Montano (1534–1621). Blunt 1979, 41–45. On Montano, see Fairbairn 1998, 540–773.

7 Although this guidebook was almost certainly intended for publication, it never appeared in print. The manuscript is held by Avery Library, Columbia University, n. AA 1115 D456. An excerpt of about two hundred pages from the 513-page document appears in Connors and Rice 1990. See Connors and Rice 1990, 79: "Boromini, qui a toujours cherché de s'écarter de la simplicité régulière de l'antique, y a tant fait de tourner et de retourner de lignes convexes et concaves que la veue qui se fatique à trouver là-dedans de la suitte n'est pas tout à fait contente, quoi que d'abord il y ait dans cette manière un je ne scais quoi qui surprend et qui empesche de l'improuver entièrement." The verbs *tourner* and *retourner* evoke a comparison between Borromini's curved forms and those produced in turning, the focus of chapter 6. If a delicate piece of virtuoso ivory turnery eludes facile visual understanding, so do Borromini's mostly white facades (and interiors), which are contemporary to the fashion for turned ivory in Europe.

8 In this study, I choose not to use the Protestant-perspective label "Counter Reformation" but instead employ the designation "Catholic Reformation." By this term, what I really mean is "Catholic Reformation Era" or "Reformation Era Catholicism"; I use the term "Catholic Reformation" for the sake of concision and to move away from partisan era labels. On these names and their benefits and drawbacks, see O'Malley 2000.

9 Clément 1867, 161.

10 Joseph Connors has contrasted the architectural methods of Bernini with those of Borromini, whose plans are made not only from centers, but epicenters, that is, points on the boundary of a geometrical form that create other forms in an extended space (Connors 1982, 25 and 33).

11 Connors 1982, 25.

12 Clément 1867, 168.

13 Clément 1867, 162.

14 On Albertina 173, see Connors 1999, 480.

15 On the plan, see Wittkower 1999, vol. 2, 199. See also the following discussion of Borromini's design that stresses other aspects besides geometric units: Raspe 1994.

16 The term "enigma" is borrowed from Joseph Connors, who writes that in San Carlo "we are in an enigma" (*siamo all'interno di un enigma*). Connors 1999, 459.

17 The friar's dates are not known. Fra Juan di San Bonaventura, *Relatione del Convento di S. Carlo alle 4.o fontane di Roma*, Rome, n.d. [1650–1652], MS in the Archivio di S. Carlo alle Quattro Fontane in Rome, vol. 77a, fols. 25 f. Cited in Wittkower 1999, 203.

18 Fra Juan di San Bonaventura, *Relatione del Convento di S. Carlo alle 4.o fontane di Roma*, Rome, n.d. [1650–1652], MS in the Archivio di S. Carlo alle Quattro Fontane in Rome, vol. 77a, fol. 50. Vienna, Albertina, Borromini Zeichnungen, 168, 169, 170, 173, 175, 176. Thelen 1958, 11. Portoghesi 1967, plates 9 and 17. See also Connors 1995, 590.

19 Steinberg 1977, 17–43.

20 Clément 1867, 162: "Il y a dans cette église seize colonnes dont les chapiteaux sont composés si bizarrement que les volutes retournent en dedans au lieu de retourner en dehors."

21 On the tradition of inverted volutes and Borromini's use of them, see Steinberg 1977, 208–17; and Beck 1971, 225–34.

22 As stressed by Steinberg 1977, 214–15. See also Martin Raspe's discussion of the capitals and his broader emphasis on the inner coherence of

Borromini's architectural system of principles, Raspe 1994, 39–42.

23 Raspe (1994) has stressed that Borromini did not overturn traditional architectural rules, but rather reinterpreted them.

24 Fra Juan di San Bonaventura, *Relatione del Convento di S. Carlo alle 4.o fontane di Roma*, Rome, n.d. [1650–1652], MS in the Archivio di S. Carlo alle Quattro Fontane in Rome, vol. 77a, fols. 45–46: "Ogni giorno vidiamo gente di queste nationi intrare in questa chiesa, mosse a vederla della fama et rumor che arrivò a li lor paessi; et quando stano in chiesa altro non fanno che guardar allo alto et voltarsi per tutta la chiesa, perché tutte le cosse d'essa sonno in tal modo disposte che una chiama alla altra, e la una picca al che guarda acciò veda la altra; et cossì noi molte volte dalle tribune et zelosie della chiesa vediamo a questi nationali fare queste attioni senza saper partirsi né dir cossa alcuna per un pezzo: et quello che più admira è che, sempre che si guarda, questa chiesa dà più gusto, et pare che si vede di nuovo, et lascia apetito di ritornarla a veder, perché vediamo li stessi huomini venir molte volte a vederla. Perché? Perché talmente la vedono che dà tal gusto che non dà fastidio, ma sì desiderio di più vederla. Mi pare che questo tenga qualche cossa di imitatione (in quanto si può dire) di divinità. … La fabrica di questa chiesa quando è vista di questi nationali pare che suspende il lor intelletto perché per un pezzo si vedono non far motto e parte nessuna altro che vederla; et doppo di vista ritornono a vederla; et questo non solo molte volte in una ocasione, se non moltissime volte in molti diversi giorni. Si havesse saciato la prima volta, ed havesse datto fastidio come dano le cosse mundane doppo di vista, non ritornariano molte volte a vederla." The passage was transcribed and discussed (though not translated) by Leo Steinberg in his thesis on San Carlo, written in the 1950s and published in 1977. Steinberg 1977, 328–29. For a transcription of the passage in the manuscript (with some modifications), see García 1999, 71–73.

25 On the *je ne sais quoi*, see Scholar 2005.

26 Bouhours 1671, 322: "Il est bien plus aisé de le sentir, que de le connnoistre … sa nature est d'estre incomprehensible, et inexplicable."

27 As stressed by Scholar 2005, 4.

28 Bouhours 1671, 326: "En effet c'est quelque chose de si délicat, & de si imperceptible, qu'il échappe à l'intelligence la plus penétrante, & la plus subtile."

29 Scholar 2005, 64.

30 On conformity, see Hyman 2022, 588–89; and Hyman 2021.

31 Ossa-Richardson 2019, chapter 1. Sluiter 2016, 39.

32 Argan 1954, 257–64. Anthony Blunt, for instance, asserts, referring to baroque art: "If this new art is to be described by a single epithet, it could be called *rhetorical*" (emphasis in original; Blunt 1988, 10).

33 Demonstrating the commitment to rhetorical principles in seventeenth-century Italian painting, see the discussion of clarity of gestural expression, composition, and color in Guercino's (1591–1666) work in Ebert-Schifferer 1992, 102.

34 Schroeder 1960, 484: "Illud vero diligenter doceant episcopi, per historias mysteriorum nostrae redemptionis, picturis vel aliis similitudinibus expressas, erudiri et confirmari populum in articulis fidei." Translation from Klein and Zerner 1990, 120–21.

35 Gregory the Great 1982, 768 [2.12–14]. Chazelle 2012, 138–53.

36 On Paleotti, see Prodi 2022.

37 Barocchi 1960–1962, vol. 2, 214: "Un altro effetto che deriva dalle cristiane pitture, molto notabile e principale, il qual a guisa degli oratori è dirizzato al persuadere il popolo e tirarlo col mezzo della pittura ad abbracciare alcuna cosa pertinente alla religione." Translation from Paleotti 2012, 110 (book 1, chapter 21).

38 Barocchi 1960–1962, vol. 2, 215: "Il fine di esse principale sarà di persuadere le persone alla pietà et ordinarle a Dio." Translation from Paleotti 2012, 111 (book 1, chapter 21). Schroeder 1960, 484: "Postremo tanta circa haec diligentia et cura ab episcopis adhibeatur, ut nihil inordinatum, aut praepostere et tumultuarie accommodatum …" Translation from Klein and Zerner 1990, 121.

39 Barocchi 1960–1962, vol. 3, 112: "Nihil operis, qualequale sit, statuatur, fiat, inscribatur, effingatur exprimaturve … quod deforme … sit."

40 Translation from Borromeo 2010, 10–11: "Sciendum est licere quidem Pictori sive Sculptori historias ornare et illustrare quantum velint, minime autem licere suscipere aliquam cum ipsarum veritate pugnam et inveteratam de una quaque re famam obscurare sive inficiari."

41 Barocchi 1960–1962, vol. 2, 408: "Una delle principali laudi che sogliono darsi ad uno autore o professore di qualche scienza, è ch'egli sappia chiaramente esplicare i suoi concetti, e le materie, se bene alte e difficili, renderle col suo facil modo di parlare intelligibili a tutti e piane. Il medesimo possiamo affermare in universale del pittore, e tanto più, quanto l'opere sue servono principalmente per libro degli idioti, alli quali bisogna sempre parlare aperto e chiaro." Translation from Paleotti 2012, 251 (book 2, chapter 33).

42 Barocchi 1960–1962, vol. 2, 408: "A che non avertendosi da diversi, accade che ogni giorno si veggono in varii luoghi, e massimamente nelle chiese, pitture così oscure et ambigue, ch'ove doveriano, illuminando l'intelletto, eccitare insieme la divozione e pungere il cuore, elle con la loro oscurità confondono per modo la mente, che la distraeno in mille parti e la tengono occupata in disputare tra sé stessa quale sia quella figura, non senza perdita della divozione." Translation from Paleotti 2012, 251 (book 2, chapter 33).

43 John W. O'Malley (1991, 190) has stressed that most Jesuits likely never saw the decrees of the Council of Trent.

44 On this mnemonic poem and the term "baroque," see Panofsky 1995, 19. Translation from Pascal 1948, 211. Pascal 1991, 428: "Ce n'est pas *barbara* et *baralipton* qui forment le raisonnement."

45 Translation from Pascal 1948, 211. Pascal 1991, 428: "Et l'une des raisons principales qui éloignent autant ceux qui entrent dans les connaissances du véritable chemin qu'ils doivent suivre est l'imagination qu'on prend d'abord que les bonnes choses sont inaccessibles, en leur donnant le nom de grandes, hautes, élevées, sublimes. Cela perd tout. Je les voudrais nommer basses, communes, familières. Ces noms-là leur conviennent mieux, Je hais ces mots d'enflure ... "

46 O'Malley 2000, 19.

47 O'Malley 2000, 19. Von Ranke 1839–1847.

48 Pütter 1776, 10. O'Malley 2000, 20.

49 Levy 2015, 66.

50 As observed by O'Malley 2000, 1. On the history of the term "Reformation," see Wolgast 1984, vol. 5, 313–60 (esp. 321–35). See also Eire 2016.

51 O'Malley 2000, 17–18 and 131. O'Malley has stressed that the Jesuits typically did not refer to the reform of the Church, although they often employed *reformatio* to refer to two practices. At times, it denoted their endeavors in assisting convents in the election of excellent superiors and handling "their morale and religious observance." More commonly, it denoted the "change of heart effected in individuals through the *Spiritual Exercises* and other ministries in which the Jesuits were engaged." O'Malley 1991, 182.

52 Enders 1890, vol. 1, 132. Cited in Wolgast 1984, 326.

53 Wrede 1901, 397–98: "Nos ... pontificatum hunc nunquam ambivimus, ... subiecimus igitur colla summe dignitati, non ob dominandi libidinem neque ad ditandos propinquos nostros, sed ad divine voluntati parendum, ad deformatam eius sponsam ecclesiam catholicam reformandam, ad subveniendum oppressis, et doctos ac virtute preditos, qui multo iam tempore neglecti iacuerunt, erigendum et ornandum, et denique ad omnia alia agendum, que bonum pontificem et legittimum beati Petri successorem agere oportet; quamquam nemo mirari debebit, si non statim omnia errata et abusus omnes per nos emendatos viderit: inveteratus nimium morbus est, nec simplex, sed varius et multiplex pedetentim in eius curam) procedendum est et prius gravioribus magisque periculosis occurrendum, ne omnia pariter reformare volentes omnia perturbemus; omnes subite mutaciones (inquit Aristotiles) in republica periculose sunt; qui nimis emungit, elicit sanguinem." Referenced in Wolgast 1984, 326n101. The passage cites Proverbs 30:33: "Qui vehementer emungit, elicit sanguinem" ("he that violently bloweth his nose, bringeth out blood"). Here and elsewhere, I have used the English of the Douai-Rheims Bible. See Aristotle, *Politics*, 1269a14–18.

54 Pallavicino 1656, vol. 1, 66.

55 Pallavicino 1656, vol. 1, 66–67.

56 *Vocabolario degli Accademici della Crusca* 1612, 262: "Guastar la forma, tor la bellezza."

57 *Vocabolario degli Accademici della Crusca* 1612, 708: "Riordinare, e dar nuova, e miglior forma."

58 *Vocabolario degli Accademici della Crusca* 1612, 708: "Ma riforma, e rifortifica in isperanza l'animo suo."

59 My discussion of this trope draws on the comments of Freedberg 1989, 184–87; Smith 2002, 47–49; and Melion and Dekoninck 2019, 527.

60 David 1601. See Melion and Dekoninck 2019, 527.

61 On this image, see Smith 2002, 48–49; and Dekoninck 2005, 194–96.

62 On spiritual discernment, see Zierholz 2021, 50; Clark 2009, 204–35; Copeland and Machielsen 2013; and Sluhovsky 2007.

63 David 1601, 353: "Quemadmodum praecellens aliquis pictor, omni sua industria ad vivum conatur exprimere, quod sibi ex arte imitandum praesumserit; ita illa homini Christiano incumbit cura, ut Christum Saluatorem nostrum in vita sanctaque conversatione imitetur, & in se ad vivum quasi delineatum exhibeat ... ita tandem in omni virtutum genere proficiat, ut Christum veluti prototypon suum in se ipse velut in tabula repraesentet." Cited in Freedberg 1989, 184. On the phrase *ad vivum* ("from the life"), see Swan 1995, 353–72.

64 Sucquet 1630, 501: "Est considerare cum animo suo, & velut in corde depingere mysterium, aut doctrinam aliquam e vita Christi, vel denique perfectionem Dei cum circumstantiis personae, actionum, verborum, loci ac temporis." Translation from Freedberg 1989, 184–85.

65 Jerome Nadal's prologue, which was written around 1567, is transcribed in Italian translation in Ignatius of Loyola 1966, 6–7. As noted in Levy 2004, 122.

66 Barocchi 1960–1962, vol. 2, 119: "Catolici, i quali, ritenendo l'uso delle imagini, hanno nondimeno in varii modi corretta e difformata la dignità loro." Translation from Paleotti 2012, Proem, 47.

67 Barocchi 1960–1962, vol. 2, 131: "La intenzione nostra in questo trattato è di riformare il catolico." Translation from Paleotti 2012, 57 (book 1, chapter 1).

68 Barocchi 1960–1962, vol. 2, 267: "Ora, per cominciare a discorrere intorno a questi abusi, acciò, scoperte le sordidezze che tengono diformata la bellezza delle imagini, più facilmente si possa reintegrare la purità loro." Translation from Paleotti 2012, 159 (book 2, chapter 2).

69 Here and elsewhere translations from Bellori are by Alice Sedwick; see Bellori 2005, 57. Bellori 1976, 13–14: "Perpetuamente giustissimi e vaghissimi ... soggetti alle alterazioni ed alla bruttezza." On the distinction between the indestructible celestial realm and the sublunary realm, see Aristotle 1984, vol. 1, *On the Heavens*, 283b26–38. Aristotle 1552, 42^{r}–43^{r}.

70 Bellori 2005, 57. Bellori 1976, 14: "Nulladimeno per l'inequalità della materia, si alterano le forme, e particolarmente l'umana bellezza si confonde, come vediamo nell'infinite deformità e sproporzioni che sono in noi."

71 Bellori 2005, 57–61; and Bellori 1976, 13–23.

72 Bellori 2005, 62. Bellori 1976, 23: "Pur troppo la deformano quelli che con la novità la trasmutano, mentre alla bellezza sta vicina la bruttezza, come li vizii toccano le virtú."

73 Concerning conformity, see Hyman 2022, 588–89; and Hyman 2021.

74 On deformed animal kinds in Aristotle, see Witt 2012. Aristotle, *Parts of Animals*, 684b1–2.

75 Aristotle, *Generation of Animals*, 737a27–29. See also Witt 2012, 93.

76 For Maignan's use of the term *ars deformandi* ("the art of deforming") and variants to denote his endeavors, see Maignan 1648, 438 and 442. Or consider how Niceron writes on his titlepage for *La perspective curieuse* (*Curious Perspective*, 1638) that his book will teach how to make and construct all sorts of *figures difformes* ("deformed figures"). For uses of *difforme* and *difformes* in his book, see also Niceron 1638, evi^{r}, 49–52, 54–59, 66–67, 69, 70–71, 83, 88, 92, 94, 96, 98–99, and 105. The terms *deformationes* ("deformations") and *reformationes* ("reformations") abound in Mario Bettini's writing on optical devices; see, for instance, Bettini 1642, vol. 1, *Apiarium Quintum, Progymnasma Primum*, 3.

77 In the last thirty years, historians have reframed labels for the period ranging from around 1400 to 1800. Many have focused especially on the increasing interrelations among diverse areas of the globe during this time and have moved away from Eurocentric perspectives. Concerning the limitations of the term "baroque," see Kaufmann 2011, 83–98; Farago 2011, 99–122; Hills 2011; Levy 2004; and Levy and Baumgarten 2008.

78 For the convenience of contemporary readers, I occasionally use the modern category of "science" and its derivatives as stand-ins for lengthier but more precise terms including "natural philosophy," "mathematical natural philosophy," and "natural history." Concerning the relationship between baroque visual art and the rise of the "new" science, see Gal and Chen-Morris 2013.

79 Levy and Baumgarten 2008, 39–40.

80 On the continued importance of the baroque for scholars, see Levy and Baumgarten 2008, 42.

81 Translation from Wölfflin 2015, 300. Wölfflin 1918, 237: "So vielgestaltig die nach-klassischen Stilabwandlungen sein mögen, sie haben alle die merkwürdige Eigenschaft, daß die Erscheinung sich irgendwie der völligen Faßbarkeit entzieht."

82 Wölfflin 1964, 15; and Wölfflin 1888, 1. Translation from Wölfflin 2015, 274–75. Wölfflin 1918, 211: "Das 17. Jahrhundert hat in der Dunkelheit, die die Form verschlingt, eine Schönheit gefunden." On the formless in Wölfflin, see also Levy 2015, 109–11 and 159nn53–54.

83 I follow Jon R. Snyder (2005, 22) in his stress on the importance of classicism when talking about the baroque and the ways in which baroque artists often distorted its norms.

84 Stoichita 1993.

85 Levy 2004, 178, 182–83, and 203.

86 Hyman 2022, 588–89; and Hyman 2021, 13.

87 Panofsky 1995, 19.

88 Furetière 1690, vol. 1, 230–31: "Terme de Joüaillier, qui ne se dit que des perles qui ne sont pas parfaitement rondes."

89 Pernety 1757, 24: "Baroque qui n'est pas selon les régles des proportions, mais du caprice."

90 In theological contexts in this period, the term "discernment" could denote the ability to distinguish or discern whether a reality is divine or diabolical (or, later, man-made). In what follows, I do not deploy the term "discernment" in this sense, unless explicitly

stated. On the discernment of spirits, see Copeland and Machielsen 2013.

91 Crary 2001. Lyle Massey's *Picturing Space, Displacing Bodies: Anamorphosis in Early Modern Theories of Perspective* similarly offers a prehistory to a phenomenon that Crary identifies in late nineteenth-century France (2007, 5). For a general study of attention, see Uhl 1890. On attention in the history of science, see Daston 2003, 100–26; Daston 2001; and Daston 2004, 434–40. Concerning attention and absorption in art historical sources, see Fried 1988. See also the Centre for Attention Studies at King's, accessed March 9, 2022, https://attentionstudies.org.

92 Michel Foucault (2009, 230–31) relates an emerging investment in directing attention to a broader interest in conduct in sixteenth- and seventeenth-century Europe. Foucault's "government of souls" is discussed in Väliaho 2022, xiii, 90, and 118–19.

93 Freedberg 2002. With the term "naturalism," I refer to efforts in the visual arts to represent details of nature as they are (rather than in a stylized fashion). With "empiricism" I refer to the assertion that sense experience and immediate observation (as opposed to abstract reasoning) is how we know reality. In regard to members of the Accademia dei Lincei, David Freedberg treats them as categorically "modern," while Irene Baldriga instead situates their scientific breakthroughs in relation to orthodox Catholicism and stresses the crucial role of previous endeavors in the evolution of their ideas. Baldriga 2002. See also Findlen 2004b.

94 Havelange 1998. Clark 2009.

95 1 Corinthians 13:12.

96 Romans 1:20. Concerning medieval discussions of this passage, see Hamburger 2000, 353–409.

97 Pereda 2018, 10, 27n72, and 138. The bibliography on *desengaño* is extensive; see, for instance, Rosales 1966; Green 1970, 141–70; Schulte 1969; and Casado 1988, 567–95. The *Tesoro de la lengua castellena o española* (*Treasure of the Castilian or Spanish Language*, 1611) by the polyglot Sebastián de Covarrubias Horozco (1539–1613) presents this definition of *desengañar*: "To pull out of deceit [illusion] one who is in it." Translation from Davenport 2010, 19. De Covarrubias Horozco 2006, 687: "Sacar de engaño al que está en él."

98 Niceron opens his preface to *La perspective curieuse* by accentuating how his and his era's work in optics demonstrates advantages as compared to the accomplishments of antiquity. See Niceron 1652, 1.

99 Martin Kemp has dubbed the period from about 1630 through 1680 "the golden age of the perspective treatise" in France, citing works by Jean-Louis Vaulezard (first published in 1630), Girard Desargues (1636), Niceron (1638 and 1646), Jean du Breuil (1642, 1647, and 1649), Jacques Aleaume and Estienne Mignon (1643), Abraham Bosse (1648, 1653, 1665, and 1667), René Gaultier de Maignannes (1648), Jacques Le Bicheur (1660), Charles Bourgoing (1661), Grégoire Huret (1670), and Sébastien Le Clerc (1682). Kemp 1990, 119–20. Seventeenth-century accomplishments in the study of perspective built on the achievements of the previous centuries by such thinkers as Piero della Francesca, Leonardo da Vinci, Jean Pélerin (known as Viator), Albrecht Dürer, Giacomo Barozzi da Vignola, Sebastiano Serlio, Daniele Barbaro, and Jean Cousin the Elder.

100 Bosse 1665, 108. On this debate and its implications, see Damisch 1994, 49–50; and Duro 1997. Writing nine years before Bosse's expulsion, Niceron appears to side with Bosse when he underlines the importance of the maxims of optics to painting: "All the faults which one notices in the pictures of a number of painters come from their ignorance of these principles." ("Toutes les fautes que l'on remarque dans les tableaux de plusieurs peintres viennent de l'ignorance de ces principes.") Translation from Niceron 2019, 62. Niceron 1652, 2.

101 Scholars have questioned traditional accounts of a struggle between the Catholic Church and the "new" science suggesting instead an uneasy balance between Catholicism and "novel" scientific findings. See, for instance, Findlen 2004a; Feist 2013, 71–72; Gorman 2020; and Feingold 2002. On the financial and social aid for research in astronomy offered by the Roman Catholic Church, see Heilbron 2001, 3. On baroque visual art and the "new" science, see Gal and Chen-Morris 2013. Christopher G. White stresses the interrelations between scientific and religious thought: "Scientific and religious ideas always come braided together … they influence and shape each other to a degree that has gone unnoticed for several reasons, including the fact that until recently cultural elites and other commentators have held simplistic, reified views about science and religion and have often understood them as implacable enemies." White 2018, 13. For studies that negate the notion that scientific and religious thought are inimical, see works mentioned in Gregory 2008, 519, 497n3. Cited in Eire 2023, 460n20. See also Peter Harrison's critical study of the terms "science" and "religion" that challenges the notion of a longstanding conflict between these concepts (2015).

102 At this point, the great variation in early modern Jesuit

art must also be acknowledged explicitly. Although I often consider works associated with this Society, it has for a long time been shown that "baroque" art was far from the only mode employed by Jesuits, who adapted to the tastes and methods of the areas where they proselytized. Moreover, just as Jesuits did not impose a singular mode, Catholic Reformation art more generally was not uniform. Kaufmann 2004, 245–46. Bailey 1999, 38–89. Howard Hibbard noted the eclectic artistic preferences of the reforming communities long ago (1972, 40).

103 Juan Caramuel 1678 [–1679], vol. 2, treatise 6, 13.

CHAPTER 1: THE CARDINAL'S PERSPECTIVE

1 For the will from December 1661, see ASR, Fondo Spada, vol. 364, A.3.M.71. Concerning this document, see Neppi 1983, 106. On different vantage points on the colonnade in Palazzo Spada, see, Cecili, Farroni, and Mancini 2022, 81.

2 The colonnade is about 29.5 feet (9 meters) in length, but it appears to extend about 124.7 feet (38 meters).

3 On the irregularity of other colonnades by Borromini, see Raspe 1994, 52.

4 On the *prospettiva*, see Neppi 1975, 175–87; Neppi 1983; and Capriotti and Roca De Amicis 2022. The colonnade is discussed in relation to baroque illusionism in Camerota 2003, 34–39. Righi's dates are unknown.

5 The Spada family's private archives entered the public domain in 1971 and this chapter draws on the foundational studies and transcriptions of primary source documents in this material authored by Lionello Neppi in 1975 and Minna Heimbürger Ravalli in 1977. I have checked their transcriptions against the original documents, many of which are translated into English here for the first time. Heimbürger Ravalli, 1977; and Neppi 1975.

6 On Gibbes, see Tabarrini 2016, 62–63.

7 ASV, Fondo Card. B. Spada, vol. 18, c. 689: "Non so se prima che V[ostra] S[ignoria] partisse di qua fusse cominciata una certa loggetta lavorata di stucchi con numero di colonne che il P[adre] Bitonto Augustiniano ha fatta qui in Casa di contro la stanza dove io soglio dare audienza; e veramente per cosa piccola che ella è, forma una prospettiva longhissima, onde ha dato occasione a un epigramma morale, ch'io le mando." Transcription from Neppi 1975, 280.

8 The suggestion that the letter is addressed to Cardinal Francesco Barberini is drawn from unpublished work on the Spada Colonnade by Joseph Connors, whom I thank for sharing his writings with me.

9 This translation is modified from one by Connors. BAV, Barb. lat. 1005, c. 102: "Mole sub exigua spectatur porticus ingens; / Cernitur in spatio semita longa brevi; / Quoque magis distant tanto maiora videntur / Quae sunt in proprio corpora parva loco. / Artis opus mirae; mundi fallentis imago; / Magna suis offert ipse pusillus, inops. / Magna, sed in speciem, capienti parvula fiunt; / Grandia sub coelo non nisi spectra manent." Transcription from Neppi 1975, 280.

10 I thank Giulio Pertile and Frederic Clark for discussing this poem with me.

11 Benardino spent 87,500 scudi to purchase and renovate Palazzo Spada. FSV, D.3.50. The construction of San Carlo alle Quattro Fontane (not counting the facade) cost only 18,939 scudi; by comparison, the Palazzo Borghese cost 275,000 scudi, and Borromini's work on San Giovanni in Laterano between 1646 and 1650 cost 127,563.06 scudi. See Connors 1980, 60.

12 The account that follows concerning the Spada family's social elevation draws on Cavallo and Storey 2013, 49–61.

13 Finocchiaro 1999, 17. ASR, FSV, vol. 463, cap. 3.

14 O'Malley 1993, 222–24.

15 For studies of Jesuit educational theater, see O'Malley 1993, 422n118.

16 Guazzo 1596, 30: "Che la conversatione civile, sia honesta, lodevole, & virtuosa." On civil conversation, see Findlen 1994, 104–46; Patrizi 1990. See ASR, Notai Tribunali, A.C. vol. 5933, f. 773^{r}. In seventeenth-century England, virtue marked the trustworthiness of a knowledge source, as Steven Shapin has argued. Shapin 1994. We can deduce from Guazzo's handbook that the same principle applied in early modern Italy.

17 Lionello Neppi (1975, 137) writes that Palazzo Spada's garden briefly featured a space in which Bernardino's grandchildren could engage in the aristocratic pastime of turning, which I will discuss further in chapter 6. Concerning Virgilio's efforts to refine Orazio's understanding of architecture, see Butler 1994, 63. This chapter builds on the work of historians of architecture, who have likewise stressed that Bernardino's lifelong building projects became crucial instruments for expanding the social prestige and grandeur of his family, by highlighting the importance of deformation in particular in increasing that prestige. Concerning the relationship between the expansion of Spada commissions and the family's quick social ascent, see Connors 1979, 194.

18 Bourdieu 1984.

19 In discussing the connections between collecting and power, I build on such pathbreaking studies as Findlen 1994.

20 See Cavallo and Storey 2013, 51.

21 Casanova 1987, 56–78.

22 On the military arts and the interest in this subject among nobles, see Long 2001, 192–209.

23 See "Anno 1633," in ASR, Fondo-Spada-Veralli, vol. 285. Cited in McPhee 2002, 310n34.

24 FSV, D.3.50. Neppi 1975, 289. On the Spada family fascination with building, see Connors 1979, 194.

25 See FSV, D.3.50; and transcription from Neppi 1975, 289. On staircases in Palazzo Spada both realized and unrealized, see Tabarrini 2016, 35–37, 48–54, and 175–79.

26 Abramson 2016, 12–16. On obsolescence, see also Cairns and Jacobs 2014, 103–33.

27 Trachtenberg 2010. Translation from Alberti 1988, 156.

28 On the imaginative power of building sites and the potential imbedded into ruins, in particular, see Hill 2019.

29 Altrini 2000, 1. For important religious and civil buildings in Rome, see Lombardi 1992, 548–49, 551–52; and Lombardi 1993.

30 On ephemeral festive architecture, see Dell'Arco and Carandini 1977–1978; and Dell'Arco 1997.

31 FSV, D.3.50. Transcription in Neppi 1975, 289. Virgilio valued frugality: when he made a list of qualities a patron should consider in choosing an architect, he included skill in preparing cost estimates in advance. Connors 1989, 82.

32 FSV, D.3.50: "Della quale compra hebbe gran gusto il P[ad]re Vergilio suo fratello, poiché sapendo l'humor peccante del Cardinale in fabbricare e variare tutte le cose hebbe a dire: *mi rallegro con V[ostra] E[minenza] che ha travato palazzo così finito, che non ci potrà porre un mattone di vantaggio, quando anche volesse.* ... Ma a pena entrato il Card[ina]le ad habitarvi, cominciò a chimerizarvi sopra, et ha continuato trenta e più anni a farvi lavorare, a segno che tutta l'economia in haver fatta la prima compra con qualche vantaggio ha giovato poco mentre oltre li scudi 31500 si trova haver speso fino al giorno della morte altri scudi 50000, e vi saranno debiti con maestranze forse per altri scudi 6000." Transcription from Neppi 1975, 289.

33 FSV, D.3.50: "Con soffitti dorati, e del primo oro, per quanto dicevano, venuto dall'Indie." Transcription from Neppi 1975, 289.

34 The literature on cabinets of curiosities is extensive. Notable contributions include Bredekamp 1995; Findlen 1994; and Pomian 1990.

35 As argued by Daston and Park 1998, 118 and 304. Aristotle, *Metaphysics*, 982b12–14.

36 See *De la Recherche de la vérité* (*The Search after Truth*, 1674–1675), in Malebranche 1963, vol. 2, 205. "Il n'y a rien de si difficile que de s'appliquer long-tems à une chose, lorsque ne l'admirant point, les esprits animaux ne se portent pas facilement aux endroits nécessaires pour se la representer. On a beau nous dire que nous soyons attentifs: nous ne pouvons pas l'être, ou nous ne pouvons pas l'être longtems; quoique d'ailleurs nous soyons persuadez d'une certaine persuasion abstraite ... que la chose mérite fort nôtre application. Il est nécessaire que nous trompions nôtre imagination pour réveiller nos esprits, et que nous nous representions d'une manière nouvelle le sujet que nous voulons méditer, afin d'exciter en nous quelque mouvement d'admiration." Translation from Malebranche 1997, 385–86.

37 On a painting by Niccolò Tornioli (1606–1651) that was displayed in Palazzo Spada, see Berger and Schechner 2021.

38 FSV, D.3.50. Neppi 1975, 289–90.

39 FSV, D.3.50. Neppi 1975, 290. On the collapse of the staircase, see Tabarrini 2016, 51–54.

40 FSV, D.3.50: "E non essendo sufficiente così gran sito alla vastità de suoi pensieri, et al prurito di fabricare, applicò l'animo ad ornare le muraglie d'una casa d'un vicino incontro la facciata del proprio palazzo." Transcription from Neppi 1975, 290.

41 As noted in Connors 1992, 58–62.

42 FSV, vol. 463, #24, para. 1: "È assioma ben ricevuto che il valore degli huomini si raccoglie dalle fabriche e dai matrimoni." Translation from McPhee 2002, 88.

43 Concerning baroque Roman palaces and how greeting guests worked to exhibit rank, see Waddy 1990, 3–13.

44 See the following handbooks, which are discussed by Waddy 1990: Evitascandolo 1598; da Bibbiena 1621; Lunadoro 1635; Tantouche 1623; and Leti 1675.

45 Concerning the popularity of the guidebook tradition in the seventeenth century, see Connors and Rice 1990, xviii.

46 See also Beckmann 1807–1809; Dumesnil 1865; Pescarzoli 1957; Schudt 1930; Schudt 1959; and Blennow and Rota 2019. For a guidebook that mentions Palazzo Spada, see Totti 1638, 210–11.

47 See Connors and Rice 1990. See also, for example, Martinelli 1650.

48 I draw here on the discussion of the progression of guests through Roman palaces in Waddy 1990, 4–5.

49 The letter is dated September 17, 1656. FSV, E.2.50: "Mercordì, se mal non mi ricordo, discorrendo col P[ad]re Abbate Ughelli nella mia camera dell'audienza incontro alla prospettiva del giardino segreto, con occasione ch'egli lodava la detta prospettiva io venni dicendogli il pensiero che qualche volta m'era passato per fantasia di farne una in qualche chiesa." Transcribed in Neppi 1975, 281.

50 A third entrance to the garden—the service entrance—was located behind a spiral staircase, beyond the site of the current ticket office for the Galleria Spada, likely at the location now occupied by a small elevator. We can deduce from various plans that by the mid-nineteenth century, the original doors in both the *Sala dei Rilievi* and the *Sala del Sole* had been replaced with windows. I thank Adriana Capriotti for her generous guidance on these matters.

51 This aspect of the tori is also noted by Roca De Amicis 2022, 42–43.

52 FSV, E.3.107: "Se io tratto di fabricare, si comincia col dire che non ce n'è bisogno, e poi si finisce col mettere in dubbio se basta; tanto ch'io spendo con poca gratia altrui e doppo speso non posso disporne con lor sodisfatione." Transcription from Neppi 1975, 128n33.

53 FSV, vol. 487, fol. 229, December 20, 1673. Cited in Tabarrini 2016, 8–9.

54 FSV, vol. 487, fol. 229, December 20, 1673: "E certo io posso affermare che il prurito di fabbricare l'ho grandissimo, ma l'intendere qualche cosa di questa professione me ne fa astenere più che non farei." Cited in Tabarrini 2016, 8–9.

55 FSV, D.3.50. Neppi 1975, 125n14.

56 Virgilio's mention that the painters came from the Dentoni workshop appears in a document titled *Descrizione delle stanze del Palazzo a Capo di Ferro fatta da Mons.r. Virgilio Spada* (*Description of the Rooms of the Palazzo Capodifero Done by Mr. Virgilio Spada*, 1662). See FSV, A.4.X.72. On the *Salone*, see Tita 2019. The room is known today as the *Sala di Pompeo* because of the ancient statue believed in the seventeenth century to represent Pompey (106–48 BCE) that stands in the space. It was also referred to in the seventeenth century as the *sala grande* or the *sala principale*.

57 FSV, D.3.50: " … facevano apparire rilievi nelle volte e fregi delle stanze dove non erano, a segno che molti per chiarirsene vollero sino con le scale salirvi, e toccarle." Transcription (with minor modification) from Neppi 1975, 125n14.

58 Pliny 1518, CCLXV[r]; and Pliny 1952, 306–311 (35.36.61–66).

59 FSV, D.3.50: " … furono chiamati da tutti quei principi di Lombardia, ma in Francia dal Re." Transcription from Neppi 1975, 125n14.

60 Tabarrini 2021, 25–31.

61 ASV, SS. Bologna, vol. 279, c. 239. Transcription from Neppi 1975, 125.

62 I thank Robert Kirk for sharing his knowledge of ornithology.

63 FSV, 747 (formerly FSV, F.1.9): " … per haver depinto la faciata del giardino fata con colone capitelli base et pilastri fato a bogni e l'ornamento alle finestre dalla testata di deta faciata con la cornice che gira intorno con li fianchi di pilastri fati a bogni et acomodato il cori cori sopra a deta faciata, sc[udi] 30." Transcription from Neppi 1975, 270.

64 FSV, 747. Neppi 1975, 272.

65 FSV, 747. Neppi 1975, 270. Further bills for fictive frescoes painted in 1650 and 1658 by Giovanni Battista Laurenti, whose dates are unknown, survive under the same call number.

66 FSV, D.3.50: "E s'ingegnò di ricevere da un vicino certo sito accanto ad un giardinetto segreto per fare in tal sito una prospettiva di materia conforme che haveva veduto per opra del Cavaliero Borromino architetto nell'oratione delle quaranthore" ("He managed to receive from a neighbor a certain site next to a secret small garden to create in that site the perspective, of a material conforming to the one that he had seen by the architect Cavaliero Borromino in the Forty Hours' Devotion"). Transcription in Neppi 1975, 290. Concerning the Forty Hours' Devotion, see Weil 1974, 218–48; Mochizuki 2019, 449–86; and Diez 1997, vol. 2, 84–97.

67 Tabarrini 2016, 14n57.

68 Weil 1974, 219.

69 For the pamphlet describing this *apparato*, see *Apparato delle Solenni Quarant'ore celebrate nel Giesù di Roma l'Anno del Giubileo 1650* 1650, 4–8. Gigli 1958, 351. Neppi 1975, 281n4; Dell'Arco 1997, 344–46; and Dell'Arco and Carandini, 1977–1978, 138–40. On the large group of laborers and artists who contributed to this *apparato*, see Horn 2019b, 245.

70 Gigli 1958, 351.

71 Bösel and Insolera 2010, 238–39.

72 Johnson 2017, 136–51. For other stage designs that employ accelerated perspective, see Furttenbach 1640, 59–70 and pls. 21–23; Sabbattini 1638, 44–48; and Troili 1672, 110–13. Cited in Baltrušaitis 1977, 5.

73 On the perspectival scenes, see Cafà 2004.

74 I thank E. J. Johnson for generously sharing this image with me. Scolari's dates are unknown.

75 Neppi 1975, 176.

76 The Spada family purchased the *jus patronatus* of San Paolo Maggiore in 1634. Heimbürger Ravalli 1977, 52–55 (esp. 54); and Neppi 1975, 183.

77 FSV, D.3.50: "Riuscì cosa molto vaga, ma il costo eccedé di assai il gusto." Transcription from Neppi 1975, 290.

78 Beyond "focusing devotion," Davies also associates perspective in tabernacles with such functions as "revealing the hidden," "enhancing the size of the holy," "distancing the heavenly," and "radiating holiness." Davies 2013, 898–921.

79 Davies 2013, 898–921.

80 For a discussion of how perspective focuses the gaze of the faithful, see Davies 2013, 898–921.

81 See the discussion of "enhancing the size of the holy," in Davies 2013, 910.

82 See the account of "revealing the hidden," in Davies 2013, 909–10.

83 See Davies 2013, 911–19.

84 For the connection between perspectival tabernacles and Bramante's architecture, see Davies 2013, 898–921.

85 The large-scale restoration of the church in the nineteenth century makes it hard to ascertain what the floor looked like in Bramante's time. As noted by Davies 2013, 916.

86 As argued by Davies 2013, 917. On the painting of the miraculous image of the Madonna, see McCall 2019, 1–53.

87 Marder 1997, 216–24. Lavin 1980, 93–98.

88 Neppi 1975, 177. Capriotti 2022, 28n26.

89 FSV, 747: "1653. Per haver dipinto un paese grande in tela quale è servito per modello nell'ingresso della Prospettiva nel giardino secreto, sc[udi] 6; per haver dato di colore verde alle spalliere del giardino finto in detta prospettiva quale fu dato doi volte, sc[udo] 1" ("1653. 6 scudi for having painted a large country scene on canvas which has served as a model at the entrance of the Perspective in the secret garden; 1 scudo for coloring in green, with two coats of paint, the back of the fake garden in the same Perspective"). Transcription from Neppi 1975, 177n16.

90 An inscription was once painted in gilt and enameled letters onto an area over the *prospettiva*; that inscription may well have contained lines from Bernardino's epigram. For the bill for gilding and enameling the letters, see FSV, 747 (transcribed in Neppi 1975, 178n22).

91 Harris 1990, 408.

92 FSV, vol. 463, chapter 2, line 23: "Ne voleva sempre a tavola tre, o quattro, e sempre gli eccitava con interrogazioni, e risposte equivoche per dargli occasione di ratiocinare" ("He always wanted three or four at the table and always excited them with equivocal questions and answers to give them an opportunity to reason"). This passage is quoted (with minor modifications) in Neppi, who does not include a reference to the archival source (1975, 127).

93 Neppi 1975, 181.

94 Bayle 1734, vol. 4, 58: "MAIGNAN (Emanuel) l'un des plus grans Philosophes du XVIIe Siecle."

95 FSV, D.3.50: "Nella volta d'una Galleria che trovò fatta vi fece fare un horologgio da ecelente matematico vedendosi le hore in luogo coperto e dove non giunge raggio di sole se non per reverbero, e cotal raggio in luogo di gnomone mostra l'hore." Transcribed (with minor modification) in Neppi 1975, 290.

96 Neppi 1975, 189–206. Weston, 2016a, 309–11. Weston 2016b, 120–23. Feist 2016, 271–86. For a technical analysis of the sundial, see Farroni 2019. See also Feist 2013, 16–73.

97 My thanks to Sara J. Schechner for sharing her erudite knowledge of sundials with me and for helping to draft this section.

98 The azimuth is the horizontal angle or compass bearing measured from north along the horizon. Different parts of Europe and different trades used different systems of hours to mark the passage of time. The system of common hours, also known as astronomical hours, divides the day into twenty-four equal hours, which are counted one to twelve starting at midnight and midday. The system of Italian hours counts equal hours from one to twenty-four beginning at sunset. The system of Babylonian hours counts equal hours from one to twenty-four beginning at sunrise. The system of unequal hours, also known as temporal hours or canonical hours, divides both the day and the night into twelve equal parts, each counted one to twelve, no matter what the season. By definition, the sixth hour of the day in the system of unequal hours in Europe and the West is at noon, and that of the night at midnight. Consequently, the daylight hours grow fat in the summer (when the night hours are thinner) and thinner in the winter (when the night hours are fatter). The hours of the day and night are equal in length only on the equinoxes.

99 In the system of planetary hours, a system of unequal hours, the seven classical planets are given rulership over particular hours of each day in the Aristotelian order of Saturn, Jupiter, Mars, the Sun, Venus, Mercury, and the moon. The ruling planet of the first hour of each weekday gives us the name of that day (Saturday, Sunday, Monday, etc.).

100 A volvelle is a rotating disk engraved with one or more concentric scales and sometimes notched or pierced with windows to reveal information underneath.

101 FSV, F.1.9: "Per aver dipinto sette putini coloriti di grandezza del naturale li quali fingano di reggere il panno di razza dove è dipinto l'orologgio." Transcribed in Neppi 1975, 275.

102 A reflective sundial had been produced by Nicolaus Copernicus (1473–1543) in the previous century in a castle in Olsztyn.

103 Maignan 1648.

104 Maignan 1648, sigs. +3^{v}-+4^{r}: "non modo Romam orbis caput, sed & cum capite orbem totum, communicatis tuae laudis ac verae gloriae radiis illustras: quod etiam nostrum Ordinem Minimorum, praeter alias plurimas religiosas familias, ipsa tuae illustris Protectionis praerogativa foves; inde est, ut sponte sua & nativo ingenio Pespectiva mea suo cum lucis radio triplici, te respiciat, atque ad eundem te perfugiat velut ad suum Solem."

105 Biblioteca Vallicelliana, Indice dei libri del P. Virgilio Spada, Z. 109, pp. 31 and 73; ASR, Inventario dei libri, Fondo Archivistico Spada Veralli, vol. 494, p. 36. Virgilio's personal library contained 850 printed books and 80 manuscripts. Finocchiaro 1999, 20–21, 117. See also Fantoni 1989, 156–66; Neppi 1975, 201–02; Feist 2016; and Rohr 1990, 5–11.

106 Saguens 1703, vol. 4, 11: "Inter aequissimos illos rerum ad artes, doctrinasque attinentium aestimatores eminebat titulo eoque multiplici purpurae, nobilitatis, peritiae, authoritatis, & summae in rebus Ecclesiasticis decernendis, administrandis, fortunandis experientiae Eminentissimus Cardinalis Spada totius Ordinis nostri amantissimus Protector, qui postquam a Maignano obtinuisset artefactum eiusdem generis praeclarum, quod in amplissimarum aedium suarum porticu splendissima delineari & appensis absolutissimae picturae tabellis exornari curavit; hanc rependere voluit magnificentiam omnibus profuturam, qua sumptibus cum effusa largitione suppeditatis Perspectivae Horariae Magnanaeae editionem acceleraret."

107 Maignan 1648, 392: "Non potuit etiam tabella haec aenea adeo exigua lineas omnes Astrolabii simul capere; quapropter ad vitandam confusionem multitudinis, praecipuae sunt exaratae, reliquae numero longe maiori sunt praetermissae." The author calls the device an astrolabe because the sundial is "astrolabe-like" 1) in its functions, and 2) in its "furniture," by which he means that the sundial has the same lines of the celestial sphere and local altazimuth coordinates that one finds on an astrolabe, but here they are projected onto a vaulted ceiling.

108 Consentino 1970, 212. For pathbreaking studies of Jesuit mathematical thesis booklets, see Collacciani and Roux 2017b, 89–113; Collacciani and Roux 2017a, 51–84; and Collacciani and Roux 2021, 79–139.

109 Lukács 1986, vol. 5, 374–77, 402: "Singulis aut alternis saltem mensibus ab aliquo auditorum magno philosophorum theologorumque conventu illustre problema mathematicum enodandum curet; posteaque, si videbitur, argumentandum." Cited and translated in Collacciani and Roux 2021, 79–139.

110 For sources on illustrated thesis prints, see Gieben 1993, 273–74n2; and Rice 1999, 165n1.

111 Henri 1639, 4: "Item quo artificio sit delineandum horologium solare in Aulae laqueari, ut horas indicet non styli umbra, sed exiguo orbe lucido, qui laqueare percurrat." Copies of this booklet are held in the BnF, Bibliothèque municipale (Lyon), and the Bibliothèque Mazarine (Paris).

112 De Cornouaille and Manchon 1638, 10: "HOROLOGIA varia exhibere tum ad indicandum quota sit hora hic, & ubivis gentium, tum ad dispensandum tempus; qualia sunt ea, quorum artificium pendet ex luce, vel umbra Solis, aut Lunae, imo & candelae si lubet." Copies of this booklet are held in the BnF, Bibliothèque municipale (Lyon), Bibliothèque Mazarine (Paris), and British Library (London).

113 On this work, see Chossat 1896, 235–37; Hornet 1982, 266. For potential traces of the sundial, see Oudet 1995, 79–94.

114 I thank Hélène Lo Iacono for kindly sharing her photographs of this instrument with me.

115 For a discussion of these epigrams, see Feist 2016, 280–81.

116 "Aemula naturae manus hic depingere coeles / Tentavit, Solem pingere non potuit. / Ne tamen incepto desit Sol / ipse labori / Seque suumque suo lumine pingit iter." Translation from Feist 2016, 281.

117 "Si quid in Orbe novum, si quid sit in Urbe vetustum, / Huius in hac specimen nobile cerne domo. / Dum diversa aliis miranda videbis in aulis: / O quales, dices, Orbis et Urbis opes! / Hic ubi diversas Sol ipse videbitur horas, / Et signare suam luce vibrante viam: / Dic, age, quis potuit Solem deducere coelo? / O quale est istud mentis et artis opus!" Maignan 1648, 396; Neppi 1975, 190. Translation (with slight modification) from Connors 1992, vol. 2, 36. Concerning another epigram by Bernardino that calls attention to the way the Sun enters his palace, see Feist 2016, 281; for the epigram, see Maignan 1648, 395.

118 On the identities of these figures, see Gagnaire 2003, 21–31. See also De Rosa 2019b, 128–29.

119 Niceron 1646, 179; and Kircher 1646, 649: "Hiusmodi Astrolabium anacampticum, non ita pridem in Palatio suo sane magnificentissimo, delineari curavit Eminentissimus Cardinalis Spada; quod revera uti opus est oppido ingeniosum, ita nullum non in summam admirationem rapit." See also Kircher 1635 (for Kircher's prior interest in reflecting sundials).

120 Kircher 1646, 649. Saguens 1703, 10.

121 See bill dated July 24, 1645, FSV, vol. 747 (previous call number F.1.9): "serve per vedere chi passa per strada à stare in stanza con fenestre chiuse." Cited in Neppi 1975, 275–76. Neppi cites this as F.1.9.

122 See bill dated February 24, 1644, in FSV, vol. 747, line 14 (formerly under call number F.1.9). Cited in Neppi 1975, 275. Neppi cites this as F.1.9.

123 Findlen 1994, 107.

124 As explained by Waddy 1990, 4–5.

CHAPTER 2: MINIM DEFORMATION AND REVELATION

1 For a classic study of anamorphoses, see Baltrušaitis 1977.

2 In a related argument, Jasmin Mersmann (2008, 29) has stressed how anamorphoses turn observers into visionaries.

3 According to Edgar R. Samuel, the skull may have been examined head-on with a correcting glass tube. If Samuel is correct, the skull may have appeared less skewed when observed via this device. Samuel 1963, 436–41.

4 Schott 1657, 101–69. For the term *deformation* and its variants, see Maignan 1648, 438 and 442; Niceron 1638, titlepage, evi[r], 49–52, 54–59, 66–67, 69, 70–71, 83, 88, 92, 94, 96, 98–99, and 105; and Bettini 1642, vol. 1, *Apiarium Quintum, Progymnasma Primum*, 3.

5 Translation from Niceron 2019, 66. Niceron 1652, 5: "Le mot de Mage n'est ny Grec, ny Latin, mais Persan; … il signifie en cette langue la mesme chose que le nom de Prophete, chez les Hebreux." Daniel J. Sheffield writes regarding the Old Persian word *maguš*, from which the word *magus* is derived: "In most of the Zoroastrian tradition the word is used for a Zoroastrian ritual expert and is conventionally translated as 'priest.'" Personal communication with the author, February 2024. I thank Sheffield for discussing this notion with me.

6 On seventeenth-century Jesuit and Minim anamorphoses, see Baltrušaitis 1977; De Rosa 2013; De Rosa 2019a; and Massey 2007. Throughout his landmark study of optical marvels, *La perspective curieuse* (1638), Niceron describes viewing anamorphic images as a transition from a state of confusion to one of clarity. Consider, for instance, his title for a section on producing anamorphic images: "To give the method to describe all sorts of figures … that seem confused in appearance and from a certain point represent perfectly a proposed object." Niceron 1638, 52: "Donner la methode de descrire toutes sortes de figures … qui semblent confuses en apparence & d'un certain poinct representent parfaictement un objet proposé."

7 Milan, Biblioteca Ambrosiana, Codex Atlanticus, fol. 35[v-a]. It has been estimated that the Codex dates from c. 1478–1518. Jean Paul Richter suggests the drawings were likely produced around 1510. See Richter 1977, vol. 2, 384; and Pedretti 1957, 68–76.

8 For the dating of the *Considerazioni*, see Favaro 1890–1909, vol. 9, 7–28 and 12–13. See also Panofsky 1954; and Panofsky 1956, 3–15.

9 Translation (with slight modifications) from Panofsky 1954, 13. Galilei 1890–1909, vol. 9, 129–30: "Riguardate in scorcio da un luogo determinato mostrino una figura umana … vedute in faccia e come naturalmente e comunemente si guardano le altre pitture, altro non rappresentano che una confusa e inordinata mescolanza di linee e di colori, dalla quale anco si potriano malamente raccapezare imagini di fiumi o sentier tortuosi, ignude spiaggie, nugoli o stranissime chimere."

10 Panofsky 1956, 9.

11 For evidence that Galileo knew of Kepler's work on elliptical planetary orbits, see Drake 1978, 190.

12 Locke 1975, 365.

13 Locke 1975, 365.

14 Locke 1975, 365.

15 Locke 1975, 366.

16 Locke 1975, 366.

17 This paragraph draws on the discussion in Stuart 2013, 236.

18 Locke 1975, 366.

19 In 1600, thirty-eight Minim houses existed in France; that number rose to 112 by 1623. This exponential growth reflects the state's support of Catholicism and religious orders in the wake of the Wars of Religion. Whitmore 1967, 14. Kent 2023, 62.

20 On the Minims, see Whitmore 1967.

21 See Whitmore 1967, 216.

22 On Saint Francis's tumor, see du Vivier 1622, 7–8;

and Dondé 1671, 10–12. See also Cousinié 1994, 48n51.

23 On the lunettes, several of which were repainted in later centuries, see Moretti 2016. In 1584, the lunettes' designs were copied in engravings by Giovanni Ambrogio Brambilla (active c. 1569–c. 1595).

24 On the convent and church, see Di Matteo 2015.

25 Lorrain was buried in the church of Trinità dei Monti.

26 Dondé's dates are unknown.

27 Dondé 1664, plate 7, n. 20: "Sainct Francois de Paule, avec Sa Salive, forme la bouche et les yeux, a un Enfant monstreux de naißance." The inscription recalls John 9:6–7: "When he had said these things, he spat on the ground, and made clay of the spittle, and spread the clay on his eyes, And said to him: Go, wash in the pool of Siloe, which is interpreted, Sent. He went therefore, and washed, and he came seeing."

28 Dondé 1664, plate 8, n. 23: "S. F. de Paule donne la veuë a une fille aveugle née, außi tost qu'il lui eut appliqué d'une herbe sur les yeux."

29 Dondé 1664, plate 12, n. 41: "S. F. de Paule faisant oraison, est rendu invisible a cinquante hommes d'armes que le Roi de Naples avoit envoyez pour le prendre."

30 Dondé 1664, "Avis sur le dessein de cet ouvrage": "Chacune de ces Images est un Prédicateur muet qui nous exhorte a la vertu, puis qu'elles ne nous representent autre chose, que l'exemple ou la recompense de la mesme vertu: si quelqu'un y est porté en les considerant, j'aurai la fin de mon travail, qui n'est autre que la gloire de Dieu dans l'exaltation et l'imitation de S. F. de Paule." The following financial details are presented under the year 1662 of the *Annales des Minimes de la Province de France* … (MS Mazarine No. 2429): "We gave 400 livres to friar Dondé, the alms collector of the convent, to engrave the plates of the miracles of Saint Francis of Paola … then we gave him 100 livres in 1668 for the same subject" ("On a donné 400 livres au Fr. Dondé, Quêteur du Couvent pour graver les planches des Miracles de Saint François de Paule … plus on lui a donné 100 livres en 1668 pour le même sujet"). Cited by Whitmore 1967, 247.

31 Camille 1985, 26–49.

32 On devotional prints produced in the sixteenth and seventeenth centuries to organize and deepen meditation on the soul's relation to God, see Melion 2009; and Melion, Dekoninck, and Guiderdoni-Bruslé 2012.

33 The portrait of Niceron in Figure 64 was sold in a Christie's auction in New York on April 23, 2021.

34 I thank Kévin Daligault for sharing his MA dissertation, "Les commandes de peintures dans les couvents Minimes de la Place Royale, de Vincennes et de Passy (XVIe-XVIIe s.)." For another anamorphosis of a saint, see the *Saint Jerome*, in the Central Museum, Utrecht.

35 The anamorphosis of Saint Francis, along with three others by Niceron—two of King Louis XIII (1601–1643) and a third showing a scene of married life—were acquired by the Italian Ministero dell'Educazione in 1937 and are held today in the Palazzo Barberini in Rome. Their inventory numbers are 1953 (*Louis XIII Before a Crucifix*), 1954 (*Portrait of Louis XIII*), 1955 (*Saint Francis of Paola*), and 1956 (*Scene of Married Life*). On the acquisition of these works, see Onori and Adamo 1989, 112–14; and Onori and Adamo 2007, 296. Concerning these paintings and two further specular anamorphoses (now lost) showing Henry IV (1553–1610) and Louis XIII, see De Rosa 2013, 10 and 70 notes 40, 41, and 42.

36 The print is signed "Friar Jean-Francois Niceron was designing [this]." ("F. Ioannes Franciscus Niceron delineabat.") Niceron 1638, table 19. The engraver is uncredited.

37 Niceron 1638, table 19: "le petit cercle FGHI est la grosseur du Cylindre[;] Et le grand cercle KLMNO &c represente sa base."

38 Pathbreaking work has been conducted on these frescoes by Italian scholars, led by Agostino De Rosa. I am grateful to De Rosa for his generosity in sharing with me photographs of frescoes from Trinità dei Monti. On the restoration, see the Appendix in Fratini and Moriconi 2010.

39 Terki 2006, 22.

40 On the nineteenth-century color additions, see Di Matteo 2015, 70 and 223.

41 For the Minim regulation that ordinarily prevented women from entering the convent, see Francis of Paola 1632, 20.

42 On the fresco of Saint Francis, see Terki 2006; Liva 2019, 76–94; and Julien 2005, 45–71.

43 Terki 2006, 22.

44 Maignan's Saint Francis is based on either Charles Mellin's (c. 1597–1649) *Saint François de Paule en prière* (*Saint Francis of Paola in Prayer*, 1597), now in the Musée Lorrain, Nancy, or Charles Audran's (1594–1674) engraving after Mellin's painting made in 1630. As noted in De Rosa 2019a, 73n16.

45 Schott 1657, 138: "Vidi saepissime Romae, aliisque monstravi, in Monte Pincio, in celebri conventu Fratrum Minimorum … dissipatam in longissimo cuiusdam ambulacri muro … effigiem S. Francisci de Paula, factam a doctissimo dicti ordinis Viro Emanuele Magnano, ea arte, ut ex solo determinato loco in ambulacri extremitate constituto appareat, extra vero illum locum, & praesertim ex adverso

inspecta refert arborum frondes, homines, animalia, aliaque similia."

46 Schott 1657, 147: "Manus … curavit inter productiores deformatae imaginis tractus depingi, minutioribus figuris, S. Francisci de Paula insigniora virtutum & operum monumenta, ut hac iucunda fallacia intuens ex adverso detineatur, nec quid suspicetur inibi ulterius esse, quod ars occultaverit."

47 Maignan 1648, 442: " … in iis monticulis, viis, aedificiis, &c. levi penicillo minutiores figurae pingantur, quae in prospectu sit, & coram cernantur: hae namque iucunda fallacia intuentem detinent, ne quid addubitet inibi ulterius esse, quod ars occultaverit. Ego in hac a me deformata, Santissimi Francisci nostri effigie; eiusdem Patris insigniora virtutum, & operum monimenta, minutis huiusmodi figuris pingenda curavi."

48 Dondé 1664, plate 12, n. 37. Regio 1582, 54^{v}. Gabriella Liva likewise discusses this detail in relation to Dondé's engraving (2019, 83).

49 Dondé 1664, plate 12, n. 37: "S. F. de Paule, paße le destroit de Messine sur son monteau; l'entrée du vaißeau lui ayant esté refusée, a cause de sa pauvreté."

50 Maignan 1648, 444: "Tota huius operis, ad visum decipiendum compositi … "

51 A number of documents exist pertaining to this unrealized project. A surviving letter addressed to Cardinal Pamphilj containing a proposal for the villa with marginal annotations said to be in Virgilio's hand; it describes the villa as a "study of applied mathematics" (*uno studio di matematica pratica*). For the letter, see BAV, Vat. Lat. 11258, fol. 14^{r-v}. Regarding the claim that the letter is annotated by Virgilio, see Camerota 2000a, 297; and Camerota 2000b, 612. Architectural drawings for the villa by amateur and professional hands are held in the BAV and the ASR; although scholars disagree about their attributions, Heimbürger Ravalli and Connors ascribe an amateur plan to Virgilio. For the drawings, see BAV, Vat. lat. 11257, fols. 201^{v} and 202^{r}; Vat. lat. 11258, f. 15; and FSV, vol. 551, p. 375. Whereas Heimbürger Ravalli supports the attribution of the professional sketches to Girolamo Rainaldi (1570–1655)—first suggested by Heinrich Thelen—Camerota follows Portoghesi in ascribing the professional drawings to Borromini instead. See Heimbürger Ravalli 1977, 271 and figs. 182–85; Camerota 2000b, fig. 3; Portoghesi 2000, 122–29; and Connors 1980, 136n15. Two copies of another document elaborating the project, titled "Matematica per adornare il Giardino dell'Em.mo Sig. Card.le Pamfilio" ("Mathematics for decorating the garden of the most eminent Cardinal Pamphilj"), are found in Virgilio's papers in the ASR. One copy appears in a volume that Virgilio titled "Miscellanea de negotii passati per mani mie sub Innocentio PP X": ASR, Archivio Spada, vol. 235, pp. 631–34. The second is held in a collection of Virgilio's writings on architecture, among other topics, titled "Libro di scritture diverse fatte da me Virgilio Spada in diversi tempi, et occasioni distinte in quattro parti, cioè Spirituale, Ecconomica, Architettonica, Varia": ASR, Archivio Spada, vol. 186, pp. 1083–89. Cited and discussed in Heimbürger Ravalli 1977, 270–74, 272nn84–85. A Latin version of this text, titled "Mathematica Pamphilianos hortos exornam," survives alongside one of the copies in Italian. ASR, Archivio Spada, vol. 235, pp. 627–30. Cited in Filippo Camerota 2000a, 310n2. See also Portoghesi 2000, 122–29. Filippo Camerota has argued in his extensive study of the villa project that this Latin manuscript was written before the Italian versions, because the Italian reads as an imperfect translation of the Latin; Camerota ascribes the "Mathematica Pamphilianos hortos exornam" to Maignan and matches the handwriting of the Latin manuscript to other manuscripts attributed to Maignan in the Bibliothèque Municipale de Toulouse. Camerota 2000a, 297. For an imagined digital reconstruction of this project, see Bortot 2020, 118–254.

52 Connors dates the project to 1644–45; Mirka Beneš dates it to November 14, 1644 through mid-February 1645. Connors 1980, 136n15; Beneš 1989, 594–95.

53 ASR, Archivio Spada, vol. 235, pp. 627–30: "In utroque latere Orientali et Occidentali aulae maioris depingetur arte Optica, effigies hinc v.g. S.ti Petri, inde ipsius S.tiss.i D.ni vel alia ut placuerit, tali quidem modo ut existentibus in ipsa aula tales effigies non appareant sed aliud quid ad placitum, appareant vero existentibus extra aulam in porticu septentrionali et per foraminulum ad hoc praeparatum spectantibus. etc." Translation and transcription from Camerota 2000b, 616–17.

54 On Saint Peter and Roman Catholic claims to authority, see Bruhn 2002, 36–37.

55 ASR, Archivio Spada, vol. 235, pp. 627–30. See also Camerota 2000b, 618.

56 Niceron 1638, 115.

57 An important model for Daniele's painting (though not for the optical effect cited by Niceron) must have been Rosso Fiorentino's (1495–1540) *Deposition from the Cross* (1521), which likewise features an enormous cross; he could have studied this painting in his hometown in the Chapel of the Campagnia della Croce di Giorno in San Francesco. Valone 1990,

80. Smith 1976, 67–70. The *Descent from the Cross* no longer appears in its original location, in the third chapel on the left in Santissima Trinità dei Monti; it is now in the second chapel on the left. On the fresco, which was restored to some controversy, see Romani 2003, 52–53n69; and Albers 2005, 42–47.

58 Niceron 1638, 72: "Un tableau peint à frais, en une Chapelle de nostre Convent de la Trinité du Mont Pincius à Rome, auquel est representée une descente de Croix, ou le Christ, qui en est la principale figure, est tellement disposé, qu'estant veu du costé gauche, il semble couché & incliné sur le travers du tableau, & son pied droit faire une faillie du mesme costé; & veu de l'autre costé, tout son corps paroist presque droit, beaucoup plus dans le racourcissement, & ce pied qui paroissoit faire sa faillie du costé gauche, semble avancer vers le droit." Translation (with slight modification) from Baltrušaitis 1977, 49.

59 The refectory paintings were restored from 2001 to 2002. Bösel and Insolera 2010. On the attribution to Pozzo, see Di Matteo 2015, 77.

60 McCormick 1990.

61 Di Matteo 2015, 234.

62 For monuments created outside Trinità dei Monti to celebrate the births of Spanish and French heirs, the eradication of the Calvinist Church, and the healing of Louis XIV, see Dell'Arco 1997, 274, 306–08, 518–21, and 532–34.

63 Dell'Arco 1997, 407–12.

64 Bernini's sketches for the church's dolphin decoration survive in three autograph drawings in Leipzig's Museum der bildenden Künste.

65 Translation from Niceron 2019, 62. Niceron 1652, 2: "L'Astronomie l'appelle aussi à son secours, pour bien juger de la hauteur, & du mouvement des Planetes, par le moyen des Astrolabes, & des autres instrumens qui conduisent le rayon visuel"; and "La Philosophie naturelle verifie la plus part de ses experiences par son moyen."

66 This sundial is cited by Di Matteo 2015, 227.

67 As stressed by Kent 2023, 64–66.

68 Hrodej established the dating of these journeys, which had been misreported by earlier authors (1997, 100).

69 According to José Beltrán, the papers amount to over 9,100 pages. Beltrán 2017, 6.

70 As estimated by Pietsch 2017, 34. The Muséum national d'histoire naturelle now holds most of Plumier's archive, although some documents are also kept in the BnF and elsewhere. See, in particular, Muséum national d'Histoire naturelle, Direction générale des collections, Bibliothèque centrale, ms Res 1-37, 1176, 1335, 2753, 3338, 3355, 5031; BnF, Bibliothèque de l'Arsenal, ms 2502, 2875, 3078, 3104; BnF, Département des Estampes, Est JD-18-FOL; RES YE-27-PET FOL; and RES YE-1.

71 Lister 1699, 73. Emphasis in original.

72 See Muséum national d'Histoire naturelle, Direction générale des collections, Bibliothèque centrale, ms Res 37, fols. 12 and 14.

73 Although Plumier is credited with designing this print, the engraver remains unacknowledged. The same engraving is integrated into Plumier 1705, plate 51.

74 Lefèvre d'Ormesson 1860, vol. 1, 333: "Le dimanche 26 novembre … aux Minimes, où le père Niceron me fit voir une carte faite à Bruxelles d'un monde dedans la lune, où toutes les terres, montagnes, isles, estoient nommées des noms des rois et grands de ce siècle, et fort proche celui de la reine de celui de Mazarin." As first reported by Whitmore 1967, 160–61.

75 Whitmore 1967, 161. See also De Rosa 2013, 39–41. There are two impressions of this print in the département des Cartes et plans of the BnF (under call numbers GE D-17925 and GE D-12817). I have not been able to find a connection between these impressions and the Minim convent in Paris. Catherine Hofmann, email communication with the author, February 19, 2020.

76 Translation from Galilei 1989, 42. Galilei 1890–1909, vol. 3.1, 62–63: " … non perpolitam … sed, contra, inaequalem, asperam, cavitatibus tumoribusque confertam, non secus ac ipsiusmet Telluris facies, quae montium iugis valliumque profunditatibus hinc inde distinguitur."

CHAPTER 3: JESUIT DEFORMATION AND MEDITATION

1 Translation (modified) from Pascal 1966, 74–75 (#21). Pascal 1954, 205: "Si on considère son ouvrage incontinent après l'avoir fait, on en est encore tout prévenu; si trop longtemps après, on [n']y entre plus. Ainsi les tableaux, vus de trop loin et de trop près. Et il n'y a qu'un point indivisible qui soit le véritable lieu: les autres sont trop près, trop loin, trop haut ou trop bas. La perspective l'assigne dans l'art de la peinture. Mais dans la vérité et dans la morale qui l'assignera?" See also, Damisch 1994, 49.

2 See "The Purposes and Ministries of the Society of Jesus According to the Official Documents, 1539–50," in O'Malley 2006, xxxiv–xxxvi (xxxiv–xxxv).

3 Much has been written on the Jesuit investment in images and architecture, and the tremendous diversity of Jesuit art and architecture is well known. Notable studies on Jesuit imagery include works by Walter S. Melion, Alexander Gauvin Bailey, Ralph Dekoninck, Christine Göttler, Mia Mochizuki, Jeffrey Chipps Smith, Jeffrey Muller, and Evonne Levy. For an introduction to Jesuit image theory, see Melion 2016, 1–49.

4 See Ignatius of Loyola, 1548, Niii[r]-Oii[v]. Note that Ignatius uses the verb *discernere* in this passage. On the discernment of spirits in Ignatius, see Pegon 1957, vol. 3, cols. 1266–81 (1266–75). On the rules for discernment, see O'Malley 1993, 41–43.

5 Ignatius of Loyola, 1548, Di[v]: "Primum praeludium est, ratio quaedam componendi loci, pro qua notandum est, quod in quavis meditatione, sive contemplatione, de re corporea, ut puta de Christo, Effingendus erit nobis, secundum visionem quandam imaginariam, locus corporeus, id quod contemplamur, repraesentans, veluti templum, aut mons, in quo reperiamus Christum Jesum, vel Mariam Virginem, et caetera quae spectant, ad contemplationis nostrae argumentum." Translation from Ignatius of Loyola 1992, 40.

6 Sucquet 1630, 498: "Juvare hae poterunt omnes in ariditate animi, simpliciores vero semper ad attentionem utilioremque meditationis usum. … quae post Adae peccatum omnino effrenis, facit animam vagam toto orbe discursare, nec piis cogitationibus inhaerere permittit: ut vero illa coerceatur piis cogitationibus & imaginibus (ut ita loquar) eam affigere expedit: quod cum adeo S. Ignatio parenti nostro probaretur, mandavit P. Natali, in hanc rem incumberet, ut vita Christi imaginibus pulcris expressa, meditantium insigni commodo ederetur. Idque quam verum sit experientia docet; & sacrosancta Ecclesia S. Spiritu edocta, in qua plurimus imaginum usus est, sua auctoritate confirmat." Translation from Freedberg 1989, 185.

7 On *curiositas*, see Carruthers 1998, 82–94.

8 St. Augustine, *Confessions*, 10.35.

9 Nadal 1595.

10 On the *Adnotationes et meditationes*, see Melion 2003, 1–96.

11 My account of this image draws on Melion 2003, 23.

12 Translation from Melion 2003, 26. Nadal 1595, 77: "Adeste omnes quotquot luce orationis, & rerum spiritualium cognitione estis destituti, audite Evangelicum caecum, aperientur oculi cordis vestri."

13 Translation from Melion 2003, 26. Nadal 1595, 77: "Audio quidem te, sentio tua mysteria, sed Domine non video, lux mihi deest celestis, eam peto o lux infinita. Intelligo benigne IESU, respicio laetum lucis tuae radium: te sequar, te laudabo in aeternum."

14 My discussion of this image draws on Melion 2003, 26–27.

15 See Melion 2003, 27.

16 As argued by Melion and Dekoninck 2019, 521.

17 On this emblem, see Rice 2007, 200.

18 On *enargeia*, see, for instance, Ginzburg 1985, 3–19; Vickers 1983, 69–102; and Preston 2007, 115–29.

19 Translation by W. Rhys Roberts in Aristotle 1984, vol. 2, 2252 (*Rhetoric* 1411b25).

20 Translation by D. A. Russell in Quintilian 2014, vol. 3, 374–75 (8.3.62).

21 See Melion 2003, 1–96.

22 Nadal 1595, 577–82. See Melion 2001, 52–55.

23 Ignatius founded the Society with the aim of active ministry. Concerning mobility in ministry and evangelization, see O'Malley 2013, 147–64.

24 The artists, carvers, and printers responsible for the woodblock prints in these Chinese editions remain unidentified. See Spence 1984, 59–64, 128–32; Chen 2004; Reed and Demattè 2007, 168–71, cat. 13; Criveller 2003, vol. 2, 437–60; Kleutghen 2015, 36–40; and Porras 2023, 17–50.

25 See, for instance, *imagines* XV, XXX, and XXXI in the *Via vitae aeternae*.

26 On representations of the *Tabula Cebetis*, see Schleier 1973. For early seventeenth-century philosophical thesis prints that feature pathways to knowledge, see Berger 2017. Concerning botanical gardens and other spaces of learning, see Findlen 2006, 272–89.

27 On this image, see Berger 2014, 343–66.

28 Campbell 2003.

29 See Gage 2008, 1177–83.

30 Giambullari 1549, 9: "Come non meno per servire alla sanità del corpo, che alla recreazione della mente." Cited and translated in Gage 2008, 1178.

31 As noted by Gage 2008, 1182. See also Wycherley 1962, 2–21.

32 Laertius 1545, 98: "Elesse ne'l Liceo il peripato ed ivi … passegiando disputava e parlava de la filosofia con suoi discepoli" ("He chose in the Lyceum the colonnade and walking there … disputed and spoke of philosophy with his disciples"). Cited by Gage 2008, 1182.

33 Cited and translated by Gage 2008, 1183n77. Mercuriale 1601, 37: " … platanos Athenis in Academiae ambulatione celebratas fuisse. …" See also Siraisi 2003, 231–51.

34 I thank Bernadette Russo-Amoros and Jacques Hancy for making it possible for me to view this painting in person.

35 Julien argues the work was produced before 1714, as it features graffiti with this date. Julien 2000, 19.

36 During the Last Supper, Christ foretold that Peter would betray him three times before the rooster crowed the following morning; the same gospel accounts subsequently report that Peter did in fact deny he knew Jesus. Gospel of Matthew 26:34; Gospel of Mark 14:30; and Gospel of Luke 22:34.

37 Henri 1639, 5: "EXPLICARE quo sint artificio delineandae imagines illae, quae in toto longissimae porticus laqueari, aut pariete, imo & pavimento depictae, etsi sint deformes, & monstris similes ob immoderatam longitudinem retenta vera latitudine; si tamen ex destinato loco, ut ex aditu spectentur, appareant perfectae, ac iustam hominis effigiem repraesentent."

38 BnF, Département des manuscrits, Ms Lat. 17862, fols. 839bis[v], 843bis[v], and 845bis[r], in-4. Mercier's notes are cited in Collacciani and Roux 2017b, 104; and Le Dividich 2003, 61.

39 Barthes 1989, 66.

40 Translation (with minor modification) by Golvers 1993, 115. Verbiest 1687, 76: "In horto domus nostrae PeKinensis, qui quatuor muris circumclauditur, P. Philippus Grimaldi in singulis muris suam imaginem, quae juxta longitudinem parietis usque ad 50. circiter pedes excurrit, juxta leges opticas ita delineavit, ut ex adverso quidem spectata nil nisi montes, silvas, venationes &c. exhibeat; si quis autem ex puncto determinato aspiciat, figuram humanam, aut unius dumtaxat animalis exactissime delineatam oculis objiciat. Singulas hasce imagines Imperator, dum ad domum nostram divertit, diu multumque contemplatus est, & adhuc quotidie Magnates & Mandarini praecipui ad haec & eiusmodi alia videnda, domum nostram frequentant, nec est aliquis, qui non miretur artem, praesertim cum facies parietum in tantam longitudinem excurrant, & valde irregulares sint, atque aliquae ex iis sint portis & fenestris interruptae, adeo ut imagines multum distrahant & interrumpant, etsi ex adverso tantum spectentur." The emperor visited the residence on July 12, 1675. Golvers 1993, 171n5.

41 Du Halde's compilation is drawn from the printed works and manuscripts of Jesuit missionaries. Du Halde 1736, vol. 3, 333: "Le P. Grimaldi donna un autre spectacle des merveilles de l'Optique dans le jardin des Jésuites de *Peking*, qui étonna fort tous les Grands de L'Empire. Il fit sur les quatre murailles quatre figures humaines, chacune de la longueur de la muraille qui étoit de cinquante pieds. Comme il avoit parfaitement gardé les règles de l'Optique, on n'y voyoit de front que des montagnes, des forêts, des chasses, & autres choses de cette nature. Mais d'un certain point on y appercevoit la figure d'un homme bien fait & bien proportionné. / L'Empereur honora la Maison des Jésuites de sa présence, & considéra ces figures fort longtems & avec admiration. Les Grands & les principaux Mandarins qui y venoient en foule, étoient dans la même surprise. Mais ce qui les frappoit davantage, c'étoit de voir des figures si régulieres & si exactes sur des murailles très-irrégulieres & entrecoupées de plusieurs portes & de fenêtres./
Il seroit trop long de rapporter toutes les figures tracées confusément, & que l'on voyoit distinctement d'un certain point, ou que l'on redressoit avec des miroirs coniques, cylindriques, pyramidaux, & tant d'autres prodiges de l'Optique que le P. Grimaldi présentoit aux plus beaux esprits de la Chine, & qui attiroient également leur surprise & leur admiration." Also cited in Musillo 2016, 110.

42 Gell 1992, 40–63.

43 Gell 1992, 44.

44 Gell 1992, 52.

45 Concerning the essential role of the mathematical sciences in the spread of the Catholic faith, see Golvers 1993, 21–23.

46 Transcribed and translated in Golvers 1999, 470 and 470n103: "Si non adornat … R.V. porticum illam suae domus Xamhaianae perspectivis et aliis curiositatibus, sicut ego hic feci, arguam defectum aliquem zeli."

47 Concerning the interrelations between optics and religion later in the seventeenth century, Martin Kemp has argued that Pozzo conceived the Sant'Ignazio nave's perspective construction in a precise metaphoric fashion; Thomas Frangenberg says the evidence does not support this interpretation. Kemp 1987, 262. Frangenberg 2000, 100n32. On the vault frescoes in Sant'Ignazio, see also Horn 2019a, 315–23.

48 On the Sant'Ignazio fictive dome, see Wilberg 1996, 215–23; Horn 2019a, 309–15.

49 See Frangenberg 2000, 94n9. Pascoli 1981, 696. See also Horn 2019a, 322. Massey 2021, 237–38.

50 See the following heading in the *Perspectiva horaria* concerning how to create anamorphoses on a wall: "Mirum, ac fidelissimum artificium tradere, quo expressa coloribus in tabella effigies, potest in superficie quacunque parietis, fornicis &c. facillime, & expeditissime deformari tali modo, ut procul, atque ex assignato puncto in latus prospecta, optice reformetur; appareatque nitida, concinna, & prototypo simillima: propius autem, & ex adverso inspecta dissipetur; & aliud quippam a se distinctum, belle tamen expressum ostendat" ("To relate the amazing and very reliable technique by which a color image produced on a panel can most easily and readily be deformed

onto any surface of a wall, vault, etc. in such a way that seen sideways at a distance and from a certain point, it may be optically reformed and appear neat, orderly, and very like its prototype, but seen up close and straight on it is scattered and shows something else, different from itself but neatly ordered"). Maignan 1648, 438. Translation (with minor modification) from Massey 2007, 153n89. Like Pozzo's dome, which even from zones beyond the marble disk still reads as a dome, it is an exaggeration to say that the anamorphic wall paintings of Maignan and Niceron resolve into coherent images only from a single point; rather, these works resolve into relative coherence when examined from one general area of the corridor and then as observers move forward they dissolve into incoherence followed by another relatively coherent image that comes into view once the observer moves into another general area at the far end of the corridor.

51 Vermeir has also emphasized that anamorphoses show a hidden order and a divine communication, if considered from a Catholic viewpoint (2005, 145). Relevant here is Vittorio de Feo's account that Pozzo's work would activate in the observer a series of concatenated psychological experiences that lead toward belief (1988, 13). Concerning the problem of a single vantage point in the observation of Pozzo's art, see also Burda-Stengel 2001. On perspective and the region of the visionary, see Panofsky 1980, 126.

52 Horn 2019b; and Horn 2019a.

53 Horn 2019b, 246.

54 Massey 2021, 237–38.

55 Massey 2021, 238.

56 Concerning the rooms on the first floor of the convent, see Di Matteo 2015, 123.

57 Pozzo 1702, vol. 1, text accompanying figure 101: "E persuadetevi, che simili opere, accioche possino facilmente ingannar l'occhio devono havere un punto stabile, e determinato, onde siano rimirate, accioche non appariscano al risguardante quelle deformità, e storcimenti, che la curvità & irregolarità delle volte suole far nascere." Cited and translated in Frangenberg 2000, 104. On the single view position in Pozzo, see Camerota 2011, 25–36.

58 O'Malley 2013, 147–64.

59 For Pozzo's discussion of the nave fresco perspective, see Pozzo 1693, vol. 1, books 1 and 2. An amended edition of this text appeared in Rome in 1702. See also Pozzo's letter to the Prince of Liechtenstein: Pozzo 1694; transcribed in Horn 2019a, 389–90.

60 Pozzo 1702, vol. 1, "Al lettore, Studioso di Prospettiva": "Cominciate dunque o mio Lettore allegramente il vostro lavoro; con risolutione di tirar sempre tutte le linee delle vostre operationi al vero punto dell'occhio che è la gloria Divina."

61 *Lettres édifiantes et curieuses, escrites des Missions Étrangères* 1819, vol. 10, 3: "Le plafond est tout à fait peint. Il est divisé en trois parties: le milieu représente un dôme tout ouvert, d'une riche architecture; ce sont des colonnes de marbre qui portent un rang d'arcades surmonté d'une belle balustrade ... on voit au-dessus le Père éternel assis dans les nues sur un groupe d'Anges, et tenant le monde en sa main." The letter is dated August 20, 1704.

62 Finlay 2007, 172–73.

63 Kircher 1646, 808: "Locus est ad Mare Australe vulgo del Zur, dictus Arauco, ubi mons Smaragdis, Turchesiis, similibusque pretiosis lapidibus ita refertus est, ut eminus se variorum huiusmodi lapidum nativo splendore depictum non sine intuentium voluptate spectandum praebeat. In huius montis concavo, ex certo tamen, & constituto puncto, Imago Beatissimae Virginis cum filio in brachiis tanta colorum varietate depicta cernitur, ut nemo sibi fere id ludentis naturae opus persuadere possit, sed omnes apparitionem divinam mordicus teneant: mirum enim videbatur indigenis spectaculum, cum illud non nisi ex certo loco sese conspiciendum praeberet; a quo si vel minimum recesseris, iam evanescente figura, praeter rupes, & praecipitia nihil amplius te videre existimes, ut proinde locus cum tempore mirum in modum frequentari fuerit solitus" ("There is a place near the Southern Sea commonly known as del Zur, called Arauco, where the mountain is so full of emeralds, turquoises, and similar precious stones, that from afar it shows itself painted with the native splendor of the various stones of this kind, not without the pleasure of the onlookers gazing (at it). In the hollow of this mountain, however, from a certain and established point, the Image of the Blessed Virgin with her Son in her arms is seen painted in such a variety of colors that almost no one can convince himself that it is the work of a playful nature, but all hold the divine apparition in check: for the spectacle seemed strange to the natives since it could only be seen from a certain place; from which, if you withdraw even a little, the shape is already disappearing, you think you can see nothing more except rocks and precipices, so that consequently the place has become accustomed to be strangely frequented with time"). Also discussed in Väliaho 2022, 49.

64 Kircher 1646, 808: "... nihil in rerum natura tam casuale, ac fortuitum apparere posse, quod sub occulta divinae providentiae dispositione omnium moderatrice non lateat." Also cited in Väliaho 2022, 49.

65 De Ovalle 1646, between 328 and 329: "Vera efigies alterius, quae in rupis cavitate in modum capellae concameratae, non ab opifice aliquo delineata, sed ab auctore naturae ex Jaspide varii coloris constructa candido vultu sub nigro capillo, palio interius caeruleo exterius rubro veste rosis contexta conspicitur."

66 Rosignoli 1708, 250: "Ma per i scoprire bene tali belle figure era di mestieri porsi in una determinata distanza: perche da vicino non si vedeva altro, che varie lividure e macchie senza veruna distinzione."

67 Kircher 1636; Kircher 1641; Kircher 1646; Kircher 1652–1655; Kircher 1665; Kircher 1671a; and Kircher 1667. Findlen 2004a.

68 Vermeir 2007, 372 and 385.

69 Kircher 1646; Schott 1657; Bettini 1642; du Breuil 1663–1669.

70 See, for example, "Vesuvius observed by Kircher in 1638," in *Mundus Subterraneus*, vol. 1, Biblioteca Nazionale Centrale "Vittorio Emanuele II," Rome, Fondo Gesuitico, ms. 562, fol. 590. Further sketches by Kircher are held in the Museo di Geologia Università degli Studi di Roma "La Sapienza"—Dipartimento di Scienze della Terra—Rome. For reproductions of some of Kircher's drawings, see Morello 2001, 179–96. On Kircher's limited skills as a draftsman, see Rowland 2002, 10–14.

71 As argued by Melion and Dekoninck 2019, 539–40.

72 On thesis prints, see Rice 1999, 148–69; and Rice 2007, 195–246.

73 Manning and van Vaeck 1999; Dimler 2005; Dimler 2007; and Enenkel, Melion, and de Boer 2016.

74 David 1601; David 1605; David 1607.

75 Hugo 1624. See also Leach 1987; and Praz 1975, 143.

76 Porteman 1996.

77 Rice 2007, 197.

78 Aguilonius 1613. On the *Opticorum libri sex*, see Dupré 2008, 53–66.

79 Niceron stresses the pleasurable aspects of his studies in optical deformations, writing of "very pleasant diversions for the pleasure of vision" (*de tres-agreables divertissemens pour la satisfaction de la veue*). Translation from Niceron 2019, 61. Niceron 1652, 2.

80 Hanneke Grootenboer describes anamorphoses' inherent construction of concealment and revelation as allegorical of God's omnipotence. I add to Grootenboer's claim by stressing the role of the Catholic Church in this movement from obfuscation to divine revelation. Grootenboer 2005, 102 and 132.

81 Bossuet 2001, vol. 2, 206: "Quand je considère en moi-même la disposition des choses humaines, confuse, inégale, irrégulière, je la compare souvent à certains tableaux, que l'on montre assez ordinairement dans les bibliothèques des curieux comme un jeu de la perspective. La première vue ne vous montre que des traits informes et un mélange confus de couleurs, qui semble *être* ou l'essai de quelque apprenti, ou le jeu de quelque enfant, plutôt que l'ouvrage d'une main savante. Mais aussitôt que celui qui sait le secret vous les fait regarder par un certain endroit, aussitôt, toutes les lignes inégales venant à se ramasser d'une certaine façon dans votre vue, toute la confusion se démêle, et vous voyez paraître un visage avec ses linéaments et ses proportions, où il n'y avait auparavant aucune apparence de forme humaine. C'est, ce me semble, messieurs, une image assez naturelle du monde, de sa confusion apparente et de sa justesse cachée, que nous ne pouvons jamais remarquer qu'en le regardant par un certain point que la foi en Jésus-Christ nous découvre." On this passage, see Leplatre 2002, 38–46.

82 Bossuet 2001, 211: "Dieu fait souvent du bien aux méchants, il laisse souffrir de grands maux aux justes" ("God often does good to the wicked, he lets the just suffer great harms").

83 On Bossuet's emphasis on the Church's regulation of the interpretation of Scripture, see Blum and Günther 2019, 41–42.

84 Bossuet 2001, 207: "Quelle est la confusion de ce tableau! et ne semble-t-il pas que ces couleurs aient été jetées au hasard, seulement pour brouiller la toile ou le papier, si je puis parler de la sorte? … Peut-être que vous trouverez que ce qui semble confusion est un art caché; et si vous savez rencontrer le point par où il faut regarder les choses, toutes les inégalités se rectifieront, et vous ne verrez que sagesse où vous n'imaginiez que désordre."

85 On Bossuet and anamorphosis, see also Leplatre 2002.

86 Baltrušaitis questions the attribution to Niceron by noting how unusual this secular subject matter is in his oeuvre. Baltrušaitis 1977, 180n17.

87 In *La perspective curieuse*, Niceron writes in his dedication to the Bishop of Ascoli Satriano Giorgio Bolognetti (1595–1680) that his book will contribute to his dedicatee's *divertissemens* ("diversions"). Niceron 1638, avi^{r}.

88 At least one Protestant, Gottfried Wilhelm Leibniz (1646–1716), who engaged in an extended correspondence with Bossuet, also analogized the process of comprehending "the apparent deformities of our little worlds" to viewing certain "devices of perspective." Leibniz 1885, vol. 6, 197–98: "Mais Dieu, par un art merveilleux, tourne tous les défauts de ces petits mondes au plus grand ornement de son grand monde.

C'est comme dans ces inventions de perspective, où certains beaux desseins ne paroissent que confusion, jusqu'à ce qu'on les rapporte à leur vrai point de veue, ou qu'on les regarde par le moyen d'un certain verre ou miroir. … Ainsi les deformités apparentes de nos petits mondes se reunissent en beautés dans le grand et n'ont rien qui s'oppose à l'unité d'un principe universel infiniment parfait: au contraire, ils augmentent l'admiration de sa sagesse, qui fait servir le mal au plus grand bien." ("God, by a wonderful art, turns all the errors of these little worlds to the greater adornment of his great world. It is as in those devices of perspective, where certain beautiful designs look like mere confusion until one restores them to the right angle of vision or one views them by means of a certain glass or mirror. … Thus the apparent deformities of our little worlds combine to become beauties in the great world, and have nothing in them which is opposed to the oneness of an infinitely perfect universal principle: on the contrary, they increase our wonder at the wisdom of him who makes evil serve the greater good.") Cited in Väliaho 2022, 27. Translation from Leibniz 1985, 216.

89 On the levels of biblical exegesis, see de Lubac 1959–1964.

90 For the impact of knowledge in Europe of Greek, Syriac, Arabic, Hebrew, and Aramaic on the interpretation of Christian Scripture, see Dunkelgrün 2012, 9.

91 Dunkelgrün 2012, 9. See also Saenger and van Kampen 1999; and Dunkelgrün, 2024.

92 Regarding visual exegesis's impact on biblical interpretation from the fifteenth through the seventeenth centuries, see Melion, Clifton, and Weemans 2014.

93 By describing an embodied exegesis, I expand upon Massey's helpful reinsertion of the body into perspectival history and theory (2007). I thank Michael Koortbojian for sharing his thoughts on actualized allegory and embodied exegesis with me.

94 In this way, they build upon the claim articulated by the French abbot Suger of Saint-Denis (1081–1151) that some artworks operated in a manner analogous to the exegetical analysis of Scripture. Suger claimed this art was not meant to cater to the illiterate, but only to the educated monk; as Conrad Rudolph has stressed, these works catered to not only the educated monk but also "the increasingly better-educated public." See Rudolph 2011, 418. See also Horn's discussion of the "broad appeal" of perspectival illusionism which "does not require any particular education or training in order to engage the curious observer and produce the nearly universal experience of astonishment." Horn 2019a, 4–5.

95 Unlike most other religious orders that had built female arms, Jesuits, following some initial testing, refused entry into the Society to women. Russell 2022, 186n28.

96 Ignatius of Loyola 1970, 161. Cited in Russell 2022, 23; and Goujon 2019, 178–79.

97 Based on the names of members of the order, P. J. S. Whitmore has postulated that Minims were drawn primarily "from the Bourgeoisie and Artisan classes." Whitemore 1967, 27. Kent 2023, 62.

98 Kent 2023, 68. It is challenging to present a precise number of books because surviving catalogues present distinct information. See also Vernet et al. 2008, 432.

CHAPTER 4: OBSCURITY JUSTIFIED

1 Koen Vermeir also considers Jesuit illusionistic performances in relation to ideas espoused in Roman rhetoric. See Vermeir 2005, 127–59.

2 Translation from Augustine 1996, 62–63 (2.6.14–15): "Qui enim prorsus non inveniunt quod quaerunt, fame laborant; qui autem non quaerunt, quia in promptu habent, fastidio saepe marcescunt. In utroque autem languor cavendus est. 15. Magnifice igitur et salubriter spiritus sanctus ita scripturas sanctas modificavit, ut locis apertioribus fami occurreret, obscurioribus autem fastidia detergeret. Nihil enim fere de illis obscuritatibus eruitur quod non planissime dictum alibi reperiatur." Cited in Ossa-Richardson 2019, 138–39.

3 Erasmus 1969–, vol. 1.6, 66 (book 1, chapter 18). Cited in Ossa-Richardson 2019, 199.

4 Barocchi 1960–1962, vol. 2, 411: "Alle volte si cercano di esprimere cose di sua natura non fattibili, tanto recondite e difficili, che non si può commodamente renderne capace il popolo, sì come sono l'operazioni delle intelligenze, i secreti della providenza divina, i misteri della predestinazione, e simili." Translation from Paleotti 2012, 253 (book 2, chapter 33).

5 Barocchi 1960–1962, vol. 2, 411: "L'astenersene più che si può, acconsentendoli però interiormente con la vera fede." Translation from Paleotti 2012, 253 (book 2, chapter 33). This chapter builds on the important work of Pasi Väliaho, who has shown that for Jesuits, the optical projections of the magic lantern and the camera obscura came to uproot chronology, such that a virtual projection could denote a visual effluence of God's sovereignty or divine providential plans (2022, 42).

6 On Bettini, see Sommervogel 1890, vol. 1, 1425.

7 Maignan 1653. Maignan 1672. Bettini 1642.

8 See Bettini 1642, vol. 1, *Apiarium Quintum, Progymnasma primum*, 7 and 8.

9 Bettini 1642, vol. 1, *Apiarium Quintum, Progymnasma Primum*, 7: "Telam, vel papyrum, ut AB, circa ligneum cylindrum incurvato."

10 Bettini 1642, vol. 1, *Apiarium Quintum, Progymnasma Primum*, 8: "quasi cribrum papyrum."

11 Bettini 1642, vol. 1, *Apiarium Quintum, Progymnasma Primum*, 8: "Lumen tu aliquod cavae papyro perforatae postponito, ut in E. Ex E radii traiecti per foramina oculi CD circumquaque perforati proiicient in subiectum planum intercisam, ac dissipatissimam quandam formam informem, qualis est FGHI."

12 On how-to books, see Leitch 2015, 189–208; Eamon 1994, 112–20; and Smith 2022. In addition to Bettini's confusing instructions cited above for creating a deformed eye such that its image in a column-like mirror would be lucid, see also, for instance, his account of a secretive mechanism that enables someone closed in a bedroom to know what happens outside, in the streets, squares, fields, or at sea: Bettini 1642, vol. 1, *Apiarium Quintum, Progymnasma secundum*, 37.

13 Bettini 1642, vol. 1, *Apiarium Quintum, Progymnasma Primum*, 21: "Imaginem in speculo erectam, & emendatam, quam idem oculus videbat ex eodem puncto per directam visionem in plano subiecto inversam, ac deformatam."

14 Bettini 1642, vol. 1, *Apiarium Quintum, Progymnasma Primum*, 3: "Arcana semiarcane prodenda esse, ac propterea consulto alicubi, post alia exemplis perspicuis explicata, alia, quae tantum indicanda censebimus, abstracte, ac sine exemplis quasi sub velo innuemus."

15 On secrecy in writings from technical, craft, and practical sources from antiquity to the early modern period, see Long 2001.

16 Drake 2014, 815–42. Bezner 2005.

17 Bettini 1642, vol. 1, *Apiarium Quintum, Progymnasma Primum*, 23: "Rosam, & oculum in cylindro exhibueram pro argumento, in quo versabar, idest optico, & pro tempore, quo haec scribebam, idest verno, & rosarum feraci. Paucis post mensibus fama pertulit praecellentissimum Dominum Hieronymum Columnam sacra purpura donatum, & paullo post Ecclesiae huic Bononiensi Archiepiscopum adventare." ("I showed a rose, and an eye in a cylinder, as an argument, which I was occupied with, that is, an optic argument, and on account of the time in which I was writing these things, that is, Spring, also rich in roses. After a few months, a report announced that the most eminent Lord Girolamo Colonna, was given the sacred purple, and a little after, the archbishopric of the Church of Bologna came to him.") As Colonna was elevated to cardinal in 1627 and was appointed archbishop of Bologna in 1632, Bettini likely authored his demonstrations before these promotions were announced. Cardella 1793, vol. 6, 282–84. The publication of the *Apiaria* was delayed by difficulty in finding an appropriate patron. See Gorman 2003, 4.

18 Bettini 1642, vol. 1, *Apiarium Quintum, Progymnasma Primum*, 23: "Sane optica scientia praevidit in tempore ac praesignavit fausta haec meae patriae omina." A marginal note on the same page refers to "The optical prognostication in praise of the most eminent Cardinal Colonna" (*Opticum praesagium in laudem Eminentissimi Cardinalis Columnae*).

19 Bettini 1642, vol. 1, *Apiarium Quintum, Progymnasma Primum*, 23: "Dum ergo, mi Lector, apud nos in speculo columnari vides oculum, prospice animo in Eminentissimo Cardinali Columna pastoralem providentiam, & ex hisce nostris opticis reformationibus agnosce futuram saluberrimam animorum, ac morum in Bononiensi Ecclesia (si qua forte deformatio fuerit) reformationem. Agnosce, dum rosam in speculo columnari vides, sacram Antistitis nostri purpuram, & lectissimam in eo erga suum populum charitatis rosam. Agnosce in eodem sacro heroe Columna quasi columnare quoddam speculum, in quo mirifice appareant, ac splendeant praecipua lumina earum omnium virtutum, quae in sacro Principe purpurato, & in optimo animorum Pastore requiruntur. In hoc fausto praesagio optico finis esto prioris Progymnasmatis."

20 On the mirrors of princes, see Nederman 1998, 18–38; and Gottardo 2011, 293–97.

21 Tesauro 1670, 460: "Egli è dunque una segreta ed innata delitia dell'Intelletto humano, l'avvedersi di essere stato *scherzevolmente ingannato*: peroche quel trapasso dall'inganno al disinganno, è una maniera d'imparamento, per via non aspettata; & perciò piacevolissima." Emphasis in original. Cited and translated (with modifications) in Schwenger 1976, 67.

22 Del Monte 1600, frontispiece. Concerning the connection to del Monte's frontispiece, see Fratini and Moriconi 2010.

23 Del Monte 1600, 4: "De varia igitur visibilium apparentia, & de eo videndi modo, qui arte quadam visum decipere videtur, quamvis mathematicis demonstrationibus, quae falli non possunt, fallax omnis tollatur apparentia, sermonem facturus."

24 See chapter 2 n. 50.

25 Niceron 1638, ē6ᵛ and 11: "Perspective ... nous fait cognoistre & discerner plus parfaictement les beaux ouvrages de la nature & de l'art"; "L'optique ... est une science, qui enseigne à discerner & bien juger des objets de la veuë."

26 Niceron 1638, 52.

27 In addition, according to an inscription dated March 16, 1640, in the *Livre des Conclusions Capitulaires de ce convent de la S.te Trinitè Du Mont* from the years spanning 1620 to 1649, Niceron was asked to produce a perspectival wall painting of King Louis XIII for a corridor above the cloister in Rome. See Archivio Generalizio dei Minimi, Rome, *Livre des Conclusions Capitulaires de ce convent de la S.te Trinitè Du Mont* (5-X-1620-26-IX-1649), (T3), XVII sec., p. 191. Concerning this project, see De Rosa 2019a, 65; and Fratini and Moriconi 2010.

28 For an eighteenth-century reference to the frescoes, see de Dargenville 1757, 285: "Above the cloister are two big galleries which run its entire length. We admire there two optic pieces by Father Niceron: one is Mary Magdalene in contemplation in the cave of Sainte Baume; the other is Saint John the Evangelist writing his Apocalypse in the island of Pathmos." ("Au-dessus du Cloître sont deux grandes Galeries qui régnent sur toute sa longueur. On y admire deux morceaux d'optique du *Pere Niceron*: l'un est la Madeleine en contemplation dans la grotte de la Sainte Baume, l'autre S. Jean l'Evangéliste écrivant son Apocalypse dans l'île de Pathmos.") Whitmore 1967, 166.

29 Niceron 1646, 177: "Sylvas, agros, nemora, aedes, hortos, arborumque, herbarum & florum insignes varietates, antra, cavernas, animantia, bestiolas, sinuosos marium anfractus, ripas, littora, naves, incendia, lacus, piscatorum cymbas, naves onerarias, & triremes ad pugnam compositas effingebamus."

30 Niceron 1646, 177: "Sic nos in obscuris & umbrosis viridis tunice sinubus, silvas densas & frequentissimis arboribus consita nemora ponebamus; in lucidioribus eiusdem tunicae partibus sive eminentiis, flavas segetes & messi iam maturas; in zona candida labentes fluminis aut fontis aquas; in candidis aperti libri foliis ingentem lacum, in eoque portus, littora, naves, piscatores, &c. In capite cavernas, antra, praecisas rupes, saxa, aedificia, immo totius civitatis Babylonis ruinam & casum, cui etiam angelos tuba canentes apponebam."

31 Niceron 1646, 177: "Memini me delineatae in Romano conventu proiecturae, sic alludere verbis graecis quae ad ipsum, quem prae manibus tenet, Evangelistae librum apponebam, & spectatoribus legenda proponebam ταυτῆς ὀπτικῆς ἀποκάλυψις, ὁ τῆς ἀποκαλύψεως αὐτοπτής."

32 This inscription is also considered by De Rosa 2013, 45. I thank Anthony Ossa-Richardson for discussing this annotation with me.

33 My thanks to Ioannis Mylonopoulos for this observation that the phrase was likely intended to be read aloud.

34 Concerning the Pan-European fascination with Ottoman culture from 1650 through 1750, see Bevilacqua and Pfeifer 2013, 75–118.

35 Niceron 1638, 115: "La [figure] est à peu pres la coppie, au moins le dessein d'un tableau que je traçai & fis peindre il y a 2 ou 3 ans, & qui se garde encore à present en la Bibliotheque de nostre Convent de la place Royale à Paris."

36 On the Florentine device, see De Rosa 2013, 26.

37 De Rosa 2013, 26. The tube disappeared in the flood of 1966.

38 Niceron 1638, 115: "Ce dessein est fait à propos de la Prophetie ... que Mahomet laissa autresfois à ses successeurs. Leur recommandant de ne jamais offencer la Monarchie Françoise; parce que leur empire ne seroit jamais ruiné que par la puissance de quelqu'un de ses Rois. Sur ce voulant montrer que l'honneur de cette conqueste n'appartient point à d'autre qu'à Louis le Juste, nous faisons que la plus part de ces Empereurs en ce tableau lui rendent hommage, en sorte qu'ils contribueut chacun quelque partie de soi pour former son image, comme s'ils se despoüilloient eux-mesmes pour honorer son triomphe."

39 Postel 1969. See also Fleischer 2018, 23–24. I thank Alexander Bevilacqua and Helen Pfeifer for drawing my attention to Fleischer's article.

40 Fleischer 2018, 23.

41 Cited and translated in Fleischer 2018, 24. Postel 1969, 102: "Les Turcs ont davantaige en especiale authorité et quasi non moindre que leur *Al Coran* un livre de profeties, là où il est expressement escript que le prince et le peuple de la jaulne couleur doibt destruire et tant les Turcs (54v) comme touts les aultres Ismaelites ou Muhamedikes, qui vulgairement se disent Mahometains. Et de cecy se peult faire foy indubitable, combien que lesdicts Turcs cachent autant qu'ils peuvent aulx estrangers ladicte profetie."

42 Fleischer 2018, 24.

43 Niceron 1638, 118: "On peut prendre au vieil testament toutes les figures d'une mesme signification, & faire qu'estant peintes & disposees au plan selon les regles prescrites, elles ne represente par la lunette que la chose figuree. Jai encore eu la pensee plusieurs fois,

de peindre quelques Prophetes de ceux, qui ont parlé plus expressement de la Vierge & de l'Incarnation, chacun avec un liteau volant, où fussent escrits les mots de sa Prophetie comme, par exemple, Isaie avec ces motes, ECCE VIRGO CONCIPIET ET PARIET FILIUM, … & faire que par la lunette on ne vit que la Vierge avec cette inscription: ECCE ANCILLA DOMINI, &c."

44 Concerning the terms *figures* ("figures") and *figurée* ("the figured thing"), see Erich Auerbach's classic discussion of *figura* and its changing meanings from early Latin literature through Dante (1265–1321). Auerbach 1938, 436–89.

45 Plumier 1705, i: "De jetter leurs regards sur la terre, pour y admirer ces rares prodiges où sa divine providence éclatte en mille & mille façons."

46 Plumier 1705, i–ii: "Tout ce que la terre produit benisse le nom du Seigneur."

47 Plumier 1705, ii: "Nous sommes obligez de les regarder comme autant d'illustres prodiges, que Dieu a posez sur la terre." Plumier repeats this notion later in the preface (1705, xiii): "Ne sommes nous pas obligez de regarder toutes ces plantes comme autant de prodiges que Dieu a mis sur la terre pour nous faire admirer sa grandeur" ("are we not obliged to look at all these plants as so many wonders that God put on the earth to make us admire his greatness").

48 Acosta 2002, 162 (book 4, chapter 1). See Prieto 2011, 168–71.

49 Ignatius of Loyola 1970, 361. See also Prieto 2011, 3.

50 Concerning the history of the passionflower, see Bleichmar 2017, 82–88; Cañizares-Esguerra 2006, 118–19; Terrada 2000, 163–64; López 2014, 195–203; Kugler and King 2004, 15–26; and Pimentel 2014, 109–23. I thank Daniela Bleichmar for drawing my attention to the passionflower.

51 Translation from Acosta 2002, 219 (book 4, chapter 27).

52 Nieremberg 1635, 299–300.

53 Parkinson 1629, 396.

54 Parkinson 1629, 394–95.

55 Parkinson 1629, 396.

56 Castelli 1625, 49–56.

57 The engravings are after Aldini 1620. As noted by Bleichmar 2017, 203n100.

58 These prints have strong thematic resonances with both the anonymous image showing miraculous pictures reportedly found in an apple tree and Pieter Saenredam's (1597–1665) and Cornelis Korning's *Print to Belie Rumors about the Images Found in an Apple Tree* discussed by Alpers 1983, 80–82. Copies of these prints are held in the Municipal Archives in Haarlem. Korning's dates are unknown.

59 As stressed by Bleichmar 2017, 83.

60 d'Eremita 1619. Four drawings of the flower from the 1620s were in the possession of Cassiano dal Pozzo. See Elliott, Guerrini, and Pegler 2015, vol. 1, 116–21, and vol. 3, 1176–77. Plumier 1693, 61–75 and plates 79–90.

61 Plumier 1693, 61: "Il y en a pourtant à qui ils donnent plus particulierement le nom de … *Fleur de la passion*, à cause qu'on croit y trouver quelques marques de la passion du Sauveur, comme les trois clous, les cinq playes, les foüets, & les cordes."

62 Plumier 1693, 67: "La colomne, les clous, & les bras où sont attachées les playes, sont blanchastres, & tous tachetez de points rouges: la teste des clous est fendue en deux, & les playes sont couvertes d'une poussiere jaune fort menue."

63 Maignan's ideas on the Eucharist and optical illusions are noted in secondary sources with only passing reference to his writings. Baltrušaitis 1977, 69. Ceñal 1952, 128–29. Malcolm 2002, 10–11. Lyle Massey presents a more detailed consideration of Maignan's ideas on this topic; she takes issue with Baltrušaitis's assertion that Maignan likens perspectival deceptions either to sensory deceptions during transubstantiation or to Christ's transfiguration as a pilgrim and gardener. Massey 2007, 102–09. Geneviève Rodis-Lewis cites relevant passages in Maignan's writings on sensory deceptions and Christ's transfiguration as a pilgrim and gardener (1956, 471–74).

64 See "Eucharist" 2005.

65 "Eucharist" 2005.

66 My account of Aristotelian scholastic theories of transubstantiation and debates around the Eucharist in the seventeenth century draws on Nadler 1988, 230–31.

67 Porphyry, *Isagoge*, 12.25–26. Aristotle 1590, vol. 1, 5.

68 Translation from Descartes 1984, 140. Descartes 1904, vol. 7, 121: "Complete intelligo quid sit corpus, putando tantum illud esse extensum, figuratum, mobile." Concerning Descartes's theology, see Watson 1987, 155–70; and Armogathe 1977.

69 Descartes 1904, vol. 7, 434. "Omnino repugnat dari accidentia realia, quia quicquid est reale, potest separatim ab omni alio subjecto existere; quicquid autem ita separatim potest existere, est substantia, non accidens." Translation from Descartes 1984, 293.

70 Translation of Arnauld from Descartes 1984, 152–53. Descartes 1904, vol. 7, 217: "*Verum, quod maxime Theologis offendiculo fore praevideo, est quod, secundum Viri C. dogmata, salva & integra remanere non posse videantur, quae de sacrosanctis altaris*

mysteriis docet Ecclesia. Fide enim credimus, ablata ab Eucharistico pane panis substantia, sola illic accidentia remanere: ea autem sunt extensio, figura, color, odor, sapor, aliaeque sensiles qualitates." Emphasis in original. Cited by Nadler 1988, 232.

71 For Descartes's response to Arnauld's concerns, see Nadler 1988, 229–46.

72 Maignan 1653, vol. 2, 527: "sed solum aliquid est a scholasticis ex principiis, ut dixi fallacibus, peripateticis explicatum" ("but only something is explained by the Peripatetic scholastics, from fallacious, as I said, principles").

73 Maignan 1653, vol. 2, 530: "Non est novum miraculum distinctum ab eo, per quod Deus facit rem apparere quae non est, nempe panem qui non est, dum facit corpus Christi verum vere ac realiter manducari sub specie panis, circumferri sub specie panis, adeoque tangi ac videri &c. sub specie panis; sicuti servata proportione faciebat videri sub specie hortulani a Magdalena, & sub specie peregrini a Discipulis euntibus Emmaum."

74 Gospel of John 20:14–15. Luke, 24:15–31.

75 Maignan 1653, vol. 2, 531: "Sola erat apparentia: id est Christus, cum esset ipsemet, alius tamen esse apparebat, nempe apparebat esse hortulanus vel peregrinus."

76 Maignan 1653, vol. 2, 531: "oculos aliosque pariter sensus inspectantium tenebat, ne se sub ea aliena occultatum specie agnoscerent."

77 Maignan 1653, vol. 2, 531: "Eodem autem modo servata proportione putandum est, id omne quod in Euch. habet aliquid realis entitatis, esse ipsummet Christum; de pane autem ac de vino prorsus nihil superesse reale, sed solam ibi esse eorum omnium apparentiam sive speciem, quae oculos aliosque sensus fidelium tenet, ne Christum sub ea sese occultantem agnoscant quasi in propria persona; sed nihilominus ipsum revera sub aliena sumant in cibum ac potum. Itaque si intelligis potuisse Christum obtendere atque exhibere speciem peregrini, vel hortulani, quamvis nihil prorsus de peregrino vel hortulano reale exhiberet, nihil inquam reale a se distinctum (non enim credis illum tunc assumpsisse vestem v.g. realem peregrini, quam secum fecerit mox evanescere) bene intelliges posse pariter speciem exhibere panis ac vini; quamvis de pane ac vino prorsus nihil reale adsit, quod exhibeat."

78 De la Grange 1675, 39–40: "Le Pere Maignan est autant opposé à la Philosophie ordinaire que les Philosophes dont j'ai parlé jusqu'à present, il se mocque perpetuellement des formes accidentelles des Paripateticiens, il croit comme Descartes & Gassendi, qu'il n'y a au monde que des esprits & des corps."

79 De la Grange 1675, 40: "Ordinairement ses opinons sont les plus mal choisies de toutes, & les plus mal debitées."

80 De la Grange 1675, 125: "Parce que supposé cette Opinion, il est vrai de dire qu'il n'y a point de Blancheur dans l'Eucharistie, qu'il n'y a point de Dureté, ny de Figure, ny de Goust, quoi qu'il nous semble y voir de la Blancheur, y sentir de la Dureté & y trouver quelque Goust; nous sommes perpetuellement trompés, lorsque nous croyons qu'il y a quelque chose de materiel dans l'Eucharistie, qui est cause des sentimens que nous en avons: de la mesme maniere que ceux qui dorment s'imaginent manger & courir, quoi qu'en effet ils ne mangent point & ne courent point."

81 De la Grange 1675, 126: "Car n'est-il pas plus facile de croire que les Perfections Accidentelles du Pain, lesquelles frappent nos sens, demeurent dans l'Eucharistie, & qu'il n'y a que la substance du Pain qui est détruite, & qui est convertie au Corps de Notre Seigneur; que de se persuader qu'il ne demeure rien de tout ce que nous appercevions dans le Pain auparavant la Consecration, & que tout ce que nos sens nous representent, est entierement faux?" ("For isn't it easier to believe that the accidental perfections of bread, which strike our senses, remain in the Eucharist, and that only the substance of bread is destroyed, and which is converted into the body of our Lord; than to convince oneself that there remains nothing of all that we perceive in the bread before the consecration, and that everything that our senses represent to us is entirely false?")

82 De la Grange 1675, 126, 128, and 129.

83 De la Grange 1675, 129: "Selon le sentiment de plusieurs Theologiens, il ne faut qu'un Miracle afin que les Accidens de l'Eucharistie subsistent sans aucune Substance … Mais l'Opinion des Cartistes suppose autant de Miracles qu'il y a de gens qui voyent le Saint Sacrement" ("according to the opinion of several theologians, it takes only one miracle so that the accidents of the Eucharist subsist without any substance … But the opinion of the Cartesians supposes as many miracles as there are people who see the Holy Sacrament").

84 Raynaud, 1665–1669, vol. 6, 467 (mispaginated as "147"): "Appendix against Maignan and others" ("APPENDIX CONTRA Magnanum & alios").

85 See Raynaud 1665, vol. 6, 467 (mispaginated as "147").

86 Raynaud 1665, vol. 6, 468 [mispaginated as "148"]: "Nulla est transsubstantiatio, cum nihil commune utrique substantiae remaneat, nec explicari

possit … vera ratio Sacramenti, ne modus existendi Sacramentalis non differat ab apparitionibus."

87 Bayle 1734, vol. 4, 58: "La maniere dont il explique la conservation des accidens sans sujet … est plus heureuse que celle de Mr. Des Cartes." ("The manner in which he explains the conservation of accidents without subject … is more felicitous than that of Mr. Descartes.")

88 Maignan 1672, vol. 2, 465: "Visibiles itaque species a speculo concavo praesertim parabolico reflexae obiectum repraesentant in spatio citra speculum. Ad hoc, fallacia, & spectrum, & praestigiae, & idolum videtur apud vulgus; cuius tamen iudicio fallacia non est, (& molto minus spectrum, &c.) quod eaedem species a speculo plano reflexae repraesentant idem obiectum, velut existens in spatio ultra speculum."

89 Maignan 1672, vol. 2, 465: "Sed sapiens quisque ridet iudicii disparitatem istam; quia non magis in primo casu fallacia est, quam in secundo."

90 Maignan 1672, vol. 2, 465: "Si namque in primo decipitur sensus hominis, aut simiae, &c. decipitur etiam in secundo sensus v.g. molossi, qui velut alium sibi similem conspicans, & suspicans adesse, exardescit in eum, allatrat, infilitque."

91 Maignan 1672, vol. 2, 465: "Iudicium vulgi circa fallaciam, quam existimat, speculi parabolici concavi, fundatur in eo quod rarius videt effectum hunc speculi concavi."

92 Maignan 1672, vol. 2, 465: "Effectum vero speculi plani … non sumit pro fallacia, eo quod iam diu assuevit similes effectus quotidie cernere; ideoque haud putat se decipi."

93 Maignan 1672, vol. 2, 465–66: "Ac proinde, cum in apparitione per speculum seu planum, seu concavum, pane ex hypothesi ultra speculum, vel citra reflexe apparente, nullus Philosophus existimet accidentia realia ulla panis ibi realiter esse ultra vel citra speculum, sed solis apparentiam hanc fieri & constare speciebus a pane emissis, a speculo reflexis, & ab oculo susceptis, una fit omnium sententia."

94 Maignan 1672, vol. 2, 466: "Mirum est, quod in Eucharistia quis ad solam apparentiam requirat accidentia; quasi non possit hic sine eis, possit vero illic apparentia panis fieri."

95 Maignan 1672, vol. 2, 466: "In Eucharistia vero apparentia, non panis existat; sed Deus, loco & vice panis emittat, seu faciat species omnino similes eis, quas ipse panis emitteret, si adesset."

96 Maignan 1672, vol. 2, 466: "nulla est difficultas, quia Deus est satis potens" ("there is no difficulty, because God is powerful enough").

97 Maignan 1672, vol. 2, 466: "Ex eo quod in speculorum dictis apparentiis potest fallacia quaecunque est unius sensus, corrigi per alium; quia potest certo agnosci obiectum non esse realiter ultra speculum. vel citra in illo catheti puncto. Id vero de pane Eucharistico post consecrationem recognosci nequit."

98 Maignan 1672, vol. 2, 466: "Deus monuit, dicente Christo *Hoc est Corpus meum*" ("God warns the faithful, as Christ says, *This is my Body*").

99 Maignan 1672, vol. 2, 466: "Et in animo fideli potentius est Christi verbum ad amovendam illusionis, fallaciae, &c. occasionem, quam sint cuilibet e vulgo, aut etiam insigni Philosopho experientiae sensatae mille. Ex his ergo iam constat nullam esse ab speciebus in Eucharistia occasionem deceptionis, praestigiarum, spectri, &c."

100 On this lens and its relation to Eucharistic imagery, see Collacciani and Roux 2017b, 89–113.

101 Du Breuil 1649, 162: "J'ai veu autre fois, un tableau semblable à cettui-ci, au milieu duquel estoit peinte une Hostie, comme celles dont on dit la Messe, & autour de cette Hostie, il y avoit quantité d'Anges en posture de devotion. Mais quand on regardoit par le trou de la lunette, on ne voyoit rien autre chose, qu'une image du petit Jesus." Cited in Collacciani and Roux 2017b, 95–97.

102 As argued by Collacciani and Roux 2021.

103 For this Eucharistic iconography, see the following thesis booklets by Bourdin: Henri 1639, 6; and Foucquet 1646, 6. These thesis booklets are also cited in Collacciani and Roux 2021.

104 For the dating of this booklet, I follow Collacciani and Roux 2017b, 94.

105 Henri 1639, 6: "I. SPECILLUM Fidei exhibere, sive tabulam in qua sit depictum sacrosanctae Eucharistiae mysterium, puta Hostia in medio, & circa illam adorantes Angeli, totaque caelestis curia in spectaculi maiestatem erecta: per specillum vero si intueare; nec Angelos videas, sed Angelorum Dominum; nec Hostiam, sed qui vere est, & sub Hostiae integumentis latet, Christum Dominum graphice repraesentatum." Cited in part in Collacciani and Roux 2017b, 95. I thank Domenico Collaciani for his generosity in helping me to access this document.

106 BnF, Département des manuscrits, ms. Lat. 17862, f. 832[r], 833[v], in-4.

107 In dating this booklet to 1640, I follow Collacciani and Roux 2017a, 59.

108 Potier 1640, 2: "Eodem die post meridiem instaurabuntur Agones circa horam 2, iisque aditum parabit Demonſtratio Specilli fidei, ac similium tabularum, quae ex principiis Refractionum descriptae ex

pluribus personis novam aliquam nec expectatam repraesentant."

109 Bourdin 1661, 9. The book's privilege notes the "great expense for the copper-plate engravings that the said exhibitor [i.e., Bourdin] had engraved in a new manner" (*la grande dépense des Planches de Taille-douce, que le dit Exposant a fait graver dans un nouvel ordre*). Bourdin 1661, 187. The notice accentuates the care and financial resources that Bourdin invested in integrating visual documents into his pedagogical practice: for Bourdin, acts of looking were integral to acts of learning.

110 Bibliothèque du CNAM, cotes 4 Y 2 Res. The volume contains twenty-two different pieces bound together. I thank Guillaume Champion for his generous assistance with details pertaining to this volume.

111 Bibliothèque du CNAM, aamrh 4 7 2 (P.14) Res.

112 Bibliothèque de la Sorbonne, shelfmark SXO 3= 11: "Pour le Reverend Pere Bourdin Professeur ès Mathématiques au College de Clairmont de la Compagnnie de Jesus: de la par de l'autheur son tres humble serviteur." This inscription is cited in Collacciani and Roux 2017b, 100n2.

113 Eire 2023, 15.

CHAPTER 5: CARAMUEL'S ARCHITECTURAL REFORMS

1 Juan Caramuel, 1678 [–1679], vol. 2, treatise 6, 13. I thank Daniela Gutierrez Flores for her assistance with Spanish translations in this chapter.

2 Perrault 1683, 108.

3 British Museum, London, 1934,0217.35: "Hic est Caramuel, toti spectabilis orbi, fama ingens clarum nomen ad astra tulit. Sculpitur os, libris mens pluribus insidet ampla, et par pluribus est; cuncta scienda tenet."

4 Tadisi 1760, v.

5 Juan Caramuel 1678 [–1679]. For the Latin edition, see Juan Caramuel 1681. Copies of the Latin and Spanish editions are held by the National Gallery of Art Library in Washington, DC. Jorge Fernández-Santos Ortiz-Iribas has argued that the *Architectura* was initially conceived as a Latin work, in part because the plates in the Spanish edition all feature Latin inscriptions: Ortiz-Iribas 2014, 144. See also Ortiz-Iribas's discussion of the need for a Spanish vernacular architectural treatise: Ortiz-Iribas 2014, 32.

6 Juan Caramuel 1678 [–1679], vol. 2, treatise 5, 56: "Se imprime mejor en la memoria lo que se mira con los ojos, que quanto con palabras se dice." Cited in Ortiz-Iribas 2014, 50.

7 Juan Caramuel 1678 [–1679], vol. 2, treatise 6, 5–6.

8 Juan Caramuel 1678 [–1679], vol. 1, dedication: "*Hoy nace una Arte Nueva ... de la qual nadie ha escrito en el Mundo.*" Emphasis in original.

9 Juan Caramuel 1678 [–1679], vol. 2, treatise 6, 1: "En obra muchas cosas he visto bien labradas, y en ellas las Leyes de la Architectura Obliqua bien executadas; he visto tambien en Reales Edificios muchos y erros; porque como no hay ningun libro, que trate de Architectura Obliqua, y son de ordinario Albañiles, los que quieren passar por Architectos, quando se offrece la occasion, en lugares obliquos encaxan fuera de proposito delineaciones rectas, suppliendo con çoquetes los vacios." Translation from Lefaivre and Tzonis 2004, 220.

10 Frangenberg 1993, 207–28.

11 Vitruvius, *De architectura libri decem*, book 3, chapter 3, 11, 13; book 3, chapter 4, 5; book 3, chapter 5, 9, 13; book 4, chapter 4, 2–3; book 6, chapter 2, 1–5; book 6, chapter 3, 11. Vitruvius 1586, 91, 110, 112, 113, 148, 223–25, and 235. Vitruvius 1983, vol. 1, 90, 176–77, 178–79, 182–83, 190–193, 194–95, and 228–29, vol. 2, 20–23 and 32–33. See also Frangenberg 1993, 208.

12 Alberti 1988, 201, 211, 215–17, 221, 268, and 362. Cited in Frangenberg 1993, 208. Ryff 1547, book 1, part 1, fols. 11ᵛ-13ʳ. Cataneo 1567, 178. Serlio 1566, fols. 8ᵛ-9ʳ. Cited in Frangenberg 1993, 208–11, 219–21.

13 Juan Caramuel 1678 [–1679], vol. 1, treatise 1, 5.

14 My account of Caramuel's life draws on details in Dvořák and Schmutz 2008, 11–17; Schmutz 2000, vol. 17, coll. 224–32; Tadisi 1760; and Ortiz-Iribas 2014. See also Sabiano and Pissavino 2012.

15 Tadisi 1760, 4 and 104.

16 For a list of Caramuel's written works, see Schmutz 2000, vol. 17, coll. 227–29. Concerning the manuscripts, see Morales 2011, 144; and Lombraña 1989, 388–415.

17 De Ureña flourished in the seventeenth century.

18 Juan Caramuel 1648. Concerning this work, see Steinberger 1906, 78–84.

19 As noted in Ortiz-Iribas 2005, 147 and 163n142.

20 Juan Caramuel 1678 [–1679], vol. 2, treatise 6, 2: "Empeçe a escribir y delinear estas Ideas alla en España, siendo mozo, año de 1624, con occasion de una hermosa capilla que en nuestro Monasterio se erigia; y ahora me hallo harto viejo, y siempre las voy perficionando. Desde el año de 1635 se han ido entallando y gravando estas Laminas; algunas en Bruselas,

en Lovayna, y Anveres. Otras en Viena de Austria. Muchas en Praga. En Roma, en Campaña y Otranto. Y ahora, despues de quarenta y tantos años se van acabando de esculpir y ordenar en Milan, y Vegeven; donde de hecho exercito esta Architectura Obliqua en el Frontispicio de mi Iglesia." ("I began to write and outline these ideas back in Spain in 1624, when I was a young man, on the occasion of a beautiful chapel that was erected in our monastery; and now I am very old, and I am always perfecting it. Since 1635, these plates have been carved and printed; some in Brussels, Louvain and Antwerp; others in Vienna of Austria, many in Prague. In Rome, in Campana and Otranto. And now, forty-some years later, the sculpting and ordering are close to being finished in Milan, and Vigevano; where in fact I exercise the oblique architecture in the facade of my Church.")

21 Giordano 2006, 105.

22 Payments for the facade from the municipality continued until 1716. On its restorations, see Giordano 2006, 96–99. See also Ortiz-Iribas 2014, 31.

23 Juan Caramuel 1678 [–1679], vol. 2, treatise 5, 80: "Estas son las que comúnmente se llaman en Latín *Placita Philosophorum*, porque donde no hay argumento que convenza, entra el libre Albedrío, y *stat pro ratione voluntas*. Deste género de opiniones están todas las ciencias llenas, porque el ingenio humano demuestra poco, y lo demás lo resuelve según las conjeturas o sospechas que tiene. ... Y destas hay no pocas en la Architectura, defendidas no solamente de Artífices modernos, sino también de los antiguos, de las quales algunas entraron por desgracia en los mismos libros de Vitruvio." Emphasis in original.

24 Juan Caramuel 1678 [–1679], vol. 2, treatise 6, 3: "El primer Architecto, que en el Cielo y la Tierra hecho lineas Obliquas, fue Dios. Porque yendo en el Cielo los dos Tropicos, y los Circulos Arctico y Antarctico parallelos a la Equinocial, hizo que el Sol con su movimiento annuo describiesse la Ecliptica, que es un circulo, que corta a la Equinocial obliquamente al Zodiaco, a la larga en dos partes iguales. ... Mando en la Tierra, que obliquamente se engriessen y erigiessen los montes: y obliquamente corriessen los rios y arroyos, por sus valles." I thank Sara J. Schechner for her generous help with translating and interpreting this section.

25 Apianus 1539, fol. VI^{v}.

26 Juan Caramuel 1678 [–1679], vol. 2, treatise 9, 74–75.

27 Juan Caramuel 1670, vol. 2, 1400: "Non eget nostra, quod habet Censuram Ecclesiae. Rejiciatur igitur Copernicanum, & duo alia sub tribunali remaneant. Systema Ptolemaicum improbabile est; nam Venerem, Mercuriumque circa Solem moveri, a nemine negari potest. Stet ergo Tychonicum." ("We have no need for that which the Church has condemned. When hence the Copernican system has been rejected, the two others remain in court. The Ptolemaic system is implausible: nobody can indeed deny that Venus and Mercury move around the Sun. Thus the Tychonic system stands.") Translation by Høyrup 2009, 63.

28 For Caramuel's relation to Copernicanism in the *Theologia moralis fundamentalis*, see Tutino 2017, 180–81.

29 Clark 2009, 270; Robbins 1998; and Robbins 2006.

30 Long and Sedley 1987, vol. 1, 479. Sextus Empiricus 1562, 35: "Idem remus intra mare quidem, fractus, extra autem, rectus."

31 There is a vast literature regarding seventeenth-century discussions of limits to visual perception and certain knowledge. See, for example, de la Flor 2009; Robbins 1998; Robbins 2006; Havelange 1998; and Clark 2009.

32 Juan Caramuel 1678 [–1679], vol. 2, treatise 9, 108: "La vista se ha de poner en la A, que es el centro. Todas las Colunas, si se huvieran de colocar en el Círculo Equante, havrían de ser iguales, y tener los mismos cortes en sus miembros: pero haviéndose de poner en una peripheria Oval, o Ellíptica, donde unas disten más de la vista, otras menos, para que *quoad nos* (en nuestros ojos) sean iguales, han de ser desiguales *quoad se* (en sus proprias medidas) y en sus Bases y miembros tener diversos cortes, como con distinción y claridad en la Lámina se representa." Emphasis in original.

33 The inscription is executed by an anonymous engraver of lettering. Juan Caramuel 1678 [–1679], vol. 3, plate 23: "Vera ovalis peristylii et multiformium illud ornantium columnarum vestigium seu ichnographia. Nota bene: Vel unam in Peristylio Plinthum, esse exacte quadratam: aut vel unius Columnae Vestigium esse perfecte sphaericum, magnus, et frequenter a celeberrimis Architectis admissus, error est." Translation (with modification) from Morales 2012, 205. The second inscription at the bottom of the print is drawn from the Book of Proverbs, 8:29.

34 Juan Caramuel 1678 [–1679], vol. 2, treatise 8, 51: "Haze una plaza Oval o Ellíptica delante del Templo de San Pedro. Es verdaderamente Obra muy sumptuosa y bella en los ojos del Vulgo, que, como no sabe los fundamentos de la Architectura, con los sentidos Exteriores discurre; pero la miran con enfado quantos saben el Arte, por haver en ella tantos errores como piedras." For an argument that Caramuel's oblique architecture factored into the design for the

colonnade in 1656–1657, see Marino 1973, 81–120. Ortiz-Iribas (2005, 143–44) argues, to the contrary, that Caramuel had no role in designing the square, as he would then have claimed credit in his *Architectura*.

35 Concerning the modifications to the square, see Connors 1982, 24n38.

36 Juan Caramuel 1678 [–1679], vol. 2, treatise 8, 61: "*Los que en aquellos siglos ignorantes y rudos fueron tenidos por Milagros hoy no se tendrían por tales, porque nuestros ojos están enseñados a veer Cosas mayores y mejores.* Y aquí vuelve a entrar la distincción quoad se & quoad nos que se puso al principio. Porque las que en sí no son más que medianas, passan por marabillas en opinión de quien no ha visto cosa grande. *Habuerunt (ista Veteres Machina) honores, quia precesserunt tempore, & in rudi seculo quidquid emersisset Novum, per ora hominum iure ferebantur Eximium.* Palabras son de Cassiodoro, y tan a mi propósito y según mi parecer tan verdaderas, que me he visto obligado a ponerlas dos vezes." Emphasis in original. For the citation, see Cassiodorus 1973, 275.

37 As discussed in Mangone 2020, 6, 22, 126–28, 145, 150, and 227–28.

38 Translation from the entry dated August 23 in a journal describing Bernini's trip to Paris in de Chantelou 1985, 139–40. De Chantelou 2001, 134: "Le Cavalier a ajouté qu'un des points le plus important était d'avoir un bon oeil pour bien juger des *contrapposti*, que les choses nous paraissent non seulement ce qu'elles sont, mais eu égard à ce qui est dans leur voisinage, qui change leur apparence."

39 Translation from de Chantelou 1985, 140. De Chantelou 2001, 134–35: "Q'il avait fait autrefois une statue, laquelle finie, la tête en paraissait petite, quoiqu'elle eût sa juste grandeur, d'une neuvième partie, comme elle doit être au corps d'un Christ … Cette tête paraissant petite, et à lui et à un de ses amis … qu'enfin, à force d'examiner et de raisonner, qu'il découvrit que c'était un morceau de draperie posé sur l'épaule de la figure qui produisait cet effet, et, l'ayant beaucoup diminué, cela changea tout à fait l'apparence de cette tête, après quoi cet ami étant revenu et voyant cette figure, dont la tête paraissait d'une juste proportion, il en fut tout étonné et ne put s'imaginer comment il y avait donné du remède." See also Mangone 2020, 126.

40 Mangone 2020, 127. The similarity between Bernini's *contrapposti* and Michelangelo's *giudizio dell'occhio* is also noted in Summers 1981, 362; and Lavin 1980, 13. Concerning *giudizio* and *giudizio dell'occhio*, see Summers 1981, 352–79; Summers 1987, 21–49 and 151–81; Cropper 1983, 157–62; and Dempsey 1977, 56–58. See also Mangone's discussion of the capacity for discernment in relation to Michelangelo's *giudizio dell'occhio* (2020).

41 Vasari 1962, vol. 1, 117: "Dicendo che bisognava avere le seste negli occhi e non in mano, perché le mani operano e l'occhio giudica." Translation from Vasari 1965, 419.

42 Lomazzo 1584, book 5, chapter 7, 262: "Soleva Michel Angelo quel grandissimo scultore, pittore e architetto dire che non valevano negli huomini tutte le ragioni ne di Geometria, ne d'Aritmetica, ne essempi di prospettiva, senza l'occhio, cioè senza l'essercitatione dell'occhio in saper vedere e far fare alla mano." Translation expanded from Mangone 2020, 219. Concerning Lomazzo and the *giudizio dell'occhio*, see also Squizzato 2004, 78–82.

43 Translation from de Chantelou 1985, 140. De Chantelou 2001, 135: "Il a allégué … le portail de l'église de Saint-Pierre de Rome et les loges, qu'il a faites de part et d'autre, afin de le faire paraître plus haut par l'opposition de ces corps beaucoup moins élevés que ce portail, et a dit que cela avait réussi, comme il l'avait imaginé."

44 Juan Caramuel 1644, 151: "*Ita debet symetriam aedificii Architectus disponere, ut statuae videantur altitudinem habere naturalem.* Demonstratur. Quia statuae non sunt hominum monstra, sed figurae; ergo tantae esse debent magnitudinis, quanta requiritur ut eodem videantur angulo, quo proceri homines. Adsequetur, qui distantiam, & altitudinem consideret. In praesenti enim schemate, si columnae altitudo sit AB. & distantia oculi CB. dico statuam futuram esse multo majorem humana; & excessum ostendo. … Dico igitur statuam AD. visam a C. apparere aequalem homini FG. quia angulo aequali conspicitur, & in magnis distantiis magnitudines rerum apparentes dependent a visionis angulo." Emphasis in original.

45 See Frangenberg 1993, 214–24.

46 De Caus 1612, chapter 30: "POUR FAIRE UNE ESCRITURE CONTRE UNE MURAILLE HAUTE D'ONT LES LETTRES PAROISTRONT DE PAREILLE GRANDEUR LES HAUTES COMME LES BAS" ("How to create a script on a tall wall whose letters appear of the same size on top as below").

47 The command "Know thyself" (in Latin: *Nosce te ipsum*) was inscribed on the temple at Delphi, where it would have been rendered in Greek ("Γνῶθι σεαυτόν"); Socrates famously took this precept as his motto.

48 De Caus 1612, chapter 30. Dürer 1538, K2ᵛ.

49 Plato, *The Sophist*, 235 and 236a; translation by

Nicholas P. White from Plato 1997, 256; Plato 1602, 166: "Si enim veram pulchrorum commensurationem simulacris praeberent, minora quam decet superiora membra, maiora vero quae infra sunt, apparerent, propterea quod illa eminur, haec cominus a nobis conspiciantur. ... Ergo artifices, veritate dimissa, non eas quae revera sunt, sed que videntur pulchrae commensurationes, simulacris ipsis accommodant."

50 Juan Caramuel 1678 [–1679], vol. 3, part 4, p. 6: "LECTORI a praejudiciis libero. COLUMELLARUM (Hispanus eas Balaustres nominat) duos ordines repraesentamus. Inferiores ineptissimae sunt et tamen in scalis ubique terrarum ab Architectis conformatae. At rejici omnino debent, Superioresque earum loco substitui; ubi enim pavimentum AB, quod censetur fundamentalis linea inclinatur; duci ad libellam aliae C et D, E et F, G et H, I et K, etc. nequeunt; quoniam sic conductae inter se non cohaerent: et, ut cohaereant debent ipsimet horizontali AB. παραλλήλως (aequali in distantia) decurrere ut in superiori ordine repraesentatio."

51 Mersmann 2013, 373.

52 These balusters were made by Bastiano and Antonio Bencivenni, who flourished in the sixteenth century. Ortiz-Iribas 2005, 144–45. See also Pasqui and Pasqui 1880, 213–19.

53 Juan Caramuel 1644, 25: "*Turres & universa aedificia quae ad perpendiculum eriguntur, latiora sunt superius quam inferius*. Ostenditur. Turrium muri non sunt paralleli, nec apropiant [*sic*] versus coelum, sed versus centrum: ergo. Probatur assumptum. Quia perpendicula dimissa ex P & ex Q in centro I coinciderent. ergo istae lineae magis distant inter se, quo magis prolongantur a centro." Emphasis in original.

54 For another image of the tower, see Juan Caramuel 1678 [–1679], vol. 3, plate 26.

55 Vitruvius 1684. Perrault 1683. Perrault 1993.

56 See Alberto Pérez-Gómez, "Introduction," in Perrault 1993, 4.

57 Perrault 1683, 96: "Elles ne doivent plus estre changées & renduës differentes dans des Edifices differens, par des raisons d'Optique, & de la difference des aspects qu'ils peuvent avoir." Perrault 1993, 154: "[Proportions] should no longer be changed or made different in different buildings for optical reasons or because of the different aspects they may have."

58 Perrault 1683, 100: "Il est pourtant vrai qu'il y a des exemples dans l'Antique & dans le Moderne, qui font voir évidemment, qu'on a quelquefois eu dessein de changer les proportions par la raison de l'Aspect: mais outre que ces sortes de changemens sont rares, il est certain qu'ils font un tres-mauvais effet." Perrault 1993, 157: "It is true, however, that there are both ancient and modern examples that show clearly that people sometimes did try to change proportions because of the aspect, but these kinds of alterations, besides being rare, undeniably create a very adverse effect."

59 Perrault 1683, 102: "Ce jugement ne manque jamais ... d'empescher qu'on ne soit trompé par les alterations & les effets desavantageux que l'on s'imagine, que l'éloignement & les differentes situations sont capables de produire." Translation from Perrault 1993, 158.

60 Perrault 1683, 102: "Le jugement dont tous les sens sont pourvûs, est une chose que l'on a sans le sçavoir & sans s'appercevoir qu'on en use, à cause de l'accoûtumance, qui estant comme une seconde nature." Translation from Perrault 1993, 158.

61 Perrault 1683, 102: "Cela fait aussi que ceux d'entre les sens dont nous usons le plus ordinairement, tels que sont la vûë & l'oüie, ont un jugement bien plus exact que les autres sens, & qu'ils se trompent bien moins dans le discernement des circonstances, dans lesquelles la tromperie peut consister. Car ce qui fait que par la vûë & par l'oüie, on juge si certainement de l'éloignement de la grandeur & de la force des objets, & que les autres sens ne peuvent discerner si facilement ces circonstances." Translation (with modification) from Perrault 1993, 158.

62 Perrault 1683, 103: "Il est tres vraisemblale [*sic*] que les animaux à leur naissance voyent mal, & qu'ils jugent les objets éloignez aussi petits que la peinture faite dans leur oeil les leur represente, & qu'il faut que l'experience leur ayant fait connoistre qu'ils se sont trompez, corrige l'erreur de ce premier jugement, & que dans la suite le jugement s'accoûtume tellement à se servir de tous les moyens qu'il peut y avoir pour se défendre de cette tromperie, qu'enfin il parvienne à la perfection, dans laquelle il se trouve lorsqu'on commence à bien voir; & cette perfection est telle qu'il n'y a personne qui croie qu'une tour éloignée qui se couvre avec le doigt mis proche de l'oeil, soit moins grande que le doigt, ny qu'un rond vû obliquement soit une ovale, ou qu'une ovale soit un rond; quoique les images de ces choses soient actuellement telles dans l'oeil." Translation from Perrault 1993, 159.

63 Perrault 1683, 105: "Ce qui prouve la justesse & l'infaillibilité du jugement de la vûë, & fait connoistre certainement, que ce sens n'est point sujet à estre surpris & trompé comme on dit, c'est la difficulté que l'Art le plus parfait & le plus ingenieux trouve à en venir à bout quand il l'entreprend." Translation from Perrault 1993, 160.

64 Perrault 1683, 105. Perrault 1993, 161.

65 Perrault 1683, 106: "Si pour empescher que dans la sculpture la grande distance ne fasse paroistre les parties des ouvrages placez enhaut, trop confuses & trop peu distinctes, on la rend rude & grossiere, l'oeil la verra rude & grossiere; parce que comparant la distance qu'il connoist, avec la confusion qu'il sçait devoir estre dans les choses qui sont à cette distance, il sera offencé s'il y trouve cette distinction qu'il juge n'y devoir pas estre." Translation from Perrault 1993, 161.

66 Perrault 1683, 106: "Où le Peintre auroit rendu les choses éloignées aussi fortes & aussi distinctes que celles qui sont proches." Translation from Perrault 1993, 161.

67 Perrault 1683, 108: "Je croi que lors qu'on y aura bien pensé, on trouvera qu'il n'y a point de raison de corrompre & de graster les proportions, pour empescher qu'elles ne paroissent corrompues, & de rendre une chose defectueuse par l'intention que l'on a de la corriger; toutes les apparences que l'éloignement & la situation produisent, qui sont prises pour des mauvais effets et des défauts, estant le vrai état & la forme naturelle des choses, que l'on ne sçauroit changer sans les rendre visiblement difformes. Car tout ce qui a esté dit & qu'on peut dire sur ce sujet, est qu'il n'est point aussi certain que l'éloignement fasse paroistre les proportions autres qu'elles ne sont, comme il est certain que le changement de la proportion est effectivement la corruption de la proportion; & qu'il y a plus de danger qu'une proportion paroisse corrompue quand elle l'est, que quand elle ne l'est pas." Translation adapted from Perrault 1993, 162–63.

68 Perrault 1683, 107. Perrault 1993, 161.

69 Perrault 1683, 107–8: "Dans ces figures d'Optique, dont les proportions sont tellement ajustées, que si lors qu'elles sont vues d'un certain endroit, elles font un bon effet, on les trouve difformes aussi-tost que l'oeil est deplacé; les proportions qu'on auroit changées dans un édifice pour leur faire faire un bon effet à l'oeil situé en un certain endroit, paroistroient aussi tres-vicieuses, lors qu'on viendroit à changer de place: parce que l'aspect qui est oblique quand on est proche, cesse de l'estre à mesure qu'on s'éloigne: Et ainsi la face d'un Larmier qu'on auroit agrandie ou inclinée, pour empescher que l'obliquité de l'aspect ne la fit paroistre trop petite, la feroit paroistre trop grande, quand le changement d'aspect auroit fait cesser cette obliquité." Translation adapted from Perrault 1993, 162.

70 Concerning free will in the moral theology of Caramuel, see Armogathe 1990, 35–40; and Tutino 2017, 218.

71 The print is by an anonymous hand. Juan Caramuel 1644, 194–99. Reprinted (with some variations) in Juan Caramuel 1670, vol. 1, plate 33, figure 4; and Juan Caramuel 1678 [–1679], vol. 3, part 2, plate 47, fig. 4. For previous discussions of this diagram, see Mersmann 2013, 368–69; Ortiz-Iribas 2014, 458–60; and Ortiz-Iribas 2008, 297–300. On geometry as a method of explaining philosophical thinking, see also Van Veen 2017.

72 Juan Caramuel 1644, 194–95: "Sol gratiae Symbolum, Deo permittente oritur super bonos & malos. ... Naturae Luna symbolum; inconstantis inconstans, variae varia. Quid Luna, ni perfusa Sole? Quid Natura, ni delibuta Gratia?"

73 Juan Caramuel 1644, 195: "Hac doctrina praemissa mystici Solis & Lunae, sphaeras & motus contemplemur. Sit A terra, GFC Lunae orbis, HEI Solis circulus. KDL Sphaera Martis. L sit oriens, M occasus, K meridies. Posito igitur oculo in B, sola Luna F videbitur; mole enim sua totum Solem E abripiens faciet, ut ille a constitutis in B videri non possit. Posito in D Martis Sphaera; nimirum in loco ipsi B opposito, Sol unus visus, Luna invisa; E enim luminare maius mole sua & radiis luminare minus F occultat. At oculo constituto in C, Lunae circulo, videbitur Sol simul & Luna, ille in P, ipsa in Q. haec antesignana, ille pedissequus. Sane diversi aspectus, quoniam diversae parallaxes; nullus verus, quoniam omnes eccentrici; omnis enim aspectus eccentricus, exorbitans."

74 Juan Caramuel 1644, 198: "Pelagiani constituti in B, aërea superbiunt vanitatis regione: Lunae naturam, seu Lunam naturae observant & speculantur; nec Solem Gratiae, nec Gratiam Solis vident: severa Manichaeorum crudelitas armis & caedibus Ecclesiam turbans in D. Martis sphaera sidens, Solem gratiae videt, Lunam naturae: (hoc est, arbitrii liberii incrementa, decrementa) non videt; Nos ad C, ipsum Lunae circulum, appellemus, reponemusque, LUNAM PRAEEUNTEM, SEQUENTEM SOLEM; NATURAM PRAEVIAM, PEDISEQUAM GRATIAM OBSERVARI."

75 Juan Caramuel 1644, 199: "Vel semel veniamus ad A circulorum omnium verum & legitimum centrum, ubi theoscopiis optimis instructi Ecclesiae primitivae Patres, Doctores, & Scholastici, accuratissime Caelum observant, & observationes suas caelestes ad tabulas & numeros reducunt."

76 On probabilism, see Deman 1936, cols. 417–619; Schuessler 2019; Tutino 2017; Pastine 1975; Pissavino 1990; Fleming 2006; Leinsle 2008, 99–116; and Garavelli 2016.

77 De Medina 1578, 308: "Si est opinio probabilis,

licitum est eam sequi, licet opposita probabilior sit." Deman 1936, col. 463–70.

78 On the term probabilism, see Fleming 2006, xxi–xxii.

79 Cited by Schuessler 2019, 117.

80 Juan Caramuel 1657, fund. 57. See Schuessler 2019, 117; Fleming 2006, 17.

81 Tutino 2017, 169–70.

82 Juan Caramuel 1663, 63: "*Probabilitatum usus est res nova. Non potest non condemnari apud Deum, qui relinquens viam tutam, insistit opinioni probabili. Opiniones, quae dicuntur aut sunt probabiles apud nos, non erunt probabiles apud Deum.*" Emphasis in original. Cited and translated by Fleming 2006, 164n1.

83 Juan Caramuel 1663, 63: "Ego impugnavi in *Theologia Regulari*, quam anno 1639. Brugis edidi … uberius fortiusque in secunda & tertia eiusdem Operis editione." Cited and translated by Fleming 2006, 26. *Theologia regularis* first appeared in 1638. Juan Caramuel 1646.

84 Since 1998, when the papacy opened the archives of the Roman Inquisition and the Congregation of the Index, researchers have been revising older narratives about Italian ecclesiastical censorship. See Tarrant 2014, 1–27; Schutte 1999, 25–28; *L'Apertura degli archivi del Sant'Uffizio romano: Giornata di studio* 1998; Godman 2000; Cavarzere 2011; and Fragnito 2001.

85 [Bianchi] 1646. Schuessler 2019, 313. This rare book is held by the Libraria Universitaria Alessandrina in Rome.

86 Pascal 1656/1657, letters V–VIII. These were written under the pseudonym Louis de Montalte. See also Deman 1936, 501–14; and Schuessler 2019, 122–23. See Ginzburg and Biasiori 2018.

87 On González and tensions within the Society of Jesus over probabilism, see Gay 2012.

88 My study builds on previous considerations of the interrelations between Caramuel's ideas in moral theology and architecture, including Morales 2013, 145–53.

89 Tutino 2017, 169–70.

90 Juan Caramuel 1640, 23: "Non simus Angeli, sed homines, vix quatuor rerum habemus evidentiam, & ideo tenemur secundum probabiles sententias operari." Cited in Fleming (2006, 28) and cited and translated in Franklin 2001, 89.

91 Juan Caramuel 1663, 94: "Inter noctem & diem; inter manifestam falsitatem, & manifestam veritatem medio, quod *Probabilitatem* appellamus. Hoc Veritatis lumen S. Ioanni in Apocalypsi sua videtur ostendi, sub libri septem sigillis clausi specie. Quandiu nullum sigillum aperitur, ambulamus in tenebris, & veritatem penitus ignoramus. Cum septimum, aut Demonstratio naturalis, aut Fides Divina in hac vita, aut in altera visio beatifica recludit, Veritatem infallibiliter assequimur, exuimurque omni errandi formidine. Quando autem aliqua referantur, & aliqua non referantur, tenebrae & luces in nostro intellectu confligunt, & per probabiles semitas conducimur ad veritatem." Emphasis in original. Cited and translated in Fleming 2006, 105.

92 Juan Caramuel 1663, 37: "*distinctio, non semper est perceptibilis & cognoscibilis*." Emphasis in original. Cited and translated (with modification) in Fleming 2006, 111.

93 Juan Caramuel 1640, 26: "Si habes unum authorem pro tua parte, contra octo alios eiusdem qualitatis, affere tuam partem adhuc esse probabilem; contra novem non afferas."

94 Juan Caramuel 1640, 26: "Unum authorem maximae dignitatis aequivalere quinque authoribus nominis magni."

95 Juan Caramuel 1646, 25. The remarks concerning the quantification of probable opinions are cited and translated in Franklin 2001, 90.

96 Juan Caramuel 1663, a3^{r}: "*Demonstratur tandem Theologos, ita centum annis ultimis constrinxisse Conscientias, ut si aliis centum eodem impetu pergere permittantur, reddetur difficillima Infidelium conversio, et apud ipsos Orthodoxos inconvenientia maxima certissime timeri poterunt.*" Emphasis in original. Cited in Høyrup 2009, 59n14.

97 See Catalano 2002, 361.

98 For a discussion of Caramuel's involvement in proselytizing among Protestants in Bohemia, see Catalano 2002.

99 Juan Caramuel 1652, 137: "*Propositio quae hodie nec est probabilis, nec improbabilis, cras esse poterit probabilis, & perendie aut evidenter vera, aut evidenter falsa. Pono exemplum; Haec Propositio, Dantur Antipodes,* Augustini & aliorum veterum Doctorum aevo non fuit positive probabiles non enim dabantur evidentes rationes, quae illos esse, aut non esse probarent. … Terras aliquas ultramarinas esse, ab eo Columbus didicit, qui illuc procella abactus, eas lustrarat; credidit probabiliter gravi testimonio fundatus. Venit ad Principes, venit ad Reges; & quia erant contraria praeoccupati ignorantia, aut sententia, fuit irrisus: tandem Isabellae Hispaniae Regine suam sententiam persuasit: Ivit, vidit, rediit, novos homines, aves, & animantes attulit, & res negata ab omnibus transivit in certam, quae hodie a nullo ignoratur. Similiter agitur in Theologia Morali, multae sunt hodie opiniones

probabiles, quae heri tales non visae; & multae heri judicatae probabiles, quae re hodie melius examinata transiverunt in certas; videlicet aut certo falsas, aut certo veras." Emphasis in original.

100 Juan Caramuel 1652, 11: "O humanae doctrinae fundamenta! sub dubiam lucem Hesperi, judicantur certissima: clarissima dum noctem obscuram mens agit: et si mane, cum lucet veritatis Phosphorus, eadem quaesiero, non subsistunt. Forte stant ipsa cum maxime vacillare videntur. Forte adfingemus rebus oculorum errores; nam si inspicillo utimur flavo, croceas esse nives dicemus. Forte nostri hallucinantur oculi nimia claritate perculsi; sed hoc ipsum ignoro; & quia neutrum evidenter cognosco, neutrum pronuncio. Et quia vellem reperire vel punctum, ubi secure pedem figam, illudque in Terris non reperio, *Ad Te confugio, & supplex tua Numina posco*." Emphasis in original.

101 Caramuel's manuscript, which is held in the *Archivio capitolare* in the Vigevano archive, was first published and transcribed in Pastine 1972, 177–221. On this text and Caramuel's reflections on the philosophy of Descartes, see Orlando 2016. For a critical edition of the text, see Agostini and Orlando 2013, 105–237.

102 For the original quotation, see Descartes 1641, 33: "Illud omne esse verum quod valde clare et distincte percipio."

103 Agostini and Orlando 2013, 152: "Multa prius ut omnino certa et manifesta admitteris, quae tamen postea deprehendisti incerta, dubia aut falsa, inde evenit, si utar antiquis terminis, quod putaveris tuas ideas cognoscere rem ut erat; si tuis, quod valde claram et distinctam cognitionem te habere putares, cum nonnisi obscuram difformem, et dissonam obiecto haberis."

104 See Tutino 2017, 2; Clark 2009; Robbins 1998; and Robbins 2006.

105 Tutino 2017, 2.

106 Juan Caramuel 1642, 380: "In hoc *nunc aeternitatis* potest ipsis Deus decernere existentiam habendam in anno 2000. Etsi hoc faceret, in hoc *nunc temporis* essent futurae." Emphasis in original. Cited in Schmutz 2001, 399.

107 Schmutz 2001.

108 Juan Caramuel 1642, 395: "Praeterita enim praesentia et futura Deo sunt praesentia objective atque intentionaliter: & in hoc sensu ab aeterno fuerunt Deo praesentia, & in aeternum erunt. … An verbi gratia ut lapis lapidi coexistit in aedificio, ita etiam praeterita & futura physice & realiter iam coexistant et sint praesentia Deo in *Nunc* Aeternitatis." Emphasis in original. See Schmutz 2001, 429.

109 Juan Caramuel 1678 [–1679], vol. 1, "Tratado Proemial," 57:"Siendo Architecto y Theologo es raçon que la Theologia me subministre materia, en que exercitar mis Architectonicas contemplaciones." Cited and translated by Morales 2012, 25.

110 Villalpando and de Prado 1596–1605. See Lara 1999, 505–21.

111 Ezekiel 40–48. For some of Caramuel's references to Villalpando in the *Architectura*, see Juan Caramuel 1678 [–1679], vol. 2, treatise 5, 45–50, 70; vol. 2, treatise 6, 22–23, 29–30, 33–34, 36; vol. 2, treatise 7, 46; and vol. 2, treatise 9, 101, 103.

112 Juan Caramuel 1678 [–1679], vol. 1, "Tratado Proemial," 17.

CHAPTER 6: PLUMIER'S IVORIES AND THE INCOMPREHENSIBLE

1 Translation from Descartes 1984, 50. Descartes 1897–1909, vol. 7, 72: "Nempe, exempli causa, cum triangulum imaginor, non tantum intelligo illud esse figuram tribus lineis comprehensam, sed simul etiam istas tres lineas tanquam praesentes acie mentis intueor, atque hoc est quod imaginari appello. Si vero de chiliogono velim cogitare, equidem aeque bene intelligo illud esse figuram constantem mille lateribus, ac intelligo triangulum esse figuram constantem tribus; sed non eodem modo illa mille latera imaginor, sive tanquam praesentia intueor."

2 Aschengreen Piacenti 1967, 37 and 155. Gennaioli 2013a, 110.

3 On Settala, see Findlen 1994, 325–34.

4 Burning mirrors, tools of navigation, clocks, and other ingenious instruments drew his attention as well. Findlen 1994, 330.

5 Terzago 1666, 206: "Gli facesse vedere che anche in Lombardia vi erano i patienti ingegni della Germania." Cited (with slight modifications) in Fogolari 1900, 89.

6 Moxon 1680, 227. Cited in Connors 1990, 223n38.

7 I build here on the work of Noam Andrews and Joseph Connors, who have both stressed the seeming impossibility of turned ivories; Connors writes, "Turnery is an art which delights in straining the limits of tool credibility." Connors 1990, 223. Andrews 2022, 236.

8 ASR, Archivio Spada, vol. 454, c. 468^{r}: "Nondimeno il medesimo Cavaliere Bernini per verità disse a me molti anni sono avanti l'altare di S. Pietro che il solo Borromino intendeva questa professione, mà che non si contentava mai, e che voleva dentro una cosa cavare un'altra, e nell'altra l'altra senza finire mai." Cited and translated in Connors 1990, 232. Martin Raspe argues that it is improbable that Bernini was

referring to turning here and suggests instead that he was characterizing "Borromini's way of joining cellular spaces, thereby creating a potentially infinite pattern, a tendency that cannot be observed in the work of his Roman colleagues." Raspe 2012, 239–40. One wonders if Bernini could have been referring to both Borromini's way of joining cellular spaces and turnery.

9 Connors 1990, 232.

10 A restoration conducted by Bettina Schindler in 2000 revealed that the Medici sculpture consists of not one but five separate pieces of ivory. Gennaioli 2013a, 110.

11 Gennaioli 2013b, 116.

12 More work is needed on Plumier's accomplishments. Concerning his biography, see Whitmore 1967, 187–98. On his natural history research, see Pietsch 2017; Hrodej 1997, 93–117; Hanomou 2008, 51–71; Beltrán 2019, 272–93; and Beltrán 2017. On Plumier's interest in turning, see Connors 1990, 224–33. Niceron 1736, vol. 33, 398; and Achard 1787, vol. 4, 95–98 (95).

13 Plumier 1701, i^{r}: "Aux gens d'Eglise dans leurs benefices, tant pour faire exercice, que pour employer une partie du temps, & fuir l'oisiveté."

14 Texts on practices of fabrication produced by craftspeople between 1400 and 1800 often attach value to training the attention. Smith 2022, 5 and 128–29.

15 For turned sculptures inscribed with invocations of the divine, see, for example, Tesoro dei Granduchi di Palazzo Pitti, Florence, Bargello inv. 1879, nos. 60, 61, 66, 68, 72, and 73.

16 Hein, 2018 10–11.

17 Friese 1708, 269: "Gott der Meister des ersten Kugel Wercks der Welt. ... Der selbst die Himmels Scheib, so zierlich drehet um." Cited and translated in Maurice 1985, 16.

18 Aschengreen Piacenti 1989, 65–96.

19 Plumier 1701, ei^{r}.

20 Heiden 1640. A copy of this text is held in the regional library of Coburg. Transcription in Aschengreen Piacenti 1964, 82–98.

21 Plumier 1701, ei^{v}: "Je m'attachai auprès de ceux que j'apris être les plus habiles, entre lesquels notre R. P. Emanuel Magnan Religieux Minime comme moi de la Province de Toulouse, connu par toute l'Europe pour son rare sçavoir, & son heureux & excellent genie, fut celui qui le premier ajoûta aux lumieres que j'avois reçuës de mon Pere qui se plaisoit à cet exercice, tout ce qu'il sçavoit de plus curieux sur le tour." ("I attached myself to those whom I took to be the most skillful, among whom our R. P. Emanuel Maignan Religious Minim like me from the Province of Toulouse, known throughout Europe for his rare knowledge, & his happy & excellent genius, was the one who was the first to add all that he knew that was most curious about the lathe to the knowledge that I had gained from my father, who enjoyed this practice.")

22 Plumier 1693, aiii^{r}.

23 Plumier 1706; and Plumier 1749.

24 Plumier 1776.

25 See Whitmore 1967, 195. Manuscript collection of the Russian Academy of Sciences in St. Petersburg, *Rospis' knigam Petra I peredannym v biblioteku Akad: Nauk*, no. 57. Cited by Pietsch 2017, 30.

26 The original letter is missing, but a copy from the early 1700s survives in the Bibliothèque municipale de La Rochelle. See extract from ms. 867, fols. 147–52. Transcribed in Whitmore 1967, 278.

27 Plumier's drawings for the seventy plates for this book survive in the Bibliothèque de l'Arsenal in Paris. See BnF, Bibliothèque de l'Arsenal, ms. 3078.

28 Plumier 1701, eiii^{v}: "J'ai representé les plans geometriques, les profils, les perspectifs, & j'ai cotté de lettres & de chiffres toutes les parties des pieces, afin que se rapportant à l'explication, on put en lisant & voyant les figures, concevoir sans peine toute la construction des machines, y en ayant même quantité qu'on peut comprendre en regardant seulement les planches."

29 See, for instance, the image of a door knocker and keys in Réaumur 1722, plate 17. On the *Kunstbüch* and the *Kunstbüchlein*, see Leitch 2015, 189–208; Eamon 1994, 112–20; and Smith 2022.

30 Boyle 1682, first fold-out before p. 1.

31 Bartisch 1583; von Gersdorff 1517. Leopold-Sophien-Bibliothek, Überlingen, (O6*). For other, analogous depictions of tools, see da Carpi 1535, fols. 156^{r}, 157^{r}, 157^{v}, 158^{r}, and 158^{r}; and della Croce 1583, 7^{r}, 7^{v}, and 8^{r}.

32 Plumier 1701, title page and i^{r}. On turning, see Maurice 1985; and von Philippovich 1961. On the readers of *Kunstbücher*, see Smith 2022, 151–76.

33 Plumier 1701, ii^{v}.

34 Plumier 1701, ii^{v}.

35 Evelyn 1955, vol. 2, 192. Cited in Connors 1990, 221.

36 Plumier 1701, i^{v}.

37 Bourdieu 1984, 6.

38 Concerning continuous drive of a lathe, see Woodbury 1961, 45–49. On foot-powered fly wheels that rendered assistants unnecessary, see Holtzapffel 1976, 31–35. Regarding a drive mechanism represented by Leonardo that made it possible for a turner to work without an assistant, see Maurice 1985, 133.

39 Further innovations made it possible for amateurs, under the watchful eyes of those more experienced in the art, to create graceful forms with relative ease. On mechanical advancements to the lathe, see Holtzapffel 1976, 29–57; Maurice 1985, 131–38.

40 Plumier 1701, aiv^{v}.

41 My discussion of ivory draws on Shalem 2004, 14–15.

42 According to Burack 1984, 29.

43 As noted by Burack 1984, 29.

44 Pliny 1952, 331 (38.78.204).

45 Mark 2007, 189–211.

46 Aschengreen Piacenti 1964, 91.

47 Moore 1931, 41–47. See also Andrews 2022, 237.

48 Andrews 2022, 239–40.

49 Plumier 1701, av^{v}: "Il est constant que cet Art est presentement dans l'Europe l'occupation la plus serieuse des gens d'esprit & de merite, & entre le divertissement & les plaisirs raisonnables, celui qui est le plus consideré par ceux qui cherchent dans quelque exercice honnête le moyen d'eviter les défauts où jette une trop grande oisiveté de la vie."

50 Plumier 1701, eiii^{r}: "bannir l'oisiveté de la vie."

51 Plumier 1701, i^{r}.

52 Plumier 1701, i^{r-v}: "Enfin les personnes curieuses y verront les machines & les inventions dont on se sert pour faire ces beaux ouvrages de tour, qui paroissent incomprehensibles à ceux qui ne les connoissent pas."

53 Plumier 1701, av^{v}. On "delicacy" and its late seventeenth- and early eighteenth-century twofold connotation of material perishability and a person's sensitivity to understated pleasures, see Wunsch 2024, 6–13, 33–35, 45–52, 53-54, 69, 71, and 113–14.

54 Hein 2018, 10–11.

55 Plumier 1701, ei^{r}: "Le Sr du Rosay Bourgeois de Lyon conserve chez lui deux chandeliers & un crucifix faits d'ivoire diversement figurez, & si artistement travaillez, qu'il semble presqu'impossible que la main d'un homme ait été capable de produire sur le tour un ouvrage si delicat."

56 *Mémoires pour l'Histoire des Sciences & des beaux-Arts* 11 (May 1702): 3–16, at 4: "une delicatesse qui surprend, & qui paroit inconcevable."

57 Plumier 1701, eii^{r}: "La beauté & la delicatesse de ses ouvrages, qui sont en divers endroits l'ornement des cabinets des Princes, & sont considerez comme inimitables" ("The beauty and delicacy of his works, which are in various places the ornaments of the cabinets of princes, & are considered inimitable").

58 On the ideal of knowledge-by-doing, see Funkenstein 2018, 37. See also Smith 2004; Smith 2022; and Bertucci 2017.

59 Plumier 1701, eiii^{r}: "Je n'ai vu jusqu'à cette heure aucun ouvrage qui ait été fait sur le tour, que je n'aie imité de ma propre main, & fait moi-même les machines necessaires pour les executer" ("I have not seen until now any work that was made on the lathe that I have not imitated with my own hands and made myself the necessary machines to create them").

60 Plumier 1701, eii^{r}. As stated earlier, Plumier traveled to see the turned objects in the cabinet of the Grand Duke of Tuscany, some of which are discussed above. He describes these sculptures in complimentary terms that again accentuate the baffling nature of the art. Plumier 1701, ei^{r}: "On voit ... plusieurs beaux ouvrages ... si finiment & delicatement travaillez, qu'ils portent l'étonnement dans les esprits de ceux qui les voyent" ("One sees ... several beautiful works ... so finely and delicately crafted that they bring astonishment to the spirits of those who see them").

61 Plumier 1701, 172.

62 Plumier 1701, 172–73: "On voit dans l'un une étoile à pointe ... mais outre ce plusieurs globes les uns dans les autres, ce qui paroit embarrassant à ceux qui ne sçavent pas le tour, mais peu difficile à ceux qui en ont l'usage."

63 Plumier 1701, 24: "La pratique & l'usage vous l'enseigneront mieux qu'aucune theorie."

64 Smith 2022.

65 De Servière 1719, aiii^{r} and eiii^{v}.

66 De Servière 1719, eiv^{r-v}.

67 De Servière 1719, 3.

68 De Servière 1719, 3.

69 De Servière 1719, 4: "La beauté de cette piece consiste en la difficulté que l'on conçoit y avoir eu de la faire d'une seule boule d'ivoire."

70 De Servière 1719, 4.

71 De Servière 1719, 4.

72 De Servière 1719, 7.

73 Plumier 1701, aiv^{v}: "Sont encor apres sa mort les preuves de la beauté de son esprit, de l'excellence de son genie & de son habileté en cet art."

74 Plumier 1701, eiii^{r}: "Subtilité d'esprit pour concevoir & executer ce qu'il y a de plus rare & de plus mysterieux en cet art."

75 Bertucci 2017, 16–17. On the early modern symbolic associations of the putto, see Dempsey 2001; and Korey 2007. See also Shapin 1989, 554–63.

76 On the relationship between art and geometry in early modern Europe, see Andrews 2022.

77 Andrews 2022.

78 Kepler 1596, table 3.

79 Two manuscripts with drawings attributed to Leonardo survive: Milan, Biblioteca Ambrosiana, Ms. 170 sup.; and Geneva, Bibliothèque de Genève, Ms. l.e. 210.

80 As noted by Andrews 2022, 37.

81 Andrews 2022, 82–99.

82 Andrews 2022, 156.

83 Kemp 1990, 119–20. Niceron 1646, plates 6–11, 13, and 17.

84 Niceron 1652, 72: "Cette Perspective semble fort

difficile à raison de la grande diversité des plans & de leurs inclinations & faillies." Translation from Niceron 2019, 134.

85 Niceron 1652, 72; Niceron 2019, 134.

86 Haag 2007, 68–69.

87 Von Philippovich 1961, 331; and von Philippovich 1960, 159–78. Cali Buckley has found over 180 works in this genre, see Buckley 2020, 64–89.

88 Doppelmayr 1730, 299–300, 310–11, and table V.

89 Bornfleth et al. 1989, 72–73.

90 This model was sold in a Bonhams auction in London on November 2, 2011.

91 Bscherer, Volkamer, and Zick 1680.

92 Weber, Wolff, and Zick 1781: "Ganz unbegreiflich scheint das menschlich Ohr und Auge / Vom feinsten Elphenbein, natürlich ausgedreht, / Vergliedert ohne Fehl zum Nutz- und Lehrgebrauche / In der Anatomie; Sind Stucke von Rarität!" Cited and translated (with modification) in Buckley 2020, 80.

93 The print is designed and engraved by J. C. Zick, whose dates are unknown.

94 Plumier 1701, 133. See also Plumier 1701, eiv[v].

95 Plumier 1701, 133: "J'appelle … le bon gout cet agrément à la vûë qui d'abord satisfait l'esprit par le seul port & aspect de l'ouvrage."

96 Plumier 1701, 133: "Veritablement il est bien difficile de pouvoir expliquer ce bon gout, & d'en établir des regles précises, puisqu'il dépend plutot de l'idée & du genie des gens que d'aucune methode certaine. L'oeil seul en doit prescrire les regles & les loix, de même que dans l'art de peindre & dans l'architecture, où toutes les regles qu'on a pu donner n'ont jamais sçu former des ouvrages aussi agreables que ceux que bien souvent la seule idée ou le seul caprice imagine." Cited in Connors 1990, 226n48.

97 For more on Borromini's rules for practice, see Raspe 1994.

98 Plumier 1701, 133: "Notre illustre Pierre Puget de Marseille grand Peintre, Grand Sculpteur & ensemble grand Architecte, voyant un Livre d'Architecture que j'avois dessiné d'après le Palladio & le Vignole, me témoigna grand plaisir de le voir; mais il m'avoua en même tems que toutes les regles de ces autheurs étoient fort peu necessaires, & qu'il falloit que l'Architecte format lui même des desseins convenans aux lieux & aux situations où l'on devoit construire."

99 Plumier 1701, 134. On the assertion that Borromini owned a treatise by Michelangelo, see Connors 1990, 227–29.

100 Plumier 1701, 134.

101 Plumier 1701, 134.

102 Plumier 1701, 134: "On peut juger de la difficulté qu'il y a d'établir des regles précises & particulieres pour les ouvrages du tour; surtout étant presque infinis, & tous de differente nature."

103 Plumier again highlights the "variety and surprising delicacy" (*varieté & la delicatesse surprenante*) of turned objects (1701, aiv[v]). On variety see, Goldgar 2007, 119–20; and Shearman 1967, 75, 86, 92, 100–101, 105, and 139.

104 Clément 1867, 160.

105 Mersenne 1634, 14: "La Musique peut estre appellée la Perspective de l'ouie, comme la Perspective la Musique de l'oeil, & ces deux sens sont destinez aux sciences, & au service de l'esprit, comme les trois autres au service du corps." Concerning Mersenne's views on the power of music to impact souls and facilitate divine communion, see Kent 2023.

106 On musical "deformation" in eighteenth-century Europe, see, for example, Hepokoski and Darcy 2006.

107 On this composition, see Palisca 1994, 323.

108 I thank David Crook, Daniele V. Filippi, Bernhold Schmid, Noel O'Regan, Craig Monson, and Arnaldo Morelli for discussing motet performances in seventeenth-century Rome with me. For the subset of the Minim library in the Place Royale convent in Paris dedicated to intricate polyphonic choral compositions, see the *Index generalissimus omnium librorum bibliothecae conventus patrum minimorum Parisiensium*, Paris, Bibliothèque Mazarine, Ms. 4149, pp. 116–18. Cited in Kent 2023, 70n47.

109 Di Lasso 2005, 28–31: "Videmus nunc per speculum aenigmate, / Tunc autem facie ad faciem. / Nunc cognosco ex parte, / Tunc autem cognoscam sicut et cognitus sum." Forney 2006, 262.

110 On the connections between music and rhetoric in seventeenth-century Europe, see Wilson, Buelow, and Hoyt 2001.

111 This pattern was widely deployed in the sixteenth and seventeenth centuries in Italy and beyond, in, for example, the wooden ceiling of the anticamera in the Palazzo della Cancelleria in Rome or in a ceiling in the oil painting on panel *A Company in an Interior* (c. 1622–c. 1624) by Bartholomeus van Bassen (1590–1652) and Esaias van de Velde (1587–1630) in the Rijksmuseum in Amsterdam. For examples of the pattern before Serlio and Borromini, see Steinberg 1977, 252–60. Borromini innovated, however, in transposing the pattern from a flat surface or a barrel-vault, as it had been deployed in the past, for instance in the examples just mentioned and in the ring vault mosaic decoration in the church of Santa Costanza in Rome, onto a dome. As observed by Steinberg 1977, 225.

BIBLIOGRAPHY

Abramson, Daniel M. 2016. *Obsolescence: An Architectural History*. New Haven, CT: Yale University Press.

Achard, Claude-François. 1787. "Plumier (Charles)." *Dictionnaire de la Provence et du comté Venaissin*. Marseille: J. Mossy.

Acosta, José de. 2002. *Natural and Moral History of the Indies*. Edited by Jane E. Mangan. Durham, NC: Duke University Press.

Agostini, Igor, and Emanuela Orlando, eds. and trans. 2013. "Johannes Caramuel Lobkowitz, Osservazioni contro le *Meditationes* cartesiane." *Rivista cistercense* 30: 105–237.

Aguilonius, Franciscus. 1613. *Opticorum libri sex*. Antwerp: Balthasar I Moretus.

Albers, Geraldine. 2005. "La *Deposizione* di Daniele da Volterra: Un capolavoro manierista." *Art e dossier* 210: 42–47.

Alberti, Leon Battista. 1988. *On the Art of Building in Ten Books*. Translated by Joseph Rykwert, Neil Leach, and Robert Tavernor. Cambridge, MA: The MIT Press.

Aldini, Tobia. 1620. *Vera e natural effigie della pianta Indiana chiamata Maraco, Grandilla, e fior della Passione D.N.S.* Rome: s.n.

Alpers, Svetlana. 1983. *The Art of Describing: Dutch Art in the Seventeenth Century*. Chicago: University of Chicago Press.

Altrini, Maria. 2000. "L'attività edilizia nella Roma barocca (1600–1650)." In *Studi in onore di Ciro Manca*, edited by Donatella Strangio. Padua: CEDAM.

Andrews, Noam. 2022. *The Polyhedrists: Art and Geometry in the Long Sixteenth Century*. Cambridge, MA: The MIT Press.

Apianus, Petrus. 1539. *Cosmographia*. Antwerp: Arnold Birckman.

Apparato delle Solenni Quarant'ore celebrate nel Giesù di Roma l'Anno del Giubileo 1650. 1650. Rome: Vitale Mascardi.

Argan, Giulio Carlo. 1954. "La rettorica e l'arte barocca." *Lettere Italiane* 6, no. 3: 257–64.

Aristotle. 1552. *De celo et mondo Tradotto di Greco in Volgare Italiano: Per Antonio Bruccioli*. Venice: Bartholomeo Imperatore.

———. 1590. *Operum Aristotelis nova editio, Graece et Latine*. Edited by Isaac Casaubon. 2 vols. Lyon: Guillelmus Laemarius.

———. 1984. *The Complete Works of Aristotle*. Edited by Jonathan Barnes. 2 vols. Princeton, NJ: Princeton University Press.

Armogathe, J. R. 1977. *Theologia Cartesiana: L'explication physique de l'Eucharistie chez Descartes et dom Desgabets*. La Haye: Martinus Nijhoff.

———. 1990. "Probabilisme et Libre-Arbitre: La théologie morale de Caramuel y Lobkowitz." In *Le meraviglie del probabile: Juan Caramuel 1606–1682, atti del convegno internazionale di studi*, edited by Paolo Pissavino. Vigevano: Comune di Vigevano.

Aschengreen Piacenti, Kirsten. 1964. " 'Beschreibung eines von Helffenbein gedrehten Kunststücks ... beneben desselben geistliche Bedeutung,' von Markus Heiden, 1640." *Anzeiger des Germanischen Nationalmuseums*: 82–98.

———. 1967. *Il Museo degli argenti a Firenze*. Florence: Cassa di Risparmio.

———. 1989. "Ivory Towers: The Ivories of Coburg." *Franco Maria Ricci* 41: 65–96.

Auerbach, Erich. 1938. "*Figura*." *Archivum Romanicum* 22: 436–89.

Augustine. 1996. *De doctrina christiana*. Edited and translated by R.P.H. Greene. Oxford: Oxford University Press.

Baglione, Giovanni. 1935. *Le vite de'pittori, scultori, et architetti: Dal pontificato de Gregorio XIII del 1572 in fino a' tempi di papa Urbano Ottavo nel 1642*. Edited by V. Mariani. Rome: R. Istituto d'Archeologia e Storia dell'Arte.

Bailey, Gauvin Alexander. 1999. " 'Le style jésuite n'existe pas': Jesuit Corporate Culture and the Visual Arts." In *The Jesuits: Cultures, Sciences, and the Arts, 1540–1773*, edited by John W. O'Malley, S.J., Gauvin Alexander Bailey, Steven J. Harris, and T. Frank Kennedy, S.J. Toronto: University of Toronto Press.

Baldinucci, Filippo. 1948. *Vita di Gian Lorenzo Bernini (1682)*. Edited by Sergio Samek Ludovici. Milan: Edizioni del Milione.

Baldriga, Irene. 2002. *L'occhio della lince: I primi Lincei tra arte, scienza e collezionismo (1603–1630)*. Rome: Accademia Nazionale dei Lincei.

Baltrušaitis, Jurgis. 1977. *Anamorphic Art*. Translated by W. J. Strachan. Cambridge, UK: Chadwyck-Healey.

Barocchi, Paola, ed. 1960–1962. *Trattati d'arte del Cinquecento fra manierismo e controriforma*. 3 vols. Bari: Gius. Laterza & Figli.

Barthes, Roland. 1989. *Sade, Fourier, Loyola*. Translated by Richard Miller. Berkeley: University of California Press.

Bartisch, Georg. 1583. *Ophthalmodouleia, das ist Augendienst*. Dresden: Matthes Stöckel the Elder.

Bayle, Pierre. 1734. *Dictionnaire historique et critique*. Amsterdam: Compagnie des Libraires.

Beck, Ingamaj. 1971. "Il capitello composito a volute invertite: Saggio su una forma antica nella struttura borrominiana." *Analecta Romana Instituti Danici* 6: 225–34.

Beckmann, Johann. 1807–1809. *Litteratur der älteren Reisebeschreibungen*. 2 vols. Göttingen: Johann Friedrich Rower.

Bellori, Giovanni Pietro. 1976. *Le vite de'pittori, scultori, e architetti moderni*. Edited by Evelina Borea. Turin: Einaudi.

———. 2005. *The Lives of the Modern Painters, Sculptors,*

and Architects: A New Translation and Critical Edition. Translated by Alice Sedwick Wohl. Cambridge, UK: Cambridge University Press.

Beltrán, José. 2017. "Nature in Draft: Images and Overseas Natural History in the Work of Charles Plumier (1646–1704)." PhD diss., European University Institute.

———. 2019. "Nature *au naturel* in Late-Seventeenth-Century France." In *Ad vivum?: Visual Materials and the Vocabulary of Life-Likeness in Europe before 1800*, edited by Thomas Balfe, Joanna Woodall, and Claus Zittel. Leiden: Brill.

Beneš, Mirka. 1989. "Villa Pamphilj (1630–1670): Family, Gardens, and Land in Papal Rome." 3 vols. PhD diss., Yale University.

Berger, Susanna. 2014. "The Invention of Wisdom in Jean Chéron's Illustrated Thesis Print." *Intellectual History Review* 24, no. 3: 343–66.

———. 2017. *The Art of Philosophy: Visual Thinking in Europe from the Late Renaissance to the Early Enlightenment.* Princeton, NJ: Princeton University Press.

Berger, Susanna, and Sara J. Schechner. 2021. "Observations on Niccolò Tornioli's *The Astronomers*." *Annals of Science* 78, no. 4: 418–62.

Bertucci, Paola. 2017. *Artisanal Enlightenment: Science and the Mechanical Arts in Old Regime France.* New Haven, CT: Yale University Press.

Bettini, Mario. 1642. *Apiaria universae philosophiae mathematicae.* 2 vols. Bologna: Giovanni Battista Ferroni.

Bevilacqua, Alexander, and Helen Pfeifer. 2013. "Turquerie: Culture in Motion, 1650–1750." *Past & Present* 221, no. 1: 75–118.

Bezner, Frank. 2005. *Vela Veritatis: Hermeneutik, Wissen und Sprache in der Intellectual History des 12. Jahrhunderts. Leiden: Brill.*

[Bianchi, Andrea]. 1646. *De opinionum praxi.* Cremona: Paulus Pueronus.

Bleichmar, Daniela. 2017. *Visual Voyages: Images of Latin American Nature from Columbus to Darwin.* New Haven, CT: Yale University Press.

Blennow, Anna, and Stefano Fogelberg Rota, eds. 2019. *Rome and the Guidebook Tradition: From the Middle Ages to the 20th Century.* Berlin: De Gruyter.

Blum, Paul Richard, and Hans-Christian Günther. 2019. *Bossuet – Artist, Intellectual and Man of Politics.* Nordhausen: Verlag Traugott Bautz.

Blunt, Anthony. 1979. *Borromini.* Cambridge, MA: The Belknap Press.

———, ed. 1988. *Baroque and Rococo: Architecture and Decoration.* Ware, UK: Wordsworth Edition.

Bornfleth, Elisabeth, Claus Pese, Arnold Rauhut, and Peter Thumann, eds. 1989. *Das Gewerbemuseum der LGA im Germanischen Nationalmuseum.* Nuremberg: Landesgewerbeanstalt Bayern.

Borromeo, Federico. 2010. *Sacred Painting. Museum.* Edited and translated by Kenneth S. Rothwell, Jr. Cambridge, MA: Harvard University Press.

Bortot, Alessio. 2020. *Emmanuel Maignan e Francesco Borromini: Il progetto di una villa scientifica nella Roma barocca.* Siracusa: Lettera Ventidue.

Bösel, Richard, and Lydia Salviucci Insolera, eds. 2010. *Mirabili disinganni: Andrea Pozzo (Trento 1642-Vienna 1709); Pittore e architetto Gesuita.* Rome: Istituto Nazionale per la Grafica. Exhibition catalogue.

Bosse, Abraham. 1665. *Traité des pratiques geometrales et perspectives.* Paris: Bosse.

Bossuet, Jacques-Bénigne Lignel. 2001. "Premier sermon pour un des jours de la deuxième semaine de Carême: Prêche à la cour; Sur la Providence." In *Sermons.* Paris: Gallimard.

Bouhours, Dominique. 1671. *Les Entretiens d'Ariste et d'Eugène.* Paris: Sebastien Mabre-Cramoisy.

Bourdieu, Pierre. 1984. *Distinction: A Social Critique of the Judgement of Tastes.* Translated by Richard Nice. Cambridge, MA: Harvard University Press.

Bourdin, Pierre. 1661. *Le cours de mathematique.* Paris: Simon Benard.

Boyle, Robert. 1682. *New Experiments Physico-Mechanicall, Touching the Spring of the Air, and Its Effects.* Oxford: Printed by Miles Flesher for Richard Davis.

Bredekamp, Horst. 1995. *The Lure of Antiquity and the Cult of the Machine: The Kunstkammer and the Evolution of Nature, Art, and Technology.* Princeton, NJ: M. Wiener Publishers.

Bruhn, Karen. 2002. "Reforming Saint Peter: Protestant Constructions of Saint Peter the Apostle in Early Modern England." *The Sixteenth Century Journal* 33, no. 1: 33–49.

Bscherer, Daniel, Johann Georg Volkamer, and Stephan Zick. 1680. *Kurtze und mechanische Beschreibung dieses Kunst-Auges.* Nuremberg: Christian Sigmund Froberg.

Buckley, Cali. 2020. "Pathos, Eros, and Curiosity: The History and Reception of Ivory Anatomical Models from the Seventeenth Century to Today." *Nuncius* 35: 64–89.

Burack, Benjamin. 1984. *Ivory and Its Uses.* Rutland, VT: Charles E. Tuttle Company.

Burda-Stengel, Felix. 2001. *Andrea Pozzo and Video Art.* Philadelphia: Saint Joseph's University Press.

Butler, David. 1994. "Orazio Spada and His Architects: Amateurs and Professionals in Late-Seventeenth-Century Rome." *Journal of the Society of Architectural Historians* 53, no. 1: 61–79.

Cafà, Valeria. 2004. "19. Interventi sul Teatro Olimpico a Vicenza (1584–1585)." In *Vincenzo Scamozzi 1548–1616*, edited by Franco Barbieri and Guido Beltramini. Venice: Marsilio, Centro Internazionale di Studi di Architettura Andrea Palladio.

Cairns, Stephen, and Jane M. Jacobs. 2014. *Buildings Must Die: A Perverse View of Architecture.* Cambridge, MA: The MIT Press.

Camerota, Filippo. 2000a. "Le bizzarrie dell'ingegno: Architettura e scienza per villa Pamphili." In *Francesco Borromini: Atti del convegno internazionale*, edited by Christoph L. Frommel and Elisabeth Sladek. Milan: Electa.

———. 2000b. "Architecture and Science in Baroque Rome: The Mathematical Ornaments of Villa Pamphilj." *Nuncius* 15, no. 2: 611–38.

———. 2003. "L'architettura illusoria." In *Storia dell'architettura italiana: Il Seicento*, edited by Aurora Scotti Tosini. Milan: Electa.

———. 2011. "Il teatro delle idee: Prospettiva e scienze matematiche nel Seicento." In *Mirabili disinganni: Andrea Pozzo (Trento 1642-Vienna 1709); Pittore e architetto gesuita*, edited by Richard Bösel and Lydia Salviucci Insolera. Rome: Artemide.

Camille, Michael. 1985. "Seeing and Reading: Some Visual Implications of Medieval Literacy and Illiteracy." *Art History* 8, no. 1: 26–49.

Campbell, Gordon. 2003. "Pilgrimage." In *The Oxford Dictionary of the Renaissance*. Oxford: Oxford University Press. Accessed online March 23, 2020: https://www-oxfordreference-com.

Cañizares-Esguerra, Jorge. 2006. *Nature, Empire, and Nation: Explorations of the History of Science in the Iberian World*. Stanford, CA: Stanford University Press.

Capriotti, Adriana. 2022. " 'Ho inteso il gusto di Vostra Eminenza': *La Prospettiva e altre illusioni del cardinal Bernardino Spada*." In *La Prospettiva di Palazzo Spada*, edited by Adriana Capriotti and Augusto Roca de Amicis. Rome: Editoriale Artemide.

Capriotti, Adriana, and Augusto Roca de Amicis, eds. 2022. *La Prospettiva di Palazzo Spada*. Rome: Editoriale Artemide.

Caramuel y Lobkowitz, Juan. 1640. *In regulam divi Benedicti commentarius*. Bruges, Belgium: Breyghels.

———. 1642. *Rationalis et realis philosophia*. Louvain: E. de Witte.

———. 1644. *Mathesis audax*. Louvain: Andrea Bouvet.

———. 1646. *Theologia regularis*. Frankfurt: Theobald Schönwetter.

———. 1648. *Sacri Romani Imperii pax licita demonstrata*. Frankfurt: J. G. Schönwetter.

———. 1652. *Theologia moralis fundamentalis*. Frankfurt: Johann Gottfried Schönwetter.

———. 1657. *Theologia moralis fundamentalis*. Lyon: L. Anisson.

———. 1663. *Apologema pro antiquissima et universalissima doctrina de probabilitate*. Lyon: Anisson.

———. 1670. *Mathesis biceps*. Campania: Episcopal Press.

———. 1678 [1679]. *Architectura civil recta y oblicua*. 3 vols. Vigevano: Juan Caramuel.

———. 1681. *Architectura civilis recta et obliqua*. Vigevano: Juan Caramuel.

Cardella, Lorenzo. 1793. *Memorie storiche de' cardinali della Santa Romana Chiesa*. Vol. 6. Rome: Stamperia Pagliarini.

Carruthers, Mary. 1998. *The Craft of Thought*. Cambridge, UK: Cambridge University Press.

Casado, Antonio Quirós. 1988. "El tema del desengaño." In *Actas del V Seminario de Historia de la Filosofía Española*, edited by A. Heredia Serrano. Salamanca: Universidad de Salamanca.

Casanova, Cesarina. 1987. "Le donne come 'risorsa': Le politiche matrimoniali della famiglia Spada." *Memoria: Rivista di storia delle donne* 21: 56–78.

Cassiodorus. 1973. *Variarum libri XII*. Edited by A. J. Fridh [with his *De anima*]. Corpus Christianorum, Series Latina 96. Turnhout: Brepols.

Castelli, Pietro. 1625. *Exactissima descriptio rariorum quarundam plantarum, que continentur Rome in Horto Farnesiano*. Rome: Giacomo Mascardi.

Catalano, Alessandro. 2002 "Juan Caramuel Lobkowicz (1606–1682) e la riconquista delle coscienze in Boemia." *Römische historische Mitteilungen* 44: 339–92.

Cataneo, Pietro. 1567. *L'architettura*. Venice: Aldus.

Cavallo, Sandra, and Tessa Storey. 2013. *Healthy Living in Late Renaissance Italy*. Oxford: Oxford University Press.

Cavarzere, Marco. 2011. *La prassi della censura nell'Itallia del seicento: Tra repressione e mediazione*. Rome: Edizioni di storia e letteratura.

Cecili, Alessandro, Laura Farroni, and Matteo Flavio Mancini. 2022. "Il rilievo della Prospettiva a Palazzo Spada." In *La Prospettiva di Palazzo Spada*, edited by Adriana Capriotti and Augusto Roca de Amicis. Rome: Editoriale Artemide.

Ceñal, Ramón. 1952. "Emmanuel Maignan: Su vida, su obra, su influencia." *Revista de estudios politicos* 46, no. 66: 111–49.

Chazelle, Celia M. 2012. "Pictures, Books, and the Illiterate: Pope Gregory I's Letters to Serenus of Marseilles." *Word & Image* 6, no. 2: 138–53.

Chen, Hui-hung. 2004. "Encounters in Peoples, Religions, and Sciences." PhD diss., Brown University.

Chossat, Marcel. 1896. *Les Jésuites et leurs oeuvres à Avignon, 1563–1768*. Avignon: François Seguin.

Clark, Stuart. 2009. *Vanities of the Eye: Vision in Early Modern European Culture*. Oxford: Oxford University Press.

Clément, Pierre, ed. 1867. *L'Italie en 1671. Relation d'un voyage du marquis de Seignelay*. Paris: Didier.

Collacciani, Domenico, and Sophie Roux. 2017a. "La querelle optique de Bourdin et de Descartes à la lumière des thèses mathématiques soutenues au collège de Clermont." In *Chemins du cartésianisme*, edited by Antonella Del Prete and Raffaele Carbone. Paris: Garnier.

———. 2017b. "Pierre Bourdin, anti-cartésien ou jésuite ordinaire?" *Historia Philosophica* 15: 89–113.

———. 2021. "The Mathematics Theses Defended at *collège de Clermont* (1637–1682): How to Guard a Fortress in Times of War." In *Teaching Philosophy in Early Modern Europe: Text and Image*, edited by Susanna Berger and Daniel Garber. Cham, Switzerland: Springer.

Connors, Joseph. 1979. Review of *Architettura scultura e arti minori*, by Minna Heimbürger Ravalli, and *Palazzo Spada*, by Lionello Neppi. *Journal of the Society of Architectural Historians* 38, no. 2: 193–96.

———. 1980. *Borromini and the Roman Oratory: Style and Society*. Cambridge, MA: The MIT Press.

———. 1982. "Bernini's S. Andrea al Quirinale: Payments and Planning." *Journal of the Society of Architectural Historians* 41, no. 1: 15–37.

———. 1989. "Virgilio Spada's Defence of Borromini." *The Burlington Magazine* 131, no. 1031: 76–90.

———. 1990. "Ars Tornandi: Baroque Architecture and the

Lathe." *Journal of the Warburg and Courtauld Institutes,* 53: 217–36.
———. 1992. "Playing the Palace." *New York Review of Books* 39, no. 21: 58–62.
———. 1995. "A Copy of Borromini's S. Carlo alle Quattro Fontane in Gubbio." *Burlington Magazine* 137, no. 1110: 588–99.
———. 1999. "Un teorema sacro: San Carlo alle Quattro Fontane." In *Il giovane Borromini: Dagli esordi a San Carlo alle Quattro Fontane,* edited by Manuela Kahn-Rossi and Marco Franciolli. Milan: Skira. Exhibition catalogue.
———. 2000. "Poussin detrattore di Borromini." In *Francesco Borromini: Atti del convegno internazionale,* edited by Christoph L. Frommel and Elisabeth Sladek. Milan: Electa.
Connors, Joseph, and Louise Rice. 1990. *Specchio di Roma Barocca: Una guida inedita del XVII secolo.* Rome: Edizioni dell'Elefante.
Consentino, Giuseppe. 1970. "Le matematiche nella *Ratio studiorum* della Compagnia di Gesù." *Miscellanea storica ligure* 2: 171–213.
Copeland, Clare, and Jan Machielsen, eds. 2013. *Angels of Light? Sanctity and the Discernment of Spirits in the Early Modern Period.* Leiden: Brill.
Cousinié, Frédéric. 1994. "Voir le sacré: Perception et visibilité du maître-autel au XVIIe siècle." *Histoire de l'art* 28: 37–49.
Crary, Jonathan. 2001. *Suspensions of Perception: Attention, Spectacle, and Modern Culture.* Cambridge, MA: The MIT Press.
Criveller, Gianni. 2003. "Christ Introduced to Late Ming China by Giulio Aleni S.J. (1582–1649)." In *The Chinese Faces of Jesus Christ,* edited by Roman Malek. Sankt Augustin: Steyler Verlag.
Cropper, Elizabeth. 1983. Review of *Michelangelo and the Language of Art,* by David Summers. *Art Bulletin* 65, no. 1: 157–62.
da Bibbiena, Francesco Sestini. 1621. *Il maestro di camera.* Florence: Zanobi Pignoni.
da Carpi, Jacopo Berengario. 1535. *Tractatus perutilis et completus de factura cranei.* Venice: Giovanni Antonio de Nicolini da Sabio.
Damisch, Hubert. 1994. *The Origin of Perspective.* Translated by John Goodman. Cambridge, MA: The MIT Press.
Daston, Lorraine. 2001. *Eine kurze Geschichte der wissenschaftlichen Aufmerksamkeit.* Munich: Carl Friedrich von Seimans Stiftung.
———. 2003. "Attention and the Values of Nature in the Enlightenment." In *The Moral Authority of Nature,* edited by Daston and Fernando Vidal. Chicago: University of Chicago Press.
———. 2004. "Disciplines of Attention." In *A New History of German Literature,* edited by David Wellbery. Cambridge, MA: Belknap Press of Harvard University Press.
Daston, Lorraine, and Katharine Park. 1998. *Wonders and the Order of Nature, 1150–1750.* New York: Zone Books.
Davenport, Randi Lise. 2010. "The Changing Shapes of Knowledge in Spain 1627–1726: From *Dreams and Discourses* to *Universal Critical Theater.*" *Sjuttonhundratal: Nordic Yearbook for Eighteenth-Century Studies* 7: 16–31.
David, Johannes. 1601. *Veridicus Christianus.* Antwerp: Jan Moretus.
———. 1605. *Occasio arrepta, neglecta.* Antwerp: Jan Moretus.
———. 1607. *Paradisus sponsi et sponsae, et Pancarpium Marianum.* Antwerp: Jan Moretus.
Davies, Paul. 2013. "Framing the Miraculous: The Devotional Functions of Perspective in Italian Renaissance Tabernacle Design." *Art History* 36, no. 5: 898–921.
de Caus, Salomon. 1612. *La perspective avec la raison.* London: Jan Norton.
de Chantelou, Paul Fréart. 1985. *Diary of the Cavaliere Bernini's Visit to France.* Edited by Anthony Blunt and George C. Bauer. Translated by Margery Corbett. Princeton, NJ: Princeton University Press.
———. 2001. *Journal de voyage du Cavalier Bernini en France.* Edited by Milovan Stanić. Paris: Macula / L'Insulaire.
de Cornouaille, Pierre, and Jacques Manchon. 1638. *Deo hominique Jesu Christo eiusque matri Virgini Mariæ deiparæ Musæum mathematicum, D.D.C. [Dicant, dedicant, consecrant] Petrus de Cornouaille, Sylvanectensis, Jacobus Manchon, Parisinus, pro annua celebritate literaria collegii Claromontani Societatis Jesu.* s.l. s.n.
de Covarrubias Horozco, Sebastián. 2006. *Tesoro de la lengua castellana o española.* Edited by Ignacio Arellano and Rafael Zafra. Madrid: Iberoamericana.
de Dargenville, Dézallier. 1757. *Voyage pittoresque de Paris.* Paris: De Bure l'aîné.
de Feo, Vittorio. 1988. *Andrea Pozzo: Architettura e illusione.* Rome: Officina Edizioni.
Dekoninck, Ralph. 2005. *Ad imaginem: Status, fonctions, et usages de l'image dans la littérature spirituelle jésuite du XVII siècle.* Geneva: Droz.
de la Flor, Fernando Rodríguez. 2009. *Imago: La cultura visual y figurativa del Barroco.* Madrid: Abada Editores.
de la Grange, Jean-Baptiste. 1675. *Les principes de la philosophie, contre les nouveaux philosophes Descartes, Rohault, Regius, Gassendi, le P. Maignan, &c.* Paris: Georges Josse.
della Croce, Giovanni Andrea. 1583. *Cirugia universale e perfetta.* Venice: Giordano Ziletti.
Dell'Arco, Maurizio Fagiolo. 1997. *La festa barocca.* Rome: Edizioni de Luca.
Dell'Arco, Maurizio Fagiolo, and Silvia Carandini. 1977–1978. *L'effimero barocco: Strutture della festa nella Roma del' 600.* 2 vols. Rome: Bulzoni.
del Monte, Guidobaldo. 1600. *Perspectivae libri sex.* Pesaro: Hieronymus Concordia.
de Lubac, Henri. 1959–1964. *Exégèse médiévale: Les quatre sens de l'Écriture.* 4 vols. Paris: Aubier.
Deman, Thomas. 1936. "Probabilisme." In *Dictionnaire de théologie Catholique,* edited by A. Vacant, E. Mangenot, and E. Amann. Paris: Librairie Letouzey et Ané.
de Medina, Bartolomé. 1578. *Expositio in primam secundae angelici doctoris d. Thomae Aquinatis.* Salamanca: Mathias Gast.

Dempsey, Charles. 1977. *Annibale Carracci and the Beginnings of Baroque Style*. Glückstadt: J. J. Augustin.

———. 2001. *Inventing the Renaissance Putto*. Chapel Hill: University of North Carolina Press.

de Ovalle, Alonso. 1646. *Historica relacion de reyno de Chile*. Rome: Francisco Cavallo.

d'Eremita, Donato. 1619. *Vera effigie della Grandiglia detta Fior della Passione*. Naples: s.n.

De Rosa, Agostino, ed. 2013. *Jean François Niceron: Prospettiva, catottrica e magia artificiale*. Rome: Aracne editrice.

———, ed. 2019a. *Roma anamorfica: Prospettiva e illusionismo in epoca barocca*. Rome: Aracne editrice.

———. 2019b. "Postfazione." In *L'Arte del disegno a Palazzo Spada: L'Astrolabium Catoptrico-Gnomonicum di Emmanuel Maignan*. Rome: De Luca Editori D'Arte.

Desargues, Girard. 1987. "The *Perspective* (1636)." In *The Geometrical Work of Girard Desargues*, edited by J. V. Field and J. J. Gray. New York: Springer.

Descartes, René. 1641. *Meditationes de prima philosophia*. Paris: Michael Soli.

———. 1897–1909. *Oeuvres de Descartes*. Edited by Charles Adam and Paul Tannery. 11 vols. Paris: Léopold Cerf.

———. 1984. *The Philosophical Writings of Descartes*. Translated by John Cottingham, Robert Stoothoff, and Dugald Murdoch. Vol. 2. Cambridge, UK: Cambridge University Press.

De Servière, Grollier. 1719. *Receuil d'ouvrages curieux de mathematique et de mecanique, ou description du cabinet de Monsieur Grollier de Servière*. Lyon: David Forey.

Diez, Renato. 1997. "Le Quarantore: Una predica figurata." *La festa a Roma dal Rinascimento al 1870*. Edited by Marcello Fagiolo. Turin: Umberto Allemandi.

di Lasso, Orlando. 2005. *The Complete Motets 13*. Edited by Rebecca Wagner Oettinger. Middleton, WI: A-R Editions, Inc.

Di Matteo, Colette. 2015. *L'église et le couvent de la Trinité-des-Monts à Rome: Les décors restaurés*. Dijon: Éditions Faton.

Dimler, G. Richard. 2005. *The Jesuit Emblem: A Bibliography of Secondary Literature with Select Commentary and Description*. New York: AMS Press.

———. 2007. *Studies in Jesuit Emblem*. New York: AMS Press.

Dondé, Antoine. 1664. *Les figures et l'abbregé de la vie, de la mort, et des miracles de Saint François de Paule, Instituteur et Fondateur de l'Ordre des Minimes*. Paris: François Muguet.

———. 1671. *Les figures et l'abbregé de la vie, de la mort, et des miracles de Saint François de Paule, Instituteur et Fondateur de l'Ordre des Minimes*. Paris: François Muguet.

Doppelmayr, Johann Gabriel. 1730. *Historische Nachricht von den Nürnbergischen Mathematicis und Künstlern*. Nuremberg: Peter Conrad Monath.

Drake, Stillman. 1978. *Galileo at Work: His Scientific Biography*. Chicago: University of Chicago Press.

Drake, Susanna. 2014. "Origen's Veils: The Askēsis of Interpretation." *Church History* 83, no. 4: 815–42.

du Breuil, Jean. 1649. *Troisième et dernière partie de la perspective pratique où se voient les beautez et Raretez de cette Science*. Paris: Vve François Langlois.

———. 1663–1669. *La perspective pratique*. 2nd ed. 3 vols. Paris: Jean du Puis.

Du Halde, Jean-Baptiste. 1736. *Description géographique, historique, chronologique, politique, et physique de l'Empire de la Chine et de la Tartarie chinoise*. The Hague: Henri Scheurleer.

Dumesnil, Antoine Jules. 1865. *Voyageurs français en Italie depuis le seizième siècle jusqu'à nos jours*. Paris: Vve Jules Renouard.

Dunkelgrün, Theodor William. 2012. "The Multiplicity of Scripture: The Confluence of Textual Traditions in the Making of the Antwerp Polyglot Bible (1568–1573)." PhD diss., The University of Chicago.

———. 2024. *The Multiplicity of Scripture: The Making of the Antwerp Polyglot Bible*. Turnhout: Brepols.

Dupré, Sven. 2008. "Aguilón, Vitruvianism and His *Opticorum libri sex*." In *Innovation and Experience in Early Baroque in the Southern Netherlands: The Case of the Jesuit Church in Antwerp*, edited by Piet Lombaerde. Turnhout: Brepols.

Dürer, Albrecht. 1538. *Underweysung der Messung*. Nuremberg: Hieronymus Andreae.

Duro, Paul. 1997. *The Academy and the Limits of Painting in Seventeenth-Century France*. Cambridge, UK: Cambridge University Press.

du Vivier, Claude. 1622. *Vie et miracles de St. François de Paule Instituteur de l'Ordre des Peres Minimes*. Douai: Baltazar Bellere.

Dvořák, Petr, and Jacob Schmutz. 2008. "Caramuel in Prague: The Intellectual Roots of *Mitteleuropa*." In *Juan Caramuel Lobkowitz: The Last Scholastic Polymath*, edited by Petr Dvořák and Jacob Schmutz. Prague: Filosofia.

Eamon, William. 1994. *Science and the Secrets of Nature: Books of Secrets in Medieval and Early Modern Culture*. Princeton, NJ: Princeton University Press.

Ebert-Schifferer, Sybille. 1992. "'Ma c'hanno da fare i precetti dell'oratore con quelli della pittura?' Reflections on Guercino's Narrative Structure." In *Guercino: Master Painter of the Baroque*, edited by Denis Mahon. Washington, DC: National Gallery of Art. Exhibition catalogue.

Eire, Carlos M. N. 2016. *Reformations: The Early Modern World, 1450–1650*. New Haven, CT: Yale University Press.

———. 2023. *They Flew: A History of the Impossible*. New Haven, CT: Yale University Press.

Elliott, Brent, Luigi Guerrini, and David Pegler. 2015. *Flora: Federico Cesi's Botanical Manuscripts*. 3 vols. London: Harvey Miller Publishers.

Enders, Ludwig, ed. 1890. *Luther und Emser: Ihre Streitschriften aus dem Jahre 1521*. 2 vols. Halle: Max Niemeyer.

Enenkel, Karl, Walter S. Melion, and Wietse de Boer, eds. 2016. *Jesuit Image Theory, 1540–1740*. Leiden: Brill.

Erasmus, Desiderius. 1969–. *Opera omnia*. Amsterdam: North Holland.

"Eucharist." 2005. In *The Oxford Dictionary of the Christian Church*, edited by F. L. Cross and E. A. Livingston. Oxford: Oxford University Press. Accessed online February 14, 2020: https://www-oxfordreference-com.

Evelyn, John. 1955. *The Diary*. Edited by E. S. de Beer. Vol. 2. Oxford: Clarendon Press.

Evitascandolo, Cesare. 1598. *Il maestro di casa*. Rome: Vullieti.

Fairbairn, Lynda, ed. 1998. *Italian Renaissance Drawings from the Collection of Sir John Soane's Museum*. Vol. 2. London: Azimuth Editions.

Fantoni, Girolamo. 1989. "Le grandi meridian a riflessione." *Orologi: Le misure del tempo* 3, no. 4: 156–66.

Farago, Claire. 2011. "Reframing the Baroque: On Idolatry and the Threshold of Humanity." In *Rethinking the Baroque*, edited by Helen Hills. Surrey: Ashgate.

Farroni, Laura, ed. 2019. *L'arte del disegno a Palazzo Spada: L'Astrolabium Catoptrico-Gnomonicum di Emmanuel Maignan*. Rome: De Luca Editori D'Arte.

Favaro, Antonio. 1890–1909. "Avvertimento." In Galilei Galileo, *Opere*, edited by Favaro. 20 vols. Florence: G. Barbèra.

Feingold, Mordechai, ed. 2002. *Jesuit Science and the Republic of Letters*. Cambridge, MA: The MIT Press.

Feist, Ulrike. 2013. *Sonne, Mond und Venus: Visualisierungen astronomischen Wissens im frühneuzeitlichen Rom*. Berlin: Akademie Verlag.

———. 2016. "The Reflection Sundial at Palazzo Spada in Rome: The Mirror as Instrument, Symbol, and Metaphor." In *The Mirror in Medieval and Early Modern Culture: Specular Reflections*, edited by Nancy M. Frelick. Turnhout: Brepols.

Findlen, Paula. 1994. *Possessing Nature: Museums, Collecting, and Scientific Culture in Early Modern Italy*. Berkeley: University of California Press.

———, ed. 2004a. *Athanasius Kircher: The Last Man Who Knew Everything*. London: Routledge.

———. 2004b. "Science, Art, and Knowledge in Seventeenth-Century Rome." *Metascience* 13: 275–302.

———. 2006. "Anatomy Theaters, Botanical Gardens, and Natural History Collections." In *The Cambridge History of Science: Early Modern Science*, edited by Katherine Park and Lorraine Daston. Cambridge, UK: Cambridge University Press.

Finlay, John R. 2007. "The Qianlong Emperor's Western Vistas: Linear Perspective and Trompe l'Oeil Illusion in the European Palaces of the Yuanming Yuan." *Bulletin de l'École française d'Extrême-Orient* 94: 159–93.

Finocchiaro, Giuseppe. 1999. *Il Museo di curiosità di Virgilio Spada: Una raccolta romana del Seicento*. Rome: Fratelli Palombi Editori.

Fleischer, Cornell H. 2018. "A Mediterranean Apocalypse: Prophecies of Empire in the Fifteenth and Sixteenth Centuries" *Journal of the Economic and Social History of the Orient* 61: 18–90.

Fleming, Julia. 2006. *Defending Probabilism: The Moral Theology of Juan Caramuel*. Washington, DC: Georgetown University Press.

Fogolari, Gino. 1900. "Il Museo Settala: Contributo per la storia della coltura in Milano nel secolo XVII." *Archivio storico lombardo* 14: 58–126.

Forney, Kristine K. 2006. "The Netherlands, 1520–1640." In *European Music, 1520–1640*, edited by James Haar. Woodbridge, UK: Boydell Press.

Foucault, Michel. 2009. *Security, Territory, Population: Lectures at the Collège de France, 1977–78*. Edited by Michel Senellart. Translated by Graham Burchell. Basingstoke: Palgrave Macmillan.

Foucquet, Yves. 1646. *Exercitatio mathematica ad agones panegyricos. Cum Deo et Beata Virgini propugnator Yvo Foucquet, Parisinus, in Claromontano Parisiensis Societatis Jesu collegio*. Paris: s.n.

Fragnito, Gigliola. 2001. *Church, Censorship and Culture in Early Modern Italy*. Translated by Adrian Belton. Cambridge, UK: Cambridge University Press.

Francis of Paola. 1632. *Les règles des frères, des soeurs, et des fidèles d'un & de l'autre sexe de l'Ordre des Minimes*. Paris: Edme Martin.

Frangenberg, Thomas. 1993. "Optical Correction in Sixteenth-Century Theory and Practice." *Renaissance Studies* 7, no. 2: 207–28.

———. 2000. "Andrea Pozzo on the Ceiling Paintings in S. Ignazio." In *Pictorial Composition: From Medieval to Modern Art*, edited by Paul Taylor and François Quiviger. London: The Warburg Institute.

Franklin, James. 2001. *The Science of Conjecture: Evidence and Probability before Pascal*. Baltimore: Johns Hopkins University Press.

Fratini, Giulio, and Francesco Moriconi. 2010. "Datazione e attribuzione dell'anamorfosi di San Giovanni a Pathmos presso il Convento della Trinità dei Monti a Roma." *Mélanges de l'École française de Rome: Italie et Méditerranée modernes et contemporaines* 122, no. 1. Accessed online February 2, 2020: http://journals.openedition.org/mefrim/563.

Freedberg, David. 1989. *The Power of Images: Studies in the History and Theory of Response*. Chicago: University of Chicago Press.

———. 2002. *The Eye of the Lynx: Galileo, His Friends, and the Beginnings of Modern Natural History*. Chicago: University of Chicago Press.

Fried, Michael. 1988. *Absorption and Theatricality: Painting and Beholder in the Age of Diderot*. Chicago: University of Chicago Press.

Friese, Friedrich. 1708. *Der vornehmsten Künstler und Hand-wercker Ceremonial-Politica....* Leipzig: Groschuff.

Funkenstein, Amos. 2018. [1986]. *Theology and the Scientific Imagination from the Middle Ages to the Seventeenth Century*. Princeton, NJ: Princeton University Press.

Furetière, Antoine. 1690. *Dictionaire universel*. Rotterdam: Arnout & Reinier Leers.

Furttenbach, Joseph. 1640. *Architectura recreationis*. Augsburg: Johann Schultes.

Gage, Frances. 2008. "Exercise for Mind and Body: Giulio Mancini, Collecting, and the Beholding of Landscape

Painting in the Seventeenth Century." *Renaissance Quarterly* 61, no. 4: 1167–1207.

Gagnaire, Paul. 2003. "Le cadran solaire à réflexion du Père Maignan, à la Trinité des Monts." *Bulletin de l'ANCAHA* 97: 21–31.

Gal, Ofer, and Raz Chen-Morris. 2013. *Baroque Science*. Chicago: University of Chicago Press.

Galilei, Galileo. 1890–1909. *Le Opere di Galileo Galilei*. Edited by Antonio Favaro. 20 vols. Florence: G. Barbèra.

———. 1989. *Sidereus nuncius, or, The Sideral Messenger*. Translated by Albert van Helden. Chicago: University of Chicago Press.

Garavelli, Bianca. 2016. *Caramuel, Vescovo eclettico*. Bergamo: Moretti & Vitali.

García, Juan María Montijano, ed. 1999. *San Carlo alle Quattro Fontane di Francesco Borromini: Nella "Relatione della fabrica" di Fra Juan de San Buenaventura*. Milan: Il Polifilo.

Gay, Jean-Pascal. 2012. *Jesuit Civil Wars. Theology, Politics, and Government under Tirso González (1687–1705)*. Aldershot: Ashgate.

Gell, Alfred. 1992. "The Technology of Enchantment and the Enchantment of Technology." In *Anthropology, Art, and Aesthetics*, edited by Jeremy Coote and Anthony Shelton. Oxford: Clarendon Press.

Gennaioli, Riccardo. 2013a. "16." In *Diafane passioni: Avori barocchi dalle corti europee*, edited by Eike D. Schmidt and Maria Sframeli. Florence: Sillabe.

———. 2013b. "18." In *Diafane passioni: Avori barocchi dalle corti europee*, edited by Eike D. Schmidt and Maria Sframeli. Florence: Sillabe.

Giambullari, Pier Francesco. 1549. *Origine della lingua fiorentina altrimento Il Gello*. Florence: Lorenzo Torrentino.

Gieben, Servus. 1993. "Iconografia antoniana in due fogli di tesi del Museo Francescano di Roma." *Il Santo* 2, no. 3: 273–98.

Gigli, Giacinto. 1958. *Diario Romano (1608–1670)*. Edited by Giuseppe Ricciotti. Rome: Tumminelli.

Ginzburg, Carlo. 1985. "Ekphrasis and Quotation." *Tijdschrift voor Filosofie* 50: 3–19.

Ginzburg, Carlo, and Lucio Biasiori, eds. 2018. *A Historical Approach to Casuistry: Norms and Exceptions in a Comparative Perspective*. London: Bloomsbury Publishing.

Giordano, Luisa. 2006. "La facciata del duomo." In *Lo zelantissimo pastore e la città: Vigevano nell'età del vescovo Caramuel*. Pisa: ETS.

Godman, Peter. 2000. *The Saint as Censor: Robert Bellarmine between Inquisition and Index*. Leiden: Brill.

Goldgar, Anne. 2007. *Tulipmania: Money, Honor, and Knowledge in the Dutch Golden Age*. Chicago: University of Chicago Press.

Golvers, Noel. 1993. *The* Astronomia Europaea *of Ferdinand Verbiest, S.J. (Dillingen, 1687): Text, Translation, Notes, and Commentaries*. Nettetal: Steyler Verlag.

———. 1999. *François de Rougement, S.J., Missionary in Ch'ang-Shu (Chiang-Nan): A Study of the Account Book (1674–1676) and the Elogium*. Leuven: Leuven University Press.

Gombrich, Ernst. 1966. *Norm and Form. Studies in the Art of the Renaissance*. London: University of Chicago Press.

Gorman, Michael John. 2003. "Mathematics and Modesty in the Society of Jesus: The Problems of Christoph Grienberger." *Archimedes* 6: 1–120.

———. 2020. *The Scientific Counter-Revolution: The Jesuits and the Invention of Modern Science*. London: Bloomsbury Academic.

Gottardo, Ketty. 2011. "Cardinal Francesco Barberini and the *Specula Principum* Tradition." *Print Quarterly* 28, no. 3: 293–97.

Goujon, Patrick. 2019. "Elites and the Constitution of Jesuit Identity." In *The Oxford Handbook of the Jesuits*, edited by Ines G. Županov. Oxford: Oxford University Press.

Green, Ottis. 1970. "*Desengaño*." In *The Literary Mind of Medieval and Renaissance Spain*. Lexington: The University Press of Kentucky.

Gregory, Brad. 2008. "No Room for God? History, Science, Metaphysics, and the Study of Religion." *History and Theory* 47: 495–519.

Gregory the Great. 1982. *S. Gregorii Magni: Registrum epistularum libri VIII-XIV, Appendix*. Edited by Dag Norberg. Turnhout: Brepols.

Grootenboer, Hanneke. 2005. *The Rhetoric of Perspective*. Chicago: University of Chicago Press.

Guazzo, Stefano. 1596. *La civil conversazione*. Venice: Domenico Imberti.

Haag, Sabine. 2007. *Meisterwerke der Elfenbeinkunst*. Vol. 8. Vienna: KHM-Museumsverband.

Hamburger, Jeffrey F. 2000. "Speculations on Speculation: Vision and Perception in the Theory and Practice of Mystical Devotion." In *Deutsche Mystik im abendländischen Zusammenhang: Neu erschlossene Texte, neue methodische Ansätze, neue theoretische Konzepte*, edited by Walter Haug and Wolfram Schneider-Lastin. Tübingen: Max Niemeye Verlag.

Hanomou, Alice. 2008. "Les voyages du Père Charles Plumier (1646–1704): Des dessins de la faune et de la flore des îles françaises d'Amérique (Martinique, Guadeloupe et Saint Domingue) à la fin du XVIIe siècle." *Bulletin d'histoire et d'épistémologie des sciences de la vie* 15, no. 1: 51–71.

Harris, Neil. 1990. "The Operational Aesthetic." In *Popular Fiction: Technology, Ideology, Production, Reading*, edited by Tony Bennett. New York: Routledge.

Harrison, Peter. 2015. *The Territories of Science and Religion*. Chicago: University of Chicago Press.

Havelange, Carl. 1998. *De l'oeil et du monde: Une histoire du regard au seuil de la modernité*. Paris: Fayard.

Heiden, Markus. 1640. *Beschreibung eines von Helffenbein gedrehten Kunststuecks … Beneben desselben geistliche Bedeutung*. Coburg: Johann Eirich.

Heilbron, J. L. 2001. *The Sun in the Church: Cathedrals as Solar Observatories*. Cambridge, MA: Harvard University Press.

Heimbürger Ravalli, Minna. 1977. *Architettura scultura e arti minori nel barocco italiano: Ricerche nell'Archivio Spada*. Florence: Leo S. Olschki.

Hein, Jørgen. 2018. *Ivories and Narwhal Tusks at Rosenborg*

Castle: Catalogue of Carved and Turned Ivories and Narwhal Tusks in the Royal Danish Collection 1600–1875. Translated by James Manley. Copenhagen: Musuem Tusculanum Press.

Henri, Yves. 1639. *Augustæ caeli reginæ Mariæ Virgini Deiparæ eiusque integerrimo sponso D. Josepho Palatium mathematicum D.D.C. [Dicat, dedicat, consecrat] Yvo Henri Briocensis, pro annua celebritate literaria collegii Claromontani Societatis Jesu*. s.l. s.n.

Hepokoski, James A., and Warren Darcy. 2006. *Elements of Sonata Theory: Norms, Types, and Deformations in the Late-Eighteenth-Century Sonata*. Oxford: Oxford University Press.

Hibbard, Howard. 1972. "*Ut picturae sermones*: The First Painted Decorations of the Gesù." In *Baroque Art: The Jesuit Contribution*, edited by Rudolf Wittkower and Irma B. Jaffe. New York: Fordham University Press.

Hill, Jonathan. 2019. *The Architecture of Ruins: Designs on the Past, Present and Future*. London: Routledge.

Hills, Helen, ed. 2011. *Rethinking the Baroque*. Surrey: Ashgate.

Holtzapffel, John Jacob. 1976. *Hand or Simple Turning: Principles and Practice*. New York: Dover Publications.

Horn, Andrew. 2019a. *Andrea Pozzo and the Religious Theatre of the Seventeenth Century*. Philadelphia: Saint Joseph's University Press.

———. 2019b. "Andrea Pozzo and the Jesuit 'Theatres' of the Seventeenth Century." *Journal of Jesuit Studies* 6: 213–48.

Hornet, Jean-Marie. 1982. *Astronomie et astronomes en Provence, 1680–1730*. Aix-en-Provence: Edisud.

Høyrup, Jens. 2009. "Baroque Mind-Set and New Science: A Dialect of Seventeenth-Century High Culture." *Sartoniana* 22: 47–73.

Hrodej, Philippe. 1997. "Saint-Domingue en 1690: Les observations du père Plumier, botaniste provençal." *Revue française d'histoire d'outre-mer* 84, no. 317: 93–117.

Hugo, Herman. 1624. *Pia Desideria*. Antwerp: Henrik Aertssens.

Huret, Grégoire. 1670. *Optique de portraiture et peinture*. Paris: Grégoire Huret.

Hyman, Aaron M. 2021. *Rubens in Repeat: The Logic of the Copy in Colonial Latin America*. Los Angeles: Getty Publications.

———. 2022. "Baroque Assumptions." *21: Inquiries into Art, History, and the Visual* 3: 565–626.

Ignatius of Loyola. 1548. *Exercitia spiritualia*. Rome: Antonio Blado.

———. 1966. *Il racconto del Pellegrino: Autobiografia di sant'Ignazio di Loyola*. Edited by Roberto Calasso. Milan: Adelphi.

———. 1970. *The Constitutions of the Society of Jesus*. Translated by George E. Ganss. St. Louis, MO: Institute of Jesuit Sources.

———. 1992. *The Spiritual Exercises of Saint Ignatius*. Translated by George E. Ganss. Chicago: Loyola University Press.

Johnson, Eugene J. 2017. *Inventing the Opera House: Theater Architecture in Renaissance and Baroque Italy*. Cambridge, UK: Cambridge University Press.

Julien, Pascal. 2000. "L'anamorphose murale du collège jésuite d'Aix-en-Provence: jusqu'à Lisbonne par la barbe de saint Pierre." *Revue de l'art* 130: 17–26.

———. 2005. "Anamorphoses et visions miraculeuses du père Maignan (1602–1676)." *Mélanges de l'École française de Rome: Italie et Méditerranée* 117, no. 1: 45–71.

Kaufmann, Thomas DaCosta. 2004. *Toward a Geography of Art*. Chicago: University of Chicago Press.

———. 2011. "Discomfited by the Baroque: A Personal Journey." In *Rethinking the Baroque*, edited by Helen Hills. Surrey: Ashgate.

Kemp, Martin. 1987. "Perspective and Meaning: Illusion, Allusion and Collusion." In *Philosophy and the Visual Arts. Seeing and Abstraction*, edited by Andrew Harrison. Dordrecht: Springer.

———. 1990. *The Science of Art: Optical Themes in Western Art from Brunelleschi to Seurat*. New Haven, CT: Yale University Press.

Kent, Emily. 2023. "'Minimizing' Marin Mersenne: Tracing the Intellectual Culture of a Seventeenth-Century Parisian Convent." *History of Universities* 36, no. 2: 58–79.

Kepler, Johannes. 1596. *Prodomus dissertationum cosmographicarum, continens mysterium cosmographicum*. Tübingen: Georgius Gruppenbachius.

Kircher, Athanasius. 1635. *Primitiae gnomonicae catoptricae, hoc est, horologiographiae novae specularis*. Avignon: J. Piot.

———. 1636. *Prodromus coptus sive aegyptiacus*. Rome: Sacra Congregatio de Propaganda Fidei.

———. 1641. *Magnes, sive de arte magnetica*. Rome: Hermann Scheus.

———. 1646. *Ars magna lucis et umbrae*. Rome: Ludovico Grignani and Hermann Scheus.

———. 1650. *Musurgia universalis sive ars magna consoni et dissoni*. Vol. 2. Rome: Ludovio Grignani.

———. 1652–1655. *Oedipus Aegyptiacus*. Rome: Vitale Mascardi.

———. 1665. *Mundus subterraneus in XII libros digestus*. Amsterdam: Joannis Janssonius a Waesberge and Elizeus Weyerstraten.

———. 1667. *China monumentis qua sacris qua profanis nec non variis naturae et artis spectaculis, aliarumque rerum memorabilium argumentis illustrata*. Amsterdam: Jacob à Meurs.

———. 1671a. *Iter exstaticum coeleste*. Würzburg: Johann Andreas Endter and Heirs of Wolfgang Junior.

———. 1671b. *Ars magna lucis et umbrae*. Amsterdam: Joannis Janssonius a Waesberge and the Heirs of Elizaeus Weyerstraten.

Klein, Robert, and Henri Zerner. 1990. *Italian Art 1500–1600: Sources and Documents*. Evanston, IL: Northwestern University Press.

Kleutghen, Kristina. 2015. *Imperial Illusions: Crossing Pictorial Boundaries in the Qing Palaces*. Seattle: University of Washington Press.

Korey, Alexandra M. 2007. "Putti, Pleasure, and Pedagogy in

Sixteenth-Century Italian Prints and Decorative Arts." PhD diss., University of Chicago.
Kugler, Emil, and Leslie King. 2004. "A Brief History of the Passionflower." In *Passiflora: Passionflowers of the world*, edited by Torsten Ulmer and John MacDougal. Portland: Timber Press.
Laertius, Diogenes. 1545. *Le vite de gli illustri filosofi*. Venice: Vincenzo Valgrisi.
L'Apertura degli archivi del Sant'Uffizio romano: Giornata di studio. 1998. Rome: Accademia nazionale dei Lincei.
Lara, Jaime. 1999. "God's Good Taste: The Jesuit Aesthetics of Juan Bautista Villalpando in the Sixth and Tenth Centuries B.C.E." In *The Jesuits: Cultures, Sciences, and the Arts, 1540–1773*, edited by John W. O'Malley, S.J., Gauvin Alexander Bailey, Steven J. Harris, and T. Frank Kennedy, S.J. Toronto: University of Toronto Press.
Lavin, Irving. 1980. *Bernini and the Unity of the Visual Arts*. New York: Oxford University Press.
Leach, Mark Carter. 1987. *The Literary and Emblematic Activity of Herman Hugo, S.J. (1588–1629)*. Ann Arbor, MI: University Microfilms.
Le Dividich, Aude. 2003. "Afficher, distribuer: Usages éphémères de l'imprimé dans l'enseignement des sciences mathématiques en France au XVIIe siècle." *Revue de la Bibiothèque nationale de France* 14: 56–63.
Leeman, Fred. 1975. *Hidden Images: Games of Perception, Anamorphic Art, Illusion*. Translated by Ellyn Childs Allison and Margaret L. Kaplan. New York: Harry N. Abrams.
Lefaivre, Liane, and Alexander Tzonis. 2004. *The Emergence of Modern Architecture: A Documentary History from 1000 to 1810*. London: Routledge.
Lefèvre d'Ormesson, Olivier. 1860. *Journal d'Olivier Lefèvre d'Ormesson. 2 vols. Paris: Imprimerie Impériale.*
Leibniz, Gottfried Wilhelm. 1885. Die philosophischen Schriften. Edited by Carl Immanuel Gerhardt. Vol. 6. Berlin: Weidmannsche Buchhandlung.
———. 1985. *Theodicy: Essays on the Goodness of God, the Freedom of Man, and the Origin of Evil*. Translated by E. M. Huggard. La Salle, IL: Open Court.
Leinsle, Ulrich Gottfried. 2008. "Probabilismus im Kloster: Caramuels *Theologia regularis*." In *Juan Caramuel Lobkowitz: The Last Scholastic Polymath*, edited by Petr Dvořák and Jacob Schmutz. Prague: Filosofia.
Leitch, Stephanie. 2015. "Visual Acuity and the Physiognomer's Art of Observation." *Oxford Art Journal* 38: 189–208.
Leplatre, Olivier. 2002. "Spiritualité de l'anamorphose: Le Carême du Louvre, Bossuet." *L'information littéraire* 54, no. 4: 38–46.
Leti, Gregorio. 1675. *Itinerario della corte di Roma o vero teatro historico, cronologico e politico della sede apostolica dataria e cancellaria Romana*. Valenza: Pietro Francesco Guerini.
Lettres édifiantes et curieuses, escrites des Missions Étrangères. 1819. Lyon: J. Vernarel.
Levy, Evonne. 2004. *Propaganda and the Jesuit Baroque*. Berkeley: University of California Press.
———. 2015. *Baroque and the Political Language of Formalism (1845–1945): Burckhardt, Wölfflin, Gurlitt, Brinckmann, Sedlmayr*. Basel: Schwabe Verlag.
Levy, Evonne, and Jens Baumgarten. 2008. "Our Baroque Confection and the Baroque Survey." *Revista Canadiense de Estudios Hispánicos* 33, no. 1: 39–64.
Lister, Martin. 1699. *A Journey to Paris in the Year 1698*. London: Jacob Tonson.
Liva, Gabriella. 2019. "San Francesco di Paola: L'anamorfosi muraria di padre E. Maignan." In *Roma anamorfica: Prospettiva e illusionismo in epoca barocca*, edited by Agostino De Rosa. Rome: Aracne editrice.
Locke, John. 1975. *An Essay Concerning Human Understanding*. Oxford: Clarendon.
Lomazzo, Gian Paolo. 1584. *Trattato dell'arte della pittura*. Milan: Paolo Gottardo Pontio.
Lombardi, Ferruccio. 1992. *Roma. Palazzi, palazzetti, case: Progetto per un inventario 1200–1870*. Rome: Edilstampa.
———. 1993. *Roma. Chiese, conventi, chiostri: Progetto per un inventario 313–1925*. Rome: Edilstampa.
Lombraña, Julian Velarde. 1989. *Juan Caramuel, vida y obra*. Oviedo: Pentalfa.
Long, A. A., and D. N. Sedley, eds. 1987. *The Hellenistic Philosophers*. 2 vols. Cambridge, UK: Cambridge University Press.
Long, Pamela O. 2001. *Openness, Secrecy, Authorship: Technical Arts and the Culture of Knowledge from Antiquity to the Renaissance*. Baltimore: Johns Hopkins University Press.
López, José Ramón Marcaida. 2014. *Arte y ciencia en el Barroco español: Historia natural, coleccionismo y cultura visual*. Seville: Fundación Focus-Abengoa.
Lukács, Ladislao, ed. 1986. *Ratio atque institutio sudiorum Societatis Jesu (1586, 1591, 1599), Monumenta paedagogica Societatis Iesu*. Rome: Institutum historicum Societatis Iesu.
Lunadoro, Girolamo. 1635. *Relatione della Corte di Roma, e de' riti da osservarsi in essa, e de' suoi magistrati, e offitii: Con la loro distinta giurisdittione*. Padua: Paolo Frambotto.
Maignan, Emanuel. 1648. *Perspectiva horaria sive de horographia gnomonica tum theoretica, tum practica libri quatuor*. Rome: Rubeus.
———. 1653. *Cursus philosophicus concinnatus ex notissimis cuique principiis*. 4 vols. Toulouse: Raymundus Bosc.
———. 1672. *Sacra philosophia sive entis supernaturalis*. Lyon: Joannis Gregoire.
———. 1673. *Cursus philosophicus*. Lyon: Joannis Gregoire.
Malcolm, Noel. 2002. *Aspects of Hobbes*. Oxford: Clarendon.
Malebranche, Nicolas. 1963. *Oeuvres de Malebranche*. Edited by Geneviève Rodis-Lewis. Vol. 2. Paris: J. Vrin.
———. 1997. *The Search after Truth*. Edited and translated by Thomas M. Lennon and Paul J. Olscamp. Cambridge, UK: Cambridge University Press.
Mangone, Carolina. 2020. *Bernini's Michelangelo*. New Haven, CT: Yale University Press.

Manning, John, and Marc van Vaeck, eds. 1999. *The Jesuits and the Emblem Tradition*. Turnhout: Brepols.
Marder, T. A. 1997. *Bernini's Scala Regia at the Vatican Palace: Architecture, Sculpture, and Ritual*. Cambridge, UK: Cambridge University Press.
Marino, Angela Guidoni. 1973. "Il colonnato di piazza S. Pietro: Dall'architettura obliqua di Caramuel al 'classicismo' berniniano." *Palladio* 23: 81–120.
Mark, Peter. 2007. "Towards a Reassessment of the Dating and the Geographical Origins of the Luso-African Ivories, Fifteenth to Seventeenth Centuries." *History in Africa* 34: 189–211.
Martinelli, Fioravente. 1650. *Roma ricercata nel suo sito & nella scuola di tutti gli Antiquarii*. Rome: Biagio Deversin.
Massey, Lyle. 2007. *Picturing Space, Displacing Bodies: Anamorphosis in Early Modern Theories of Perspective*. University Park, PA: Penn State Press.
———. 2021. "Thaumaturgic Visions: Andrea Pozzo's St Ignatius Corridor." *Word & Image* 73, no. 3: 229–44.
Maurice, Klaus. 1985. *Sovereigns as Turners: Materials on a Machine Art by Princes*. Translated by Dorothy Ann Schade. Zürich: Verlag Ineichen.
McCall, Timothy. 2019. "Bramante's *coro finto*, Ex-Votos, and Cult Practice in Sforza Milan." *Renaissance Quarterly* 72, no. 1: 1–53.
McCormick, Thomas J. 1990. *Charles-Louis Clérisseau and the Genesis of Neo-Classicism*. Cambridge, MA: The MIT Press.
McPhee, Sarah. 2002. *Bernini and the Bell Towers: Architecture and Politics at the Vatican*. New Haven, CT: Yale University Press.
Melion, Walter S. 2001. "*Cordis circumcisio in spiritu*: Imitation and the Wounded Christ in Hendrick Goltzius's *Circumcision* of 1594." *Nederlands Kunsthistorisch Jaarboek* 52: 31–77.
———. 2003. "The Art of Vision in Jerome Nadal's *Adnotationes et meditationes in Evangelia*." In *Jerome Nadal, Annotations and Meditations on the Liturgical Gospels*, translated by Frederick A. Homann, S.J. Philadelphia: Saint Joseph's University Press.
———. 2009. *The Meditative Art: Studies in the Northern Devotional Print, 1550–1625*. Philadelphia: Saint Joseph's University Press.
———. 2016. "Introduction: The Jesuit Engagement with the Status and Functions of the Visual Image." In *Jesuit Image Theory*, edited by Wietse de Boer, Karl A. E. Enenkel, and Walter S. Melion. Leiden: Brill.
Melion, Walter S., James Clifton, and Michel Weemans, eds. 2014. *Imago Exegetica: Visual Images as Exegetical Instruments, 1400–1700*. Leiden: Brill.
Melion, Walter S. and Ralph Dekoninck. 2019. "Jesuit Illustrated Books." In *The Oxford Handbook of the Jesuits*, edited by Ines G. Županov. Oxford: Oxford University Press.
Melion, Walter S., Ralph Dekoninck, and Agnès Guiderdoni-Bruslé, eds. 2012. *Ut Pictura Meditatio: The Meditative Image in Northern Art, 1500–1700*. Turnhout: Brepols.
Mémoires pour l'Histoire des Sciences & des beaux-Arts. 1702. 11: 3–16.
Mercuriale, Girolamo. 1601. *De arte gymnastica*. Venice: Giunta.
Mersenne, Marin. 1634. *Questions inouies ou recreation des scavans*. Paris: Jaques Villery.
Mersmann, Jasmin. 2008. "Wandelgänge. Monumentale Anamorphosen im 17. Jahrhundert." In *Der entstellte Blick: Anamorphosen in Kunst, Literatur und Philosophie*, edited by Kyung-Ho Cha and Markus Rautzenberg. Leiden: Brill.
———. 2013. "Schieflagen. Die *Architectura obliqua* des Juan Caramuel y Lobkowitz." In *Mobile Eyes: Peripatetisches Sehen in den Bildkulturen der Vormoderne*, edited by David Ganz and Stefan Neuner. Munich: Fink.
Milizia, Francesco. 1768. *Le Vite di più celebri architetti d'ogni nazione e d'ogni tempo*. Rome: Paolo Giunchi Komarek.
Mochizuki, Mia M. 2019. "Jesuit Visual Culture in a Machine Age." In *The Oxford Handbook of the Jesuits*, edited by Ines G. Zupanov. Oxford: Oxford University Press.
Moore, Ernst D. 1931. *Ivory: Scourge of Africa*. New York: Harper & Brothers.
Morales, Maria Elisa Navarro. 2011. "History as Storytelling in the Account of the Eleven Orders of Architecture According to Juan Caramuel de Lobkowitz." In *Chora Six: Intervals in the Philosophy of Architecture*, edited by Alberto Pérez-Gómez and Stephen Parcell. Montreal: McGill-Queens University Press.
———. 2012. "Architectura civil recta y obliqua: A Critical Reading." PhD diss., McGill University.
———. 2013. "La arquitectura oblicua de Juan Caramuel, un enfoque casuístico." In *Barroco iberoamericano: identidades culturales de un imperio*, edited by Carmen López Calderón, María de los Ángeles Fernández Valle, and María Inmaculada Rodríguez Moya. Vol. 1. Santiago de Compostela: Andavira Editora.
Morello, Nicoletta. 2001. "Nel corpo della Terra: Il geocosmo di Athanasius Kircher." In *Athanasius Kircher: Il Museo del Mondo*, edited by Eugenio Lo Sardo. Rome: Edizioni de Luca.
Moretti, Massimo. 2016. "La decorazione cinquecentesca del chiostro della Trinità dei Monti: Arte e politica nei rapporti diplomatici tra la Francia e la Roma di Gregorio XIII." In *La chiesa e il convento della Trinità dei Monti: Ricerche, nuove letture, restauri*, edited by Colette Di Matteo and Sebastiano Roberto. Rome: De Luca editori d'arte.
Moxon, Joseph. 1680. *Mechanick Exercises, or The Doctrine of Handy-Works: Applied to the Art of Turning*. London: Joseph Moxon.
Musillo, Marco. 2016. *The Shining Inheritance: Italian Painters at the Qing Court, 1699–1812*. Los Angeles: Getty Research Institute.
Nadal, Jerome. 1595. *Adnotationes et meditationes in Evangelia*. Antwerp: Martinus Nutius.
Nadler, Steven. 1988. "Arnauld, Descartes, and Transubstantiation: Reconciling Cartesian Metaphysics and Real Presence." *Journal of the History of Ideas* 49, no. 2: 229–46.

Nederman, Cary J. 1998. "The Mirror Crack'd: The *Speculum Principum* as Political and Social Criticism in the Late Middle Ages." *The European Legacy* 3, no. 3: 18–38.

Neppi, Lionello. 1975. *Palazzo Spada*. Rome: Editalia.

———. 1983. "Punti di vista sulla prospettiva Spada." *Bollettino d'arte* 68: 105–18.

Niceron, Jean-François. 1638. *La perspective curieuse*. Paris: Pierre Billaine.

———. 1646. *Thaumaturgus opticus*. Paris: Francisus Langlois.

———. 1652. *La perspective curieuse*. Paris: window of François Langlois.

———. 2019. *Curious Perspective*. With a mathematical and historical commentary by James L. Hunt, John Sharp, and Dominique Raynaud. Tempe, Arizona: Arizona Center for Medieval & Renaissance Studies.

Niceron, J. P. 1736. "Charles Plumier." In *Mémoires pour servir à l'histoire des hommes illustres dans la république des lettres, avec un catalogue raisonné de leurs ouvrages*. Paris: Briasson.

Nieremberg, Juan Eusebio. 1635. *Historia naturae, maxime peregrinae*. Antwerp: Plantin.

O'Malley, John W, S.J. 1991. "Was Ignatius Loyola a Church Reformer? How to Look at Early Modern Catholicism." *The Catholic Historical Review* 77, no. 2: 177–93.

———. 1993. *The First Jesuits*. Cambridge, MA: Harvard University Press.

———. 2000. *Trent and All That: Renaming Catholicism in the Early Modern Era*. Cambridge, MA: Harvard University Press.

———. 2006. "Introduction: The Pastoral, Social, Ecclesiastical, Civic, and Cultural Mission of the Society of Jesus." In *The Jesuits II: Cultures, Sciences, and the Arts, 1540–1773*, edited by O'Malley, Gauvin Alexander Bailey, Steven J. Harris, T. Frank Kennedy. Toronto: University of Toronto Press.

———. 2013. "To Travel to Any Part of the World: Jerónimo Nadal and the Jesuit Vocation." In *Saints or Devils Incarnate? Studies in Jesuit History*, ed. John W. O'Malley, S.J. Leiden: Brill.

Onori, Lorenzo Mochi, and Rosella Vodret Adamo. 1989. *La Galleria Nazionale d'Arte Antica: Regesto delle didascalie*. Rome: Palombi.

———. 2007. *Galleria Nazionale d'Arte Antica: Palazzo Barberini, I dipinti, Catalogo sistematico*. Florence: L'Erma di Bretschneider.

Orlando, Emanuela. 2016. "Juan Caramuel y Lobkowitz (1606–1682) lettore di Descartes: Studio delle opere a stampa e dei testi manoscritti." PhD diss., École Pratique des Hautes Études.

Ortiz-Iribas, Jorge Fernández-Santos. 2005. "Classicism *Hispanico More*: Juan de Caramuel's Presence in Alexandrine Rome and Its Impact on His Architectural Theory." *Annali di architettura* 17: 137–65.

———. 2008. "Caramuel's 'Theoscope' and the Vindication of Prudential Probabilism." In *Juan Caramuel Lobkowitz: The Last Scholastic Polymath*, edited by Petr Dvořák and Jacob Schmutz. Prague: Filosofia.

———. 2014. *Juan Caramuel y la probable arquitectura*. Madrid: Centro de Estudios Europa Hispánica.

Ossa-Richardson, Anthony. 2019. *A History of Ambiguity*. Princeton, NJ: Princeton University Press.

Oudet, J. F. 1995. "Un théâtre astronomique en Avignon – Le Planétarium à miroir de Kircher (1632–1633) – Avant-project de reconstruction." In *Astronomie et Sciences Humaines*. Strasbourg: Observatoire Astronomique.

Paleotti, Gabriele. 2012. *Discourse on Sacred and Profane Images*. Translated by William McCuaig. Los Angeles: Getty Research Institute.

Palisca, Claude V. 1994. *Studies in the History of Italian Music and Music Theory*. Oxford: Clarendon Press.

Pallavicino, Sforza. 1656. *Istoria del Concilio di Trento*. Rome: Angelo Bernabo del Verme.

Panofsky, Erwin. 1954. *Galileo as a Critic of the Arts*. The Hague: Martinus Nijhoff.

———. 1956. "Galileo as a Critic of the Arts: Aesthetic Attitude and Scientific Thought." *Isis* 147, no. 1: 3–15.

———. 1980. "Die Perspektive als 'symbolische Form.'" In *Aufsätze zu Grundfragen der Kunstwissenschaft*, edited by Hariolf Oberer and Egon Verheyen. Berlin: Volker Spiess.

———. 1995. "What Is Baroque?" In *Three Essays on Style*, edited by Irving Lavin. Cambridge, MA: The MIT Press.

Parkinson, John. 1629. *Paradisi in Sole Paradisus Terrestris; or A Garden of All Sorts of Pleasant Flowers Which Our English Ayre will Permitt to Be Noursed*. London: Humfrey Lownes and Robert Young.

Pascal, Blaise. 1656/1657. *Les Provinciales ou les Lettres escrites par Louis de Montalte, à un Provincial de ses Amis, et aux R.R.P.P. Jesuites*. Cologne: Pierre de la Vallée.

———. 1948. *Greater Shorter Works of Pascal*. Translated by Emilie Cailliet and John C. Blankenagel. Philadelphia: Westminster Press.

———. 1954. *Les Pensées de Pascal*. Edited by Francis Kaplan. Paris: Gallimard, Bibliothèque de la Pléiade.

———. 1966. *Pensées*. Translated by A. J. Krailsheimer. Harmondsworth: Penguin.

———. 1991. *OEuvres complètes*. Edited by Jean Mesnard. Vol. 3. Paris: Desclée de Brouwer.

Pascoli, Lione. 1981. *Vite de' pittori, scultori, ed architetti viventi dai manoscritti 1383 e 1743 della Biblioteca comunale "Augusta" di Perugia*. Treviso: Libreria Editrice Canova.

Pasqui, Angiolo, and Ubaldo Pasqui. 1880. *La Cattedrale Aretina e suoi monumenti*. Arezzo: Ernesto Bellotti.

Pastine, Dino. 1972. "Caramuel contro Descartes: Obiezioni inedite alle Meditazioni." *Rivista critica di storia delle filosofia* 27: 177–221.

———. 1975. *Juan Caramuel: Probabilismo ed enciclopedia*. Florence: La Nuova Italia Editrice.

Patrizi, Giorgio, ed. 1990. *Stefano Guazzo e la civil conversazione*. Rome: Bulzoni editore.

Pedretti, Carlo. 1957. *Studi Vinciani: Documenti, analisi e inediti leonardeschi*. Geneva: E. Droz.

Pegon, Joseph. 1957. "IV: Période Moderne." ["Discernement des esprits"]. In *Dictionnaire de la Spiritualité*. Paris: Beauchesne.

Pereda, Felipe. 2018. *Crime and Illusion: The Art of Truth in the Spanish Golden Age*. Turnhout: Harvey Miller Publishers.

Pérez-Gómez, Alberto. 1993. "Introduction." In Claude Perrault, *Ordonnance for the Five Kinds of Columns after the Method of the Ancients*, translated by Kagis McEwen. Santa Monica, CA: The Getty Center for the History of Art and the Humanities.

Pernety, Antoine-Joseph. 1757. *Dictionnaire portatif de peinture, sculpture, et gravure*. Paris: Bauche.

Perrault, Claude. 1683. *Ordonnances des cinq espèces de colonnes selon la méthode des Anciens*. Paris: Jean-Baptiste Coignard.

———. 1993. *Ordonnance for the Five Kinds of Columns after the Method of the Ancients*. Translated by Kagis McEwen. Santa Monica, CA: The Getty Center for the History of Art and the Humanities.

Pescarzoli, Antonio, ed. 1957. *I libri di viaggio e le guide della raccolta Luigi Vittorio Fossati Bellani: Catalogo descrittivo*. 3 vols. Rome: Edizioni di Storia e Letteratura.

Pietsch, Theodore W. 2017. *Charles Plumier (1646–1704) and His Drawings of French and Caribbean Fishes*. Translated by Fanja Andriamialisoa and Agathe Larcher-Gosha. Paris: Publications scientifiques du Muéum.

Pimentel, Juan. 2014. "Flores de la pasión: Dos alegorías del Nuevo Mundo." In *Alegorías: Imagen y discurso en la España moderna*, edited by María Tausiet. Madrid: CSIC.

Pissavino, Paolo, ed. 1990. *Le meraviglie del probabile: Juan Caramuel 1606–1682*. Vigevano: Commune di Vigevano.

Plato. 1602. *Opera omnia*. Frankfurt: Claudius Marnius.

———. 1997. *Complete Works*. Edited by John M. Cooper. Indianapolis: Hackett Publishing Company.

Pliny. 1518. *C. Plynii Secundi Naturae Historiarum Libri XXXVII....* Haguenau: Thomas Anshelm.

———. 1952. *Natural History*. Translated by Harris Rackham. Cambridge, MA: Harvard University Press.

Plumier, Charles. 1693. *Description des plantes de l'Amérique*. Paris: L'Imprimerie royale.

———. 1701. *L'art de tourner en perfection*. Lyon: Jean Certe.

———. 1705. *Traité des fougères de l'Amérique*. Paris: L'Imprimerie royale.

———. 1706. *L'art de tourner en perfection*. Paris: Pierre Aubouin, Pierre Ribou, and Claude Jombert.

———. 1749. *L'art de tourner en perfection*. Paris: Charles Antoine Jombert.

———. 1776. *Die Kunst zu drechseln, oder alle Arten von Arbeit auf der Drehbank vollkommen zu verfertigen*. Leipzig: Breitkopf.

Pomian, Krzysztof. 1990. *Collectors and Curiosities: Paris and Venice, 1500–1800*. Translated by Elizabeth Wiles-Portier. Cambridge, UK: Polity Press.

Porras, Stephanie. 2023. *The First Viral Images: Maerten de Vos, Antwerp Print, and the Early Modern Globe*. University Park, PA: The Pennsylvania State University Press.

Porteman, Karel. 1996. *Emblematic Exhibitions (affixiones) at the Brussels Jesuit College (1630–1685): A Study of the Commemorative Manuscripts (Royal Library, Brussels)*. Brussels: Brepols.

Portoghesi, Paolo. 1967. *Borromini: Architettura come linguaggio*. Rome: Ugo Bozzi Editore.

———. 2000. "Rivisitando i disegni della Villa Pamphilj." In *Francesco Borromini: Atti del convegno internazionale*. Milan: Electa.

Postel, Guillaume. 1969. *Le thrésor des prophéties de l'univers*. Edited by François Secret. The Hague: Martinus Nijhoff.

Potier, Charles. 1640. *Deo Optimo Maximo Encyclopædia mathematica, ad agones panegyricos in Claromontano Parisiensis Societatis Jesu collegio. Agonista Carolus Potier Castrotheodoricus*. Paris: Alexandre Bourdan.

Pozzo, Andrea, S.J. 1693. *Prospettiva de' pittori e architetti: Perspectiva pictorum et architectorum*. 2 vols. Rome: Giovanni Giacomo Komarek.

———. 1694. *Copia d'una Lettera scritta da Andrea Pozzo della Compagnia di Giesu Pittore all'Illustrissimo ed Eccellentissimo Principe Antonio Floriano di Liechtenstein Ambasciadore dell'Angustissimo Imperadore Leopoldo Ignazio presso la Santita di Nostro Signore Papa Innocenzo Duodecimo circa alli Significati della Volta da lui dipinta nel Tempio di Sant'Ignazio in Roma*. Rome: Giovanni Giacomo Komarek.

———. 1702. *Prospettiva de' pittori e architetti: Perspectiva pictorum et architectorum*. 2 vols. Rome: Antonius de Rubeis.

Praz, Mario. 1975. *Studies in Seventeenth-Century Imagery*. Rome: Edizioni di Storia e Letteratura.

Preston, Claire. 2007. "Ekphrasis: Painting in Words." In *Renaissance Figures of Speech*, edited by Sylvia Adamson, Gavin Alexander, and Katrin Ettenhuber. Cambridge, UK: Cambridge University Press.

Prieto, Andrés I. 2011. *Missionary Scientists: Jesuit Science in Spanish South America, 1570–1810*. Nashville: Vanderbilt University Press.

Prodi, Paolo. 2022. *Il cardinale Gabriele Paleotti (1522–1597)*. Bologna: Il Mulino.

Pütter, Johann Stephan. 1776. *Die Augsburgische confession*. Göttingen: Wittwe Vandenhoeck.

Quintilian. 2014. *The Orator's Education*. Translation by Donald A. Russell. Cambridge, MA: Harvard University Press.

Raspe, Martin. 1994. *Das Architektursystem Borrominis*. Munich: Deutscher Kunstverlag.

———. 2012. Review of *Borromini's Book: The "Full Relation of the Building" of the Roman Oratory by Francesco Borromini and Virgilio Spada of the Oratory* by Kerry Downes. *Journal of the Society of Architectural Historians* 71, no. 2: 238–40.

Raynaud, Théophile. 1665–1669. *Opera omnia*. 20 vols. Lyon: Horace Boissat and George Remeus.

Réaumur, René. 1722. *L'art de convertir le fer forgé*. Paris: Brunet.

Reed, Marcia, and Paola Demattè, eds. 2007. *China on Paper: European and Chinese Works from the Late Sixteenth to the Early Nineteenth Century*. Los Angeles: Getty Research Institute.

Regio, Paolo. 1582. *La miracolosa vita di San Francesco da Paola*. Naples: Giovanni Battista Cappelli.

Rice, Louise. 1999. "Jesuit Thesis Prints and the Festive Academic Defense at the Collegio Romano." In *The Jesuits: Cultures, Sciences, and the Arts, 1540–1773*, edited by John O'Malley, S.J., Gauvin Alexander Bailey, Steven J. Harris, and T. Frank Kennedy, S.J. Toronto: University of Toronto Press.

———. 2007. "*'Pomis Sua Nomina Servant'*: The Emblematic Thesis Prints of the Roman Seminary." *Journal of the Warburg and Courtauld Institutes* 70: 195–246.

Richter, Jean Paul. 1977. *The Literary Works of Leonardo da Vinci*. Revised edition with commentary by Carlo Pedretti. 2 vols. Berkeley: University of California Press.

Robbins, Jeremy. 1998. *The Challenges of Uncertainty: An Introduction to Seventeenth-Century Literature*. London: Duckworth.

———. 2006. *The Arts of Perception: The Epistemological Mentality of the Spanish Baroque, 1580–1720*. London: Routledge.

Roca de Amicis, Augusto. 2022. "Forma e illusione nella Prospettiva di Palazzo Spada." In *La Prospettiva di Palazzo Spada*, edited by Adriana Capriotti and Augusto Roca de Amicis. Rome: Editoriale Artemide.

Rodis-Lewis, Geneviève. 1956. "Machineries et Perspectives curieuses dans leurs rapports avec le Cartésianisme." *XVII siècle* 32: 461–74.

Rohr, René R. J. 1990. "On Reflected Ceiling Dials." *Bulletin of the British Sundial Society* 3: 5–11.

Romani, Vittoria, ed. 2003. *Daniele da Volterra: Amico di Michelangelo*. Florence: Mandragora. Exhibition Catalogue.

Rosales, Luis. 1966. *El sentimiento del desengaño en la poesía barroca*. Madrid: Ediciones Cultura Hispánica.

Rosignoli, Carlo Gregorio. 1708. *Meraviglie della natura, ammaestramenti di moralità*. Venice: Andreas Poletti.

Rowland, Ingrid. 2002. "Through a Glass, Darkly." *New York Review of Books* 49, no. 3: 10–14.

Rudolph, Conrad. 2011. "Inventing the Exegetical Stained-Glass Window: Suger, Hugh, and a New Elite Art." *Art Bulletin* 93, no. 4: 399–422.

Russell, Camilla. 2022. *Being a Jesuit in Renaissance Italy: Biographical Writing in the Early Global Age*. Cambridge, MA: Harvard University Press.

Ryff, Walther Hermann. 1547. *Der furnembsten, notwendigsten, der gantzen Architectur angehörigen Mathematischen und Mechanischen künst, eygentlicher bericht …* Nuremberg: Johan Petreius.

Sabbattini, Nicola. 1638. *Pratica di fabricar scene e machine ne'teatri*. Ravenna: Pietro de' Paoli and Giovanni Battista Giovannelli.

Sabiano, Daniele, and Paolo C. Pissavino, eds. 2012. *Un'altra modernità: Juan Caramuel Lobkowitz (1606–1682); Enciclopedia e probabilismo*. Pisa: ETS.

Saenger, Paul and Kimberly van Kampen, eds. 1999. *The Bible as Book: The First Printed Editions*. London: Oak Knoll Press.

Saenredam, Pieter, and Cornelis Korning. 1628. *Print to Belie Rumors about the Images Found in an Apple Tree*. Haarlem: Adriaen Rooman.

Saguens, Johannes. 1703. *Philosophia Maignani scholastica. …* Toulouse: Johannes Vialar.

Samuel, Edgar R. 1963. "Death in the Glass—A New View of Holbein's 'Ambassadors.'" *The Burlington Magazine* 105, no. 727: 436–41.

Schleier, Reinhart. 1973. *Tabula Cebetis oder "Spiegel Des Menschlichen Lebens / darin Tugent und untugent abgemalet ist"*. Berlin: Gebr. Mann Verlag.

Schmutz, Jacob. 2000. "Caramuel y Lobkowitz, Juan." In *Biographisch-Bibliographisches Kirchenlexikon*, edited by Friedrich Wilhelm Bautz and Traugott Bautz. Herzberg: Verlag Traugott Bautz.

———. 2001. "Juan Caramuel on the Year 2000: Time and Possible Worlds in Early Modern Scholasticism." In *The Medieval Concept of Time: Studies on the Scholastic Debate and Its Reception in Early Modern Philosophy*, edited by Pasquale Porro. Leiden: Brill.

Scholar, Richard. 2005. *The* Je-Ne-Sais-Quoi *in Early Modern Europe: Encounters with a Certain Something*. Oxford: Oxford University Press.

Schott, Gaspar. 1657. *Magia universalis*. Würzburg: Joannis Godefridi Schönwetter.

Schroeder, H. J., ed. 1960. *The Canons and Decrees of the Council of Trent*. St. Louis, MO: B. Herder.

Schudt, Ludwig. 1930. *Le guide di Roma*. Vienna: Filser.

———. 1959. *Italienreisen im 17. und 18. Jahrhundert*. Vienna: Schroll Verlag.

Schuessler, Rudolf. 2019. *The Debate on Probable Opinions in the Scholastic Tradition*. Leiden: Brill.

Schulte, Hansgerd. 1969. *El desengaño: Wort und Thema in der Spanischen Literatur des Goldenen Zeitalters*. Munich: Fink.

Schutte, Anne Jacobson. 1999. "Palazzo del Sant'Uffizio: The Opening of the Roman Inquisition's Central Archive." *Perspectives on History* 37, no. 5: 25–28.

Schwenger, Peter. 1976. "Crashaw's Perspectivist Metaphor." *Comparative Literature* 28, no. 1: 65–74.

Serlio, Sebastiano. 1566. *Libro primo d'architettura, di Sebastiano Serlio Bolognese….* Venice: Francesco Senese & Zuane Krugher Alemanno.

Sextus Empiricus. 1562. *Sexti Philosophi Pyrrhoniarum hypotyposeon libri III, Quibus in tres philosophiae partes severissime inquiritur*. Paris: Henri Estienne.

Shalem, Avinoam. 2004. *The Oliphant: Islamic Objects in Historical Context*. Leiden: Brill.

Shapin, Steven. 1989. "The Invisible Technician." *American Scientist* 77: 554–63.

———. 1994. *A Social History of Truth: Civility and Science in Seventeenth-Century England*. Chicago: University of Chicago Press.

Shearman, John. 1967. *Mannerism*. Harmondsworth: Penguin.

Siraisi, Nancy. 2003. "History, Antiquarianism and Medicine: The Case of Girolamo Mercuriale." *Journal of the History of Ideas* 64: 231–51.
Sluhovsky, Moshe. 2007. *Believe Not Every Spirit: Possession, Mysticism, and Discernment in Early Modern Catholicism*. Chicago: University of Chicago Press.
Sluiter, Ineke. 2016. "Obscurity." *Canonical Texts and Scholarly Practices: A Global Comparative Approach*. Cambridge, UK: Cambridge University Press.
Smith, G. 1976. "On the Original Location of Rosso Fiorentino's *Descent from the Cross*." *Zeitschrift für Kunstgeschichte* 39: 67–70.
Smith, Jeffrey Chipps. 2002. *Sensuous Worship: Jesuits and the Art of the Early Catholic Reformation in Germany*. Princeton, NJ: Princeton University Press.
Smith, Pamela. 2004. *The Body of the Artisan: Art and Experience in the Scientific Revolution*. Chicago: University of Chicago Press.
———. 2022. *From Lived Experience to the Written Word: Reconstructing Practical Knowledge in the Early Modern World*. Chicago: University of Chicago Press.
Snyder, Jon R. 2005. *L'estetica del Barocco*. Bologna: Il Mulino.
Sommervogel, Carlos. 1890. *Bibliothèque de la Compagnie de Jésus*. 12 vols. Paris: Alphonse Picard.
Spence, Jonathan D. 1984. *The Memory Palace of Matteo Ricci*. New York: Viking Penguin.
Squizzato, Alessandra. 2004. "Michelangelo negli scritti d'arte di Giovan Paolo Lomazzo." In *Tracce di letteratura artistica in Lombardia*, edited by Alessandro Rovetta. Bari: Pagina.
Steinberg, Leo. 1977. *Borromini's San Carlo alle Quattro Fontane: A Study in Multiple Form and Architectural Symbolism*. New York: Garland Publishing, Inc.
Steinberger, Ludwig. 1906. *Die Jesuiten und die Friedensfrage in der Zeit von Prager Frieden bis zum Nürnberger Friedensexekutionshauptrezess 1635–1650*. Freiburg: Herdesche.
Stoichita, Victor I. 1993. *L'instauration du tableau: Métapeinture à l'aube des temps modernes*. Paris: Méridiens Klincksieck.
Stuart, Matthew. 2013. *Locke's Metaphysics*. Oxford: Oxford University Press.
Sucquet, Antonius. 1630. *Via vitae aeternae*. Antwerp: Henricus Aertssius.
Summers, David. 1981. *Michelangelo and the Language of Art*. Princeton, NJ: Princeton University Press.
———. 1987. *The Judgment of Sense: Renaissance Naturalism and the Rise of Aesthetics*. Cambridge, UK: Cambridge University Press.
Swan, Claudia. 1995. "*Ad vivum, naer het leven*, from the Life: Defining a Mode of Representation." *Word & Image* 11, no. 4: 353–72.
Tabarrini, Marisa. 2016. *Borromini e gli Spada: Un palazzo e la committenza di una grande famiglia nella Roma barocca*. Rome: Gangemi Editore.
———. 2021. "Francesco Borromini e i pavimenti in cotto di Palazzo Spada a Roma." In *Palazzo Spada e i pavimenti del Seicento a Roma*, edited by Adriana Capriotti and Augusto Roca de Amicis. Roma: Artemide.
Tadisi, Jacopo Antonio. 1760. *Memorie della vita di Monsignore Giovanni Caramuel di Lobkowitz, vescovo di Vigevano*. Venice: Giovanni Tevernin.
Tantouche, F. 1623. *Traicte de tout ce qui s'observe en la cour de Rome, tant par nostre saincte Père, que par Messeigneurs les Illustrissimes Cardinaux*. Paris: Jean Ballagny.
Tarrant, Neil James. 2014. "Censoring Science in Sixteenth-Century Italy: Recent (And Not-So-Recent) Research." *History of Science* 52, no. 1: 1–27.
Terki, Florence. 2006. *L'anamorphose murale de la Trinité-des-Monts à Rome ou L'invisible intelligible*. Montpellier: Espérou.
Terrada, María Luz López. 2000. "Hernández and Spanish Painting in the Seventeenth Century." In *Searching for the Secrets of Nature: The Life and Works of Dr. Francisco Hernández*, edited by Simon Varey, Rafael Chabrán, and Dora B. Weiner. Stanford, CA: Stanford University Press.
Terzago, Paolo Maria. 1666. *Museo ò galeria adunata dal sapere, e dallo studio Del Sig. Canonico Manfredo Settala*. Translated and revised by Pietro Francesco Scarabelli. Tortona: For the heirs of Eliseo Viola.
Tesauro, Emanuele. 1670. *Il cannocchiale Aristotelico*. Torino: Bartolomeo Zavatta.
Thelen, Heinrich. 1958. *70 disegni di Francesco Borromini dalle collezioni dell'Albertina di Vienna: 19 novembre 1958–6 gennaio 1959*. Rome: Gabinetto Nazionale delle Stampe Farnesina.
Tita, Silvia. 2019. "The Spada Swords and the Barberini Bees Define the Borders of the Papal States: The Frescoes by Colonna and Mitelli in the Salone of the Palazzo Spada." *Artibus et Historiae* 40, no. 79: 191–227.
Totti, Pompilio. 1638. *Ritratto di Roma moderna*. Rome: Mascardi.
Trachtenberg, Marvin. 2010. *Building-in-Time: From Giotto to Alberti and Modern Oblivion*. New Haven, CT: Yale University Press.
Troili, Giulio. 1672. *Paradossi per pratticare la prospettiva*. Bolonga: Gioseffo Longhi.
Tutino, Stefania. 2014. *Shadows of Doubt: Language and Truth in Post-Reformation Catholic Culture*. Oxford: Oxford University Press.
———. 2017. *Uncertainty in Post-Reformation Catholicism: A History of Probabilism*. Oxford: Oxford University Press.
Tzonis, Alexander, and Liane Lefaivre. 1986. *Classical Architecture: The Poetics of Order*. Cambridge, MA: The MIT Press.
Uhl, Lemon L. 1890. *Attention: A Historical Summary of the Discussions Concerning the Subject*. Baltimore: Johns Hopkins University Press.
Väliaho, Pasi. 2022. *Projecting Spirits: Speculation, Providence, and Early Modern Optical Media*. Redwood City, CA: Stanford University Press.
Valone, Carolyn. 1990. "Elena Orsini, Daniele da Volterra, and the Orsini Chapel." *Artibus et Historiae* 11, no. 22: 79–87.
Van Veen, Otto. 2017. *Physicae et Theologicae Conclusiones:*

Conclusions de Physique et de Théologie (1621). Edited by Agnès Guiderdoni, Ralph Dekoninck, and Aline Smeesters. Turnhout: Brepols.

Vasari, Giorgio. 1962. *La vita de Michelangelo, nella redazioni del 1550 e del 1568*. Edited by Paola Barocchi. 5 vols. Milan: R. Ricciardi.

———. 1965. *Lives of the Artists Volume 1*. Translated by George Bull. London: Penguin.

Verbiest, Ferdinand, S.J. 1687. *Astronomia Europaea*. Dilingen: Bencard.

Vermeir, Koen. 2005. "The Magic of the Magic Lantern (1660–1700): An Analogical Demonstration and the Visualization of the Invisible." *British Journal for the History of Science* 38, no. 2: 127–59.

———. 2007. "Athanasius Kircher's Magical Instruments: An Essay on 'Science,' 'Religion' and Applied Metaphysics." *Studies in History and Philosophy of Science* 38: 363–400.

Vernet, André, Claude Jolly, Dominique Varry, and Martine Poulain, eds. 2008. *Histoire des bibliothèques françaises: Les bibliothèques sous l'Ancien Régime, 1530–1789*. Paris: Editions du Cercle de la Librairie.

Vickers, Brian. 1983. "Epideictic Rhetoric in Galileo's *Dialogo*." *Annali dell'Istituto e Museo di Storia della Scienza di Firenze* 8: 69–102.

Villalpando, Juan Bautista, and Jerónimo de Prado. 1596–1605. *In Ezechielem explanationes et apparatus urbis ac templi hierosolymitani*. 3 vols. Rome: C. Vulliettus, I. Ciacconius, A. Zannetti.

Vitruvius. 1586. *De architectura libri decem*. Lyon: Joannes Tornaesius.

———. 1684. *Les dix livres d'architecture de Vitruve*. Translated by Claude Perrault. 2nd ed. Paris: Jean-Baptiste Coignard.

———. 1983. *Ten Books of Architecture*. Translated by F. Granger. Loeb Classical Library. Cambridge, MA: Harvard University Press.

Vocabolario degli Accademici della Crusca. 1612. Venice: Giovanni Alberto.

von Gersdorff, Hans. 1517. *Feldtbuch der Wundtartzney*. Strasbourg: Johannes Schott.

von Philippovich, Eugen. 1960. "Anatomische Modelle in Elfenbein und anderen Materialien." *Sudhoffs Archiv für Geschichte der Medizin und der Naturwissenschaften* 44, no. 2: 159–78.

———. 1961. *Elfenbein: Ein Handbuch für Sammler und Liebhaber*. Braunschweig: Klinkhardt & Biermann.

von Ranke, Leopold. 1839–1847. *Deutsche Geschichte im Zeitalter der Reformation*. 6 vols. Berlin: Duneker und Humblot.

Waddy, Patricia. 1990. *Seventeenth-Century Roman Palaces: Use and the Art of the Plan*. Cambridge, MA: The MIT Press.

Watson, Richard. 1987. *The Breakdown of Cartesian Metaphysics*. Atlantic Highlands, NJ: Humanities Press International.

Weber, Hans, Gottlieb S. Wolff, and Johannes Christoph Zick. 1781. *Ruhmverneuertes Ehren-Maal Der Erbaren und Kunstreichen Holz-Bein-Horn-Metall-und-Silber-Drechsler*. Nuremberg: s.n.

Weil, Mark S. 1974. "The Devotion of the Forty Hours and Roman Baroque Illusions." *Journal of the Warburg and Courtauld Institutes* 37: 218–48.

Weston, Giulia Martina. 2016a. "After Galileo: The Image of Science in Niccolò Tornioli's *Astronomers*." *Art History* 39, no. 2: 302–17.

———. 2016b. *Niccolò Tornioli (1606–1651): Art and Patronage in Baroque Rome*. Rome: Editoriale Artemide.

White, Christopher G. 2018. *Other Worlds: Spirituality and the Search for Invisible Dimensions*. Cambridge, MA: Harvard University Press.

Whitmore, P.J.S. 1967. *The Order of Minims in Seventeenth-Century France*. The Hague: Martinus Nijhoff.

Wilberg, P. Vignau. 1996. "Le finte cupole e la loro recezione nella Germania meridionale." In *Andrea Pozzo*, edited by Aberta Battisti. Milan: Luni.

Wilson, Blake, George J. Buelow, and Peter A. Hoyt. 2001. "Rhetoric and Music." In *Grove Music Online*. Oxford: Oxford University Press. Accessed online November 30, 2023: https://www-oxfordmusiconline-com.

Witt, Charlotte. 2012. "Aristotle On Deformed Animal Kinds." *Oxford Studies in Ancient Philosophy* 43: 83–106.

Wittkower, Rudolf. 1999. *Art and Architecture in Italy, 1600 to 1750*. Revised by Joseph Connors and Jennifer Montague. 3 vols. New Haven, CT: Yale University Press.

Wölfflin, Heinrich. 1888. *Renaissance und Barock: Eine Untersuchung über Wesen und Entstehung des Barockstils in Italien*. Munich: Theodor Ackermann.

———. 1918. *Kunstgeschichtliche Grundbegriffe: Das Problem der Stilentwicklung in der neueren Kunst*. Munich: Hugo Bruckmann.

———. 1964. [1928]. *Renaissance and Baroque*. Translated by Kathrin Simon. Ithaca, NY: Cornell University Press.

———. 2015. *Principles of Art History: The Problem of the Development of Style in Early Modern Art*. Translated by Jonathan Blower. Los Angeles, CA: The Getty Research Institute.

Wolgast, Eike. 1984. "Reform, Reformation." In *Geschichtliche Grundbegriffe: Historisches Lexikon zur politisch-sozialen Sprache in Deutschland*, edited by Otto Brunner et al. 9 vols. Stuttgart: Klett-Cotta.

Woodbury, Robert S. 1961. *History of the Lathe to 1850: A Study in the Growth of a Technical Element of an Industrial Economy*. Cambridge, MA: The MIT Press.

Wrede, Adolf, ed. 1901. *Deutsche Reichstagsakten Jüngere Reihe: Deutsche Reichstagsakten unter Kaiser Karl V. Dritter Band*. Gotha: Friedrich Andreas Perthes.

Wunsch, Oliver. 2024. *A Delicate Matter: Art, Fragility, and Consumption in Eighteenth-Century France*. University Park, PA: Penn State University Press.

Wycherley, R. E. 1962. "The Athenian Philosophical Scene—II." *Greece & Rome* 9, no. 1: 2–21.

Zierholz, Steffan. 2021. "The Subiconographic Surface: Two *Temptations of Saint Anthony* Painted on Stone." *Art Bulletin* 103, no. 4: 36–60.

INDEX

Note: Page numbers in *italic* type indicate illustrations.

IMAGE CREDITS

© Accademia Nazionale dei Lincei (Fig. 1)
© Albertina Museum, Vienna [Alb. 173] (Fig. 7)
© Albertina Museum, Vienna [AZRom1156] (Fig. 48)
© Alinari Archives / Art Resource, NY (Fig. 47)
© Archdiocese of Bari-Bitonto, Diocesan Museum "Aurelio Marena" of Bitonto (Fig. 37)
© Bayerische Staatsbibliothek München, 2 Math.a.105, frontispiece, urn:nbn:de:bvb:12-bsb10147839-8 (Fig. 124)
© Bayerische Staatsbibliothek München, Einbl. XI,745 a, urn:nbn:de:bvb:12-bsb00095666-8 (Fig. 186)
© Bayerische Staatsbibliothek München, V.ss. 335 g, fol. 54v, urn:nbn:de:bvb:12-bsb10208848-3 (Fig. 78)
© Bibliothèque de la Sorbonne, Paris (Fig. 140)
© Bibliothèque nationale de France (Figs. 9, 60, 61, 63, 77, 92, 98, 106, 107, 108, 118, 119, 129, 134, 136, 137, 138, 145, 146, 148, 149, 150, 151, 153, 154, 164, 166, 169, 171, 176)
© Bibliothèque Sainte-Geneviève, Paris (Fig. 139)
© 2021 Christie's Images Limited (Fig. 64)
© CICRP - MAP GAMSAU - Odile Guillon - Anthony Pamart - Stephan Faraci (Fig. 103)
© Comune di Milano, tutti i diritti riservati - Gabinetto dei Disegni del Castello Sforzesco, Milano (Fig. 6)
© Courtesy of Galleria Spada, Direzione Musei Statali della città di Roma (Fig. 13)
© From the British Library collection, General Reference Collection DRT Digital Store 10007.g.24 (Fig. 144)
© From the British Library collection, L.35/4 (Figs. 127, 128)
© f Typ 625.48.550. Houghton Library, Harvard University (Fig. 69)
© Gallerie Nazionali di Arte Antica, Roma (MiC) - Bibliotheca Hertziana, Istituto Max Planck per la storia dell'arte / Enrico Fontolan (Figs. 66, 120)
© Germanisches National Museum, Nuremberg (Figs. 180, 181, 182, 183, 184)
© Heidelberg University Library, P 7584 Folio RES "Feldtbuoch der Wundtartzney," XXXIXr (Fig. 168)
© Huntington Library (Figs. 99, 100, 117, 132, 133, 165, 167)
© Institut national d'histoire de l'art (Fig. 68)
© John Carter Brown Library, Brown University, Providence, RI, box 1894, 02912 (Fig. 116)
© KHM-Museumsverband (Fig. 178)
© Mauro Coen; Pii Stabilimenti della Francia a Roma e Loreto (Figs. 65, 87, 114)
© Musei Civici di Vicenza - Gabinetto Disegni e Stampe di Palazzo Chiericati (Fig. 36)
© Museo Galileo, Firenze; photo by Franca Principe (Figs. 130, 131)
© Muséum national d'Histoire naturelle (Fig. 91)
© National Gallery, London / Art Resource, NY (Figs. 56, 58)
© Photo: A. De Rosa / Imago rerum / Iuav University of Venice; Pii Stabilimenti della Francia a Roma e Loreto (Figs. 70, 72, 123, 126, Gatefold 1, Gatefold 2)
© Photo: Alvesgaspar, Wikimedia Commons, CC BY-SA 4.0 (Fig. 187)
© Photo: Antonio Calandriello / Imago rerum / Iuav University of Venice; Pii Stabilimenti della Francia a Roma e Loreto (Figs. 80, 81, 82. 83, 84)
© Photo: Architas, Wikimedia Commons, CC BY-SA 4.0 (Fig. 2)
© Photo: author (Figs. 39, 40, 115)
© Photo: author; courtesy of Galleria Spada, Direzione Musei Statali della città di Roma (Fig. 20)
© Photo: author; Pii Stabilimenti della Francia a Roma e Loreto (Fig. 71, 74, 76, 86, 88, 89)
© Photo: Carlo Alberto Magagnini, Wikimedia Commons, CC BY-SA 4.0 (Fig. 142)
© Photo: Didier Descouens, Wikimedia Commons, CC BY-SA 4.0 (Fig. 35)
© Photo: Diliff, Wikimedia Commons, CC BY-SA 3.0 (Fig. 147)
© Photo: G. Cigolini © DeA Picture Library / Art Resource, NY (Fig. 42)
© Photo: Hélène Lo Iacono (Figs. 53, 54)
© Photo: Jastrow, Wikimedia Commons, Public Domain (Fig. 46)
© Photo: Jean-Cristophe Benoist, Wikimedia Commons, CC BY-SA 3.0 (Fig. 109)
© Photo: Marina Cotugno (Figs. 38, 110)
© Photo: Marina Cotugno; courtesy of Galleria Spada, Direzione Musei Statali della città di Roma (Figs. 16, 17, 18, 19, 21, 32, 49, 50, 51, 52)
© Photo: Paolobon140, Wikimedia Commons, CC BY-SA 4. (Fig. 43)

© Photo: private (Figs. 8, 67, 102, 104, 105, 113)
© Photo: private; courtesy of Galleria Spada, Direzione Musei Statali della città di Roma (Figs. 14, 22, 23, 24, 25, 26, 27, 28, 29, 30, 31, 33)
© Photo: private; Pii Stabilimenti della Francia a Roma e Loreto (Figs. 59, 62, 73, 75, 85, 93, 94, 95, 96, 97)
© Photo: Rabax63, Wikimedia Commons, CC BY-SA 4.0 (Fig. 5)
© Photo: Sailko, Wikimedia Commons, CC BY 3.0 (Figs. 41, 43, 112)
© Photo: Toni Spagone / RealyEasyStar (Fig. 152)
© Plan: Nick M. L. Mols (Figs. 4, 143)
© Princeton University Library, Special Collections (Figs. 10, 101, 135)
© Roma - Sovrintendenza Capitolina ai Beni Culturali - Museo di Roma (Figs. 34, 90)
© Scala / Art Resource, NY (Figs. 3, 44, 45, 79, 111)
© Scala / Art Resource, NY; courtesy of Galleria Spada, Direzione Musei Statali della città di Roma (Fig. 12)
© Scala / Ministero per i Beni e le Attività culturali / Art Resource, NY (Fig. 15)
© Staatliche Bibliothek Regensburg, 999/Bav.957 (Fig. 179)
© Staatsbibliothek zu Berlin – Preußischer Kulturbesitz, Abteilung Handschriften und Historische Drucke, shelf mark: 2" Mx 261-1: R. (Fig. 125)
© Staats- und Stadtbibliothek Augsburg: 2 Math 11-1 (Figs. 121, 122)
© University of Chicago Library, Hanna Holborn Gray Special Collections Research Center (Fig. 11)
© Tesoro dei Granduchi di Palazzo Pitti, Su concessione del Ministero della Cultura (Figs. 155, 157, 158, 159, 163)
© The Metropolitan Museum of Art, New York, Gift of Robert Gordon, 1910 (10.212.2a, b) (Fig. 170)
© The Metropolitan Museum of Art, New York, Harris Brisbane Dick Fund, 1924 (24.45.1) (Figs. 174, 175)
© The Royal Danish Collection, Rosenborg Castle (Figs. 160, 161, 162)
© The Trustees of the British Museum (Figs. 141, 177)
© Typ 625.48.550 Houghton Library, Harvard University (Fig. 55)
© Universitätsbibliothek Johann Christian Senckenberg Frankfurt am Main [Ms. germ. fol. 15 Nr. 10] (Fig. 185)
© Veneranda Biblioteca Ambrosiana / Metis e Mida Informatica / Mondadori Portfolio Codice Atlantico (Fig. 57)
© Veneranda Biblioteca Ambrosiana / Mondadori Portfolio (Fig. 156)
© Veneranda Biblioteca Ambrosiana / Mondadori Portfolio / La divina proporzione (Figs. 172, 173)

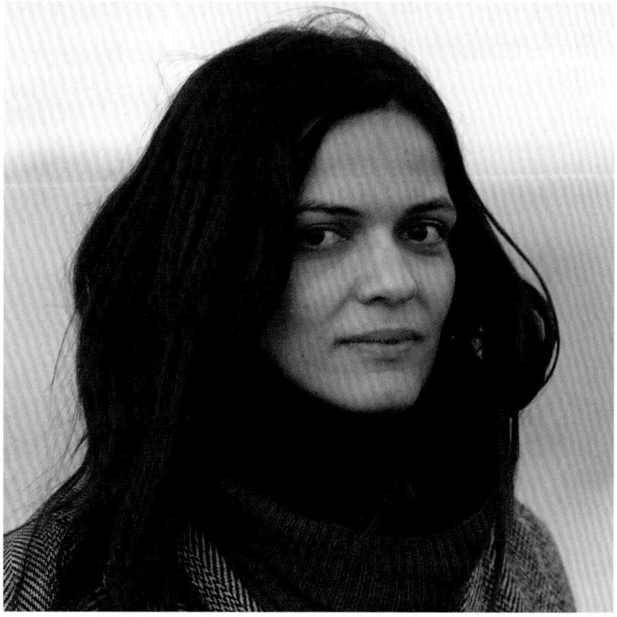

Susanna Berger is associate professor of art history and philosophy at the University of Southern California. She is the author of *The Art of Philosophy: Visual Thinking in Europe from the Late Renaissance to the Early Enlightenment* (Princeton).

Published by Princeton University Press, 41 William Street, Princeton, New Jersey 08540
In the United Kingdom: Princeton University Press, 99 Banbury Road, Oxford OX2 6JX
press.princeton.edu

Cover image: Unidentified artist, four figures inside the sundial gallery from Emanuel Maignan's *Perspectiva horaria*, 1648, between pp. 390 and 391. Engraving printed on paper, 17.32 × 11.42 in. (44 × 29 cm). Harvard University, Houghton Library, Cambridge, MA.
Back cover image: Possibly workshop of Lorenz Zick, Standing Cup with Cover (detail), mid-seventeenth century. Turned ivory, height 14 in (35.6 cm). The Metropolitan Museum of Art, New York; gift of Robert Gordon, 1910.

Author photo by Torbjørn Rødland

Page ii, detail of fig. 66

GPSR Authorized Representative: Easy Access System Europe - Mustamäe tee 50, 10621 Tallinn, Estonia, gpsr.requests@easproject.com

ISBN 9780691201016
ISBN (ebook) 9780691266121
Library of Congress Control Number: 2025930828

British Library Cataloging-in-Publication Data is available
Publication is made possible in part by a grant from the Barr Ferree Foundation Publication Fund, Department of Art and Archaeology, Princeton University.

Editorial: Michelle Komie and Annie Miller
Production Editorial: Mark Bellis
Text and Cover Design: Jeff Wincapaw
Production: Steve Sears
Publicity: William Pagdatoon
Copyeditor: Lachlan Brooks
This book has been composed in Arno Pro

Printed in China
10 9 8 7 6 5 4 3 2 1

MIX
Paper | Supporting responsible forestry
FSC™ C136333
FSC www.fsc.org